Elastic Waves in Solids II

Springer
Berlin
Heidelberg
New York
Barcelona
Hong Kong
London
Milan
Paris
Singapore
Tokyo

Advanced Texts in Physics

This program of advanced texts covers a broad spectrum of topics which are of current and emerging interest in physics. Each book provides a comprehensive and yet accessible introduction to a field at the forefront of modern research. As such, these texts are intended for senior undergraduate and graduate students at the MS and PhD level; however, research scientists seeking an introduction to particular areas of physics will also benefit from the titles in this collection.

Daniel Royer Eugène Dieulesaint

Elastic Waves in Solids II

Generation, Acousto-optic Interaction, Applications

Translated by Stephen N. Lyle
With 275 Figures, Numerous Problems and Solutions

Professor Daniel Royer
Université Denis Diderot
Laboratoire Ondes et Acoustique
ESPCI
10 rue Vauquelin
F-75231 Paris Cedex 05, France
E-mail: Daniel.Royer@espci.fr

Professor Eugène Dieulesaint
Emeritus professor
Université Pierre et Marie Curie
École Supérieure de Physique et de Chimie
10 rue Vauquelin
F-75231 Paris Cedex 05, France

Translator:
Stephen N. Lyle
8 Impasse du Coteau
F-69480 Pommiers, France

Title of the original French edition:
Ondes élastiques dans les solides.
Tome 2: Génération, interaction acousto-optique, applications.

Library of Congress Cataloging-in-Publication Data.
Royer, D. (Daniel) [Ondes élastiques dans les solides. English] Elastic waves in solids / Daniel Royer, Eugène Dieulesaint. p. cm. Includes bibliographical references and index. Contents: v. 1. Free and guided propagation - v. 2. Generation, acousto-optic interaction, applications. ISBN 3-540-65932-3 (v. 1 : hc. : alk. paper). – ISBN 3-540-65931-5 (v. 2 : hc. : alk. paper) 1. Acoustic surface waves. 2. Signal processing. 3. Elastic waves. I. Dieulesaint, E. II. Title. QC176.8.A3R6913 2000 531'.33–dc21 99-34758

ISSN 1439-2674

ISBN 3-540-65931-5 Springer-Verlag Berlin Heidelberg New York

Typesetting: Data conversion by Steingraeber Satztechnik GmbH, Heidelberg
Cover design: *design & production* GmbH, Heidelberg

Printed on acid-free paper SPIN 10681913 57/3144/di 5 4 3 2 1 0

Preface

This second of two volumes on *Elastic Waves in Solids* carries the subtitle *Generation, Acousto-optic Interaction and Applications.* To obtain maximum benefit from this book the reader should either have studied Vol. I or be familiar with the modes of propagation of elastic waves in ordinary and piezoelectric crystals. The aims of this volume are: to investigate ways of generating and detecting the various bulk and surface waves studied in the first volume; to emphasize their action on light waves; and to describe how their properties can be used to construct sensors and other components. The major methods for generating elastic waves have been known for several decades. However, in this time they have been continually improved and the dimensions and cost of the devices have both been reduced significantly in recent years. Compact filters for mobile telephones are good examples. The interaction between acoustic and optical waves is also carefully examined in this book. This interaction is responsible for the operation of devices such as modulators, spectrometers and tunable filters. These devices are encountered in a wide variety of systems, sometimes in quite unexpected situations: for example, in astronomical instruments. The presence of a gas or a liquid in contact with a surface modifies the propagation and reflection of the waves. These modifications provide a means of measuring the pressure, the temperature or other relevant quantities. Several such sensors and instruments are described here.

A number of problems are included at the end of each chapter, complete with worked solutions. These will enable readers to consolidate the material and verify that they have understood the main concepts.

The authors would like to thank friends and colleagues who provided documents and helpful remarks during the writing of this second volume: Robert Adler (Zenith, USA), Fred Hickernell (Motorola, USA), Joël Kent (ELOTouchsystems, USA), Lawrence Lynnworth (Panametrics, USA), Kimio Shibayama and Kazuhiko Yamanouchi (Tohoku University, Japan), Sir Eric Ash (Royal Society, UK), Meirion Lewis (DERA, UK), David Morgan (Impulse Consulting, UK), Walter Zingg (MicroCrystal, Switzerland), Michiel Wellekoop (Delft University, Netherlands), Mauro Varasi (Alenia, Italy), Jean-Jacques Boy (ENSMM, Besançon, France), Frank Darde (AA Saint Rémy-les-Chevreuses, France), Patrick Dierich (Meudon Observatory, France),

Jacques Simon (ESPCI, Paris), Mathias Fink (ESPCI, Paris) and all the members of the Laboratoire Ondes et Acoustique, ESPCI, Paris, especially Olivier Casula.

The authors are particularly grateful to Jacques Détaint (CNET, Bagneux, France) and Jean-Michel Hodé (Thomson Microsonics, Sophia-Antipolis, France) for their competent assistance. They would also like to thank Stephen N. Lyle who efficiently translated this text from French.

Paris *D. Royer*
September 1999 *E. Dieulesaint*

Contents

Introduction 1

1. Piezoelectric Transducers for Bulk Waves 5
1.1 Structures. One-Dimensional Model 6
1.2 Impedance Matrix 10
1.3 Equivalent Circuits 12
1.3.1 Mason and Redwood Circuits 12
1.3.2 The KLM Model 16
1.4 Low-Frequency Transducers 19
1.4.1 Emitted Power and Frequency Response 20
1.4.2 Time Response 23
1.4.3 Electrical Impedance 26
1.5 High-Frequency Transducers 30
1.5.1 Piezoelectric Crystal–Electrode-Propagating Medium 32
1.5.2 Frequency Variation of the Bandshape Factor 34
1.5.3 Electrical Tuning. Conversion Losses 38
1.6 Materials and Technology 40
Problems 46

2. Interdigital-Electrode Transducers for Surface Waves 57
2.1 Operating Principles 58
2.1.1 Secondary Effects and Possible Corrections 62
2.1.2 Spatial Distribution of Electrical Quantities 67
2.1.3 Discrete Source Method 70
2.2 Impulse Response Model 72
2.2.1 Time Responses 73
2.2.2 Radiation Conductance 76
2.3 Piezoelectric Permittivity Method 80
2.3.1 Generation. Green Functions 83
2.3.2 Reception and Reemission 91
2.4 Three-Port Circuit (Hexapole). Matrices 94
2.4.1 Scattering Matrix 95
2.4.2 Mixed Matrix 99
2.5 Coupled Mode Method 103

2.5.1 Passive Array. Reflection and Transmission Coefficients 105
2.5.2 Transducer 110
2.5.3 Unidirectional Transducers 120
2.6 Adjacent Element Analysis 123
2.6.1 *P*-Matrix Model 123
2.6.2 Harmonic Admittance. Numerical Model 127
2.6.3 Equivalent Circuits 139
2.7 Materials and Technology 143
Problems 148

3. Elastic Waves and Light Waves 157
3.1 Acousto-optic Interaction 158
3.1.1 Three Cases of Interaction 158
3.1.2 Propagation of Light Waves in Crystals 160
3.1.3 Elasto-optic Tensor 164
3.1.4 Normal Incidence Diffraction 172
3.1.5 Bragg Incidence Diffraction 177
3.1.6 Interaction with Surface Elastic Waves 193
3.2 Optical Measurement of Mechanical Displacements 196
3.2.1 Deflection and Diffraction Probes 196
3.2.2 Interferometric Probes 200
3.2.3 Applications 206
3.3 Photothermal Generation 210
3.3.1 Thermoelastic Regime 212
3.3.2 Ablation Regime 220
3.3.3 Increasing Efficiency 223
3.3.4 Experiments 224
Problems 225

4. Signal Processing Components 235
4.1 Elements and Structures 236
4.2 Delay Lines 245
4.2.1 Bulk Wave Delay Lines 245
4.2.2 Rayleigh Wave Delay Lines 249
4.3 Surface Wave Filters 251
4.3.1 Filters with Bidirectional Transducers 253
4.3.2 Filters with Multistrip Coupler 258
4.3.3 Filters with Unidirectional Transducers 259
4.3.4 Stationary Wave Filters 261
4.4 Bulk Wave Filters 269
4.4.1 Quartz Resonators 269
4.4.2 Monolithic Filters 278
4.5 Filters Matched to Signals 280
4.5.1 Response and Signal-to-Noise Ratio 281

4.5.2 Signal with Linearly Modulated Frequency 284
4.5.3 Rayleigh Wave Filters 290
4.5.4 Filters Based on Elasto-optic Interaction 301
4.6 Spectrum Analysers and Fourier Transform Processors 304
4.7 Convolvers 306
Problems 310

5. Sensors and Instrumentation 319
5.1 Quartz Resonator Sensors 320
5.1.1 Thickness Shear Mode Sensors 320
5.1.2 Vibrating Beam Sensors 324
5.1.3 Quartz Tuning Fork Thermometer 330
5.2 Guided Wave Sensors 331
5.2.1 Surface Wave Sensors 331
5.2.2 Lamb Wave Sensors 336
5.2.3 Cylindrical Guide Sensors 344
5.3 Acousto-optic Components 348
5.3.1 Modulator for Laser Printer 349
5.3.2 Stationary Wave Frequency Modulator 350
5.3.3 Spectral Line Selection by Tunable Filter 351
5.3.4 Radio Astronomy Spectrometer 354
5.4 Ultrasonic Motors and Actuators 358
5.5 Instruments and Methods 362
5.5.1 Acoustic Microscope 362
5.5.2 Ultrasonic Osteodensitometer 368
5.5.3 Time Reversal of Elastic Waves 371
5.5.4 Determination of Charge Distributions in Insulators Using Pressure Waves 376
5.5.5 Parametric Interaction Probe 379

A. Spatial Distribution of Electrical Quantities 383
A.1 Simple Transducer 383
A.2 Single-Element Charge Distribution 387
A.3 General Case. Charge Density. Current Intensity 389

B. Coupled Mode Theory 391
B.1 Formulation. Coupled Mode Equations 391
B.2 Solution 394
B.2.1 Propagation in the Same Direction 394
B.2.2 Propagation in Opposite Directions 395

C. Legendre Functions and Polynomials 399

D. Scattering Matrix and Mixed Matrix 403
D.1 Quadripole. Scattering Matrix 403

D.2 Hexapole. Mixed Matrix. Directionality 407

E. Matrix Representation of a Dioptric System 413

F. Linear Systems of Differential Equations 417

G. Stationary Phase Method 421

References 423

Index 441

Contents of Volume I

Introduction 1

1. Waves. Fluid as a Scalar Model 9
1.1 Travelling, Stationary and Guided Waves 9
1.1.1 Expressions for a Plane Wave 10
1.1.2 Total Reflection 14
1.1.3 Velocity of a Wave Packet 20
1.2 Acoustic Waves 23
1.2.1 Wave Equation for a Plane Wave 23
1.2.2 Power Flow. Poynting Vector 32
1.2.3 Attenuation 35
1.2.4 Reflection and Refraction 38
1.3 Spherical Acoustic Waves. Radiation 46
1.3.1 Solution of the Wave Equation 46
1.3.2 Radiation 48
Problems 60

2. Crystal Properties and Their Representation by Tensors 69
2.1 Crystalline Structure 70
2.1.1 Lattices, Rows, Lattice Planes and Cells 72
2.1.2 Atomic Structure 76
2.2 Point Groups of Crystals 78
2.2.1 Point Symmetry Transformations 78
2.2.2 Lattice Point Groups and the Seven Crystal Systems 83
2.2.3 The Thirty-Two Point Symmetry Classes of Crystals 88
2.3 Quasicrystals 91
2.4 Examples of Structures 94
2.4.1 Close-Packed Structures 94
2.4.2 Useful Crystals 96
2.5 Representation of Physical Properties of Crystals by Tensors 102
2.5.1 Change of Coordinate System 103
2.5.2 Definition of a Tensor 105
2.6 Effect of Crystal Symmetry on Tensor Components 108
2.6.1 Matrices for Point Symmetry Transformations 109

2.6.2 Effect of a Centre of Symmetry 110
2.6.3 Reduction of the Number of Dielectric Constants 110
Problems 112

3. Elasticity and Piezoelectricity 119
3.1 The Elasto-electric Field 120
3.1.1 Mechanical Variables 120
3.1.2 Field Equations 127
3.1.3 Boundary Conditions 131
3.1.4 Poynting's Theorem and Energy Conservation 132
3.2 Linear Behaviour of an Elastic Solid 133
3.2.1 Hooke's Law 133
3.2.2 Elastic Energy of a Strained Solid. Maxwell's Relations 136
3.2.3 Effect of Crystal Symmetry on Elastic Constants 138
3.3 Piezoelectric Solids 147
3.3.1 Physical Mechanism – One-Dimensional Model 150
3.3.2 Tensor Expressions 154
3.3.3 Effect of Crystal Symmetry on Piezoelectric Constants 158
Problems 166

4. Plane Waves in Crystals 171
4.1 Monatomic One-Dimensional Model 172
4.2 Anisotropic Solids 177
4.2.1 Wave Equation 177
4.2.2 Directions Linked to Symmetry Elements 180
4.2.3 Isotropic Solid 182
4.2.4 Energy Velocity 185
4.2.5 Characteristic Surfaces 189
4.2.6 Attenuation 211
4.3 Piezoelectric Crystals 216
4.3.1 Wave Equation 216
4.3.2 Examples of Slowness Curves 218
4.3.3 Propagation Along Directions Related to Symmetry Elements 224
4.3.4 Bulk Piezoelectric Permittivity 225
4.4 Reflection and Refraction 229
4.4.1 Polarizations and Propagation Directions of Reflected and Refracted Waves 229
4.4.2 Amplitudes. Continuity Equations 235
Problems 249

5. Guided Waves 261
5.1 Planar Waveguides 263

5.1.1 Decomposition of Equations 263
5.1.2 Mechanical Boundary Conditions 265
5.1.3 Main Types of Guided Wave 266
5.2 Power Flow 270
5.2.1 Harmonic Case 272
5.2.2 Susceptance 274
5.2.3 Free Modes 275
5.3 Rayleigh Waves 276
5.3.1 Search Procedure 276
5.3.2 Isotropic Solid. Solutions for Components 278
5.3.3 Crystals 282
5.3.4 Piezoelectric Solid 290
5.4 Transverse Horizontal Waves 303
5.4.1 Non-piezoelectric TH Waves 304
5.4.2 Piezoelectric TH Waves ($\overline{\text{TH}}$ Waves) 309
5.5 Lamb Waves in an Isotropic Plate 311
5.5.1 Dispersion Relation. Modes 313
5.5.2 Dispersion Curves. Mechanical Displacement 315
5.6 Cylindrical Waveguides 321
5.6.1 Compressional Waves 325
5.6.2 Torsional Waves 327
5.6.3 Flexural Waves 329
5.6.4 Tubular Waveguides 330
Problems 330

A. Cylindrical and Spherical Coordinates 347

B. Eigenvectors and Eigenvalues of a Matrix 353

C. Tensor Representation of a Surface Element 357

References 359

Further Reading 365

Symbols 367

Index 369

Introduction

The purpose of this volume is to investigate and explain the methods of generating and detecting bulk and surface elastic waves, to detail their interaction with light waves, and to describe how their properties can be exploited in the construction of a variety of devices. In the first chapter we will investigate the generation and detection of bulk elastic waves. The more classical approach to generating (and detecting) free waves in a solid or liquid – and also guided waves in a cylinder or plate – is to rigidly attach a piezoelectric resonator (cut monocrystal, oriented layer, polarised polymer or ceramic). This excites the required longitudinal or transverse modes. We often have recourse to an equivalent circuit diagram when describing such loaded resonators, with impedance-matching plates or layers: direct analysis is only feasible if the electromechanical coupling is much smaller than unity, even in the 1-dimensional case. Low-frequency transducers are therefore studied using an equivalent electromechanical circuit, and high-frequency transducers directly from the equations.

Chapter 2 deals with generation and detection of waves guided by a surface, and in particular, with Rayleigh waves. These waves (like horizontal transverse waves and other pseudo-surface waves) are launched by comb-shaped electrodes which are generally made on the propagating solid itself using microelectronic techniques. The latter is also the generating medium and is piezoelectric. We explain qualitatively how these interdigital-electrode transducers operate and describe the effects usually encountered in practice. It is no simple matter to include them all in one analysis. Several models exist for transducers of this type: arrays of discrete sources, impulse response, surface piezoelectric permittivity, matrix representation, coupled modes, direct use of equations via numerical techniques and equivalent circuits. Each model is presented with its weak and strong points. Piezoelectric reemission and internal reflection by the interdigital fingers are only considered in the most advanced models. However, it is not always necessary or useful to eliminate these effects. They have taken on a new significance because they can be exploited to produce unidirectional transducers and also resonator reflectors.

Chapter 3 is divided into three main parts. The action of a beam of elastic waves on a beam of light waves is discussed in the first and longest part. It can be modelled by a moving grating with modifiable parameters. A rank 4 elasto-optic tensor describes deformation of the ellipsoid of refractive indices due to the elastic wave. The result is separation into several diffracted beams or simple deviation of the incident light beam, depending on the angle of incidence. In practice, the second case is the more interesting since angle, intensity and frequency of the deviated light beam all depend on parameters of the elastic waves. Use of this interaction has led to the acousto-optic modulator, which has developed in parallel with lasers. It is an essential component in instruments such as optical interferometers and ellipsometers. The interferometer has proved extremely useful in registering mechanical displacements of surfaces produced by elastic waves. Hence the second part of this chapter is devoted to optical probe measurements of such displacements, with amplitudes in the angström range. These measurements are made without mechanical contact and are naturally related to photothermal excitation, described in the final section. This is a relatively new excitation technique using one or more lasers. It has applications in non-destructive testing of materials and will be illustrated by the results of laboratory experiments.

Chapter 4 presents elastic wave components suitable for processing electrical signals. We begin by examining their structural features and then give examples of delay lines, filters and signal compression devices, as well as devices for spectral analysis or convolution of two signals. Delay and filtering are implemented either by bulk or surface waves, although the latter provide a much wider range of structures. For example, whilst bulk wave filters are made using resonators, those for surface waves are made not only from resonators, i.e., with standing waves, but also from delay lines with travelling waves. Both types of resonator have evolved over the past ten years. However, the variety of filters based on surface wave resonators is quite astonishing. These arise from the various ways of disposing cavities relative to transducers, and then coupling them together. It is true to say that they are simply made up of metallic strips and grooves! From the few structures described, the reader is free to imagine others, preferably with applications in the fast-developing area of portable telecommunication systems. The main ideas are reviewed: relationship between time and frequency domains, response of a linear system to general signals, definition and response of a filter matched to a signal. These are illustrated by the classical filter, matched to a signal with linearly varying frequency.

Applications to detection and transformation of physical quantities, that is, to construction of sensors and instrumentation, are dealt with in Chap. 5. Our aim is to show the relevancy of elastic waves in a wide range of contexts, from geological exploration and son et lumière to radioastronomy. Most of the sensors presented here exhibit some original feature with regard to accuracy, size (miniaturisation), vibrational mode or conditions of use. Thus, we describe a pressure sensor, based on a bulk wave resonator and used for detection of oil layers, because it represents a technical innovation, involving simultaneous use of two modes, one pressure-sensitive and the other temperature-sensitive. Among those devices described are a Rayleigh wave touch-sensitive screen, a Lamb wave coordinate sensor, liquid detectors, and acousto-optic components such as a spectrometer used in radioastronomy. Optical probes like the heterodyne interferometer (developed over the past few years) are presented in Chap. 3. We therefore include the acoustic lens microscope here, in the category of instrumentation. This has not yet met with the success expected of it, but the principle remains attractive. We also describe an osteodensitometer and a parametric interaction acoustic probe. And two methods are presented: time reversal and measurement of the charge distribution in a dielectric by pressure pulse. Finally, principles and characteristics of the relatively new piezoelectric motor are discussed.

1. Piezoelectric Transducers for Bulk Waves

Chapter 4 of Vol. I, entitled Plane Waves in Crystals, discusses free propagation of elastic waves in unbounded solids. Chapter 5 of Vol. I, Guided Waves, deals with propagation in solids with at least one bounded dimension. As we understand the modes of propagation, we shall now tackle the question of how such waves can be generated, in particular, from an electrical signal. The most natural method is to use the piezoelectric effect, which couples mechanical and electrical quantities. The mechanism was described in Sect. 3.3 of Vol. I, together with conditions for its occurrence (no centre of symmetry). In the discussion of free and guided propagation, the electromechanical coupling coefficient arose in the various equations. This characterises the particular piezoelectric solid. Crystal cuts liable to generate either longitudinal or transverse bulk waves were also mentioned in Vol. I, Sect. 4.3.2b on slowness curves.

The present chapter will deal with the generation (and detection) of transverse and longitudinal bulk waves by means of a piezoelectric solid equipped with electrodes and in contact with the propagating medium. Operation of this transducer is analysed on the basis of various 1-dimensional models, all derived from the same equations but presented in different forms. Each approach selects hypotheses which bring out some particular point of view. The piezoelectric solid, essential element of the transducer, is first described by an impedance matrix. It corresponds to a hexapole, comprising one electrical and two mechanical (acoustic) ports, which can be modelled by an equivalent electrical circuit (the classic Mason or Mason–Redwood model). This piezoelectric part is then treated as an active medium in which acoustic waves obey similar equations to those governing propagation of electrical quantities in an active circuit element, that is, one with current sources. This analogy leads to the model of Krimholtz, Leedom and Matthaei. The electrodes, impedance-matching plates, with rear load (absorber) and front load (propagating medium), can also be taken into account and represented by quadripoles, circuit elements or simple impedances. Transducers with strong electromechanical coupling, used at relatively low frequencies ($f < 20$MHz) in non-destructive testing of materials and medical ultrasonic imaging (echography), are adequately described by means of equivalent circuit diagrams. The impedance of an unloaded transducer, that is, a simple resonator, is easily deduced from these diagrams. Another model is made from expressions for particle velocity waves and stresses, assuming that the electromechanical coupling coefficient for the material is much smaller than unity. This applies to transducers commonly used in high frequency delay lines ($f > 100$ MHz). The latter are composed of a thin disk or a quasi-monocrystalline active layer together with an impedance-matching layer. The influence of this intermediate layer on frequency response will be examined. The chapter closes with a section on materials and technology.

1.1 Structures. One-Dimensional Model

In its simplest version, the bulk wave transducer comprises a piezoelectric material carrying two electrodes (metallic films). The electric field of the signal across the electrodes sets the piezoelectric solid into vibration. The thickness of this solid is equal to a fraction of the elastic wavelength. The solid itself is a ceramic or a crystal, or occasionally a sheet of some piezoelectric polymer of type PVDF (see Sect. 1.6). Its polarisation or crystallographic cut is chosen so that the desired elastic mode should be preferentially excited. Vibrations propagate in the medium mechanically coupled to one face of the piezoelectric solid. This face may carry one or more impedance-matching plates (see Problem 1.6, Vol. I). The other face is free or loaded with an absorbent material.

Figure 1.1a shows a classic single element transducer for non-destructive testing or ultrasonic imaging ($f < 20$ MHz). The term 'single element' means that there is a single electrical source. The external electrode is loaded so as to damp the piezoelectric resonator, often ceramic, and thereby widen the transducer bandwidth. Figure 1.1b shows a composite structure made of piezoelectric elements embedded in a resin. This type of transducer has appeared quite recently and is better suited to liquid media (e.g., water, the human body). It can be analysed in the same way as the classic transducer, to a first approximation (see Sect. 1.6).

Figure 1.1c shows a transducer for an acoustic microscope, acousto-optic modulator or high-frequency delay line ($f > 1$ GHz). The internal electrode, in contact with the propagating medium, is crossed by the elastic wave beam, whose dimensions are imposed by the external electrode. The piezoelectric material may be a layer of zinc oxide, with its 6-fold axis of symmetry parallel to the electric field so as to excite longitudinal waves. Or it may be a thin lithium niobate monocrystal, which can generate either longitudinal or transverse waves, depending on its crystallographic cut (see Sect. 4.3.2b, Vol. I). The low-attenuation propagating medium (e.g., aluminium oxide crystal) is positioned (cut) in such a way that the elastic mode is pure, the energy propagating along the axis of the structure (see Sect. 4.2.4, Vol. I).

Transmission of longitudinal waves from one medium to another does not require any rigid connection between them. They can therefore be used

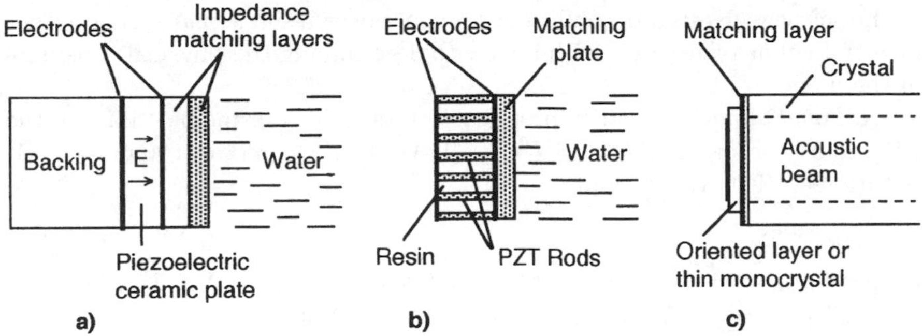

Fig. 1.1. Structure of a bulk wave single element plane transducer, used in (**a**) and (**b**) to generate waves in water or the human body ($f < 20$ MHz, e.g., in ultrasonic imaging). Diagram (**c**) shows a delay line crystal operating at high frequency ($f > 100$ MHz, e.g., in signal processing). The piezoelectric material is: (**a**) PZT-type ceramic, (**b**) composite, (**c**) monocrystal or crystal layer. It is excited by an electric field $\boldsymbol{E}$ applied across the electrodes

as primary waves for generating other waves, such as transverse waves, via oblique incidence refraction, or Lamb waves and even Rayleigh waves.

The way transducers operate can be analysed according to several models, as already mentioned. Since they are 1-dimensional, let us begin by specifying conditions for their validity in the general case of a piezoelectric crystal. An essential condition is generation of a single mode.

Figure 1.2 shows the orientation of reference axes x_1', x_2', x_3'. Unprimed coordinates x_1, x_2, x_3 are reserved for crystallographic axes. The thickness d of the piezoelectric layer, in which elastic waves are generated, is very small compared with other dimensions. The electric field is then parallel to Ox_3' and planes $x_3' = \text{Const.}$ are equiphase planes. Diffraction effects are neglected. This means that the direction of propagation, parallel to x_3', is conserved at each interface.

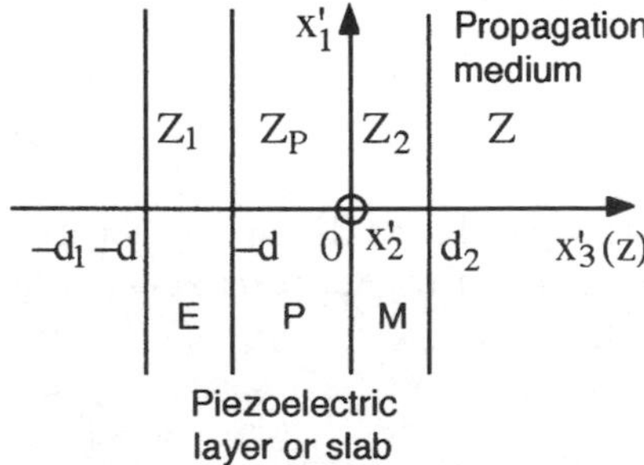

Fig. 1.2. One-dimensional model of a transducer. P is a piezoelectric crystalline plate or layer. E and M are electrodes or, if these are not taken into account in the model, the rear medium or matching plate

Let us now investigate conditions for generating a single *plane wave*, either longitudinal or transverse, with a view to determining the physical constants of the model.

A *longitudinal wave* will propagate if Ox'_3 is a principal axis of the Christoffel tensor, i.e., $\Gamma'_{13} = \Gamma'_{23} = 0$. With $n'_1 = n'_2 = 0$ and $n'_3 = 1$, equations (4.11), Vol. I, imply

$$c'_{34} = c'_{35} = 0 \,. \tag{1.1}$$

In the piezoelectric layer, $\overline{\Gamma}'_{il} = \Gamma'_{il} + \gamma'_i \gamma'_l / \varepsilon$ (see Sect. 4.3.1, Vol. I). Moreover, we must have $\gamma'_1 = \gamma'_2 = 0$. Then, from (4.75) of Vol. I,

$$e'_{34} = e'_{35} = 0 \,. \tag{1.2}$$

Since the modulus $e'_{33} \neq 0$, the wave is excited by the electric field parallel to x'_3. In the same way, conditions for propagation and excitation of a transverse wave, polarised in the x'_1 direction for example, are

$$c'_{45} = c'_{35} = 0 \,, \quad e'_{33} = e'_{34} = 0 \,, \quad e'_{35} \neq 0 \,. \tag{1.3}$$

Equations. If these conditions are satisfied, the *equation of state* (3.94), Vol. I, of the piezoelectric solid is, without indices,

$$\boxed{T = c^E \frac{\partial u}{\partial z} - eE} \,, \tag{1.4}$$

where $z = x'_3$, $E = E'_3$ and other notation is defined for longitudinal or transverse modes by:

- T is the stress T'_{33} or T'_{31}, u the displacement u'_3 or u'_1;
- c^E is the stiffness c'_{33} or c'_{55}, e the piezoelectric modulus e'_{33} or e'_{35}.

Similarly for the *second equation of state* (3.86), Vol. I,

$$\boxed{D = \varepsilon^S E + e \frac{\partial u}{\partial z}} \quad \text{with} \quad \varepsilon^S = \varepsilon'_{33} \,. \tag{1.5}$$

When there are no mechanical sources, the *equation of motion* (3.15), Vol. I, becomes, with $f_i = 0$,

$$\boxed{\frac{\partial T}{\partial z} = \rho \frac{\partial^2 u}{\partial t^2}} \,. \tag{1.6}$$

An *electrical external excitation* presupposes some charge density ρ_e, located on the electrodes in the case of an insulating material, and this appears in the second term of *Poisson's equation* (3.21), Vol. I,

$$\boxed{\frac{\partial D}{\partial z} = \rho_\mathrm{e}} \,. \tag{1.7}$$

It also implies injection of a current of intensity I. If this is uniform over the whole area A of the electrodes, the *charge conservation equation* (3.28), Vol. I, becomes, for an insulator ($J_n = 0$),

$$\boxed{\frac{\partial D}{\partial t} = J(t) = \frac{I(t)}{A}} \, , \tag{1.8}$$

where $J = -J_n'$ is the incoming current density per unit electrode area.

Equations (1.7) and (1.8), describing electrical excitation of the piezoelectric layer, bring out the importance of the normal component of electric displacement D. It is useful, therefore, to eliminate the electric field E between piezoelectric equations (1.4) and (1.5), in order to obtain the stress T in terms of D:

$$\boxed{T = c\frac{\partial u}{\partial z} - hD} \, , \tag{1.9}$$

setting $h = e/\varepsilon^S$ and $c = c^D$, where c^D is the stiffness at constant electric displacement,

$$c^D = c^E + \frac{e^2}{\varepsilon^S} = c^E\left(1 + \frac{e^2}{\varepsilon^S c^E}\right) . \tag{1.10}$$

Differentiating with respect to time and bringing in particle velocity $v = \partial u/\partial t$, the charge conservation equation (1.8) implies

$$\frac{\partial T}{\partial t} = c\frac{\partial v}{\partial z} - h\frac{\partial D}{\partial t} = c\frac{\partial v}{\partial z} - \frac{h}{A}I(t) . \tag{1.11}$$

In the same way, the dynamical equation (1.6) becomes

$$\boxed{\rho\frac{\partial^2 v}{\partial t^2} = c\frac{\partial^2 v}{\partial z^2}} \, , \tag{1.12}$$

since I is independent of z. The general solution of this equation of propagation is the sum of two waves propagating in opposite directions at velocity $V = \sqrt{c/\rho}$. In the *harmonic regime* and omitting the factor $\mathrm{e}^{\mathrm{i}\omega t}$,

$$v = a\mathrm{e}^{-\mathrm{i}kz} + b\mathrm{e}^{\mathrm{i}kz} = v_a + v_b \, , \tag{1.13}$$

where $k = \omega/V$. Given that the acoustic pressure δp in a fluid is analogous to the opposite of the stress T in a solid [compare (1.6) and (1.24), Vol. I], the results of Sect. 1.2.1.2, Vol. I, can all be transposed to the present case.

Characteristic impedance $Z = \rho V$ such that $-T_a = Zv_a$ is associated with the forward wave, i.e., the wave moving along $z > 0$, and impedance $-Z$ such that $-T_b = -Zv_b$ is associated with the wave moving along $z < 0$, where

$$Z = \rho V = \sqrt{\rho c} = \frac{c}{V} = \frac{ck}{\omega} . \tag{1.14}$$

Note that in Chap. 1 of Vol. I, c denoted phase speed in the fluid, whereas here it denotes the elastic constant of the piezoelectric material.

The stress is given by (1.11), using (1.13) for the speed,

$$T = -Z(a\mathrm{e}^{-\mathrm{i}kz} - b\mathrm{e}^{\mathrm{i}kz}) + \mathrm{i}\frac{h}{A\omega}I . \tag{1.15}$$

From the two solutions to the propagation equation in the harmonic regime, linear relations can be found between mechanical quantities (force F and particle velocity v) which are conserved at an interface, and electrical quantities (applied voltage U and injected current intensity I). These relations can be illustrated by equivalent electromechanical circuits. The Mason–Redwood equivalent circuit (see Sect. 1.3.1) follows from a matrix representation (see Sect. 1.2). The circuit due to Krimholtz et al. (the KLM circuit, Sect. 1.3.2) follows from an analysis in the form of two first order equations.

In the case of small electromechanical coupling $K^2 \ll 1$, mechanical boundary conditions on stresses and velocities at the interface (see Fig. 1.2) can be written down and the relations simplify so that the frequency response of the transducer can be calculated directly (see Sect. 1.5).

1.2 Impedance Matrix

We consider a piezoelectric slab of finite thickness $d = z_2 - z_1$ and cross-section A, as shown in Fig. 1.3a, bounded by planes $z = z_1$ and $z = z_2$. This is subject to voltage U and forces F_1, F_2 from the surrounding material. It can be represented by a three-port circuit, as shown in Fig. 1.3b. There is one electrical port (applied voltage U, injected current intensity I) and two acoustic ports, in which forces F_1, F_2 exerted on each face and incoming velocities v_1, v_2 play a role similar to voltage and current in an electrical circuit (see Sect. 1.2.1.2, Vol. I).

T corresponds to a traction per unit area exerted on the piezoelectric solid. Hence

$$F_1 = -AT(z_1) = \mathcal{Z}(a\mathrm{e}^{-\mathrm{i}kz_1} - b\mathrm{e}^{\mathrm{i}kz_1}) - \mathrm{i}\frac{h}{\omega}I\,, \tag{1.16a}$$

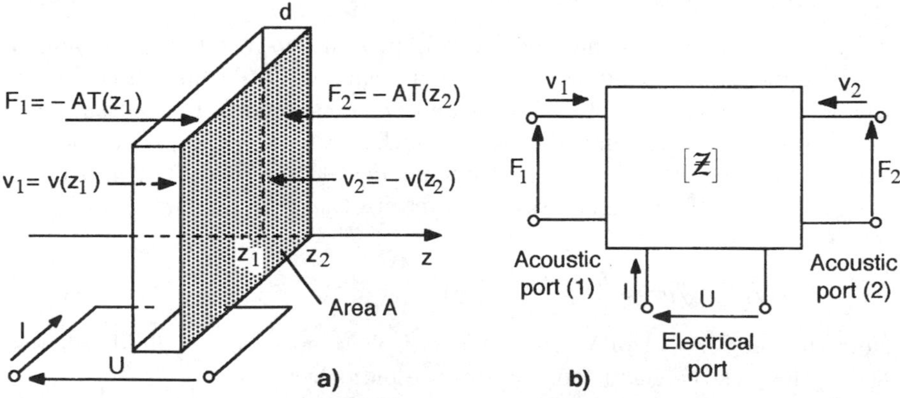

Fig. 1.3. Piezoelectric plate of cross-sectional area A. (**a**) Physical quantities: forces and velocities, voltage and current intensity. (**b**) Hexapole with two acoustic ports and one electrical port

$$F_2 = -AT(z_2) = \mathcal{Z}(a\mathrm{e}^{-\mathrm{i}kz_2} - b\mathrm{e}^{\mathrm{i}kz_2}) - \mathrm{i}\frac{h}{\omega} I\,, \tag{1.16b}$$

where

$$\mathcal{Z} = ZA = \frac{ck}{\omega} A \tag{1.17}$$

is the mechanical impedance in [kgs^{-1}]. The latter is defined as the ratio of force $F = -AT$ to vibration velocity v of a wave propagating up the z-axis, $\mathcal{Z} = F/v$. The aim is to express F_1 and F_2 in terms of (incoming) velocities v_1 and v_2 at the two faces,

$$v_1 = v(z_1) = a\mathrm{e}^{-\mathrm{i}kz_1} + b\mathrm{e}^{\mathrm{i}kz_1}\,, \tag{1.18a}$$

$$v_2 = -v(z_2) = -a\mathrm{e}^{-\mathrm{i}kz_2} - b\mathrm{e}^{\mathrm{i}kz_2}\,, \tag{1.18b}$$

which imply

$$a = \frac{v_1\mathrm{e}^{\mathrm{i}kz_2} + v_2\mathrm{e}^{\mathrm{i}kz_1}}{2\mathrm{i}\sin kd}\,, \qquad b = -\frac{v_2\mathrm{e}^{-\mathrm{i}kz_1} + v_1\mathrm{e}^{-\mathrm{i}kz_2}}{2\mathrm{i}\sin kd}\,. \tag{1.19}$$

Substituting these into (1.16a) and (1.16b), we find

$$F_1 = \mathcal{Z}\left(\frac{v_1}{\mathrm{i}\tan kd} + \frac{v_2}{\mathrm{i}\sin kd}\right) + \frac{hI}{\mathrm{i}\omega}\,, \tag{1.20a}$$

$$F_2 = \mathcal{Z}\left(\frac{v_1}{\mathrm{i}\sin kd} + \frac{v_2}{\mathrm{i}\tan kd}\right) + \frac{hI}{\mathrm{i}\omega}\,. \tag{1.20b}$$

The voltage U applied to metallised faces (see Fig. 1.3a) is calculated from the electric field E:

$$U = \int_{z_1}^{z_2} E\,\mathrm{d}z\,, \qquad E = -h\frac{\partial u}{\partial z} + \frac{D}{\varepsilon^S}\,. \tag{1.21}$$

In terms of the displacement current $I = \mathrm{i}\omega DA$ crossing the plate of area A,

$$U = h[u(z_1) - u(z_2)] + \frac{Id}{\mathrm{i}\omega\varepsilon^S A}\,. \tag{1.22}$$

Introducing speeds $v_1 = \mathrm{i}\omega u(z_1)$ and $v_2 = -\mathrm{i}\omega u(z_2)$ and the static capacitance C_0 of the rigidly attached transducer,

$$U = \frac{h}{\mathrm{i}\omega}(v_1 + v_2) + \frac{I}{\mathrm{i}\omega C_0}\,, \qquad C_0 = \frac{\varepsilon^S A}{d}\,. \tag{1.23}$$

It is useful to write (1.20a), (1.20b) and (1.23) for the piezoelectric part in matrix form:

$$\begin{pmatrix} F_1 \\ F_2 \\ U \end{pmatrix} = -\mathrm{i}\begin{pmatrix} \mathcal{Z}/\tan kd & \mathcal{Z}/\sin kd & h/\omega \\ \mathcal{Z}/\sin kd & \mathcal{Z}/\tan kd & h/\omega \\ h/\omega & h/\omega & 1/\omega C_0 \end{pmatrix}\begin{pmatrix} v_1 \\ v_2 \\ I \end{pmatrix}. \tag{1.24}$$

This involves the electromechanical impedance matrix $[\mathcal{Z}]$ of the three-port circuit in Fig. 1.3b. It is symmetric because acoustic ports 1 and 2 are equivalent. Introducing load impedances on each face of the piezoelectric plate, the transmitted acoustic power can be found, together with the frequency response and electrical impedance of the transducer (Problem 1.1 and Sect. 1.4). Another method is to associate an equivalent circuit to each part of the model in Fig. 1.2, in which forces and velocities play the roles of voltage and current intensity (see Sect. 1.2.2, Vol. I).

1.3 Equivalent Circuits

Two circuits are often used to illustrate the operation of transducers. One is composed of localised elements (the Mason circuit [1.1]), some of which may be replaced by a section of coaxial cable (Redwood [1.2]). This model is deduced from the impedance matrix. The other is a transmission line in the centre of which a current is injected, between the two conducting wires. This is the KLM model due to Krimholtz, Leedom and Matthaei [1.3]. It is based on the dynamical equation and an expression for the stress.

1.3.1 Mason and Redwood Circuits

Let us begin by examining a non-piezoelectric medium and then a piezoelectric solid.

- An infinite medium of cross-section A, in which a single progressive wave propagates along the positive z axis, is represented by its mechanical impedance $\mathcal{Z} = ZA$, ratio of the force $F = -AT$ to the vibration velocity $v = \partial u/\partial t$.
- The equivalent circuit of a (non-piezoelectric) slab of finite thickness $d = z_2 - z_1$ and cross-section A, bounded by planes $z = z_1$, $z = z_2$, follows from (1.20a) and (1.20b) with $h = 0$, relating forces F_1 and F_2 to velocities v_1 and v_2 at the two faces. Using the identity

$$\frac{1}{\tan kd} = \frac{1}{\sin kd} - \tan\frac{kd}{2}\,, \tag{1.25}$$

it follows that

$$F_1 = -\mathrm{i}\frac{\mathcal{Z}}{\sin kd}(v_1 + v_2) + \mathrm{i}\mathcal{Z}\tan\frac{kd}{2}\,v_1\,, \tag{1.26a}$$

$$F_2 = -\mathrm{i}\frac{\mathcal{Z}}{\sin kd}(v_1 + v_2) + \mathrm{i}\mathcal{Z}\tan\frac{kd}{2}\,v_2\,. \tag{1.26b}$$

These relations are also found by applying Kirchoff's laws to the circuit in Fig. 1.4a. This is the equivalent T element for the solid slab. The coaxial line element of length d and characteristic impedance $\mathcal{Z}$ in Fig. 1.4b obeys the same laws and is also an equivalent circuit for the solid layer (see Problem 1.2).

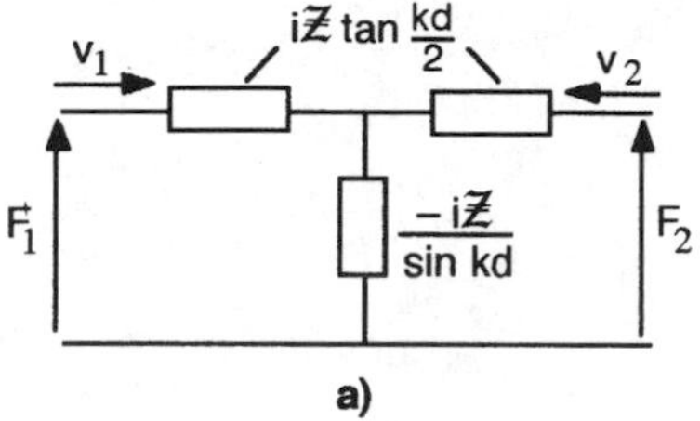

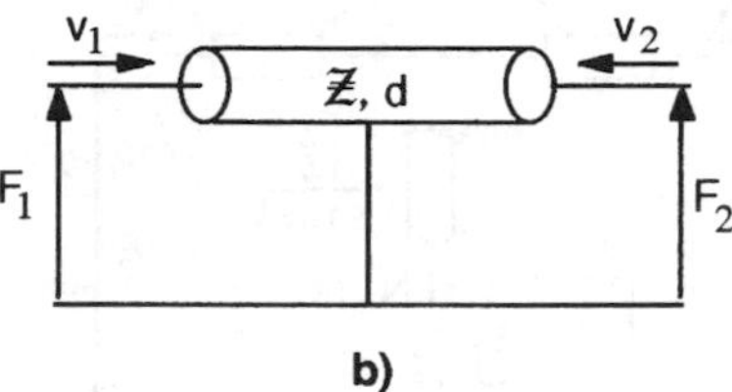

Fig. 1.4. (**a**) Equivalent T element and (**b**) equivalent coaxial line element for a (non-piezoelectric) slab of thickness d, cross-section A and elastic impedance $Z = \mathcal{Z}/A$

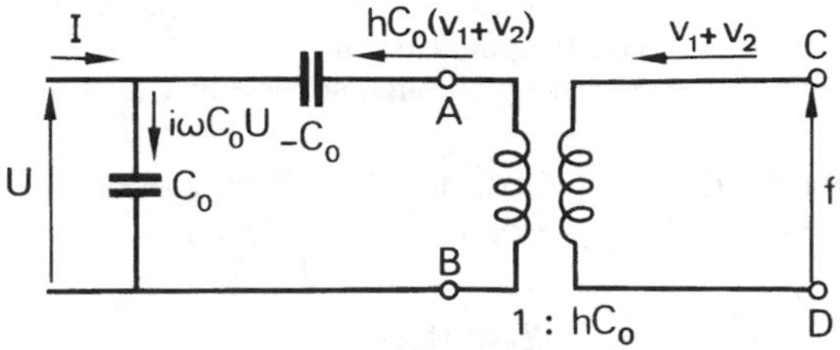

Fig. 1.5. Electromechanical transformer representing piezoelectricity

- When the material is piezoelectric, there is a further force $f = hI/\mathrm{i}\omega$ occurring as the second term in each of (1.20a) and (1.20b). Given (1.23), this force can be written

$$f = \frac{hI}{\mathrm{i}\omega} = hC_0 \left[U - \frac{h}{\mathrm{i}\omega}(v_1 + v_2) \right] . \tag{1.27}$$

This can be obtained by means of an electromechanical transformer with velocity $v_1 + v_2$ in the secondary and current $hC_0(v_1 + v_2) = \mathrm{i}\omega C_0 U - I$ in the primary, implying a transformation ratio equal to $N = hC_0$ (see Fig. 1.5). Indeed, the series capacitance $-C_0$ in the primary does create a force f between terminals C and D:

$$f = hC_0(U_\mathrm{A} - U_\mathrm{B}) = hC_0 \left[U - \frac{hC_0(v_1 + v_2)}{\mathrm{i}\omega C_0} \right] .$$

The equivalent circuit (the Mason equivalent circuit) for a piezoelectric plate with one electrical port and two mechanical ports is sketched in Fig. 1.6.

The equivalent circuit for a loaded transducer, shown in Fig. 1.7, is found by joining together in cascade the equivalent circuits for each constituent medium. If the face of the external electrode is free, as in Fig. 1.2, the corresponding terminals are practically short-circuited since the elastic impedance of air is so small.

The mean elastic power generated can be found from the amplitude of the applied force F on the interface with the propagating medium, or from the amplitude of the particle velocity v in the emitted wave (see Sect. 1.4.1):

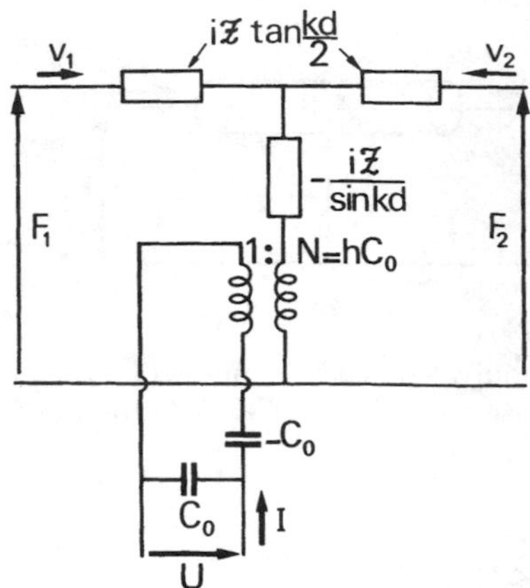

Fig. 1.6. Equivalent electromechanical circuit for a slab of piezoelectric material of thickness d, cross-section A and mechanical impedance $\mathcal{Z} = ZA$. There is one electrical port and two mechanical ports. Propagation in the slab is also represented by the line section in Fig. 1.4b

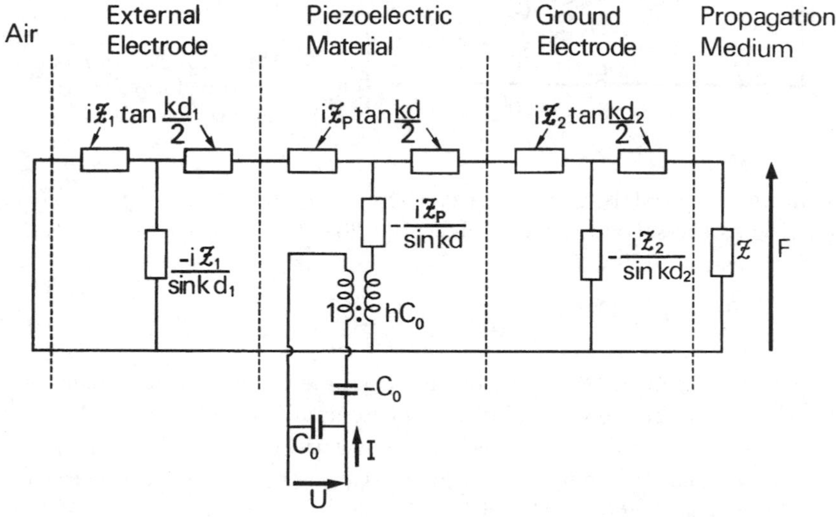

Fig. 1.7. Complete equivalent circuit for a transducer. The left-hand end, in contact with the air here, is short-circuited. The other end is loaded by the elastic impedance of the propagating medium

$$\langle P\rangle = \frac{|F|^2}{2\mathcal{Z}} = \frac{1}{2}\mathcal{Z}|v|^2 = \frac{1}{2}ZA|v|^2 \,. \tag{1.28}$$

Crossed-Field Model. The circuit in Fig. 1.6 corresponds to the case where the electric field is parallel to the direction of propagation of elastic waves (in-line field model). It must be modified if the electric field is perpendicular to the direction of propagation, as shown in Fig. 1.8. The piezoelectric rod is crystallographically oriented and metallised, so that the field $\boldsymbol{E}$, parallel to x_1, excites either a longitudinal wave (displacement u_3), or a transverse wave (displacement u_1 or u_2), propagating in the x_3 (z) direction. Equations

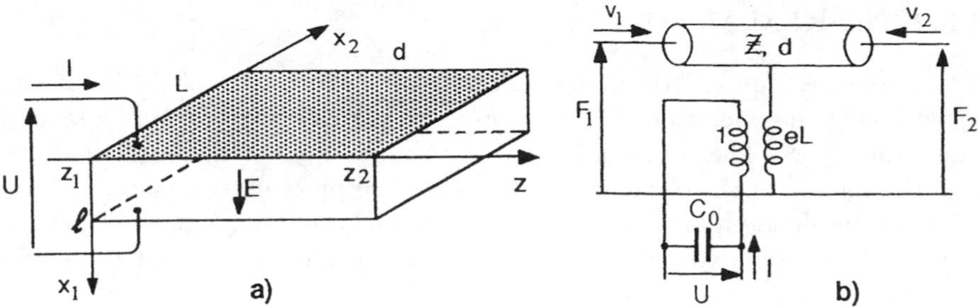

Fig. 1.8. Crossed-field model. (**a**) A piezoelectric rod of length d, excited by an electric field perpendicular to the direction of propagation of the elastic waves. (**b**) Equivalent circuit

of state for the piezoelectric solid are

$$T_\alpha = c^E_{\alpha\alpha}\frac{\partial u}{\partial z} - e_{1\alpha}E_1 \,, \tag{1.29a}$$

$$D_1 = \varepsilon_{11}E_1 + e_{1\alpha}\frac{\partial u}{\partial z} \,, \tag{1.29b}$$

where $\alpha = 3$ in the case of a longitudinal wave and $\alpha = 1$ or 2 in the case of transverse waves.

If the thickness l of the rod is small compared with other dimensions L and d, the electric field does not depend on z. Then $E_1 = U/l$ and the propagation equation

$$\rho\frac{\partial^2 u}{\partial t^2} = \frac{\partial T_\alpha}{\partial z} = c^E_{\alpha\alpha}\frac{\partial^2 u}{\partial z^2} \tag{1.30}$$

shows that the phase speed of the wave is $V = \sqrt{c^E_{\alpha\alpha}/\rho}$. The force f resulting from the piezoelectric stress $t_\alpha = -e_{1\alpha}E_1$ is the same on the two faces with abscissas z_1 and z_2 :

$$f = -At_\alpha = eEA \,,$$

setting $e_{1\alpha} = e$ and $E_1 = E$. The current intensity is given by $I = \mathrm{d}Q/\mathrm{d}t = \mathrm{i}\omega Q$, where Q is the charge on electrode $x_1 = 0$,

$$Q = \int_S D_1 \mathrm{d}s = L\int_{z_1}^{z_2} D_1(z)\mathrm{d}z = \varepsilon^S_{11}E_1Ld + eL[u(z_2) - u(z_1)] \,.$$

Now bringing in the static capacitance $C_0 = \varepsilon^S_{11}Ld/l$ and velocities $v_1 = \mathrm{i}\omega u(z_1)$, $v_2 = -\mathrm{i}\omega u(z_2)$ on the two faces, it follows that

$$I = \mathrm{i}\omega C_0U - eL(v_1 + v_2) \,, \qquad f = eLU \,. \tag{1.31}$$

The equivalent circuit in Fig. 1.8b no longer contains the capacitance $-C_0$ and the transformation ratio is $M = eL = hmC_0$, where $m = l/d$.

If the rod is loaded on either side by some material of the same characteristic impedance, operation of the transducer is described by two equal and opposite sources placed at the ends of the rod (Problem 1.3).

1.3.2 The KLM Model

This model is due to Krimholtz, Leedom and Matthaei [1.3]. It is useful for studying the operation of high bandwidth transducers containing several matching layers. The transducer is set into vibration by injecting a current into the middle of the element which models the piezoelectric plate.

Equations chosen for the 1-dimensional model are the dynamical equation (1.6), with particle velocity v in place of the displacement, and the stress equation (1.11):

$$\begin{aligned}\frac{\partial T}{\partial z} &= \rho\frac{\partial v}{\partial t}\,,\\ \frac{\partial T}{\partial t} &= c\frac{\partial v}{\partial z} - \frac{h}{A}I(t)\,.\end{aligned} \tag{1.32}$$

Hence, in the harmonic regime and introducing the acoustic impedance $Z = ck/\omega$ and mechanical impedance $\mathcal{Z} = AZ$,

$$\begin{aligned}\frac{\mathrm{d}T}{\mathrm{d}z} &= \mathrm{i}\omega\rho v = \mathrm{i}kZv\,,\\ \frac{\mathrm{d}v}{\mathrm{d}z} &= \mathrm{i}\frac{\omega}{c}T + \frac{h}{cA}I(z) = \mathrm{i}\frac{k}{Z}T + \frac{h}{\mathcal{Z}V}I(z)\,.\end{aligned} \tag{1.33}$$

In order to exhibit clearly the two modes propagating in opposite directions in the circuit, and the injected current intensity, we write

$$v = v_a + v_b\,, \qquad T = Z(v_b - v_a)\,,$$

using indices a and b to denote the corresponding physical quantities. Substituting into (1.33), this yields

$$\begin{aligned}\frac{\mathrm{d}}{\mathrm{d}z}(v_b - v_a) &= \mathrm{i}k(v_a + v_b)\,,\\ \frac{\mathrm{d}}{\mathrm{d}z}(v_b + v_a) &= \mathrm{i}k(v_b - v_a) + \alpha I(z)\,,\end{aligned} \tag{1.34}$$

where $\alpha = h/\mathcal{Z}V$. Separating the two modes by term by term subtraction and addition, we find

$$\begin{aligned}\frac{\mathrm{d}v_a}{\mathrm{d}z} + \mathrm{i}kv_a &= \frac{\alpha}{2}I(z)\,,\\ \frac{\mathrm{d}v_b}{\mathrm{d}z} - \mathrm{i}kv_b &= \frac{\alpha}{2}I(z)\,,\end{aligned}$$

or alternatively

$$\begin{aligned}\frac{\mathrm{d}}{\mathrm{d}z}(v_a\mathrm{e}^{\mathrm{i}kz}) &= \frac{\alpha}{2}I(z)\mathrm{e}^{\mathrm{i}kz}\,,\\ \frac{\mathrm{d}}{\mathrm{d}z}(v_b\mathrm{e}^{-\mathrm{i}kz}) &= \frac{\alpha}{2}I(z)\mathrm{e}^{-\mathrm{i}kz}\,.\end{aligned} \tag{1.35}$$

When there is no current, solutions of these two equations are normal modes (1.13) propagating in the piezoelectric material.

In the general case, placing the origin in the middle of the piezoelectric plate, the first equation is integrated from the end lying at $z = -d/2$ to give:

$$v_a(z) = v_a(-d/2)\mathrm{e}^{-\mathrm{i}k(z+d/2)} + \frac{\alpha}{2}\mathrm{e}^{-\mathrm{i}kz}\int_{-d/2}^{z} I(\zeta)\mathrm{e}^{\mathrm{i}k\zeta}\,\mathrm{d}\zeta\,. \tag{1.36}$$

This expression shows that the amplitude of the wave v_a in the plane through z results from the wave originating in the plane $z = -d/2$ added to the cumulative excitation through the emitting region (transducer) due to the current $I(z)$ (see Fig. 1.9). The second equation is obtained by integrating from the $z = +d/2$ end of the transducer, equivalent to making the replacement $k \to -k$, $d \to -d$:

$$v_b(z) = v_b(d/2)\mathrm{e}^{\mathrm{i}k(z-d/2)} + \frac{\alpha}{2}\mathrm{e}^{\mathrm{i}kz}\int_{d/2}^{z} I(\zeta)\mathrm{e}^{-\mathrm{i}k\zeta}\,\mathrm{d}\zeta\,. \tag{1.37}$$

Assuming intensity I to be independent of z, velocities v_a and v_b at the ends $z = +d/2$ and $z = -d/2$ are given by

$$v_a(+d/2) = v_a(-d/2)\mathrm{e}^{-\mathrm{i}kd} + \frac{hI}{\mathcal{Z}\omega}\mathrm{e}^{-\mathrm{i}kd/2}\sin\frac{kd}{2}\,, \tag{1.38a}$$

$$v_b(-d/2) = v_b(+d/2)\mathrm{e}^{-\mathrm{i}kd} - \frac{hI}{\mathcal{Z}\omega}\mathrm{e}^{-\mathrm{i}kd/2}\sin\frac{kd}{2}\,. \tag{1.38b}$$

Examining phases of the second terms, we find that the wave v_a (v_b) leaving the face of the transducer at $z = +d/2$ ($z = -d/2$) is the sum of a wave which has entered by the other face, at $z = -d/2$ ($z = +d/2$), and a wave which leaves from the centre plane at $z = 0$, excited by the current of intensity I. In order to understand how waves are generated in the centre plane, we must express the jump in speed for each wave as it crosses the plane $z = 0$, i.e., we must calculate velocities immediately to the right ($z = \varepsilon > 0$) and to the left ($z = -\varepsilon$) of this plane. Multiplying (1.38a) and (1.38b) by $\mathrm{e}^{\mathrm{i}kd/2}$ yields

$$v_a(d/2)\mathrm{e}^{\mathrm{i}kd/2} = v_a(-d/2)\mathrm{e}^{-\mathrm{i}kd/2} + \frac{hI}{\mathcal{Z}\omega}\sin\frac{kd}{2}\,, \tag{1.39}$$

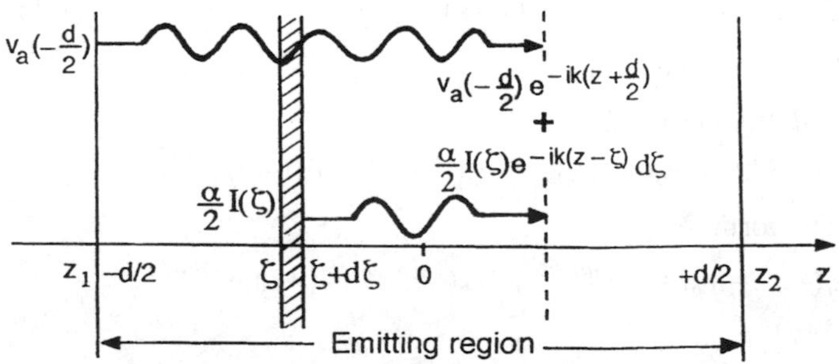

Fig. 1.9. In the abscissa plane through $z > \zeta$, the wave propagating up the z-axis is the resultant of the wave at the entry $z_1 = -d/2$ to the emitting region and those waves emitted by a continuous array of sources of amplitude $\alpha I(\zeta)\mathrm{d}\zeta/2$, which arrive at z with phase delay $k(z - \zeta)$

$$v_b(-d/2)\mathrm{e}^{\mathrm{i}kd/2} = v_b(d/2)\mathrm{e}^{-\mathrm{i}kd/2} - \frac{hI}{\mathcal{Z}\omega}\sin\frac{kd}{2}\,. \tag{1.40}$$

Given the direction of propagation of modes v_a and v_b, and the phase difference $-kd/2$ for a path of length $d/2$ in either direction between abscissa points $-d/2$ and $-\varepsilon$ for one and between $+\varepsilon$ and $+d/2$ for the other, it follows that

$$v_a(d/2) = v_a(+\varepsilon)\mathrm{e}^{-\mathrm{i}kd/2}\,, \qquad v_a(-\varepsilon) = v_a(-d/2)\mathrm{e}^{-\mathrm{i}kd/2}\,, \tag{1.41a}$$

$$v_b(-d/2) = v_b(-\varepsilon)\mathrm{e}^{-\mathrm{i}kd/2}\,, \qquad v_b(+\varepsilon) = v_b(d/2)\mathrm{e}^{-\mathrm{i}kd/2}\,. \tag{1.41b}$$

Equations (1.39) and (1.40) now become

$$v_a(+\varepsilon) = v_a(-\varepsilon) + \frac{hI}{\mathcal{Z}\omega}\sin\frac{kd}{2}\,, \tag{1.42a}$$

$$v_b(+\varepsilon) = v_b(-\varepsilon) + \frac{hI}{\mathcal{Z}\omega}\sin\frac{kd}{2}\,. \tag{1.42b}$$

The total difference in velocity $v = v_a + v_b$ from one side of the centre plane to the other, produced by injecting a current of intensity I, is

$$\begin{aligned} v(+\varepsilon) - v(-\varepsilon) &= [v_a(+\varepsilon) + v_b(+\varepsilon)] - [v_a(-\varepsilon) + v_b(-\varepsilon)] \\ &= \frac{2hI}{\mathcal{Z}\omega}\sin\frac{kd}{2}\,. \end{aligned} \tag{1.43}$$

It is just as though the current of intensity I in the primary coil of an electromechanical transformer were inducing a velocity of intensity NI in the secondary coil, with turns ratio N a function of angular frequency ω:

$$N = \frac{2h}{\mathcal{Z}\omega}\sin\frac{\omega d}{2V}\,. \tag{1.44}$$

It is the potential difference U applied across the electrodes of the crystal of capacitance C_0 which produces the current I and the jump in speed. Before finding the sum $v_1 + v_2$ in (1.23) explicitly, note that the stress $T(z)$ is not discontinuous across the centre plane since, according to (1.42a) and (1.42b),

$$\begin{aligned} T(+\varepsilon) &= -Z[v_a(+\varepsilon) - v_b(+\varepsilon)] \\ &= -Z[v_a(-\varepsilon) - v_b(-\varepsilon)] = T(-\varepsilon)\,. \end{aligned} \tag{1.45}$$

Given (1.41a) and (1.41b), the sum

$$\begin{aligned} v_1 + v_2 &= v(-d/2) - v(+d/2) \\ &= [v_a(-d/2) - v_b(+d/2)] + [v_b(-d/2) - v_a(+d/2)] \end{aligned}$$

can be written in the form

$$v_1 + v_2 = [v_a(-\varepsilon) - v_b(+\varepsilon)]\mathrm{e}^{\mathrm{i}kd/2} + [v_b(-\varepsilon) - v_a(+\varepsilon)]\mathrm{e}^{-\mathrm{i}kd/2}\,,$$

which implies

$$\begin{aligned} v_1 + v_2 = {} & [v(-\varepsilon) - v(+\varepsilon)]\cos\frac{kd}{2} + \\ & \mathrm{i}[v_a(-\varepsilon) - v_b(-\varepsilon) + v_a(+\varepsilon) - v_b(+\varepsilon)]\sin\frac{kd}{2}\,. \end{aligned}$$

Using (1.43) and (1.45),

$$v_1 + v_2 = -2\frac{hI}{\mathcal{Z}\omega}\sin\frac{kd}{2}\cos\frac{kd}{2} - 2\mathrm{i}\frac{T}{Z}\sin\frac{kd}{2}\,,$$

and the applied voltage U is obtained by substituting this sum into (1.23):

$$U = \frac{I}{\mathrm{i}\omega C_0} - \frac{h^2 I}{\mathrm{i}\mathcal{Z}\omega^2}\sin kd - \frac{2hT}{\mathcal{Z}\omega}\sin\frac{kd}{2}\,. \tag{1.46}$$

The equivalent form

$$U = \frac{I}{\mathrm{i}\omega}\left(\frac{1}{C_0} + \frac{1}{C}\right) - NTA$$

exhibits the static capacitance C_0 of the rigidly bonded transducer and the frequency dependent motional capacitance C,

$$C = -\frac{\mathcal{Z}\omega}{h^2 \sin kd} = -\frac{\mathcal{Z}V}{dh^2}\frac{\pi\omega/\omega_{\mathrm{P}}}{\sin kd}\,, \tag{1.47}$$

where $\omega_{\mathrm{P}} = \pi V/d$. The term $-NTA$, proportional to the force $-TA$, indicates the generation of elastic waves. Introducing the electromechanical coupling coefficient, defined in Sect. 4.3.2 of Vol. I by $K = e/\sqrt{\varepsilon^S c^D}$,

$$\frac{h^2 d}{\mathcal{Z}V} = \frac{e^2}{(\varepsilon^S)^2}\frac{1}{\rho V^2}\frac{d}{A} = \frac{e^2}{\varepsilon^S c^D}\frac{d}{\varepsilon^S A} = \frac{K^2}{C_0}\,,$$

so that the motional capacitance becomes

$$C = -\frac{C_0}{K^2\,\mathrm{sinc}\,x}\,,$$

where $x = kd = \pi\omega/\omega_{\mathrm{P}}$. It is infinite at the $\lambda/2$ resonance ($\omega = \omega_{\mathrm{P}}$). When ω approaches zero, it tends to $-C_0/K^2$. When $0 < \omega < \omega_{\mathrm{P}}$, it is negative with absolute value greater than C_0.

The KLM model, represented by the three-port circuit in Fig. 1.10, comprises a transmission line corresponding to propagation, and an electromechanical transformer modelling the electrical source.

The great advantage of this circuit diagram is that forces appear only at line terminals, rather than occurring partly at transformer terminals as in the Mason circuit (Fig. 1.6). Mechanical actions at the ends of the line and electrical actions localised at its centre are physically separated at the expense of having a frequency dependence in some of the circuit elements (motional capacitance and transformation ratio).

1.4 Low-Frequency Transducers

Transducers operating at low frequencies ($f < 100$ MHz, i.e., $\lambda > 50$ μm) are often made from piezoelectric ceramics (piezoceramics) such as PZT, or crystals such as lithium niobate or tantalate, which have strong electromechanical coupling coefficients. The mechanical effect of electrodes is negligible, because they are extremely thin compared with the piezoelectric plate ($\approx \lambda/2$).

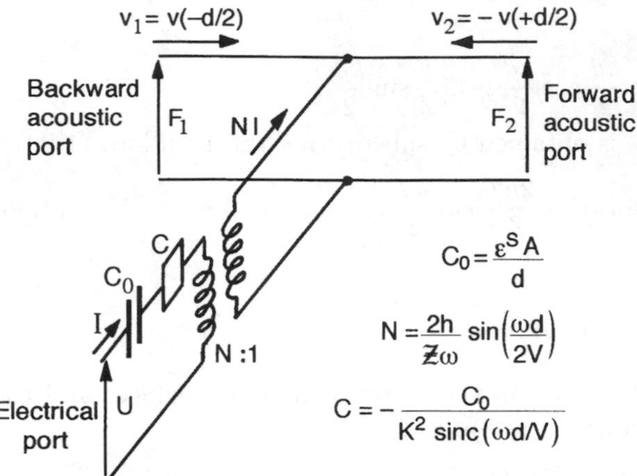

Fig. 1.10. KLM model, due to Krimholtz et al. [1.3]. Transformation ratio N and motional capacitance C are frequency dependent

1.4.1 Emitted Power and Frequency Response

In order to bring out the role of the electromechanical coupling, let us calculate the emitted power for the piezoelectric solid/propagating medium structure shown in Fig. 1.11.

Solving (1.24) with $F_1 = 0$ and $F_2 = -\mathcal{Z}v_2$ leads to an expression for particle velocity $v = -v_2$ in the propagating medium of impedance $\mathcal{Z}_2 = \mathcal{Z} = AZ$ (Problem 1.1). From (1.93) in Problem 1.1,

$$v = \frac{hC_0}{\sqrt{\mathcal{Z}\mathcal{Z}_{\mathrm{P}}}m(\theta)}U\,, \quad \text{where} \quad \theta = \frac{kd}{2} = \frac{\pi}{2}\frac{f}{f_{\mathrm{P}}}\,, \quad f_{\mathrm{P}} = \frac{V_{\mathrm{P}}}{2d}\,, \tag{1.48}$$

and we have defined

$$m(\theta) = \sqrt{\frac{Z}{Z_{\mathrm{P}}}}\left(\frac{K^2}{2\theta} - \cot 2\theta\right)\cot\theta + \mathrm{i}\sqrt{\frac{Z_{\mathrm{P}}}{Z}}\left(\frac{K^2}{\theta} - \cot\theta\right)\,. \tag{1.49}$$

Putting $C_0 = \varepsilon^S A/d$, $h = e/\varepsilon^S$ and $\mathcal{Z}_{\mathrm{P}} = AZ_{\mathrm{P}}$, this velocity can be written simply as

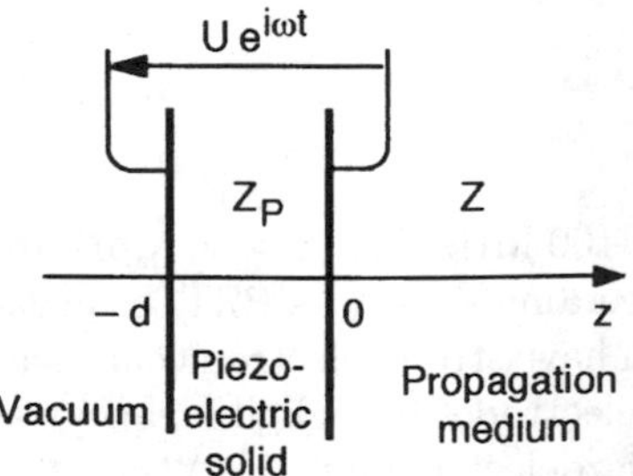

Fig. 1.11. By hypothesis, electrode thicknesses are negligible in comparison with dimensions of the piezoelectric crystal

$$v = \frac{e}{d\sqrt{ZZ_{\mathrm{P}}}} \frac{U}{m(\theta)} . \tag{1.50}$$

It depends on the ratio of the excitation frequency and half-wavelength resonance frequency f_{P} of the free piezoelectric plate, via the *frequency response* $H(f)$ of the transducer, defined by

$$H(f) = \frac{1}{m(\pi f/2f_{\mathrm{P}})} \quad \Rightarrow \quad \frac{v(f)}{U(f)} = \frac{e}{d\sqrt{ZZ_{\mathrm{P}}}} H(f) . \tag{1.51}$$

The *transmitted power* at any instant t is $P(t) = -ATv = AZv^2$. In the sinusoidal regime $U = |U|\mathrm{e}^{\mathrm{i}\omega t} \Rightarrow v = a\mathrm{e}^{\mathrm{i}\omega t}$, this has average

$$\langle P \rangle = \frac{1}{2} AZ|a|^2 = \frac{1}{2} C_0 \frac{e^2}{Z_{\mathrm{P}}\varepsilon^S d} |H(f)|^2 |U|^2 .$$

Given the relation $Z_{\mathrm{P}} = \rho_{\mathrm{P}} V_{\mathrm{P}} = c^D/V_{\mathrm{P}}$, the electromechanical coupling coefficient for thickness mode vibrations is

$$K^2 = \frac{e^2}{\varepsilon^S c^E + e^2} = \frac{e^2}{\varepsilon^S c^D} = \frac{h^2 \varepsilon^S}{c^D} . \tag{1.52}$$

Introducing the resonance frequency $f_{\mathrm{P}} = V_{\mathrm{P}}/2d$ of the piezoelectric crystal,

$$\boxed{\langle P \rangle = K^2 C_0 f_{\mathrm{P}} |H(f)|^2 |U|^2} . \tag{1.53}$$

Equation (1.53) shows that the transmitted power is proportional to the square of the electromechanical coupling factor and that it depends on frequency through the term $|H(f)|^2$. But the frequency dependence of $|H|$ is different for different values of K^2. The curves in Fig. 1.12, plotted as functions of relative frequency f/f_{P} for various values of K^2 and elastic impedance ratio Z_{P}/Z, are hardly symmetric about f_{P} once K^2 goes beyond 0.2. As K^2 increases, curves shift down towards low frequencies.

This behaviour can be understood in the following way. When the impedance of the propagating medium is small compared with that of the piezoelectric crystal, the latter vibrates at half-wavelength, its faces being practically free (curves for $Z_{\mathrm{P}}/Z = 2$). This results in a maximum elastic power produced at $f = f_{\mathrm{P}}$. Conversely, when Z is large compared with Z_{P} ($Z_{\mathrm{P}}/Z = 1/2$ or $1/4$), one of the faces of the transducer is free and the other almost completely fixed. It is then the $\lambda_{\mathrm{P}}/4$ or $3\lambda_{\mathrm{P}}/4$ vibrations which are excited. The corresponding radiated power is maximal near $f = 0.5 f_{\mathrm{P}}$ and $f = 1.5 f_{\mathrm{P}}$. The value $Z_{\mathrm{P}}/Z = 1/\sqrt{2}$, above which there is only a minimum in the power, leads to a very flat frequency response around centre frequency f_{P}. The bandwidth extends from $0.4 f_{\mathrm{P}}$ to $1.5 f_{\mathrm{P}}$ and exceeds 100% in relative value.

The *efficiency* of a piezoelectric transducer can be measured by the factor

$$G = \frac{\langle P \rangle}{|U|^2} = K^2 C_0 f_{\mathrm{P}} |H(f)|^2 ,$$

given in $[\mathrm{WV}^{-2}]$. Since $C_0 = \varepsilon^S A/d$ and $f_{\mathrm{P}} = V_{\mathrm{P}}/2d$, G can also be written

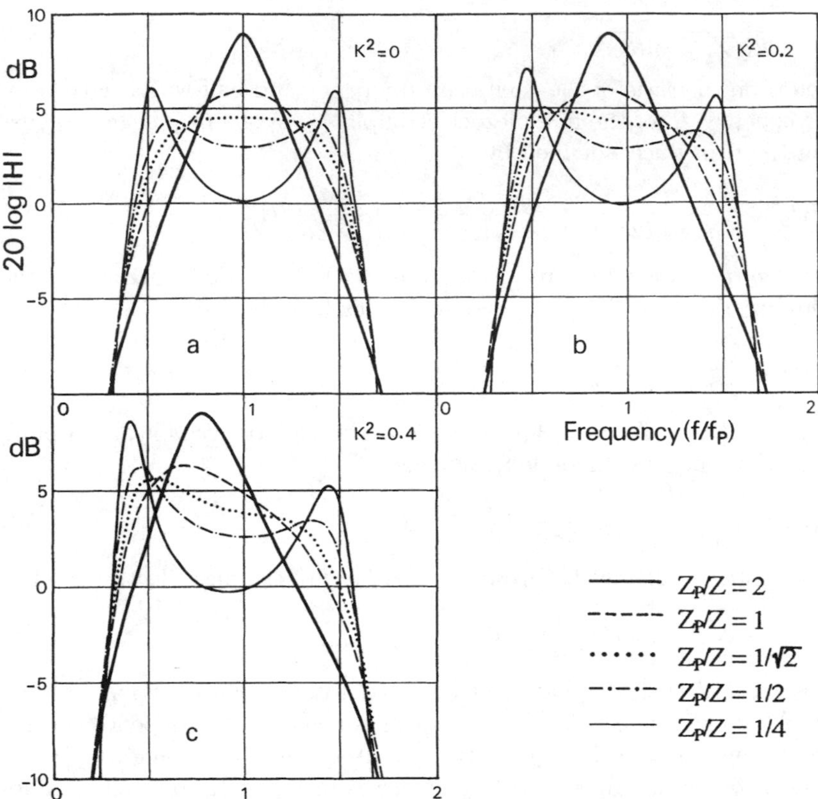

Fig. 1.12a–c. Responses of a transducer as a function of relative frequency f/f_{P} for different values of K^2 and elastic impedance ratios Z_{P}/Z. If the impedance Z of the medium is smaller than that of the crystal Z_{P}, the crystal vibrates at $\lambda/2$ ($Z_{\mathrm{P}}/Z = 2$). Otherwise, it vibrates at $\lambda/4$ and $3\lambda/4$ ($Z_{\mathrm{P}}/Z = 0.25$)

$$G = \frac{K^2}{2}\varepsilon^S V_{\mathrm{P}} r |H(f)|^2 , \tag{1.54}$$

where $r = A/d^2$ is a dimensionless ratio characterising the geometry of the transducer. The thickness mode vibration is preferentially excited provided that lateral dimensions of the piezoelectric plate are greater than ten times its thickness d, i.e., $r > 100$. Characteristics of the material enter through the coefficient $K^2\varepsilon^S V_{\mathrm{P}}$, and some of these characteristics are given in Table 1.1 for the main piezoelectric transducer materials (Sect. 1.6).

Let us compare efficiencies of a mid-frequency transducer (100 MHz), made from $(Y + 36°)$-cut lithium niobate, and a low-frequency transducer (10 MHz), made from Z-cut PZT ceramic. In both cases, the elastic impedance Z is matched to Z_{P}, for example, with an intermediate layer (see Problem 1.6, Vol. I). Curves in Fig. 1.12 show that in the transducer bandwidth, $20 \log |H| \approx 3$ dB, i.e., the factor $|H|$ is of order $\sqrt{2}$.

In the case of lithium niobate, using the information in Table 1.1,

$$K^2 = 0.24\ , \quad \varepsilon^S = 3.4 \times 10^{-10}\,\mathrm{Fm}^{-1}\ , \quad V_\mathrm{P} = 7\,340\,\mathrm{ms}^{-1}\ ,$$

and given the geometric characteristics,

$$f_\mathrm{P} = 100\,\mathrm{MHz} \Rightarrow d = 37\,\mu\mathrm{m} \quad \text{and} \quad A = 10\,\mathrm{mm}^2 \Rightarrow r = 7\,300\ ,$$

we find $G = 4.3\times10^{-3}$ WV^{-2}. A sinusoidal voltage of amplitude 1 V generates an elastic power of 4.3 mW. Characteristics of the second transducer,

$$K^2 = 0.26\ , \quad \varepsilon^S = 56 \times 10^{-10}\,\mathrm{Fm}^{-1}\ , \quad V_\mathrm{P} = 4\,560\,\mathrm{ms}^{-1}\ ,$$

and

$$f_\mathrm{P} = 10\,\mathrm{MHz} \Rightarrow d = 228\,\mu\mathrm{m} \quad \text{and} \quad A = 100\,\mathrm{mm}^2 \Rightarrow r = 1\,920\ ,$$

lead to a greater efficiency, viz., $G = 12.7 \times 10^{-3}$ WV^{-2}. A voltage of amplitude 1 V generates an elastic power of 12.7 mW.

1.4.2 Time Response

When used in devices for medical echography and non-destructive testing of materials, transducers operate in impulse mode. Axial resolution is directly related to the duration of the pulse, which is a function of the transducer bandwidth. The impulse response $h(t)$ is the response to a brief pulse $\delta(t)$ of unit area (Dirac pulse). It can be found by Fourier transformation of the frequency response $H(f)$:

$$h(t) = \int_{-\infty}^{\infty} H(f)\mathrm{e}^{-2\pi \mathrm{i} f t}\mathrm{d}f\ .$$

Making the variable change $u = f/f_\mathrm{P}$,

$$h(t) = f_\mathrm{P} \int_{-\infty}^{\infty} H(u)\mathrm{e}^{-2\pi \mathrm{i} u f_\mathrm{P} t}\mathrm{d}u = f_\mathrm{P} h_\mathrm{N}(t/\tau_\mathrm{P})\ ,$$

where $\tau_\mathrm{P} = 1/f_\mathrm{P} = 2d/V_\mathrm{P}$ is twice the time taken to cross the piezoelectric plate. The normalised response $h_\mathrm{N}(t/\tau_\mathrm{P})$ is plotted in Fig. 1.13, using (1.51), for $K^2 = 0.4$ and for several values of the elastic impedance ratio Z_P/Z. The time scale is $f_\mathrm{P}t = t/\tau_\mathrm{P}$. The shortest acoustic impulses are obtained when Z_P is close to Z (curves b and c). The response shape is then almost bipolar for $Z = Z_\mathrm{P}$. The curve f reveals spectral components at $f_\mathrm{P}/2$ and $3f_\mathrm{P}/2$.

Particle velocities are found from the convolution of the impulse response $h(t)$ and the voltage $U(t)$ applied to the transducer. Denoting the Fourier transform of $U(t)$ by $\overline{U}(f)$ and using (1.51),

$$v(f) = \frac{e}{d\sqrt{ZZ_\mathrm{P}}} H(f)\overline{U}(f) \quad \rightarrow \quad v(t) = \frac{e}{d\sqrt{ZZ_\mathrm{P}}} h(t) \otimes U(t)\ .$$

If the electrical impulse is a rectangular pulse of height U_m and width $\Theta \ll \tau_\mathrm{P}$, the transducer response is similar to $h(t)$. In fact, $v(t) = Ah(t)$, where the constant of proportionality is the area $A = U_\mathrm{m}\Theta$ of the pulse:

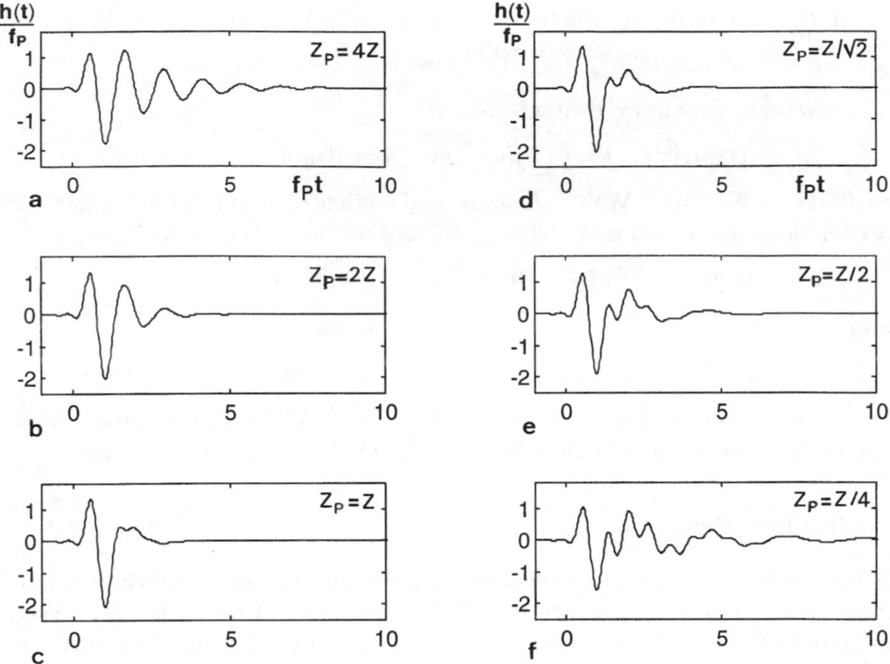

Fig. 1.13a–f. Normalised impulse responses calculated for $K^2 = 0.4$ by taking the Fourier transform of the frequency response for different values of the impedance ratio Z_P/Z. *Ordinate*: $h(t)/f_P$. *Abscissa*: $f_P t = t/\tau_P$

$$v(t) = \frac{eU_m}{d\sqrt{ZZ_P}}\Theta h(t) \qquad (\Theta f_P \ll 1)\,.$$

According to Fig. 1.13, the maximum of $h(t)$ is of order $2f_P$ (see Problem 1.4). The maximum particle velocity is therefore

$$v_m \approx \frac{2eU_m}{d\sqrt{ZZ_P}}\Theta f_P\,. \tag{1.55}$$

For example, in the case of the PZT4 ceramic transducer with centre frequency 10 MHz considered earlier, $d = 0.228$ mm, $e = 15.1$ Cm^{-2}, $Z = Z_P =$ 34.2 MRayl, with $U_m = 50$ V and $\Theta = 0.02$ µs, the peak value v_m may attain 3.9 cm/s.

These signals represent the velocity $v(t)$, assumed uniform, of the front transducer face vibrating in piston mode. In order to calculate the acoustic field radiated by a transducer with finite lateral dimensions, diffraction effects must be taken into account. It was shown in Sect. 1.3.2.3, Vol. I, that, in fluids, the velocity potential at point M is given by the convolution of acoustoelectric response $v(t)$ and diffraction impulse response $h_D(M, t)$ [see (1.106), Vol. I].

The impulse response shape can be predicted intuitively in certain cases. An example is the ideal case where the front of the transducer is loaded with a solid having the same elastic impedance, and at the same time, the electrodes are extremely thin and give rise to no reflection [1.4]. The dynamical equation, deduced from (1.6) and (1.9),

$$\rho\frac{\partial^2 u}{\partial t^2} - c\frac{\partial^2 u}{\partial z^2} = -h\frac{\partial D}{\partial z} ,$$

shows that signals are generated in the region near the electrodes, where the electric displacement is varying most rapidly.

Suppose to begin with that the transducer is backed with an absorbent medium. Supplying an electrical impulse (of short duration relative to the transducer transit time), produces an extension of the piezoelectric plate, which then returns to equilibrium. The two faces of the transducer first move apart and then come together again. Each face is the source of two elastic impulses of opposite signs, propagating in opposite directions (see Fig. 1.14a). Hence the front face emits a compression pulse propagating towards the right and an extension pulse moving towards the left. The signal progressing from one side of the transducer to the other is made up of two pulses of opposite sign, separated by the time required to cross the transducer. However, the two pulses which enter the absorber load disappear. The other two propagate into the right-hand medium.

If the transducer is loaded on just one side, as in Fig. 1.14b, with the other face remaining free, only the loaded face emits two pulses. The free face emits a single pulse of twice the height. The pulse moving towards the

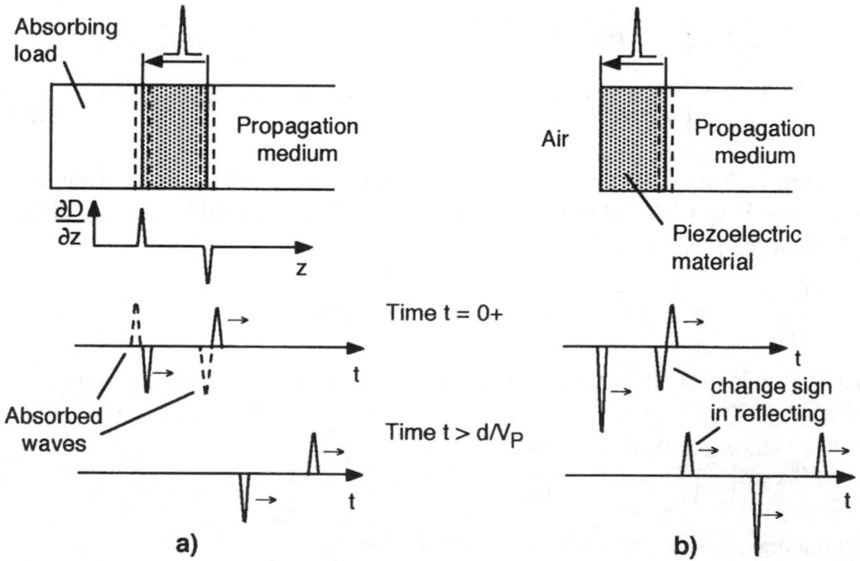

Fig. 1.14. Impulse response of a transducer when one face is loaded with a medium of the same elastic impedance and the other is (**a**) loaded with a perfectly absorbent material, (**b**) free

left, emitted by the loaded face, is reflected by the free face and changes sign. The elastic signal comprises three pulses (see Fig. 1.14b).

1.4.3 Electrical Impedance

By hypothesis, the transducer operates at low frequency and there is no need to take its electrodes into account, elastically speaking, since they are much thinner than the wavelength. Its input electrical impedance is easily calculated from the impedance matrix (1.24). The result is given by (1.94) in Problem 1.1:

$$Z_e = \frac{1}{iC_0\omega}\left(1 + \frac{K^2}{\varphi} Z_P \frac{2Z_P(1-\cos\varphi) - i(Z_1+Z_2)\sin\varphi}{-(Z_P^2 + Z_1Z_2)\sin\varphi + iZ_P(Z_1+Z_2)\cos\varphi}\right) . \tag{1.56}$$

Z_1 and Z_2 are elastic impedances of the media behind and in front of the piezoelectric material, respectively. The latter has impedance Z_P and electromechanical coupling factor K^2. The angle $\varphi = kd$ is equal to $\omega d/V_P$. Let us begin by investigating the simplest case, a free resonator, and then move on to the transducer.

1.4.3.1 Resonator. A *free resonator* is a simple piezoelectric plate whose two faces are metallised. It is an unloaded transducer, with $Z_1 = 0 = Z_2$. It can vibrate in different modes: thickness extension and shear according to the crystallographic cut (see Sect. 4.4.1). If no power is dissipated, its electrical impedance is purely imaginary:

$$\begin{aligned} Z_e &= \frac{1}{iC_0\omega}\left(1 - K^2 \frac{\tan(\varphi/2)}{\varphi/2}\right) \\ &= \frac{1}{iC_0\omega}\left(1 - K^2 \frac{\tan(\omega d/2V_P)}{\omega d/2V_P}\right) . \end{aligned} \tag{1.57}$$

Figure 1.15 shows the frequency variation of the modulus of the resonator admittance $Y = 1/Z_e$. This is zero (i.e., $Z_e = \infty$) at each odd multiple of the *antiresonance frequency* f_a:

$$\frac{\omega_a^{(n)} d}{2V_P} = (2n+1)\frac{\pi}{2} \quad \Rightarrow \quad f_a^{(n)} = (2n+1) f_a , \tag{1.58}$$

where $f_a = V_P/2d$. It is infinite (i.e., $Z_e = 0$) at *resonance frequencies* $f_r^{(n)}$ such that

$$K^2 \tan\left(\frac{\pi f_r^{(n)}}{V_P} d\right) = \pi \frac{f_r^{(n)}}{V_P} d .$$

Introducing f_a, this can be restated in the form

$$K^2 \tan\left(\frac{\pi}{2}\frac{f_r^{(n)}}{f_a}\right) = \frac{\pi}{2}\frac{f_r^{(n)}}{f_a} . \tag{1.59}$$

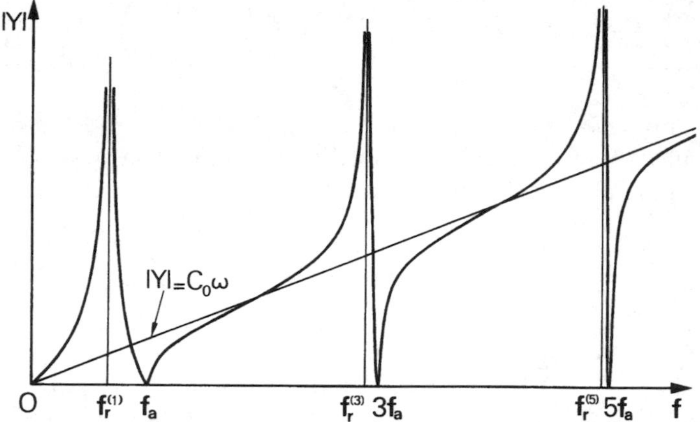

Fig. 1.15. Free piezoelectric resonator. Frequency variation of the admittance. The straight line $|Y| = C_0\omega$ is the admittance of the static capacitance C_0. In practice, at resonance frequencies, propagation and reflection losses limit the admittance to finite values. The same is true for the impedance at antiresonancces frequencies

By measuring the antiresonance frequency $f_{\rm a}$ and the first resonance frequency $f_{\rm r} = f_{\rm r}^{(1)}$, we can find the phase speed $V_{\rm P} = 2f_{\rm a}d$ and the coupling coefficient K^2 corresponding to the excited wave,

$$K^2 = \frac{\pi}{2}\frac{f_{\rm r}}{f_{\rm a}}\tan\left(\frac{\pi}{2}\frac{f_{\rm a}-f_{\rm r}}{f_{\rm a}}\right) . \tag{1.60}$$

Except for materials with large electromechanical coupling coefficient $K > 0.3$, the relative difference $(f_{\rm a} - f_{\rm r})/f_{\rm a}$ is very small, so that

$$K^2 \approx \frac{\pi^2}{4}\frac{f_{\rm a}-f_{\rm r}}{f_{\rm a}} .$$

The coupling coefficient for AT cut quartz (shear mode) is 0.08 and the relative difference between resonance and antiresonance frequencies is only 0.26%. This difference decreases as $1/n^2$ for vibrational modes of higher orders n. The latter are called partial (instead of harmonic) modes, because corresponding frequencies are not integer multiples of the resonance frequency (see Problem 1.5).

An approximate expression for the impedance can be found by expanding [1.5],

$$\frac{\tan\theta}{\theta} = \sum_{n=0}^{\infty}\frac{2}{[(2n+1)\pi/2]^2-\theta^2} , \quad \text{where} \quad \theta = \frac{\pi}{2}\frac{\omega}{\omega_{\rm a}} , \quad \omega_{\rm a} = \pi\frac{V_{\rm P}}{d} .$$

This clearly shows the poles at $\theta = (2n+1)\pi/2$. Each term corresponds to a resonance around the angular frequency $(2n+1)\omega_{\rm a}$. These do not overlap when $K^2 \ll 1$ (e.g., for quartz). Near the fundamental resonance $n = 0$, the resonator admittance is

$$Y = \frac{\mathrm{i}C_0\omega}{1 - \dfrac{8K^2/\pi^2}{1-\omega^2/\omega_\mathrm{a}^2}} = \mathrm{i}C_0\omega\frac{\omega_\mathrm{a}^2-\omega^2}{\omega_\mathrm{r}^2-\omega^2}\,, \quad \omega_\mathrm{r}^2 = \omega_\mathrm{a}^2\left(1-\frac{8K^2}{\pi^2}\right)\,. \tag{1.61}$$

The resonator is represented by an equivalent electrical circuit composed of the static capacitance C_0 of the rigidly clamped plate and a series resonant circuit connected in parallel, as shown in Fig. 1.16.

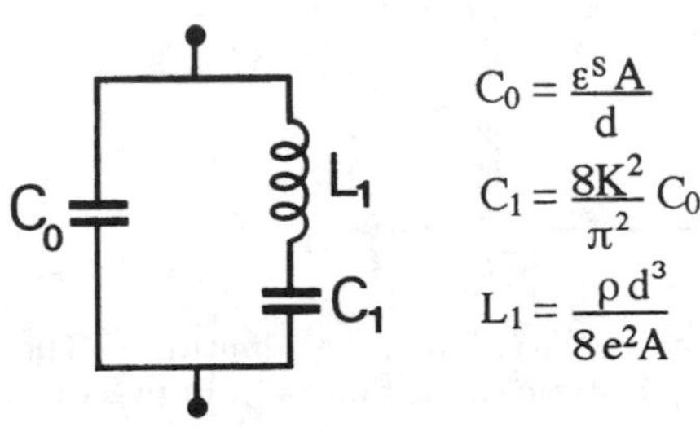

Fig. 1.16. Equivalent electrical circuit for a resonator near the resonance frequency: static capacitance C_0 of the rigidly clamped resonator and motional impedance

Values of elements in the equivalent circuit are found by identifying its admittance

$$Y = \mathrm{i}C_0\omega\left[1 + \frac{C_1/C_0}{1-L_1C_1\omega^2}\right]$$

with (1.61), and in particular, comparing series and parallel resonance angular frequencies,

$$\omega_\mathrm{s}^2 = \frac{1}{L_1C_1}\,, \qquad \omega_\mathrm{p}^2 = \omega_\mathrm{s}^2\left(1+\frac{C_1}{C_0}\right)\,, \tag{1.62}$$

with ω_r and ω_a. Motional inductance and capacitance are given by

$$\frac{C_1}{C_0} = \frac{\omega_\mathrm{a}^2-\omega_\mathrm{r}^2}{\omega_\mathrm{r}^2} = \frac{8K^2/\pi^2}{1-8K^2/\pi^2}\,, \qquad L_1 = \frac{1+C_0/C_1}{C_0\omega_\mathrm{a}^2} \approx \frac{\rho d^3}{8Ae^2}\,. \tag{1.63}$$

Hence,

$$\frac{f_\mathrm{a}-f_\mathrm{r}}{f_\mathrm{r}} \approx \frac{C_1}{2C_0}\,.$$

Order of Magnitude. For quartz, the ratio $C_0/C_1 = \pi^2/8K^2$ lies between 125 and 500, depending on the cut. The inductance L_1 is relatively high ($\approx$ 1 H at 1 MHz). Consider an AT cut crystal, vibrating in thickness shear mode at frequency 5 MHz. From the value $K = 0.08$ of the coupling coefficient, we have $C_0/C_1 = 193$, speed $V_\mathrm{P} = 3\,320\ \mathrm{ms}^{-1}$ and $d = 0.332$ mm. If the electrode surface area is $A = 50\ \mathrm{mm}^2$, then with $\varepsilon^S = 40$ pF/m, $C_0 = 6$ pF and $L_1 = 33$ mH.

In order to estimate losses due to intrinsic attenuation by the material (see Sect. 4.2.6, Vol. I) or resonator attachments, we must introduce a resistance R_1 into the equivalent circuit, placing it in series with L_1 and C_1 (see

Problem 1.6). This resistance is inversely proportional to the quality factor $Q = L_1\omega_r/R_1$ of the resonator. If losses are of viscoelastic nature, then from (4.61), Vol. I, $Q = c^D/\omega_r\eta$.

The product $Qf_r = c^D/2\pi\eta$ of the quality factor and the resonance frequency characterises material performance. It reaches 10^{13} s^{-1} in an AT cut quartz crystal vibrating in thickness shear mode with frequency close to 5 MHz.

The bulk wave resonator is an important component in filtering and time measurement (see Sect. 4.4.1).

1.4.3.2 Transducer. The electrical impedance (1.56) of a resonator loaded both in front and behind by media with mechanical impedances Z_1 and Z_2 is complex:

$$Z_e = \frac{1}{iC_0\omega} + iX_a(\omega) + R_a(\omega)\,, \quad \text{or} \quad Y = iC_p\omega + iB_a(\omega) + G_a(\omega)\,.$$

It includes, apart from the reactance of the static capacitance, a real part $R_a(\omega)$ and an imaginary part $X_a(\omega)$ (motional impedance) which are both frequency dependent (see Fig. 1.17a). The corresponding elements of the parallel equivalent circuit, shown in Fig. 1.17b, are $B_a(\omega)$ and $G_a(\omega)$. The resistance $R_a(\omega)$ and the radiation conductance $G_a(\omega)$ express the conversion of part of the electrical power supplied by the source into mechanical power, transported by emitted elastic waves. The mean value of this mechanical power is

$$\langle P\rangle = \frac{1}{2}R_a(\omega)|I|^2\,, \quad \text{or} \quad \langle P\rangle = \frac{1}{2}G_a(\omega)|U|^2\,. \tag{1.64}$$

At the centre frequency $f_P = V_P/2d$, for which $\varphi = k_P d = \pi$, the electrical impedance (1.56)

$$Z_e(\omega_P) = \frac{1}{iC_0\omega_P} + \frac{4K^2}{\pi C_0\omega_P}\frac{Z_P}{Z_1 + Z_2} = \frac{1}{iC_0\omega_P} + R_a(\omega_P)$$

is composed of a resistance in series with the static capacitance. The radiation reactance $X_a(\omega_P)$ is zero. If also $Z_1 = 0$ and $Z_2 = Z_P$ (the model shown in Fig. 1.11), the radiation resistance is

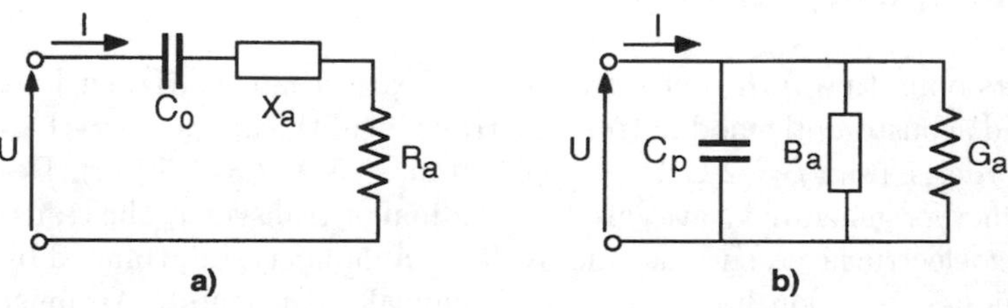

Fig. 1.17. (**a**) Series and (**b**) parallel equivalent electrical circuits for a bulk wave transducer. R_a (G_a) is the frequency dependent radiation resistance (conductance)

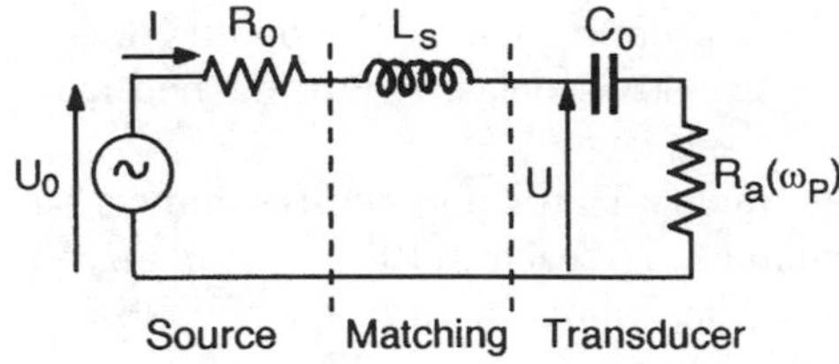

Fig. 1.18. Equivalent electrical circuit of a transducer at its centre frequency, matched with a series impedance. When R_0 and R_a are very different, matching requires a transformer

$$R_\mathrm{a}(\omega_\mathrm{P}) = \frac{4K^2}{\pi}\frac{1}{C_0\omega_\mathrm{P}} = \frac{2K^2}{\pi^2}\frac{1}{C_0 f_\mathrm{P}} \,. \tag{1.65}$$

When $K^2 \ll 1$, this is much smaller than the reactance $1/C_0\omega_\mathrm{P}$ due to the static capacitance. The latter can be compensated by a series inductance L_s, as shown in Fig. 1.18. Transducer efficiency increases at the expense of its bandwidth.

When a large bandwidth is required, it is better to increase the efficiency by choosing $1/C_0\omega_\mathrm{P}$ close to the resistance R_0 of the source. The mean emitted power $\langle P\rangle$ is a fraction of the maximal electrical power P_d available from the supply e.m.f. U_0,

$$\langle P\rangle = \frac{1}{2}\frac{R_\mathrm{a}U_0^2}{(R_0+R_\mathrm{a})^2+(1/C_0\omega_\mathrm{P})^2}\,, \qquad P_\mathrm{d} = \frac{U_0^2}{8R_0}\,. \tag{1.66}$$

The transducer efficiency η, defined as the ratio of these two powers,

$$\eta = \frac{\langle P\rangle}{P_\mathrm{d}} = \frac{4R_0R_\mathrm{a}}{(R_0+R_\mathrm{a})^2+(1/C_0\omega_\mathrm{P})^2}\,,$$

is maximum for $R_0 = \sqrt{R_\mathrm{a}^2+(1/C_0\omega_\mathrm{P})^2}$, i.e., $R_0 \approx 1/C_0\omega_\mathrm{P}$, when $K^2 \ll 1$. In practice, for a given material and frequency, this condition determines the cross-sectional area of the transducer. From (1.65), the maximum efficiency is

$$\eta_\mathrm{max} = 2R_\mathrm{a}C_0\omega_\mathrm{P} \approx \frac{8K^2}{\pi} \ll 1\,.$$

1.5 High-Frequency Transducers

In components operating with centre frequency greater than 1 GHz, such as delay lines and acousto-optic modulators, the transducer thickness is less than $\lambda/2$, and of order 1 μm ($f = 2$ GHz, $V = 6\,000$ m/s $\rightarrow \lambda/2 = 1.5$ μm). Depending on whether generated waves are longitudinal or transverse, the transducer is a piezoelectric layer, such as zinc oxide, or a monocrystal, thinned by special techniques (e.g., ion bombardment, chemical treatments). We must now take into account the role played by the electrodes, in particular, the internal electrode, placed between transducer and propagating medium. The aim of the present section is to calculate frequency responses for the transducer when it is equipped with different internal electrodes, always assuming that the electromechanical coupling factor is much less than unity.

Consider again Fig. 1.2. Expressions for velocity and stress in each part of the transducer are given by (1.13) and (1.15), the factor $\mathrm{e}^{\mathrm{i}\omega t}$ being understood:

$$v_{\mathrm{S}} = a_{\mathrm{S}}\mathrm{e}^{-\mathrm{i}k_{\mathrm{S}}z} + b_{\mathrm{S}}\mathrm{e}^{\mathrm{i}k_{\mathrm{S}}z} \,, \tag{1.67a}$$

$$T_{\mathrm{S}} = -Z_{\mathrm{S}}(a_{\mathrm{S}}\mathrm{e}^{-\mathrm{i}k_{\mathrm{S}}z} - b_{\mathrm{S}}\mathrm{e}^{\mathrm{i}k_{\mathrm{S}}z}) - hD\delta_{\mathrm{SP}} \,, \tag{1.67b}$$

where S denotes M for the earthed electrode, P for the piezoelectric solid, and E for the external electrode. The Kronecker symbol δ_{SP} indicates that the term $hD = hI/\mathrm{i}\omega A$ only occurs for the piezoelectric material. In the propagating medium, there is a travelling wave

$$v = a\mathrm{e}^{-\mathrm{i}kz} \,, \tag{1.68a}$$

$$T = -Za\mathrm{e}^{-\mathrm{i}kz} \,. \tag{1.68b}$$

The seven unknowns a, a_{M}, b_{M}, a_{P}, b_{P}, a_{E}, b_{E} are determined by the seven boundary conditions which express continuity of velocities and stresses between:

- internal electrode and propagating medium,

$$v_{\mathrm{M}}(d_2) = v(d_2) \,, \tag{1.69a}$$

$$T_{\mathrm{M}}(d_2) = T(d_2) \,, \tag{1.69b}$$

- piezoelectric solid and internal electrode,

$$v_{\mathrm{P}}(0) = v_{\mathrm{M}}(0) \,, \tag{1.70a}$$

$$T_{\mathrm{P}}(0) = T_{\mathrm{M}}(0) \,, \tag{1.70b}$$

- external electrode and piezoelectric crystal,

$$v_{\mathrm{E}}(-d) = v_{\mathrm{P}}(-d) \,, \tag{1.71a}$$

$$T_{\mathrm{E}}(-d) = T_{\mathrm{P}}(-d) \,, \tag{1.71b}$$

and on the free surface of the external electrode,

$$T_{\mathrm{E}}(-d - d_1) = 0 \,. \tag{1.72}$$

Solving this system directly gives amplitudes a, a_{S} and b_{S} as a function of the electric displacement D. However, the measurable quantity is the potential difference U applied across the electrodes,

$$U = \int_0^{-d} -E(z)\,\mathrm{d}z \,.$$

According to (1.7), the electric displacement D is constant for an insulating medium. Integrating (1.5) between the ends 0 and $-d$ of the piezoelectric solid, it follows that

$$-Dd = -\varepsilon^S U + e[u_{\mathrm{P}}(-d) - u_{\mathrm{P}}(0)] \,. \tag{1.73}$$

Hence, since $u = v/\mathrm{i}\omega$ and replacing v_{P} by the expression from (1.67a),

$$D = \varepsilon^S \frac{U}{d} + \mathrm{i}\frac{e}{\omega d}\left[a_{\mathrm{P}}\left(\mathrm{e}^{\mathrm{i}k_{\mathrm{P}}d} - 1\right) + b_{\mathrm{P}}\left(\mathrm{e}^{-\mathrm{i}k_{\mathrm{P}}d} - 1\right)\right] .$$

Substituting into the expression for the stress T_{P} and using $h = e/\varepsilon^S$, it follows that

$$T_{\mathrm{P}} = -Z_{\mathrm{P}}\left\{a_{\mathrm{P}}\left[\mathrm{e}^{-\mathrm{i}k_{\mathrm{P}}z} + \mathrm{i}\frac{e^2}{\varepsilon^S \omega Z_{\mathrm{P}} d}\left(\mathrm{e}^{\mathrm{i}k_{\mathrm{P}}d} - 1\right)\right] - b_{\mathrm{P}}[\mathrm{c.c.}]\right\} - e\frac{U}{d} \,,$$

where c.c. denotes complex conjugate. The factor $e^2/\varepsilon^S \omega Z_{\mathrm{P}} d$ contains the square of the electromechanical coupling coefficient $K^2 = e^2/\varepsilon^S c_{\mathrm{P}}^D$. The thickness d of the piezoelectric plate is of the order of one half-wavelength, $k_{\mathrm{P}} d \approx \pi$. The ratio

$$\frac{e^2}{\varepsilon^S \omega Z_{\mathrm{P}} d} = \frac{e^2}{\varepsilon^S c_{\mathrm{P}}^D k_{\mathrm{P}} d} = \frac{K^2}{k_{\mathrm{P}} d} \approx \frac{K^2}{\pi}$$

is much smaller than 1 for an averagely piezoelectric solid ($K < 0.3 \Rightarrow K^2 < 0.1$). The stress T_{P} is given in terms of the potential difference U by

$$T_{\mathrm{P}} = -Z_{\mathrm{P}}\left(a_{\mathrm{P}}\mathrm{e}^{-\mathrm{i}k_{\mathrm{P}}z} - b_{\mathrm{P}}\mathrm{e}^{\mathrm{i}k_{\mathrm{P}}z}\right) - e\frac{U}{d} \,. \tag{1.74}$$

Knowing the amplitude $a = |v|$ as a function of U, the elastic power transported by the beam of cross-section A in the propagating medium of elastic impedance Z can be found from (1.28).

We can apply this method to the relatively simple structure of a high-frequency delay line transducer ($f > 1$ GHz). This is usually a thin zinc oxide layer deposited by sputtering onto the propagating crystal. The internal electrode cannot be ignored. For a given piezoelectric material and propagating crystal, we shall now analyse the influence of electrode type and thickness on transducer frequency response.

1.5.1 Piezoelectric Crystal–Electrode-Propagating Medium

For this structure, boundary conditions are expressed by (1.69a), (1.69b), (1.70a), (1.70b) and the relation $T_{\mathrm{P}}(-d) = 0$. Using also (1.74) and putting

$$\varphi_{\mathrm{M}} = k_{\mathrm{M}} d_2 \,, \qquad \varphi_{\mathrm{P}} = k_{\mathrm{P}} d \,,$$

this leads to the system

$$\begin{aligned}
0 &= a_{\mathrm{M}}\mathrm{e}^{-\mathrm{i}\varphi_{\mathrm{M}}} + b_{\mathrm{M}}\mathrm{e}^{\mathrm{i}\varphi_{\mathrm{M}}} - a\mathrm{e}^{-\mathrm{i}kd_2} \,, \\
0 &= Z_{\mathrm{M}}\left(a_{\mathrm{M}}\mathrm{e}^{-\mathrm{i}\varphi_{\mathrm{M}}} - b_{\mathrm{M}}\mathrm{e}^{\mathrm{i}\varphi_{\mathrm{M}}}\right) - Za\mathrm{e}^{-\mathrm{i}kd_2} \,, \\
0 &= a_{\mathrm{P}} + b_{\mathrm{P}} - a_{\mathrm{M}} - b_{\mathrm{M}} \,, \\
eU/d &= -Z_{\mathrm{P}}(a_{\mathrm{P}} - b_{\mathrm{P}}) + Z_{\mathrm{M}}(a_{\mathrm{M}} - b_{\mathrm{M}}) \,, \\
eU/d &= -Z_{\mathrm{P}}\left(a_{\mathrm{P}}\mathrm{e}^{\mathrm{i}\varphi_{\mathrm{P}}} - b_{\mathrm{P}}\mathrm{e}^{-\mathrm{i}\varphi_{\mathrm{P}}}\right) \,.
\end{aligned} \tag{1.75}$$

Solving these leads to (see Problem 1.8)

$$a = \frac{eU}{Zd} \frac{e^{ikd_2}}{M(\varphi_P, \varphi_M)} , \tag{1.76}$$

where the *bandshape factor* $M(\varphi_P, \varphi_M)$ is defined by

$$M(\varphi_P, \varphi_M)(\cos\varphi_P - 1) = \cos\varphi_P \cos\varphi_M - \frac{Z_P}{Z_M} \sin\varphi_P \sin\varphi_M \tag{1.77}$$
$$+ i \left(\frac{Z_M}{Z} \cos\varphi_P \sin\varphi_M + \frac{Z_P}{Z} \sin\varphi_P \cos\varphi_M \right) .$$

This function M depends on frequency through the angles $\varphi_P = k_P d$ and $\varphi_M = k_M d_2$, and also the $\lambda/2$ resonance frequencies f_P, f_M of the piezoelectric layer and internal electrode, given by

$$f_P = V_P/2d , \qquad f_M = V_M/2d_2 . \tag{1.78}$$

The *frequency response*, defined by (1.51), is equal to

$$H(f) = \sqrt{\frac{Z_P}{Z}} \frac{e^{ikd_2}}{M(\varphi_P, \varphi_M)} , \quad \varphi_P = \pi \frac{f}{f_P} , \quad \varphi_M = \pi \frac{f}{f_M} . \tag{1.79}$$

The *elastic power* supplied to the delay line follows from (1.28):

$$\langle P \rangle = \frac{1}{2} A Z |a|^2 = K^2 C_0 f_P \frac{Z_P}{Z} \frac{U^2}{|M|^2} . \tag{1.80}$$

The *radiation conductance*

$$G_a = \frac{2 Z_P K^2 C_0 f_P}{Z |M|^2} \tag{1.81}$$

is inversely proportional to the square of the modulus of the bandshape factor M.

When the internal electrode is not taken into account, we retrieve the results of Sect. 1.4.1. Indeed, if $d_2 = 0$ then $\varphi_M = 0$, so putting $\varphi_P = \varphi$,

$$M(\varphi) = \frac{\cos\varphi + i \dfrac{Z_P}{Z} \sin\varphi}{\cos\varphi - 1} .$$

The particle velocity amplitude is given by

$$a = \frac{eU}{d} \frac{\cos\varphi - 1}{Z\cos\varphi + iZ_P \sin\varphi} = \frac{eU}{\sqrt{Z Z_P} d m_0(\varphi)} , \tag{1.82}$$

where

$$m_0(\varphi) = \frac{\sqrt{Z/Z_P}\cos\varphi + i\sqrt{Z_P/Z}\sin\varphi}{\cos\varphi - 1} , \tag{1.83}$$

and $\varphi = k_P d = \pi f / f_P$. This formula can be obtained from the more general relation (1.49) for the factor $m(\theta)$, with θ replaced by $\varphi/2$ and $K = 0$. The effect of the elastic impedance ratio Z_P/Z on transmitted power is illustrated in Fig. 1.12a as a function of normalised frequency f/f_P.

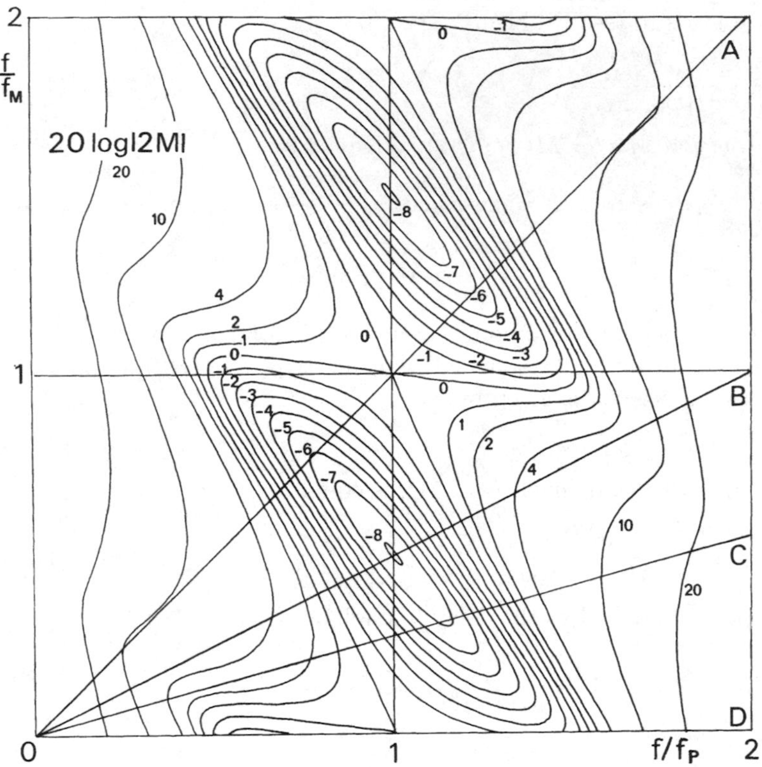

Fig. 1.19. Contours of the CdS-Al-Al_2O_3 bandshape factor M for the longitudinal wave. A transducer with $f_P/f_M = 0.25$ is represented by the vertical plane through OC

1.5.2 Frequency Variation of the Bandshape Factor

Since the bandshape factor $M(\varphi_P, \varphi_M)$ takes the value 1/2 for $f = f_P$ and $f = f_M$, the normalised frequency response is represented in the following figures by plotting $20 \log |2M|$. From the symmetries of the 2-variable function $M(f/f_P, f/f_M)$, it suffices to calculate its values when f/f_P and f/f_M vary over the range 0 to 1. In Figs. 1.19 and 1.21, the function $20 \log |2M|$ is represented by contours of constant M value. These contours have been plotted for two structures which are quite typical with regard to elastic impedances [1.6]. They contain either an aluminium electrode or a gold electrode, between a cadmium sulfide layer and a sapphire crystal (propagating medium). Cadmium sulfide (CdS) is rarely used today. It has been superceded by zinc oxide (ZnO) and aluminium nitride (AlN), but it has lower elastic impedance than these two materials ($Z_{CdS} = 21.5$ MRayl, $Z_{ZnO} = 35$ MRayl, $Z_{AlN} = 34$ MRayl). In the present case, this is an advantage in bringing out the effect of the internal electrode impedance on frequency response. All three materials belong to the same crystal class, viz., $6mm$. The 6-fold axis is par-

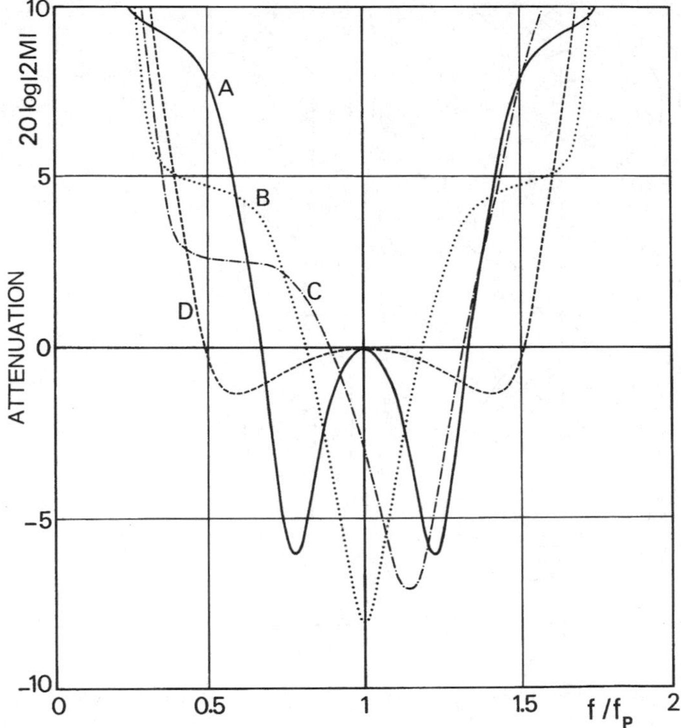

Fig. 1.20. CdS-Al-Al_2O_3 structure (longitudinal wave). The attenuation has one or two troughs depending on the thickness of aluminium

allel to the electric field and hence only the longitudinal wave is excited. It propagates along the 3-fold axis of the sapphire crystal. The one-dimensional model can therefore be applied.

The behaviour of a given transducer is found from a cross-section of the surface $M(f/f_P, f/f_M)$ by a vertical plane passing through the origin. This is because the ratio

$$\frac{f/f_M}{f/f_P} = \frac{f_P}{f_M}$$

is constant, being determined by layer thicknesses and propagation speeds. Straight lines A, B, C and D in Figs. 1.19 and 1.21 are projections of such planes. The corresponding frequency responses are given in Fig. 1.20 for the CdS-Al-Al_2O_3 sequence and Fig. 1.22 for the CdS-Au-Al_2O_3 sequence. The shapes of these curves can be explained physically in the following way. The elastic impedance of cadmium sulfide (Z_P = 21.5 MRayl) is significantly lower than that of sapphire (Z = 44.5 MRayl). Consequently, for zero metal thickness (curve D), the transducer has one face free and the other more or less rigidly bonded, so that $\lambda/4$ (f/f_P = 0.5) and $3\lambda/4$ (f/f_P = 1.5)

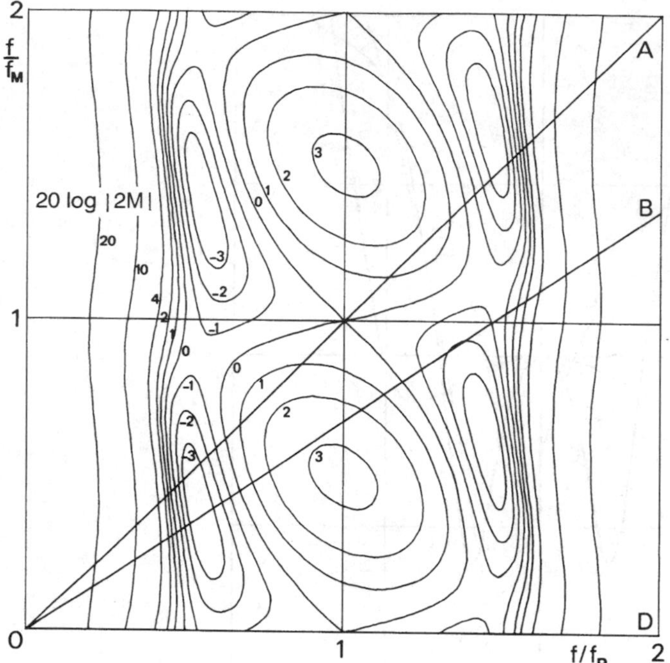

Fig. 1.21. Contours of the CdS-Au-Al_2O_3 bandshape factor (longitudinal wave). A transducer with given f_P/f_M is represented by a vertical plane like the one through OB

vibrations are set up. The bandshape factor M exhibits two troughs around these points, corresponding to two elastic power maxima.

Adjoining a gold film of impedance $Z_M = 64$ MRayl, of any thickness, deepens the two troughs (curves A and B in Fig. 1.22). The metallic film on the sapphire crystal is not monocrystalline. However, the 3-fold axis of sapphire favours growth in the [111] direction. The elastic impedance takes on an intermediate value, lying between the values for directions [100] and [111]. In contrast, the aluminium layer of elastic impedance $Z_M = 17.5$ MRayl interposed between cadmium sulfide and sapphire plays the role of an *impedance transformer* when it has the right thickness, equal to $\lambda/4$. It brings the impedance on the inner face of the transducer ($z = 0$) to the value (see Problem 1.6, Vol. I)

$$Z_T(z=0) = \frac{Z_M^2}{Z} = 6.9\,\text{MRayl}\,,$$

so that this face is in fact almost free. The transducer vibrates at $\lambda/2$ ($f = f_P$) and the frequency response only exhibits one trough (curves B and C in Fig. 1.20). It is easy to follow the way the bandshape factor evolves from the two-trough shape to the single-trough shape by pivoting the projection of the characteristic plane around the origin.

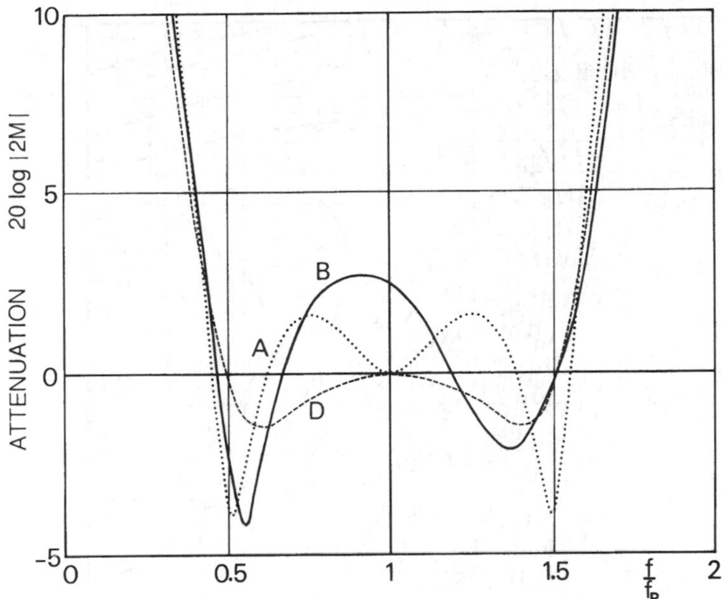

Fig. 1.22. CdS-Au-Al_2O_3 structure (longitudinal wave). Attenuation always has two troughs

The bandshape factor for the ZnO-Pt-Al_2O_3 structure (Fig. 1.23) is similar to that for CdS-Au-Al_2O_3, having an analogous sequence of elastic impedances:

$$Z_{\mathrm{ZnO}} = 35\,\mathrm{MRayl}\,, \quad Z_{\mathrm{Pt}} = 87\,\mathrm{MRayl}\,, \quad Z_{\mathrm{Al_2O_3}} = 44.5\,\mathrm{MRayl}\,.$$

In the above, the influence of the external electrode has not been taken into account. It is generally made from a layer of aluminium or silver. Provided it remains thin compared with the piezoelectric layer ($< 10\%$), it tends to shift the bandshape curves slightly towards lower frequencies.

Order of Magnitude. For the ZnO-Pt-Al_2O_3 structure, at frequency $f_\mathrm{P} =$ 1 GHz, the thickness of the layer is

$$d = \frac{V_\mathrm{P}}{2f_\mathrm{P}} = 3.2\,\mu\mathrm{m}\,.$$

The static capacitance of a transducer of cross-sectional area $A = 1\ \mathrm{mm}^2$ is $C_0 = \varepsilon_{33}^S A/d = 26$ pF. Given that the coupling coefficient for the longitudinal wave is

$$K^2 = \frac{e_{33}^2}{\varepsilon_{33}^S c_{33}^E + e_{33}^2} = 0.073\,,$$

the radiation conductance (1.81) is $G_\mathrm{a} = 3 \times 10^{-3} |M|^{-2}\,\Omega^{-1}$. A voltage of magnitude $|U| = 1$ V and frequency such that $M = 0.5$ generates elastic power

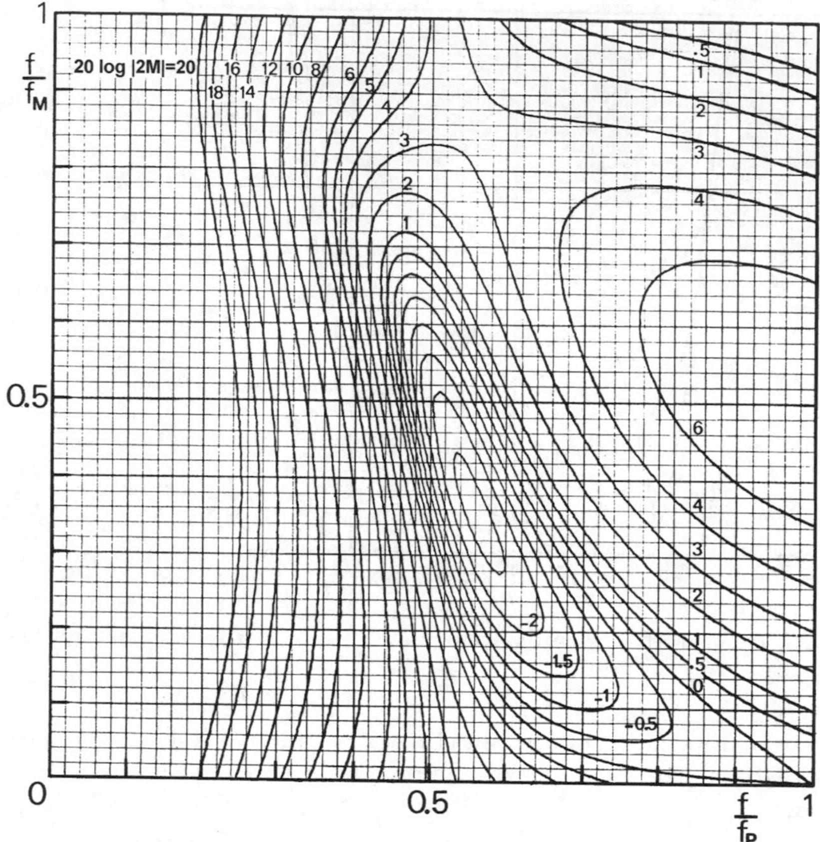

Fig. 1.23. Contours of the ZnO-Pt-Al_2O_3 bandshape factor (longitudinal wave). f/f_P and f/f_M vary between 0 and 1. Note the similarity with Fig. 1.21

$$P = \frac{1}{2} G_a |U|^2 \approx 6\,\mathrm{mW} .$$

1.5.3 Electrical Tuning. Conversion Losses

Assuming the piezoelectric material to be an insulator and perfect dielectric, the only electrical current is the displacement current of intensity $I = \mathrm{i}\omega DA$. The electric displacement is given by (1.73) and the electrical impedance of the transducer contains two terms:

$$Z_e = \frac{U}{I} = \frac{d}{\mathrm{i}\omega\varepsilon^S A} + \frac{e}{\varepsilon^S I}[u_P(-d) - u_P(0)] . \tag{1.84}$$

The first is just the impedance of the static capacitance $C_0 = \varepsilon^S A/d$. The second leads to the radiation resistance R_a in the case of a moderately piezoelectric material $K^2 \ll \pi$ (see Sect. 1.4.3.2).

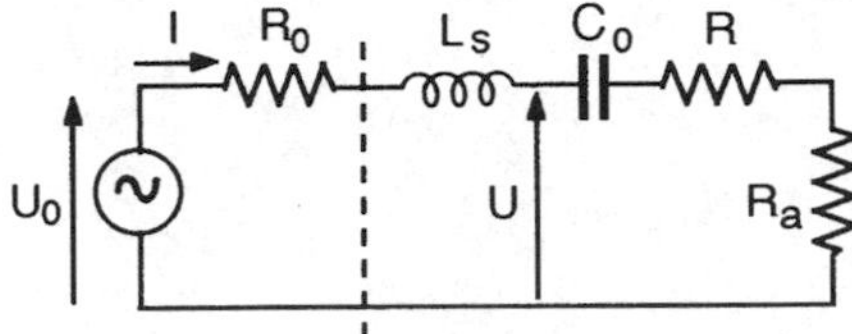

Fig. 1.24. Equivalent circuit for a transducer. Resistance R represents all losses of electrical and elastic origin in the various layers

In practice, a resistance R must be added to the equivalent circuit to represent electrical power dissipation by mechanisms other than generation of elastic waves (see Fig. 1.24). In high-frequency transducers, these losses can be imputed to the following causes:

- piezoelectric materials (e.g., ZnO, AlN) are semiconductors and not insulators;
- dielectric losses are high for frequencies above 1 GHz;
- resistivity of metallic layers is not zero;
- the various (non-monocrystalline) layers attenuate elastic waves.

Experience shows that the equivalent resistance R for these losses is greater than the radiation resistance R_a.

The electrical power supplied to the transducer by a source of e.m.f. U_0 and internal resistance R_0 is maximal when impedances are matched, i.e., when capacitance C_0 is tuned to the centre frequency f_P with a series inductance L_s and $R_0 = R_a + R$. The transducer efficiency η is, by definition, the ratio of the mean elastic power $\langle P \rangle$ produced in the crystal to the maximal available power from the source:

$$\eta = \frac{\langle P \rangle}{P_d}, \quad \langle P \rangle = \frac{1}{2} R_a |I|^2, \quad P_d = \frac{U_0^2}{8R_0}. \tag{1.85}$$

Since impedances are matched, this means $I = U_0/2(R_a + R)$ and

$$\eta = \frac{R_a}{R + R_a}.$$

This definition of efficiency is only useful if impedances can be matched within the bandwidth of the transducer. Since R_a is frequency dependent, this means that R must remain greater than R_a and also

$$(R_a + R_0 + R)C_0\omega_P > \left(\frac{\Delta f}{f}\right)_M,$$

where $(\Delta f/f)_M$ is the relative bandwidth of the bandshape factor M, which is of order 0.5 (see Figs. 1.20 and 1.22). As the term

$$R_a C_0 \omega_P \approx 4K^2/\pi$$

is less than 1, the condition is satisfied when $R \gg R_a$. The source is then matched to the resistance R and the efficiency, which is small, is inversely proportional to the square of the modulus of the bandshape factor M. From (1.80) and (1.85),

$$\eta \approx 2K^2 \frac{Z_P}{Z} \frac{R_0 C_0 f_P}{|M|^2} . \tag{1.86}$$

1.6 Materials and Technology

The essential element in a transducer is the piezoelectric material, converting electrical energy from some external source into usable elastic energy in a chosen medium, in the form of bulk waves here (and conversely). Unfortunately, only a very few piezoelectric materials can actually be used. The main ones are given together with their most relevant characteristics in Table 1.1. Other characteristics, such as crystal class, stiffnesses, density and piezoelectric constants, have already been given in Volume I (see Figs. 3.10, 3.15, 4.36, 5.27 of Vol. I). In some cases (e.g., lithium niobate and zinc oxide), the slowness (reciprocal speed) curves have been plotted (see Figs. 4.33 and 4.35, Vol. I).

Table 1.1. Main constants (V, Z, K and ε^S) characterising piezoelectric materials used as transducers

Material	Cut	Polar-isation	Speed V [m/s]	Imped. Z [MRayl]	K	$\varepsilon^S \times 10^{11}$ [F/m]
$LiNbO_3$	$Y + 36°$	$\sim$ L	7340	34.5	0.49	34.3
$LiNbO_3$	Z	L	7316	34.4	0.16	25.7
ZnO	Z	L	6400	36.3	0.28	7.8
AlN	Z	L	10400	34	0.17	7.5
Quartz	X	L	5747	15.2	0.092	3.92
PZT4 ceramic	Z	L	4560	34.2	0.51	560
P(VDF-TrFE)	Z	L	2400	4.5	0.30	5.3
$LiNbO_3$	$Y + 163°$	T$\sim\perp X$	4560	21.4	0.62	37.8
Quartz	AC	T $\parallel X$	3317	8.8	0.10	4.0
Quartz	BC	T $\parallel X$	5105	13.5	0.037	4.0
Quartz	AT	T $\parallel X$	3320	8.8	0.08	4.0
PZT4 ceramic	Z	T $\perp Z$	2600	19.5	0.71	650

Note: Some authors define a piezoelectric coupling coefficient κ by the relation $\kappa^2 = e^2/\varepsilon^S c^E$, denoting by $K_T^2 = e^2/\varepsilon^S c^D$ the effective coupling coefficient that we have called K^2. For weakly piezoelectric solids (e.g., quartz, zinc oxide, cadmium sulfide), the difference between κ and K is very slight. Indeed, from the formula $c^D = c^E + e^2/\varepsilon^S$, we have

$$\frac{c^D}{c^E} = 1 + \kappa^2 , \qquad K_T^2 = \frac{\kappa^2}{1+\kappa^2} .$$

The difference becomes noticeable in PZT-type ceramics. For example, $\kappa^2 = 0.4$ implies $K_T^2 \approx 0.29$.

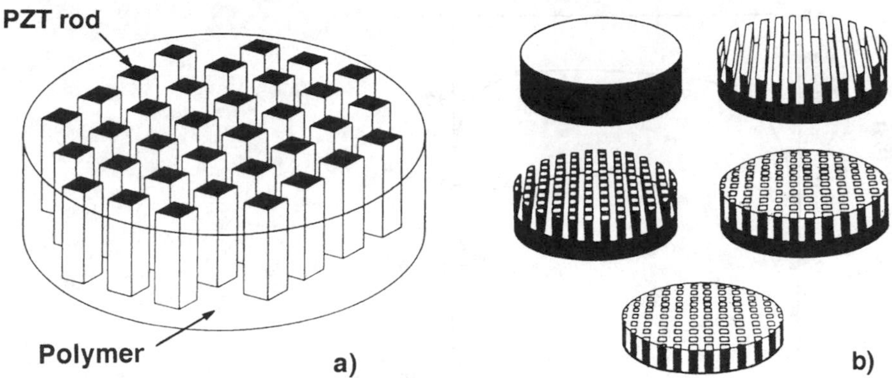

Fig. 1.25. (**a**) Structure of a piezocomposite transducer. (**b**) Production stages [1.7]. ©1991 IEEE

Naturally, the choice of material depends on the frequency of the waves we hope to generate (or detect), all the more so because its thickness is generally a fraction of the length λ of the wave, and also on the medium (solid or liquid) to which it is mechanically bonded.

Piezoceramics. In the low-frequency domain ($f < 25$ MHz, $\lambda/2 < 0.1$ mm if $V = 5000$ m/s), the piezoelectric material is very often a ceramic plate with metallised faces, cemented onto the (solid) propagating medium with a relatively thick binding agent, like an epoxy resin. However, transducers comprising a single piezoceramic plate are often unsuitable, especially if the propagating medium is a liquid (e.g., in underwater acoustics, medical ultrasound imaging). The problem here is that their acoustic impedance tends to be too large ($>$ 20 MRayl) relative to that of the medium (e.g., 1.5 MRayl for water and the human body) into which they send waves in the form of impulses or wideband signals. Matching transducer to propagating material is accomplished by interposing intermediate layers. The aim is an efficient conversion of electrical energy into mechanical energy when impulses are emitted, and conversely when echos are received. One way of reducing the number of matching layers is to use a composite transducer.

Piezocomposite Materials. The transducer is composed of ceramic active elements, in the form of rods, for example, embedded in a passive resin (polymer), as shown in Fig. 1.25. The latter has a much smaller intrinsic impedance (3 MRayl). The quantity of high-impedance material is much smaller than in a simple plate, so that the overall impedance of the transducer is also smaller. Although it may seem surprising, the electromechanical coupling coefficient is larger than for a single plate. This can be interpreted as follows: for the same supply voltage, a rod surrounded by relatively soft material vibrates more easily than one embedded within material of the same type.

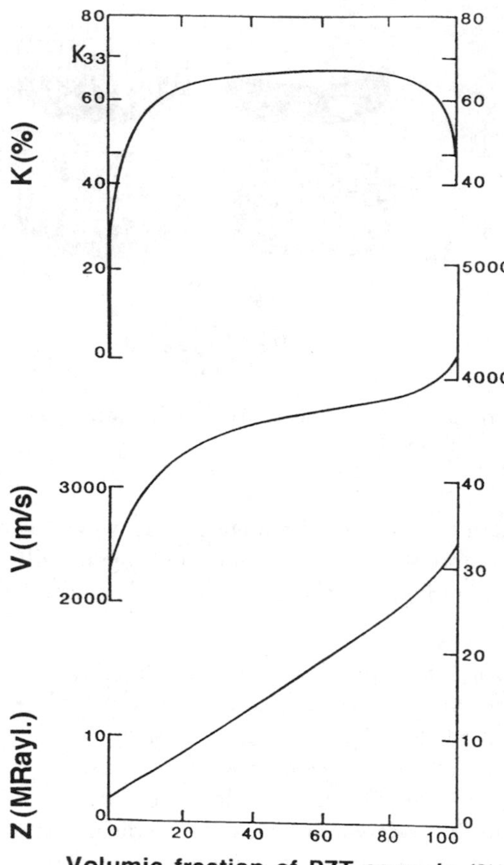

Fig. 1.26. Elastic impedance Z, speed V of longitudinal waves and electromechanical coupling coefficient K for a composite made from PZT5 in an epoxy resin (Dow) as a function of the volume fraction of PZT5. K almost reaches the value K_{33} of the coupling coefficient for extensional vibration of a free rod [1.7]. ©1991 IEEE

If the rods are distributed in the polymer with small period relative to its thickness, Smith and Auld [1.7] have shown that elastic and piezoelectric constants of the composite can be calculated in terms of those for the ceramic and the polymer, and the volume fraction of each constituent. Density and dielectric constant increase linearly from 0 to 100% with the volume of ceramic. Almost the same is true for the impedance (up to 90%). The coupling coefficient rises very rapidly from 0 (the polymer being passive) to reach a plateau anywhere between 20 and 80%, before falling again towards the normal value K for the ceramic plate. This is shown by the curves in Fig. 1.26, for PZT5 ceramic rods and a rather soft polymer.

We shall see in Chap. 3 (Fig. 3.31) that it is instructive to examine the vibrations of the various parts of the transducer for different frequencies. This can be done with the help of an optical probe.

Piezopolymers. The advantage in using composite transducers is the strong electromechanical coupling, but operating frequencies are rather restricted

($f < 20$ MHz). Another type of transducer is based on polymers and copolymers such as P(VDF-TrFe). This is made from vinylidene fluoride (VDF) and trifluoroethylene (TrFE) and can operate up to frequencies of 100 MHz [1.8]. No mechanical processing is needed for these copolymers as they can be dissolved in a solvent and then spread onto the substrate. Specific properties of such materials are:

- flexibility, allowing a wide range of shapes;
- low acoustic impedance, removing the need for any matching layers;
- low cost, light and freely available.

These advantages have given rise to applications in medicine and NDT (non-destructive testing [1.9]).

Monocrystals. In higher frequency ranges, with $\lambda/2$ lying between 10 and 100 μm, one technique is to solder a thick monocrystal (≥ 0.1 mm) by indium metallic diffusion, and then to thin it down to the required value by classic polishing. Thicknesses of just a few μm can be attained by ion bombardment. Indium soldering involves vacuum evaporation of a chromium fixing layer (150 to 300 Å), followed by a gold layer (1 500 Å) and finally an indium layer (2 500 Å), onto the face of each element. Once the indium has been deposited (maintaining the vacuum so as to avoid oxidation), crystals are brought into contact and submitted to a pressure of around 200 kg/cm^2, by means of a hydraulic jack, for around 20 minutes.

Thin Layers. If $\lambda/2$ is of order 1 μm, the piezoelectric material is deposited in the form of a thin layer. Such layers are deposited onto the substrate when it has already been coated with a metallic film, which constitutes the internal electrode. The useful fraction of the piezoelectric thin layer (diameter of order 1 mm) is fixed by dimensions of the external electrode. The latter is deposited by vacuum coating or sputtering. It is not easy to control the crystallographic orientation of the piezoelectric layer. It is not monocrystalline and growth conditions depend on the state of the underlying metal, which itself depends on the crystallographic cut of the substrate. This is a major difference with the previous technique, in which the cut of the monocrystalline transducer is completely independent of the substrate cut, so that the type of elastic wave can be freely chosen.

Zinc oxide films deposited on a parallelipiped (e.g., $20 \times 6 \times 6$ mm^3) or cylindrical (diameter 3 mm) alumina monocrystal (sapphire) have received much attention, whilst the CdS layers mentioned in Sect. 1.5.2 to illustrate the effect of impedance on frequency response have fallen into disuse. In order to produce a longitudinal wave which propagates along the 3-fold axis of the sapphire crystal, the 6-fold axis A_6 of the ZnO layer must be parallel to A_3. A platinum electrode is often associated with ZnO. An underlying layer of titanium or chromium improves fixing of the metal. Chromium has elastic impedance 41.2 MRayl which is very close to that of sapphire (44.5 MRayl) and hence has little effect on frequency response, whatever its thickness. The

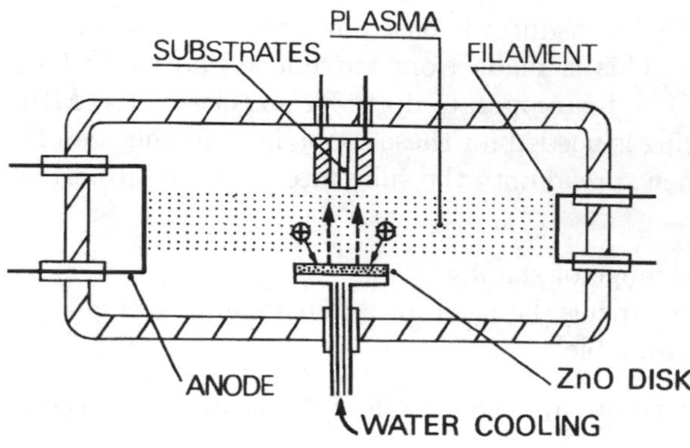

Fig. 1.27. Zinc oxide sputtering with triode setup. Zinc oxide in the target is bombarded by ions from the plasma and coats the substrate (sapphire rods) at a rate of 200 Å/min.

thickness of these metals is checked during deposition by frequency variation of a quartz oscillator placed very close to the substrate.

Zinc oxide is deposited by sputtering using a diode or triode setup, as shown in Fig. 1.27. A ZnO (sintered powder) or Zn disk is bombarded by ions extracted from an oxygen/argon column. The A_6 axis of ZnO deposited in this way is always perpendicular to the surface and the layer generates only longitudinal waves. The external electrode is generally aluminium.

This technique was first studied in 1965 [1.10] for generation and detection of bulk waves with frequency greater than 1 GHz. Since then it has undergone many improvements during the development of surface wave technology. Use of a piezoelectric layer on a commonplace substrate like glass gave rise to very low-priced (< £1) filters (e.g., for televisions). Because these components operate at frequencies of order 50 MHz, they require greater thicknesses (and areas) of ZnO, and therefore higher deposition rates. These rates have increased from a fraction of 1 μm per hour to more than 10 μm per hour. This subject will be further discussed in Sect. 2.7, which deals with surface wave technology.

Let us return to high-frequency bulk waves with an example of the kind of results which can be achieved. Figure 1.28 shows the frequency response of a ZnO-Pt-Al_2O_3 transducer. The thickness of zinc oxide is 2.6 μm and the thickness of platinum is 0.4 μm. This frequency response is plotted from 0.5 to 4.5 GHz and reveals resonances at $\lambda/4$, $3\lambda/4$, $5\lambda/4$ and $7\lambda/4$.

There are, of course, other types of transducer. Two further examples are:

- Concave transducer. This can be simple or composite. It concentrates waves and the energy they carry at its focus.
- Directional transducer. This is made up of elements with phase shifted voltages, which emit waves in a direction depending on those phase shifts. Waves guided by a plate, a rectangular or cylindrical rod or a tube, are excited by a bulk wave resonator fixed to one end of the guide, with or without intermediates. Figure 1.29 shows the possible arrangements of a

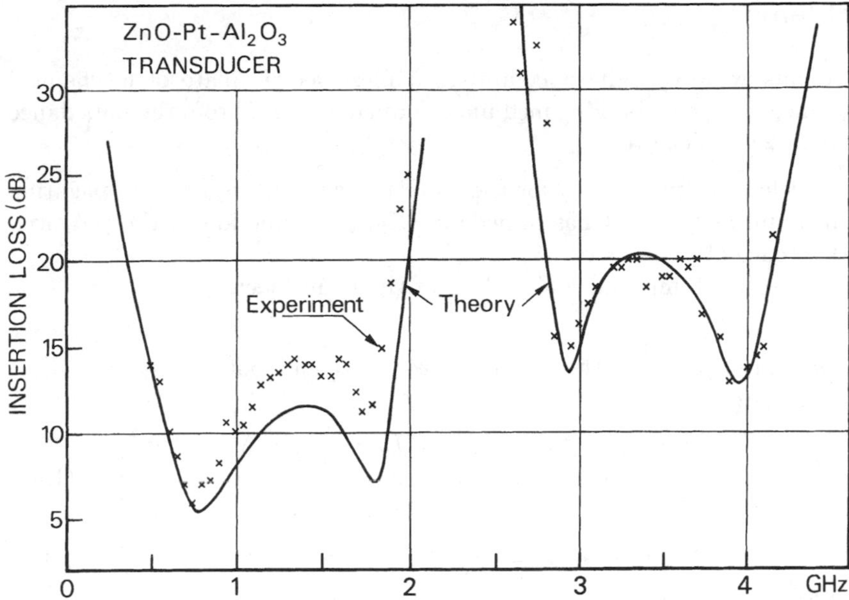

Fig. 1.28. Frequency response of a thin film (2.6 μm) zinc oxide transducer launching or detecting longitudinal bulk waves

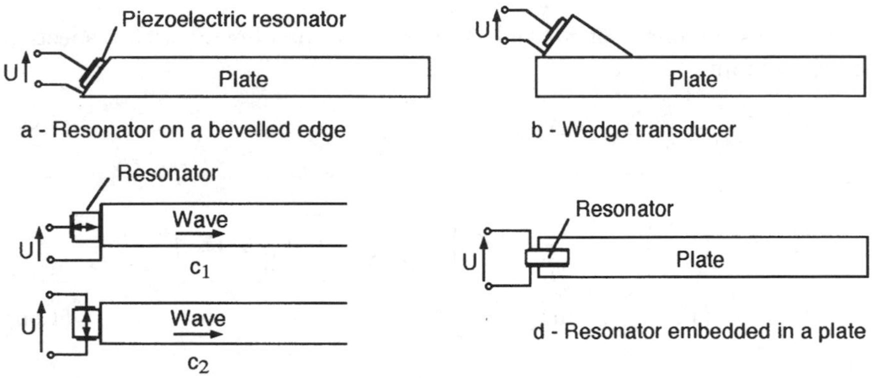

Fig. 1.29a–d. Generation of Lamb waves in a homogeneous plate. The transducer drives one of the two components. The other arises during reflections from free faces

resonator generating Lamb waves. Only one component need be excited, the other arising at the first reflection.

The technology of free bulk wave resonators is important in frequency stabilisation and filtering, and will be discussed in Sect. 4.4.1.

Problems

1.1 A bulk wave transducer comprises a piezoelectric plate of mechanical impedance $\mathcal{Z}_\mathrm{P}$, and a backing medium of impedance $\mathcal{Z}_1$. From the impedance matrix (1.24), calculate:

(a) particle velocity $v = -v_2$ for the wave transmitted into the propagating medium, assuming it has impedance $\mathcal{Z}_2$, as a function of the potential difference U;

(b) the electrical impedance $Z_\mathrm{e} = U/I$ of the transducer.

Solution. Multiplying both sides of (1.24) by i and putting $F_1 = -\mathcal{Z}_1 v_1$, $F_2 = -\mathcal{Z}_2 v_2$ give,

$$\begin{pmatrix} \mathrm{i}\mathcal{Z}_1 + \mathcal{Z}_\mathrm{P}/\tan\varphi & \mathcal{Z}_\mathrm{P}/\sin\varphi & h/\omega \\ \mathcal{Z}_\mathrm{P}/\sin\varphi & \mathrm{i}\mathcal{Z}_2 + \mathcal{Z}_\mathrm{P}/\tan\varphi & h/\omega \\ h/\omega & h/\omega & 1/\omega C_0 \end{pmatrix} \begin{pmatrix} v_1 \\ v_2 \\ I \end{pmatrix} = \begin{pmatrix} 0 \\ 0 \\ \mathrm{i}U \end{pmatrix}, \tag{1.87}$$

where $\varphi = kd$. Denoting by $Y = (y_{ij})$ the inverse of the symmetric matrix $\mathcal{Z} = (z_{ij})$:

(a) The particle velocity $v = -v_2$ in the propagating medium is given by

$$v_2 = \mathrm{i}y_{23}U\,, \quad y_{23} = -\frac{\Delta_{23}}{\Delta} \quad \Rightarrow \quad v = \mathrm{i}\frac{\Delta_{23}}{\Delta}U\,,$$

where Δ_{23} is the minor of z_{23} and $\Delta = \det \mathcal{Z}$. Expanding Δ with respect to the last column,

$$\Delta = \frac{\Delta_{33}}{C_0\omega} + \frac{h}{\omega}(\Delta_{13} - \Delta_{23})\,, \tag{1.88}$$

with

$$\Delta_{13} = \frac{h}{\omega}\left(\frac{\mathcal{Z}_\mathrm{P}}{\sin\varphi} - \frac{\mathcal{Z}_\mathrm{P}}{\tan\varphi} - \mathrm{i}\mathcal{Z}_2\right) = \frac{h}{\omega}\left(\mathcal{Z}_\mathrm{P}\tan\frac{\varphi}{2} - \mathrm{i}\mathcal{Z}_2\right),$$

$$\Delta_{23} = \frac{h}{\omega}\left(\mathrm{i}\mathcal{Z}_1 - \mathcal{Z}_\mathrm{P}\tan\frac{\varphi}{2}\right), \tag{1.89}$$

and

$$\begin{aligned} \Delta_{33} &= \left(\mathrm{i}\mathcal{Z}_1 + \frac{\mathcal{Z}_\mathrm{P}}{\tan\varphi}\right)\left(\mathrm{i}\mathcal{Z}_2 + \frac{\mathcal{Z}_\mathrm{P}}{\tan\varphi}\right) - \frac{\mathcal{Z}_\mathrm{P}^2}{\sin^2\varphi} \\ &= -\mathcal{Z}_1\mathcal{Z}_2 - \mathcal{Z}_\mathrm{P}^2 + \mathrm{i}(\mathcal{Z}_1 + \mathcal{Z}_2)\frac{\mathcal{Z}_\mathrm{P}}{\tan\varphi}\,. \end{aligned} \tag{1.90}$$

Grouping together real and imaginary terms,

$$\begin{aligned} \Delta = \frac{1}{C_0\omega}\Bigg[& -\mathcal{Z}_1\mathcal{Z}_2 - \mathcal{Z}_\mathrm{P}^2 + 2\frac{h^2C_0}{\omega}\mathcal{Z}_\mathrm{P}\tan\theta \\ & +\mathrm{i}(\mathcal{Z}_1 + \mathcal{Z}_2)\left(\mathcal{Z}_\mathrm{P}\cot 2\theta - \frac{h^2C_0}{\omega}\right)\Bigg], \end{aligned}$$

where $\theta = \varphi/2$. Since

$$K^2 = \frac{e^2}{\varepsilon^S c^D} , \quad \mathcal{Z}_P = \rho_P V_P A = \frac{c^D A}{V_P} ,$$

we have

$$\frac{h^2 C_0}{\omega} = \frac{e^2}{\varepsilon^S} \frac{A}{kdV_P} = \frac{K^2}{2\theta} \mathcal{Z}_P , \tag{1.91}$$

and the particle velocity is given by

$$v = \frac{hC_0(\mathrm{i}\mathcal{Z}_1 - \mathcal{Z}_P \tan\theta)U}{(\mathcal{Z}_1 + \mathcal{Z}_2)\mathcal{Z}_P \left(\cot 2\theta - \dfrac{K^2}{2\theta}\right) + \mathrm{i}\left(\mathcal{Z}_1\mathcal{Z}_2 + \mathcal{Z}_P^2 - \dfrac{K^2}{\theta}\mathcal{Z}_P^2 \tan\theta\right)} , \tag{1.92}$$

where $\theta = kd/2$. When the rear face of the transducer is free, $\mathcal{Z}_1 = 0$ and a factor of $\mathcal{Z}_P$ can be cancelled to give

$$v = -\frac{hC_0 U}{\mathcal{Z}_2 \left(\cot 2\theta - \dfrac{K^2}{2\theta}\right) \cot\theta + \mathrm{i}\mathcal{Z}_P \left(\cot\theta - \dfrac{K^2}{\theta}\right)} . \tag{1.93}$$

(b) The intensity of the current entering the transducer is

$$I = \mathrm{i}y_{33}U , \quad \text{where} \quad y_{33} = \frac{\Delta_{33}}{\Delta} ,$$

and the impedance of the transducer is given by (1.88):

$$Z_\mathrm{e} = \frac{U}{I} = \frac{\Delta}{\mathrm{i}\Delta_{33}} = \frac{1}{\mathrm{i}C_0\omega}\left(1 + hC_0 \frac{\Delta_{13} - \Delta_{23}}{\Delta_{33}}\right) .$$

From the expressions for minors Δ_{13} and Δ_{23} and the relation

$$\frac{h^2 C_0}{\omega} = \frac{K^2}{\varphi} \mathcal{Z}_P ,$$

this can be rewritten

$$Z_\mathrm{e} = \frac{1}{\mathrm{i}C_0\omega}\left(1 + \frac{K^2}{\varphi}\mathcal{Z}_P \frac{2\mathcal{Z}_P \tan\varphi/2 - \mathrm{i}(\mathcal{Z}_1 + \mathcal{Z}_2)}{-(\mathcal{Z}_P^2 + \mathcal{Z}_1\mathcal{Z}_2) + \mathrm{i}\mathcal{Z}_P(\mathcal{Z}_1 + \mathcal{Z}_2)/\tan\varphi}\right) .$$

Replacing $\tan\varphi/2$ by $(1 - \cos\varphi)/\sin\varphi$,

$$Z_\mathrm{e} = \frac{1}{\mathrm{i}C_0\omega}\left(1 + \frac{K^2}{\varphi}\mathcal{Z}_P \frac{2\mathcal{Z}_P(1 - \cos\varphi) - \mathrm{i}(\mathcal{Z}_1 + \mathcal{Z}_2)\sin\varphi}{-(\mathcal{Z}_P^2 + \mathcal{Z}_1\mathcal{Z}_2)\sin\varphi + \mathrm{i}\mathcal{Z}_P(\mathcal{Z}_1 + \mathcal{Z}_2)\cos\varphi}\right) , \tag{1.94}$$

where $\varphi = kd$.

1.2 Show that the T element in Fig. 1.4a, corresponding to a 1-dimensional solid slab, can be replaced by the coaxial line element in Fig. 1.4b. One way of establishing equivalence is to calculate the mechanical impedance $\mathcal{Z}_1 = F_1/v_1$ of the slab loaded by a medium of impedance $\mathcal{Z}_2 = -F_2/v_2$.

Solution. Setting $h = 0$, for a non-piezoelectric solid, into (1.20b),

$$\left(\mathcal{Z}_2 + \frac{\mathcal{Z}}{\mathrm{i}\tan kd}\right) v_2 = \frac{\mathrm{i}\mathcal{Z}}{\sin kd} v_1 \,.$$

Substituting into (1.20a),

$$F_1 = \left(\frac{\mathcal{Z}}{\mathrm{i}\tan kd} + \frac{\mathcal{Z}^2}{\sin^2 kd} \frac{\mathrm{i}\tan kd}{\mathcal{Z} + \mathrm{i}\mathcal{Z}_2 \tan kd}\right) v_1 \,.$$

The mechanical impedance at $z = z_1$,

$$\mathcal{Z}_1 = \frac{F_1}{v_1} = \frac{\mathcal{Z}}{\mathrm{i}\tan kd}\left(1 - \mathcal{Z}\frac{1+\tan^2 kd}{\mathcal{Z} + \mathrm{i}\mathcal{Z}_2 \tan kd}\right) = \mathcal{Z}\frac{\mathcal{Z}_2 + \mathrm{i}\mathcal{Z}\tan kd}{\mathcal{Z} + \mathrm{i}\mathcal{Z}_2 \tan kd} \,,$$

is found by applying equation (1.66) of Vol. I, which gives the transformed impedance for a transmission line element of length $d = z_2 - z_1$.

1.3 Show that when the equivalent electromechanical circuit for a rod, depicted in Fig. 1.8, is closed on the characteristic impedance $\mathcal{Z}$ of the material, it is equivalent to two equal and opposite forces acting on each end of the rod, i.e.,

$$F_1 = F_2 = \frac{1}{2} eLU(1 - \mathrm{e}^{-\mathrm{i}kd}) \,.$$

Solution. When the two acoustic ports in the circuit of Fig. 1.8b are loaded by the characteristic impedance $\mathcal{Z}$ of the material, it is symmetrical and $F_1 = F_2$, $v_1 = v_2$, with

$$F_1 = -\mathcal{Z}v_1 \,, \quad eLU = 2\mathrm{i}\frac{\mathcal{Z}}{\sin\varphi} v_1 - \mathcal{Z}\left(1 + \mathrm{i}\tan\frac{\varphi}{2}\right) v_1 \,,$$

and $\varphi = kd$. Replacing $-\mathcal{Z}v_1$ by F_1, this implies

$$F_1 = \frac{eLU}{1 + \mathrm{i}(\tan\varphi/2 - 2/\sin\varphi)} \,.$$

From the identity

$$\frac{2}{\sin\varphi} - \tan\frac{\varphi}{2} = \cot\frac{\varphi}{2} \,,$$

it follows that

$$F_1 = \frac{eLU \sin\varphi/2}{\sin\varphi/2 - \mathrm{i}\cos\varphi/2} = \mathrm{i}eLU\frac{\sin\varphi/2}{\exp(\mathrm{i}\varphi/2)} = \frac{1}{2} eLU[1 - \exp(-\mathrm{i}\varphi)] \,.$$

1.4 Show by Parseval's theorem (2.35) that the amplitude h_m of the impulse response $h(t)$ of a transducer with $Z = Z_\mathrm{P}$ or $Z_\mathrm{P}\sqrt{2}$ is of order $2f_\mathrm{P}$. The response is approximated by a half sine wave of width $\tau_\mathrm{P} = 1/f_\mathrm{P}$ (see Fig. 1.13b, c).

Solution. We express the energy in the time and frequency domains:

$$E = \int_{-\infty}^{\infty} h^2(t)\,\mathrm{d}t = \int_{-\infty}^{\infty} |H(f)|^2 \mathrm{d}f = 2\int_0^{\infty} |H(f)|^2 \mathrm{d}f \,.$$

When $Z \approx Z_{\mathrm{P}}$, the frequency response (see Fig. 1.12) is almost a rectangle of width f_{P} and height $\sqrt{2}$ (3 dB). Hence,

$$\int_0^{\infty} h^2(t)\,\mathrm{d}t \approx 4f_{\mathrm{P}} \,, \quad \text{or} \quad \int_0^{\infty} h^2(f_{\mathrm{P}}t)\,\mathrm{d}(f_{\mathrm{P}}t) \approx 4f_{\mathrm{P}}^2 \,.$$

For a sine wave arch of amplitude h_{m},

$$\frac{h_{\mathrm{m}}^2}{2} \approx 4f_{\mathrm{P}}^2 \quad \Rightarrow \quad h_{\mathrm{m}} \approx 2\sqrt{2}f_{\mathrm{P}} \,.$$

The effective value is slightly smaller ($\approx 2f_{\mathrm{P}}$) because not all the energy is contained in the first sinusoidal oscillation.

1.5 How does the relative separation of resonance and antiresonance frequencies vary with order n (odd)?

Solution. Let ε_n be the relative separation between $f_{\mathrm{a}}^{(n)} = nf_{\mathrm{a}}$ (where $n = 2p+1$) and $f_{\mathrm{r}}^{(n)}$, so that $f_{\mathrm{a}}^{(n)} - f_{\mathrm{r}}^{(n)} = nf_{\mathrm{a}}\varepsilon_n$. Then since

$$\tan\frac{\pi}{2}\frac{f_{\mathrm{r}}^{(n)}}{f_{\mathrm{a}}} = \cot\left(\frac{\pi}{2}\frac{f_{\mathrm{a}}^{(n)} - f_{\mathrm{r}}^{(n)}}{f_{\mathrm{a}}}\right) = \cot\left(n\frac{\pi}{2}\varepsilon_n\right) \approx \frac{2}{n\pi\varepsilon_n} \,,$$

for $\varepsilon_n \ll 1$, relation (1.59) implies

$$\frac{2K^2}{n\pi\varepsilon_n} \approx \frac{\pi}{2}n \quad \Rightarrow \quad \varepsilon_n = \frac{f_{\mathrm{a}}^{(n)} - f_{\mathrm{r}}^{(n)}}{f_{\mathrm{a}}^{(n)}} \approx \frac{4K^2}{\pi^2}\frac{1}{n^2} \,. \tag{1.95}$$

1.6 Resonator with losses. Generalise (1.61) to the case of a resonator with viscoelastic losses. Attenuation is characterised by a coefficient $\eta = c^D\tau$. Draw the Nyquist plot representing electrical impedance in the complex plane, when $\omega\tau \ll 1$.

Solution. Expanding $\tan\theta/\theta$ ($\theta = kd/2$) in the resonator impedance formula (1.57), we find that

$$Z_{\mathrm{e}} = \frac{1}{\mathrm{i}C_0\omega}\left[1 - \frac{8K^2}{\pi^2}\sum_{n>0,\,\mathrm{odd}} \frac{1}{n^2 - (k'd/\pi)^2}\right] = \frac{1}{\mathrm{i}C_0\omega} + Z_{\mathrm{m}} \,,$$

where $k' = k/(1+\mathrm{i}\omega\tau)^{1/2}$ is a complex wave number taking viscoelastic losses into account, as in Vol. I, Sect. 1.2.3. Since

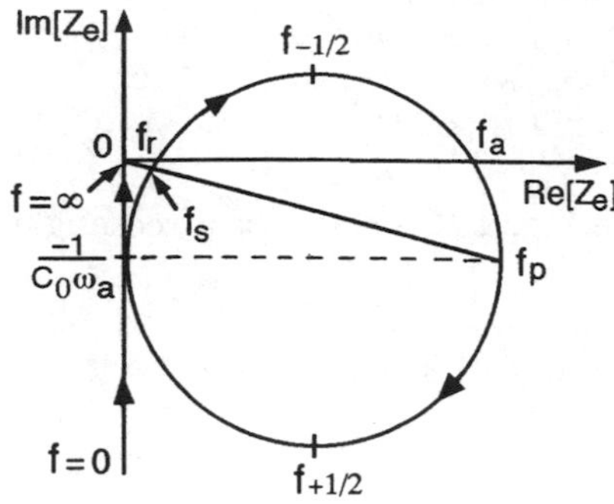

Fig. 1.30. Near resonance, the curve representing the impedance of an electromechanical resonator is a circle in the complex plane. f_{s} and f_{p} are the series and parallel resonance frequencies

$$\left(\frac{k'd}{\pi}\right)^2 = \frac{(kd/\pi)^2}{1+\mathrm{i}\omega\tau} = \frac{\omega^2}{\omega_{\mathrm{a}}^2(1+\mathrm{i}\omega\tau)} ,$$

where $\omega_{\mathrm{a}} = \pi V/d$, the motional impedance Z_{m} is given by

$$Z_{\mathrm{m}} = \frac{8K^2}{\pi^2 C_0 \omega} \sum_{n>0,\,\mathrm{odd}} \frac{\mathrm{i}(1+\mathrm{i}\omega\tau)\omega_{\mathrm{a}}^2}{\omega_n^2 - \omega^2 + \mathrm{i}\omega\tau\omega_n^2} ,$$

where $\omega_n = n\omega_{\mathrm{a}}$. For low levels of attenuation $\omega\tau \ll 1$, this implies

$$Z_{\mathrm{m}} = \frac{8K^2}{\pi^2 C_0 \omega} \sum_{n>0,\,\mathrm{odd}} \frac{\omega_{\mathrm{a}}/n\omega}{\omega_n\tau + \mathrm{i}x} , \quad x = \frac{\omega}{\omega_n} - \frac{\omega_n}{\omega} . \tag{1.96}$$

Near the resonant angular frequency $\omega \approx n\omega_{\mathrm{a}}$, each term in the sum is represented by a circle of diameter $D_n = 1/\omega_n\tau n^2$ in the complex plane. The bandwidth of the resonance, defined by cutoff angular frequencies $\omega_{-1/2}$ and $\omega_{1/2}$ such that $x(\omega_{\pm 1/2}) = \pm\omega_n\tau$, is $\Delta\omega_n = \tau\omega_n^2 = \tau\omega_{\mathrm{a}}^2 n^2$. The complex electrical impedance Z_{e} is shifted down by $1/C_0\omega$ and the diameter of the circle multiplied by $8K^2/\pi^2 C_0\omega$, as shown in Fig. 1.30. Cutoff angular frequencies $\omega_{-1/2}$ and $\omega_{1/2}$ correspond to the minimum and maximum of the imaginary part of the impedance, respectively.

1.7 Applying the method described in Vol. I, Sect. 5.2.1, express the ratio of mechanical to electrical energy stored in a lossy resonator (see previous problem) in terms of the coupling coefficient K and quality factor Q.

Solution. Damping of resonator vibrations due to attenuation is represented by a complex angular frequency $\omega + \mathrm{i}\delta\omega$. The imaginary part $\delta\omega$ is inversely proportional to the quality factor,

$$Q = \frac{1}{\omega\tau} \gg 1 \quad \text{and} \quad \delta\omega = \frac{\omega}{2Q} .$$

The complex power is given by (5.26) in Vol. I, where E_{k} and E_{p} are kinetic and potential energies:

$$\frac{1}{2}UI^* = \frac{1}{2}Z_e(\omega + i\delta\omega)|I|^2 = 2i\omega\langle E_k - E_p\rangle - 2\delta\omega\langle E_k + E_p\rangle \ . \tag{1.97}$$

Using (1.57) for the resonator impedance, with

$$\theta = \frac{\pi}{2}\frac{\omega + i\delta\omega}{\omega_a} \approx \frac{\pi}{2}\left(1 + \frac{i}{2Q}\right) \Rightarrow \tan\theta = -\frac{1}{\tan(i\pi/4Q)} \approx i\frac{4Q}{\pi} \ ,$$

at the resonance frequency $\omega \approx \omega_a$, it follows that

$$Z_e(\omega + i\delta\omega) \approx \frac{1}{iC_0\omega} - \frac{\delta\omega}{C_0\omega^2} - \frac{2K^2}{C_0\omega}\frac{4Q}{\pi^2} \ .$$

Identifying the real parts on both sides of (1.97), we obtain the mean value of the total stored energy $E_k + E_p$ in the resonator ($2\delta\omega = \omega/Q$):

$$\langle E_k + E_p\rangle = \frac{1}{4}\frac{|I|^2}{C_0\omega^2} + \frac{4K^2Q^2}{\pi^2 C_0\omega^2}|I|^2 = \langle E_{el}\rangle + \langle E_m\rangle \ .$$

The first term corresponds to the electrical energy $C_0|U|^2/4$, and the second to mechanical energy:

$$\frac{\langle E_m\rangle}{\langle E_{el}\rangle} = \left(\frac{4KQ}{\pi}\right)^2 \quad \text{and} \quad \langle E_m\rangle = \left(\frac{2KQ}{\pi}\right)^2 C_0|U|^2 \ . \tag{1.98}$$

Order of Magnitude. In the case of an AT cut quartz resonator ($K = 0.08$), with static capacitance $C_0 = 8$ pF and quality factor $Q = 10^5$, driven by an alternating voltage of amplitude $U = 5$ V, the ratio of mechanical to electrical energies is 10^8 and the stored mechanical energy is 5 mJ.

1.8 Prove (1.76) and derive the expression for the bandshape factor M.

Solution. Putting $a_P + b_P = a_M + b_M = x$, $a_P - b_P = y$ and $a_M - b_M = z$, the system of equations (1.75) becomes

$$\begin{aligned} a e^{-ikd_2} &= x\cos\varphi_M - iz\sin\varphi_M \ , \\ \frac{Z}{Z_M} a e^{-ikd_2} &= -ix\sin\varphi_M + z\cos\varphi_M \ , \\ \frac{eU}{Z_P d} &= -y + \frac{Z_M}{Z_P}z \ , \\ \frac{eU}{Z_P d} &= -y\cos\varphi_P - ix\sin\varphi_P \ . \end{aligned}$$

The determinant of the coefficients of unknowns x and z in the first two equations is equal to 1, and these equations imply

$$\begin{aligned} x &= \left(\cos\varphi_M + i\frac{Z}{Z_M}\sin\varphi_M\right) a e^{-ikd_2} \ , \\ z &= \left(\frac{Z}{Z_M}\cos\varphi_M + i\sin\varphi_M\right) a e^{-ikd_2} \ . \end{aligned}$$

Eliminating y between the last two equations leads to

$$z\frac{Z_{\mathrm{M}}}{Z_{\mathrm{P}}}\cos\varphi_{\mathrm{P}} + \mathrm{i}x\sin\varphi_{\mathrm{P}} = \frac{eU}{Z_{\mathrm{P}}d}(\cos\varphi_{\mathrm{P}} - 1)\,.$$

Substituting in x and z, the result is

$$\frac{eU}{Z_{\mathrm{P}}d}(\cos\varphi_{\mathrm{P}} - 1)\mathrm{e}^{\mathrm{i}kd_2} = a\left[\left(\frac{Z}{Z_{\mathrm{P}}}\cos\varphi_{\mathrm{P}}\cos\varphi_{\mathrm{M}} - \frac{Z}{Z_{\mathrm{M}}}\sin\varphi_{\mathrm{P}}\sin\varphi_{\mathrm{M}}\right) + \mathrm{i}\left(\cos\varphi_{\mathrm{M}}\sin\varphi_{\mathrm{P}} + \frac{Z_{\mathrm{M}}}{Z_{\mathrm{P}}}\cos\varphi_{\mathrm{P}}\sin\varphi_{\mathrm{M}}\right)\right], \tag{1.99}$$

as required.

1.9 Coupling of vibration modes in a rod. Consider an infinite rod along the x_3 axis, of width w in the x_2 direction and height h in the x_1 direction, as shown in Fig. 1.31. In the two limiting cases $w \ll h$ and $w \gg h$, express the vibration frequencies for thickness modes and longitudinal extensional modes in the rod as functions of longitudinal wave speeds $V_1 = \sqrt{c_{11}/\rho}$ and $V_2 = \sqrt{c_{22}/\rho}$ and plate speeds $V_{1\mathrm{P}}$, $V_{2\mathrm{P}}$ of the Lamb mode S_0 propagating in directions x_1, x_2. Check that these frequencies satisfy the coupled mode equation [1.11]

$$\left[1-\left(\frac{2fh}{V_1}\right)^2\right]\left[1-\left(\frac{2fw}{V_2}\right)^2\right] = \frac{c_{12}^2}{c_{11}c_{22}} = \Gamma^2\,. \tag{1.100}$$

Plot the dependence of $y = 2fh/V_1$ on $x = hV_2/wV_1$ in the general case (i.e., for arbitrary values of w and h).

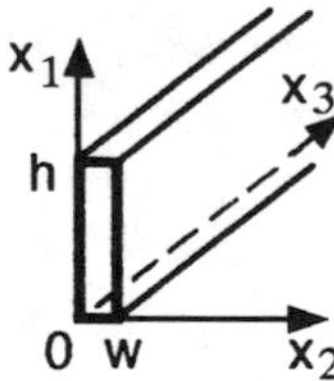

Fig. 1.31. Infinite rod

Solution. (a) The case $w \ll h$. The frequency of the thickness-mode vibration of the rod satisfies

$$2fw = V_2 = \sqrt{\frac{c_{22}}{\rho}}\,,$$

whilst the frequency of the longitudinal extension mode along the x_1-axis of the plate satisfies

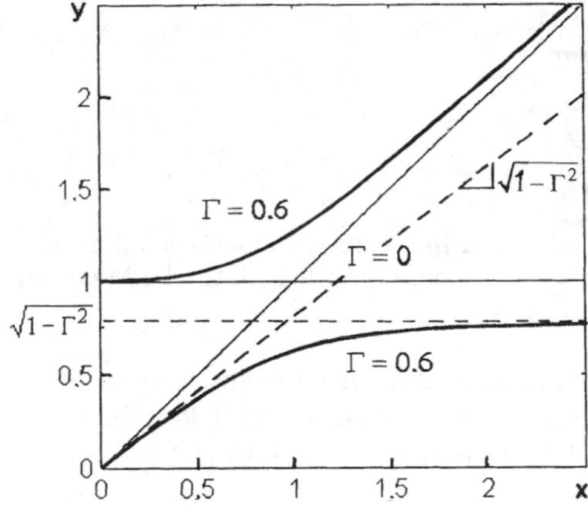

Fig. 1.32. Resonance frequencies of a PZT4 ceramic rod ($\Gamma = 0.6$) as a function of the height/thickness ratio

$$2fh = V_{1\mathrm{P}} = \sqrt{\frac{c_{11}}{\rho}\left(1 - \frac{c_{12}^2}{c_{11}c_{22}}\right)}.$$

$V_{1\mathrm{P}}$ is the phase speed of the S_0 mode propagating in the x_1 direction in a thin plate [see (5.123), Vol. I, with $c_{22} \neq c_{11}$].
(b) The case $w \gg h$. The roles of x_1 and x_2 are reversed. For the thickness mode,

$$2fh = V_1 = \sqrt{\frac{c_{11}}{\rho}},$$

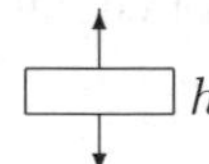

and for the longitudinal extension mode in the x_2 direction,

$$2fw = V_{2\mathrm{P}} = \sqrt{\frac{c_{22}}{\rho}\left(1 - \frac{c_{12}^2}{c_{11}c_{22}}\right)}.$$

w

These frequencies are solutions of (1.100) when $w \to 0$ and when $h \to 0$.
(**c**) Using the above notation, equation (1.100) becomes

$$(1-y^2)(x^2-y^2) = \Gamma^2 x^2 \Rightarrow y^4 - y^2(x^2+1) + (1-\Gamma^2)x^2 = 0\,.$$

The two solutions

$$y^2 = \frac{1+x^2}{2} \pm \frac{1}{2}\sqrt{(1-x^2)^2 + 4\Gamma^2 x^2}$$

are shown in Fig. 1.32.
Some significant values are:

- $x \ll 1$, $y \approx 1 \Rightarrow 2fh \approx V_1$ and $y \approx \sqrt{1-\Gamma^2}x \Rightarrow 2fw \approx V_{2\mathrm{P}}$;
- $x = 1$, $y \approx \sqrt{1 \pm \Gamma}$ and coupling pushes the resonance frequencies apart;
- $x \gg 1$, $y \approx x \Rightarrow 2fw \approx V_2$ and $y \approx \sqrt{1-\Gamma^2} \Rightarrow 2fh \approx V_{1\mathrm{P}}$.

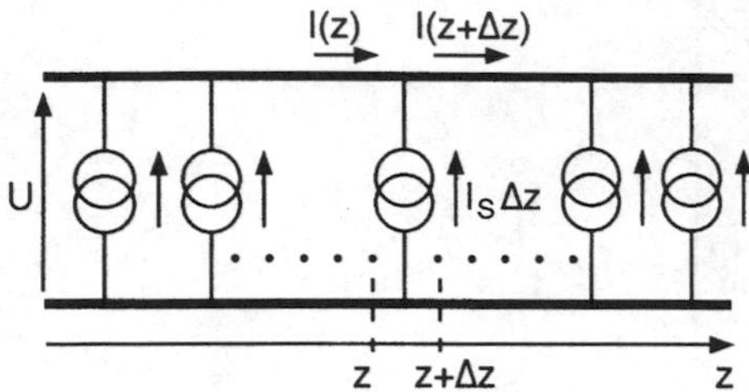

Fig. 1.33. Transmission line element equivalent to the KLM circuit

1.10 The circuit in Fig. 1.33 represents a section of transmission line with a distributed current supply of constant intensity I_S per unit length. Show that this circuit is equivalent to the KLM circuit in Fig. 1.10 [1.12].

Solution. Two circuit diagrams are equivalent when they are described by the same equations. The line shown satisfies (1.32) in Sect. 1.2.1.2 of Vol. I:

$$\frac{\partial U}{\partial z} = -L\frac{\partial I}{\partial t} \,. \tag{1.101}$$

It also obeys (1.33), Vol. I, modified by addition of I_S,

$$\frac{\partial I}{\partial z} = -C\frac{\partial U}{\partial t} + I_\mathrm{S} \quad \Rightarrow \quad \frac{\partial U}{\partial t} = -\frac{1}{C}\frac{\partial I}{\partial z} + \frac{1}{C}I_\mathrm{S} \,. \tag{1.102}$$

These two equations have the same form as (1.32), potential difference U corresponding to stress and current intensity to particle velocity. They therefore lead to the KLM circuit of Fig. 1.10 and the circuit with localised constants in Fig. 1.6.

1.11 Find an expression for the electrical impedance of a resonator with each face loaded by a layer of elastic impedance $Z' = \rho' V'$ and thickness d'. What is the relative variation of the antiresonance frequency when $k'd' \ll 1$?

Solution. Since $Z_1 = Z_2$, putting $r = Z_1/Z_\mathrm{P}$, equation (1.56) implies

$$\mathrm{i}C_0\omega Z_\mathrm{e} = 1 - 2\frac{K^2}{\varphi}\frac{1-\cos\varphi - \mathrm{i}r\sin\varphi}{(1+r^2)\sin\varphi - 2\mathrm{i}r\cos\varphi} \,, \qquad \varphi = kd = 2\theta \,.$$

Hence,

$$\begin{aligned}\mathrm{i}C_0\omega Z_\mathrm{e} &= 1 - \frac{K^2}{\theta}\frac{(\sin\theta - \mathrm{i}r\cos\theta)\sin\theta}{(1+r^2)\sin\theta\cos\theta + \mathrm{i}r(\sin^2\theta - \cos^2\theta)} \\ &= 1 - \frac{K^2}{\theta}\frac{\sin\theta}{\cos\theta + \mathrm{i}r\sin\theta} \,.\end{aligned}$$

Equation (1.66), Vol. I, giving the transformed impedance, leads to

$$Z_1 = \mathrm{i}Z'\tan k'd' = \mathrm{i}Z'\tan 2\theta' \,,$$

because the end impedance Z_L is zero. The electrical impedance of the resonator loaded on both sides is then

$$Z_e = \frac{1}{iC_0\omega}\left(1 - \frac{K^2/\theta}{\cot\theta - (Z'/Z_P)\tan 2\theta'}\right) . \tag{1.103}$$

At antiresonance frequencies ω_a, Z_e is infinite:

$$\cot\frac{\omega_a d}{2V_P} = \frac{Z'}{Z_P}\tan\frac{\omega_a d'}{V'} .$$

If the layers are very thin $k'd' \ll 1$, the frequency decreases slightly, according to $\omega_a = \pi V_P(1-\varepsilon)/d$, where

$$\cot\left[\frac{\pi}{2}(1-\varepsilon)\right] = \frac{Z'}{Z_P}\tan\left(\pi\frac{d'}{d}\frac{V_P}{V'}\right) \Rightarrow \varepsilon \approx \frac{\pi^2}{2}\frac{\rho' d'}{\rho_P d} .$$

The relative variation of the antiresonance frequency is proportional to the relative mass variation:

$$\frac{\Delta f_a}{f_a} = -\frac{\pi^2}{2}\frac{\rho' d'}{\rho d} = -\frac{\pi^2}{2}\frac{\Delta m}{m_P} . \tag{1.104}$$

2. Interdigital-Electrode Transducers for Surface Waves

This chapter deals with generation and detection of surface waves. In principle, the term 'surface wave' can be applied to any deformation which propagates by vibrating only a thin layer of matter close to the surface. The thickness is assessed in terms of the wavelength. The term therefore covers Rayleigh waves, Bleustein–Gulyaev waves and Love waves, all described in Sects. 5.3 and 5.4 of Vol. I. The Bleustein–Gulyaev wave is held near the surface by piezoelectricity of the medium, whilst the Love wave is retained by its inhomogeneity, produced by a thin layer deposited on a substrate. In fact, there are other ways of holding a transverse wave close to a surface. For example, a grating made from notches or metallic strips will have this effect. Rigidity is reduced near the surface, just as it is by the thin layer holding a Love wave, and the speed of any wave of wavelength longer than the grating period is decreased. These waves, together with other longitudinal pseudo-surface waves, have grown in importance over the past few years as a result of the increased demand for devices operating at higher and higher frequencies. Indeed they are faster than Rayleigh waves and technology is simplified because technical difficulties decrease as wavelength increases. These waves are excited and also detected by the transducers we shall describe, provided the cut is suitably chosen. The Rayleigh wave, which is our main concern, is a complex wave propagating at the surface of any solid. In the simplest case, the mechanical displacement has a longitudinal component and a transverse component which are $\pi/2$ out of phase, and which cancel totally beyond a depth of order twice the wavelength. However, this complexity is compensated by a considerable advantage. The Rayleigh wave can be excited and detected on a piezoelectric substrate by a transducer made from two comb-shaped electrodes (interdigital electrodes). This simple technique is extremely well-suited to tasks other than electromechanical conversion.

We begin by examining the operating principles of this transducer, and then analyse it according to several different models. The first signal processing devices based on Rayleigh waves were designed by representing the transducer as an ideal transversal filter whose impulse response was a replica of the comb shape. This model is useful because it can easily be inverted to determine the weighting function which leads to the required frequency response. Its main weakness is that it does not take into account internal reflections on electrodes. Solutions were therefore proposed to reduce these reflections, considered as mere side effects. For this reason, the discrete array model (see Sect. 2.1.3) and also the most developed impulse response model (see Sect. 2.2) are applicable to transducers equipped with electrodes whose geometrical disposition aims to reduce internal reflection. They have finite impulse response. Frequency response and electrical admittance of a simple transducer are calculated.

In each of the above models, one quantity is idealised. In the first, it is the distribution of charges, treated as highly localised spikes or delta functions. In the second, it is distribution of the potential according to a sinusoidal variation. However, a closer approximation to reality is made possible through the idea of piezoelectric permittivity. Quantities characterising the transducer (e.g., generation factor, admittance) are then expressed in terms of the real charge distribution. This method, which also has the advantage of explaining piezoelectric reemission, will be discussed in Sect. 2.3.

Piezoelectric reemission is the phenomenon whereby a loaded receiver transducer behaves in part as an emitter. Any incident wave induces a voltage across the load terminals and this voltage in turn generates two waves which leave the transducer on either side. One of these moves towards the transducer which launched the incident wave. The latter then reemits a new wave, and the result is a series of spurious echos. The largest of these is the triple-transit signal, which can have a significant adverse effect on the desired response of a two-transducer configuration. This reemission leads to a matrix representation of the transducer considered as a hexapole (see Sect. 2.4).

There is another reason why a wave travelling under a loaded or unloaded interdigital transducer continually generates a wave moving in the opposite direction as it advances: reflection from the sequence of fingers. This is, of course, valid for any wave arising under the active fingers of the transducer. The coupled mode method described in Appendix B applies equally well to passive transducers, i.e., reflectors, and active transducers (see Sect. 2.5).

Three other methods are given in Sect. 2.6 under the title: Adjacent Element Analysis. The first is based on the P-matrix of an elementary cell, the second uses the idea of harmonic admittance and involves numerical techniques, and the third relies on equivalent circuits, derived from the classic Mason equivalent circuit, but adapted to take reflections into account.

The demand for structures, preferably unidirectional and operating at frequencies above 1 GHz, requires more and more sophisticated technology (see Sect. 2.7).

2.1 Operating Principles

The interdigital-electrode transducer has two comb-shaped electrodes, placed on a piezoelectric substrate (e.g., by photoengraving a vacuum-deposited metallic film). In its simplest version (see Fig. 2.1a), the space d between

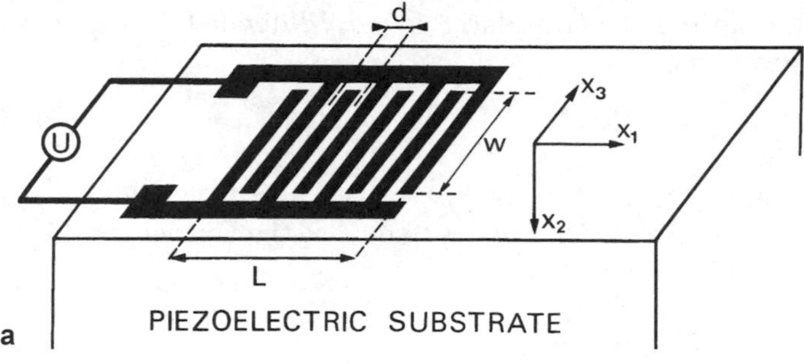

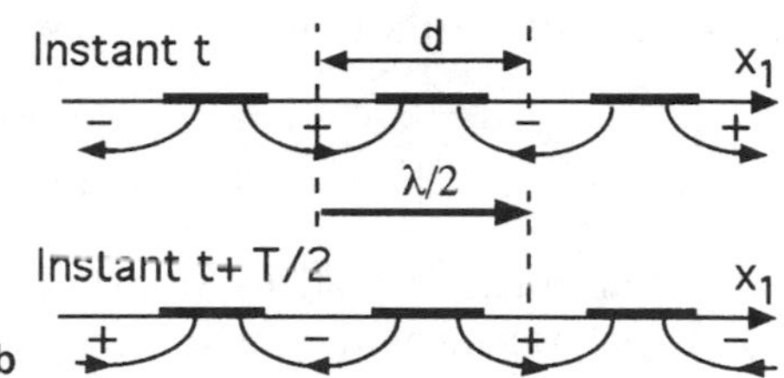

Fig. 2.1. (**a**) Simple transducer with interdigital electrodes. (**b**) The voltage applied across the two comb-shaped electrodes produces stresses near the surface of the solid

electrode fingers and the overlap width w of any two adjacent fingers are both constant [2.1].

The voltage U across the electrodes produces an electric field which generates compressions and expansions near the surface. These give rise to various elastic waves. With regard to the Rayleigh wave emitted perpendicularly to the electrode fingers, the transducer behaves like an array of ultrasound sources. When the supply voltage is sinusoidal, vibrations add constructively only if the distance d equals half the elastic wavelength. Indeed, any stress produced at time t by a pair of fingers, for a given polarity of the voltage, travels the distance $\lambda/2$ during the half-period $T/2$ at the speed V_{R} of the Rayleigh wave (see Fig. 2.1b). At time $t + T/2$, the stress arrives under the neighbouring pair of fingers, just when the voltage, which has changed sign, produces a stress with the same phase. The stress due to the second pair of fingers adds constructively to the first. The frequency $f_0 = V_{\mathrm{R}}/2d$ corresponding to this cumulative effect is called the *synchronism frequency* or *resonance frequency.* If the frequency shifts away from this value, interference between elastic signals generated by the various pairs of fingers is not totally constructive and the resulting signal is reduced. It follows that the bandwidth of such a transducer is narrower the more fingers the electrodes have.

The *frequency response* of a simple transducer with N pairs of fingers can be deduced from its impulse response. A pulse applied to the electrodes, which has short duration compared with the time required for the Rayleigh wave to pass between two fingers, simultaneously excites the various parts of the transducer. Since the electric field is reversed at each interval between fingers, the transmitted elastic signal has spatial period $2d$. It lasts a time Θ

equal to the active length of the transducer $L = 2Nd$ divided by the speed V_R of the Rayleigh wave:

$$\Theta = \frac{2dN}{V_R} = \frac{N}{f_0} . \tag{2.1}$$

Intuitively assimilating this impulse response of width Θ to a sine wave of frequency $f_0 = V_R/2d$, the frequency response, equal to the Fourier transform of this response, is a $\sin X/X$ curve, where

$$X = \pi\Theta(f - f_0) = N\pi\frac{f - f_0}{f_0} . \tag{2.2}$$

The 3 dB bandwidth ($X \approx \pm 0.885\pi/2$) is inversely proportional to the number of pairs of fingers N,

$$\frac{\Delta f}{f_0} \approx \frac{0.885}{N} . \tag{2.3}$$

In practice, the impulse response of a transducer is read by transforming the emitted elastic wave train into an electrical signal using a receiver transducer, e.g., another interdigital transducer (the response is read without mechanical contact using an optical probe, as described in Sect. 3.2). The electric field accompanying the elastic wave induces a time-varying potential difference when the elastic wave passes beneath the electrodes. The signal shape depends on the number of fingers in the receiver. Let us consider two typical cases:

- Receiver with a single pair of fingers (Fig. 2.2a). Reception is localised on a line parallel to the wave fronts. The electrical signal faithfully reproduces the elastic signal as it passes under the receiver.
- Receiver identical to emitter (Fig. 2.2b). The received voltage amplitude increases from the time $\tau = l/V_R$ when the first wave front of the elastic wave train reaches the first interdigital interval of the receiver. It goes through a maximum at time $\tau + \Theta$ when the various elements of the receiver are simultaneously excited, and then falls off. The overall impulse response, with triangular envelope and duration 2Θ, is the autocorrelation function of the impulse response of one of the transducers. When the transducers are identical, the frequency response of the two-transducer configuration is therefore a $[\sin X/X]^2$ curve, with 3 dB bandwidth equal to

$$\frac{\Delta f}{f_0} \approx \frac{0.635}{N} . \tag{2.4}$$

Figure 2.3 shows the impulse response for a configuration of two identical 60 finger transducers, made of quartz (Y-cut, propagation in the X-direction), with centre frequency $f_0 = 17$ MHz. The diamond-shaped envelope and width $2\Theta = 3.5\,\mu\text{s}$ confirm the $\sin X/X$ frequency response for each transducer ($\Theta = L/V_R = N/f_0 = 1.75\,\mu\text{s}$).

This agreement between experiment and the predictions of such a simple model is indeed satisfying. It stems from the fact that the electromechanical

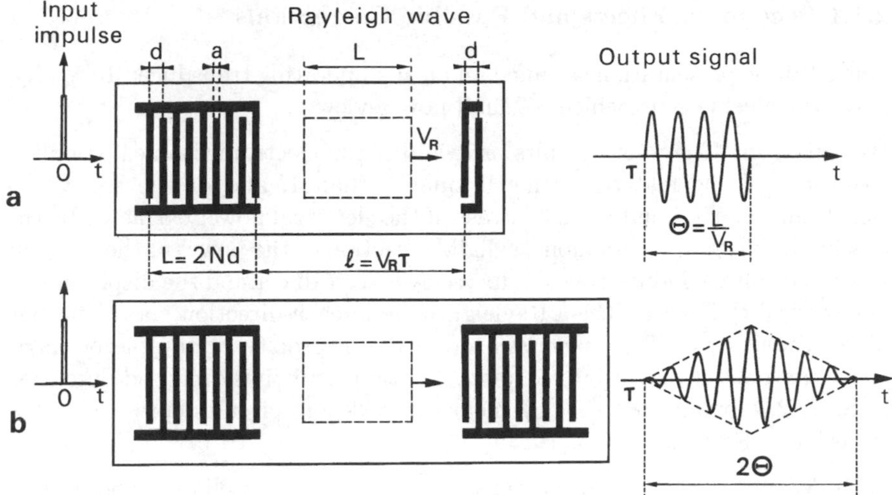

Fig. 2.2. Impulse response for a configuration of two interdigital transducers. (**a**) Two-finger receiver. (**b**) Receiver identical to emitter. The substrate carrying the transducers is piezoelectric

coupling coefficient for quartz is very small ($K^2 = 0.22\%$ for this particular cut) and that aluminium electrodes represent only a small load on the substrate. For a strongly piezoelectric material such as lithium niobate with thick electrodes (constituting a non-negligible fraction of the wavelength), the results are much less satisfactory, all the more so that other phenomena must then be taken into account. These other phenomena are the subject of the next section. In addition, the structure of the transducer is more complex. Fingers of width $a \neq \lambda_0/4$, playing the role of reflectors, for example, are then introduced between the fingers of width $\lambda_0/4$.

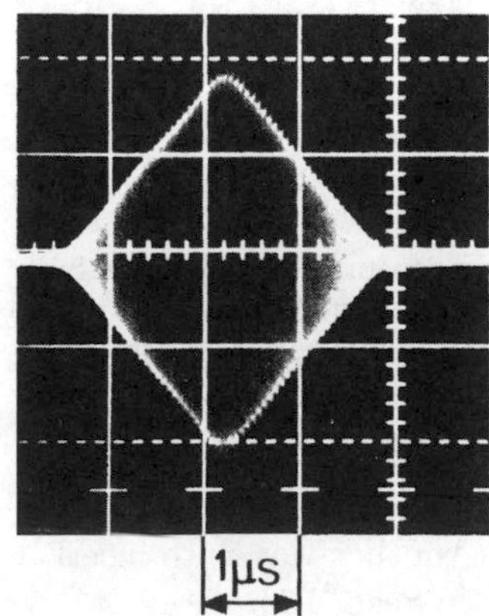

Fig. 2.3. Impulse response of a quartz delay line with two identical transducers and centre frequency 17 MHz

2.1.1 Secondary Effects and Possible Corrections

Our intuitive presentation assumed an ideally operating transducer. In reality, parasitic effects occur, which we shall now review.

Reemission. The above results are valid if the electromechanical coupling coefficient of the material is much smaller than 1, and/or the receiver is electrically short-circuited. In contrast, if the electrical power extracted by the passing Rayleigh wave is a non-negligible fraction of the power of the incident wave, the voltage induced across the receiver transducer and the displacement of electrical charge produce a Rayleigh wave in each direction via the inverse piezoelectric effect. This *reemission* due to reciprocity of the piezoelectric effect can be analysed using the piezoelectric permittivity method described in Sect. 2.3, or alternatively by means of a 3-port circuit representing the transducer (Sect. 2.4).

Bulk Waves. The transducer generates bulk waves as well as surface waves. Indeed, the grating structure favours emission of waves towards the solid core, in directions $\pm\theta$ which are symmetrical about the normal to the surface (see Fig. 2.4) and for which constructive interference conditions are satisfied:

$$2d \sin\theta = \lambda_B = V_B/f \ . \tag{2.5}$$

λ_B and V_B are the wavelength and speed of bulk waves in the (x_1, x_2) plane. This effect only occurs for frequencies greater than a cutoff value:

$$f\frac{\sin\theta}{V_B(\theta)} = \frac{1}{2d} = \frac{f_0}{V_R} \quad \Rightarrow \quad f > \frac{V_{B\min}}{2d} \ . \tag{2.6}$$

Since the minimal speed for bulk waves is generally greater than V_R, the frequency band in which such waves are generated is located well beyond the synchronism frequency f_0 of the Rayleigh wave transducer. These spurious waves, partially reflected on the lower face of the plate, are a major problem in the design of high bandwidth devices.

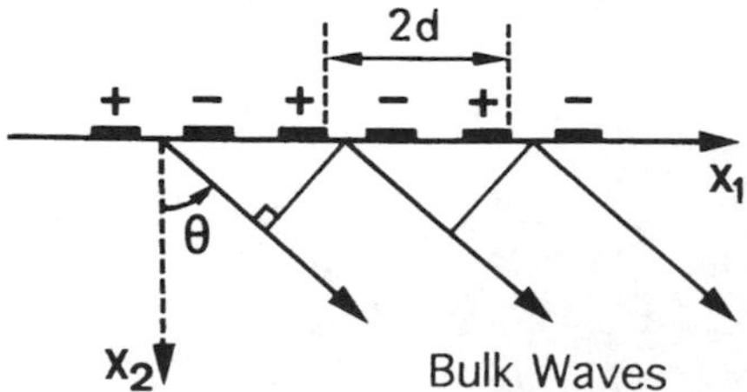

Fig. 2.4. Far from the transducer, bulk waves are preferentially emitted in directions for which interference from the various sources is constructive

Most bulk waves can be eliminated by sanding the opposite face of the crystal or covering it with an absorbent material. Another way is to insert a multistrip coupler between emitter and receiver (see Sect. 4.1).

Diffraction. The Rayleigh wave beam widens when it leaves the emitter. Such diffraction (see Sect. 1.3.2, Vol. I) is relevant if the width of the transducer is not more than a few times the wavelength. The receiver then detects only part of the beam. In different applications, the finger overlap width must be varied (see Fig. 2.5). Waves from the smallest sources are then diffracted. This effect can be partially compensated by anisotropy of the material (see Problem 2.1).

Phase Speed Variations. The Rayleigh wave front emitted by the transducer is not parallel to the fingers. Examining the beam perpendicularly to the direction of propagation, it is found that the wave front is distorted when the lengths of the fingers vary. This is especially true if the electromechanical coupling coefficient is large, as in the case of $LiNbO_3$. Such distortion arises because the transducer is not homogeneous. The speed of Rayleigh waves across the surface of a piezoelectric solid is affected by the presence of a metallic film (see Sect. 5.3.4.2, Vol. I). When the length of the electrode fingers is not constant, waves produced at the various points of a linear source travel paths of unequal metallised lengths, thereby undergoing different phase shifts (lens effect). An example in which the interdigital distance also varies is given in Fig. 2.5a [2.2]. The wave front was registered with a capacitive probe consisting of a tungsten needle in contact with the surface. The phase delay for waves coming from the central zone with maximal acoustic path is as great as 120° relative to waves coming from lateral zones. Wave fronts emitted in opposite directions are not distorted equally, because the transducer is not symmetric. Such large phase differences give rise to unwanted interference in straight-fingered receiver transducers, but are considerably reduced when *inactive fingers* are built in (see Fig. 2.5b). These 'dummy' fingers are made inactive by maintaining them at the same potential as their neighbours. Their role is to provide a more uniform distribution of metal over the surface (see Fig. 2.6a).

In general, the relative variation of phase speed comprises two terms:

$$\frac{\Delta V}{V_R} = D_p \frac{K^2}{2} + D_m \frac{h}{\lambda} \quad \text{provided that} \quad \frac{h}{\lambda} \ll 0.2\% \,. \tag{2.7}$$

The first term, proportional to the square of the electromechanical coupling coefficient, describes the piezoelectric effect. The second, proportional to the thickness h of the layer relative to the wavelength, arises from the mechanical load exerted by the electrode. The coefficient D_p depends on the extent of metallisation, given by the ratio of finger width to interdigital distance, $\eta = a/d$, and also on the way the electrodes are connected [2.3]. When all fingers are at the same potential (short-circuited transducer or array), and at the synchronism frequency f_0 such that $d = \lambda/2$,

$$D_p^{sc}(\eta) = -\frac{1}{2}\left[1 + \frac{P_{1/2}(-\cos\eta\pi)}{P_{-1/2}(-\cos\eta\pi)}\right] , \tag{2.8}$$

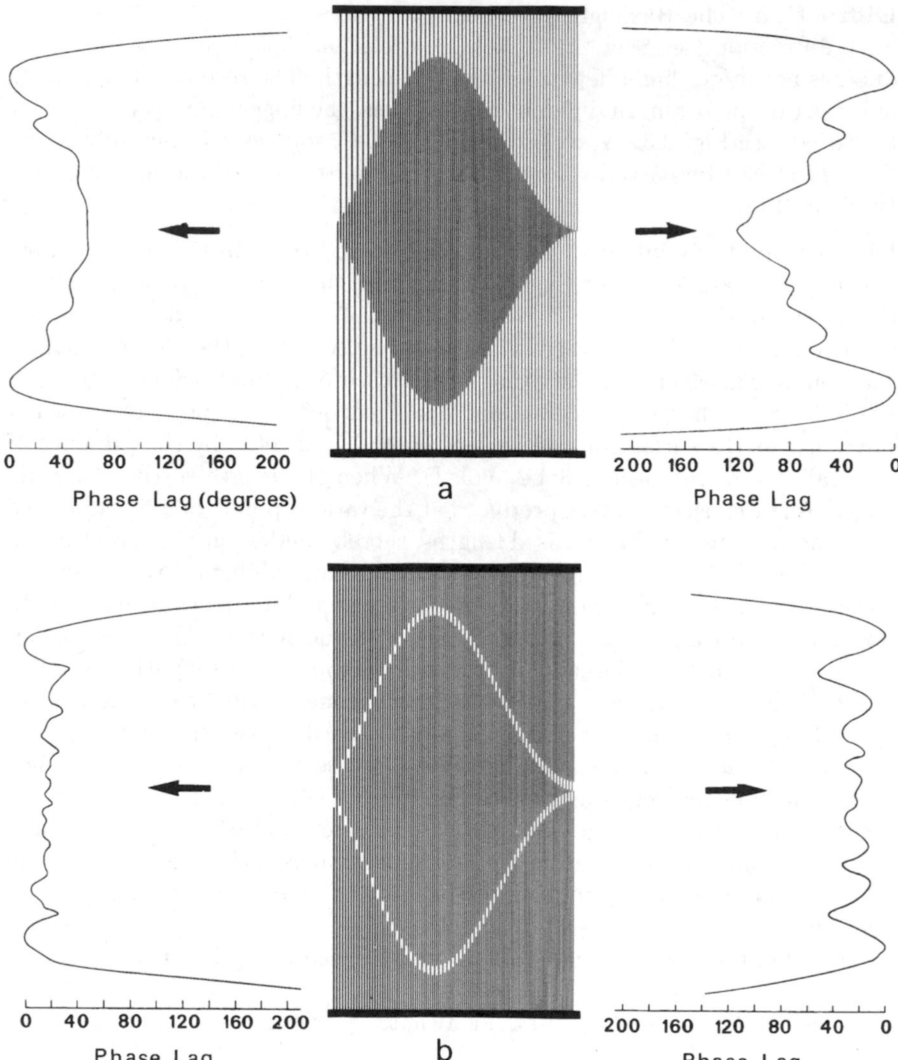

Fig. 2.5. Rayleigh wave front on a lithium niobate crystal. (**a**) Distortion by inhomogeneous metallisation of the surface. (**b**) Correction by means of inactive fingers. The excitation frequency, 57 MHz, is the resonance frequency of sources on the longest fingers. The phase is measured at 0.89 cm from the centre of the transducer. By kind permission of R.H. Tancrell [2.2]

where P_ν is the Legendre function of degree ν (see Appendix C). When fingers are in open circuit (e.g., in a grating for which the electrodes are not connected),

$$D_{\mathrm{p}}^{\mathrm{oc}}(\eta) = -\frac{1}{2}\left[1 - \frac{P_{1/2}(\cos\eta\pi)}{P_{-1/2}(\cos\eta\pi)}\right] . \tag{2.9}$$

Given the value of the Legendre functions of degree $\pm 1/2$, and for metallisation ratio $\eta = 0.5$, coefficients D_{p} are equal to

$$D_{\mathrm{p}}^{\mathrm{sc}} = -0.73 , \qquad D_{\mathrm{p}}^{\mathrm{oc}} = -0.27 .$$

The coefficient D_{m} depends on elastic constants and densities of the metal used for the electrodes and substrate. An analytic expression can be deduced from perturbation methods due to Auld [2.4].

Reflections. Variations in electrical and mechanical impedance at the surface, due to the metallic fingers, give rise to reflections whenever an elastic wave encounters a transducer. These reflections are all the more significant as the electromechanical coupling of the substrate is strong and the mechanical impedance of the material used in the fingers is very different from that used in the substrate. They depend not only on wave frequency, but also on transducer load. Let us note in passing that the term representing reemission can include reflection from the fingers. Structures have been invented to reduce reflection. For example, splitting a finger into two parts, as shown in Fig. 2.6b. Indeed, waves reflected from corresponding edges of two ordinary neighbouring fingers of width $\lambda/4$, separated by an interval equal to $\lambda/4$, add

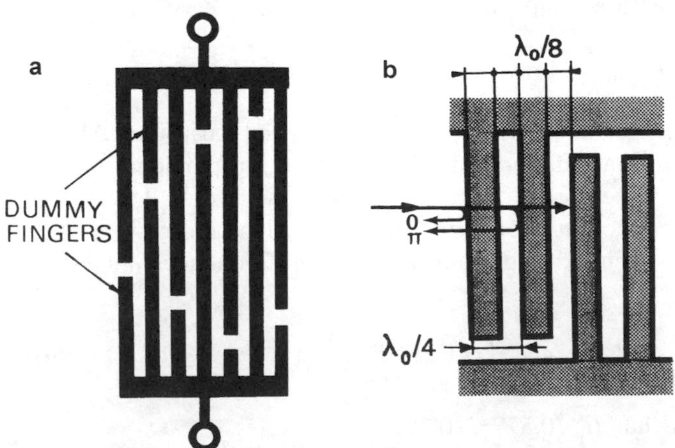

Fig. 2.6. Variants. (**a**) Transducer with inactive fingers, maintained at the same potential as their neighbours. The path difference due to non-uniform distribution of metal in an ordinary transducer is thereby compensated. (**b**) Fingers split into two parts reduce reflection effects within the bandwidth

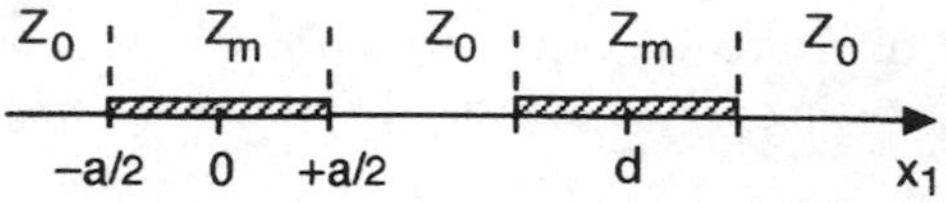

Fig. 2.7. Reflection by finger edges

constructively because the path difference causes a phase shift of 2π. In principle, these reflections are suppressed when each finger is composed of two strips of width $\lambda/8$. The distance between axes of two neighbouring fingers is still $\lambda/2$, so there is destructive interference [2.5].

These reflections have been the subject of much analysis and experiment [2.6, 2.7, 2.8] and it turns out that, in many cases, they can actually be exploited to produce unidirectional transducers, indispensable in the design of low-loss filters. The transducer is then considered as an array of free zones of impedance Z_0 and metallised zones of impedance $Z_\mathrm{m} = Z_0 + \Delta Z$, as shown in Fig. 2.7.

When $\Delta Z \ll Z_0$, the reflection coefficient (Sect. 1.2.4.2, Vol. I) at one edge of a finger is $\Delta Z/2Z_0$ and at the other $-\Delta Z/2Z_0$. Taking the centre of a finger as origin, and putting a for the finger width, its reflection coefficient r_e is

$$r_\mathrm{e} = \frac{\Delta Z}{2Z_0}\left(\mathrm{e}^{\mathrm{i}\theta} - \mathrm{e}^{-\mathrm{i}\theta}\right) = \mathrm{i}\frac{\Delta Z}{Z_0}\sin\theta\,, \quad \theta = ka = \eta\pi\frac{f}{f_0}\,, \tag{2.10}$$

where $\eta = a/d$ is once again the degree of metallisation and $f_0 = V_\mathrm{R}/2d$. This coefficient is purely imaginary. The total reflection coefficient of the transducer is found by summing (see Problem 2.2). Datta and Hunsinger have established an expression for $\Delta Z/Z_0$ when fingers are insulated, short-circuited, or loaded with an impedance:

$$\frac{\Delta Z}{Z_0} = R_\mathrm{p}\frac{K^2}{2} + R_\mathrm{m}\frac{h}{\lambda} \quad \text{provided that} \quad \frac{h}{\lambda} \ll 0.2\%\,. \tag{2.11}$$

It consists of a piezoelectric term proportional to speed variations $\Delta V/V \approx K^2/2$ caused by metallisation, and a mechanical term proportional to finger thickness relative to wavelength.

R_p is an analytic function depending on the metallisation ratio η. As an example, for a short-circuited finger and at the stop frequency f_0 such that $d = \lambda/2$,

$$R_\mathrm{p}^\mathrm{sc}(\eta) = -\frac{\pi/2}{\sin\eta\pi}\left[\cos\eta\pi + \frac{P_{1/2}(-\cos\eta\pi)}{P_{-1/2}(-\cos\eta\pi)}\right]\,. \tag{2.12}$$

For $\eta = 0.5$, we find that $R_\mathrm{p}^\mathrm{sc}(0.5) = -0.72$.

R_m is a function depending on mechanical properties of the substrate and the metal used for the electrodes. For example, it takes the value -0.51 for an aluminium electrode on an ST-cut quartz substrate (X propagation). Any reflection gives rise to a certain proportion of stationary waves (see Sect. 1.1.2.1, Vol. I), to evanescent waves and hence also to some storage of energy [2.6]. Datta and Hunsinger [2.9] have shown that a term in $(h/\lambda)^2$ should be added to the expression for $\Delta Z/Z_0$ (see Sect. 2.6.1). This term may be greater than the sum of the other two terms if the substrate is only slightly piezoelectric and electrodes are thin, with impedance close to that of the substrate (such is the case for quartz with aluminium electrodes).

Electromagnetic Coupling. From the point of view of the source, each transducer behaves as a capacitance in parallel with a radiation resistance (see Sect. 2.2.2). The two capacitances are close on an electromagnetic scale and are coupled. The receiver directly picks up a fraction of the voltage applied to the emitter, so that shielding is necessary.

2.1.2 Spatial Distribution of Electrical Quantities

We now return to the simple transducer shown in Figs. 2.1 and 2.2. The intuitive description of its operating features is useful but incomplete. Although the elastic signal emitted in response to a short pulse does indeed have periodicity and duration imposed by the structure, its variations are not necessarily sinusoidal. The frequency response is therefore made up of $\sin X/X$ curves with centre frequencies f_0, $3f_0$, $5f_0$, ... Moreover, it has been assumed that the transducer emits a Rayleigh wave. This in turn supposes that strains making up this wave are indeed coupled to components of the electric field by piezoelectric constants of the material. In order to carry out a more careful analysis of the generation of elastic waves by comb-shaped transducers, we must begin by investigating the distribution of the electric field, and then solve the propagation equation implied by this distribution. Such an analysis is, in fact, rather difficult and we shall not attempt it here. However, the static distribution of electric charges and potential, which does not take into account the electric field induced by generated waves, is readily obtained for simple transducers, such as the emitter transducer shown in Fig. 2.2. These calculations, due to Engan [2.10], are reproduced in Appendix A.

Figure 2.8 shows physically predictable spatial distributions at the solid surface, of potential, charge density and electric field, for a simple transducer subject to a constant voltage U.

When the number of finger pairs is large enough ($N \geq 10$) to ensure that edge effects can be neglected, electrical quantities are periodic, of period $2d$. They can be decomposed in Fourier series. Choosing the abscissa origin at the centre of a finger, the electric potential is an even function of x_1:

$$\Phi(x_1, x_2) = \sum_{m=0}^{\infty} \frac{UF_m}{(2m+1)\pi} \exp(-r\chi_m x_2)\cos(\chi_m x_1) ,$$

where $\chi_m = (2m+1)\pi/d$. This decreases exponentially on either side of the surface $x_2 = 0$:

$$r = \begin{cases} \varepsilon_{11}/\varepsilon_{22} & \text{for } x_2 > 0 \quad \text{(piezoelectric solid)} , \\ -1 & \text{for } x_2 < 0 \quad \text{(vacuum)} . \end{cases}$$

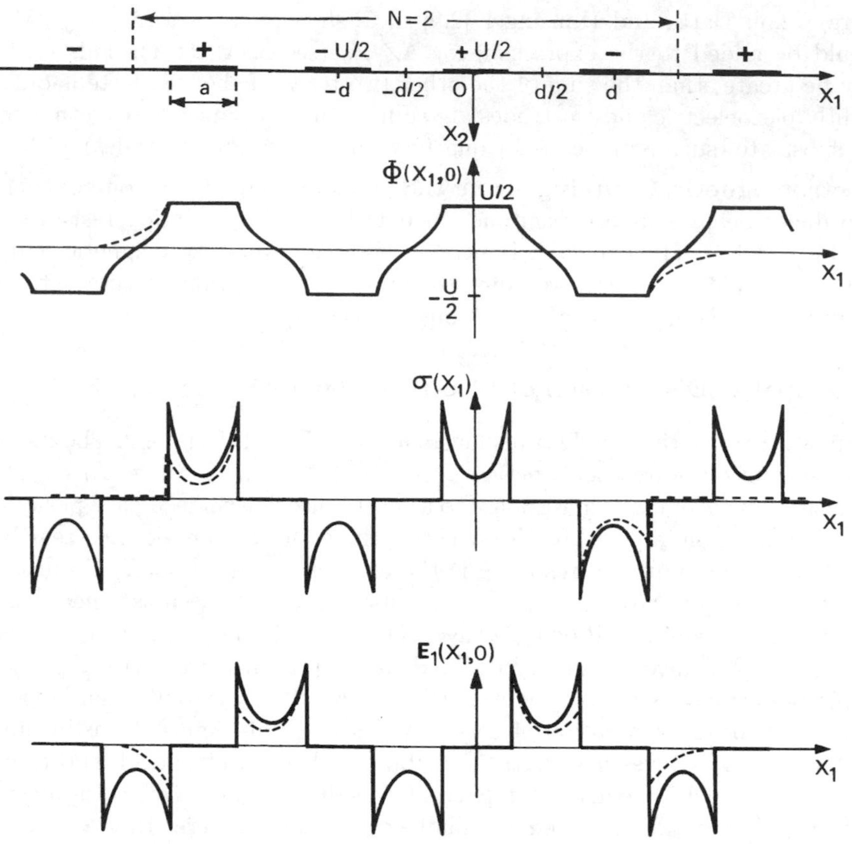

Fig. 2.8. Static spatial distribution, at the solid surface, of potential Φ, charge density σ and field component E_1 for a simple Rayleigh wave transducer (equidistant fingers of equal lengths). *Dashed line* corresponds to $N = 2$ and *continuous line* to $N = \infty$

Coefficients F_m of the harmonics are given in terms of Legendre polynomials P_m of the variable $\cos\eta\pi$ and the elliptic integral $K(\cos\eta\pi/2)$, all of which are defined in Appendix C:

$$F_m = \pi\frac{P_m(\cos\eta\pi)}{K(\cos\eta\pi/2)} . \tag{2.13}$$

They are strongly dependent on the metallisation ratio η, as can be seen from Fig. 2.9. Amplitudes of all harmonics except the first may be zero. In particular, there are no 3, 7, ..., $4p + 3$ harmonics when the finger width equals the interdigital distance ($\eta = 1/2$).

Examination of the charge or electric field distribution suggests idealising the peaks. Operation of the transducer is then described by an array of discrete sources (the delta function method, Sect. 2.1.3). If only the fundamental

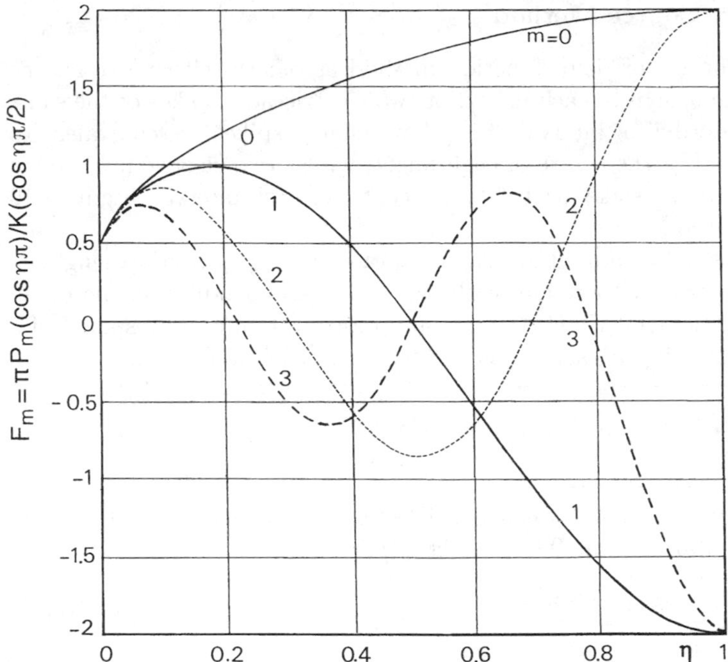

Fig. 2.9. Dependence of the amplitude of harmonic terms F_m, $m = 0, 1, 2, 3$, in the electric potential on the relative width $\eta = a/d$ of electrode fingers

term in the potential curve is retained, any response to a step variation in the supply voltage U can be taken as a sine curve of limited duration. This was the simplification adopted in Sect. 2.2. However, the characteristic features of a transducer can be expressed in terms of the charge density σ, through the notion of piezoelectric permittivity (see Sect. 2.3).

Another characteristic quantity associated with a transducer is its static capacitance $C_{\mathrm{T}} = NC_1 w$, where w is the finger overlap length and C_1 the capacitance per finger pair and per unit length. The latter depends on the metallisation ratio η (see Appendix A). It is very simply expressed when $\eta = 0.5$:

$$C_1 = \varepsilon_0 + \varepsilon_{\mathrm{p}} \,, \quad \varepsilon_{\mathrm{p}} = \left[\varepsilon_{11}^T \varepsilon_{22}^T - (\varepsilon_{12}^T)^2\right]^{1/2} \,, \tag{2.14}$$

where ε_{p} is the effective permittivity of the piezoelectric substrate (see Sect. 5.3.4.1, Vol. I).

The discrete source model, derived from examination of the charge distribution, and the impulse response model, derived from variation of the potential, have a common feature: finite duration of the impulse response. Applying an electrical pulse across the two comb-shaped electrodes drives a wave train of duration given by the ratio of the transducer length to the Rayleigh wave speed.

2.1.3 Discrete Source Method

The discrete source, or delta function method associates either two charge spikes of the same sign to each finger, or two electric field spikes of the same sign to each interdigital interval [2.11]. Two charge spikes are equivalent to one spike located at the centre of each finger, whilst two electric field spikes are equivalent to one spike located at the centre of each interval. Figure 2.10 illustrates the second case.

Each source n has amplitude A_n proportional to the overlap length w of the two fingers on either side, with sign + or − depending on the direction of the electric field. Under these assumptions, the impulse response of a transducer comprising N_s sources is a succession of Dirac functions:

$$\boxed{h(t) = \sum_{n=0}^{N_\mathrm{s}-1} s_n A_n \delta(t - t_n)} . \tag{2.15}$$

s_n is a sign, equal to $(-1)^n$ when the direction of the electric field reverses at each interdigital interval. The frequency response,

$$H(\omega) = \int_{-\infty}^{\infty} h(t) \mathrm{e}^{-\mathrm{i}\omega t} \mathrm{d}t , \tag{2.16}$$

is then

$$H(\omega) = \sum_{n=0}^{N_\mathrm{s}-1} s_n A_n \mathrm{e}^{-\mathrm{i}\omega t_n} . \tag{2.17}$$

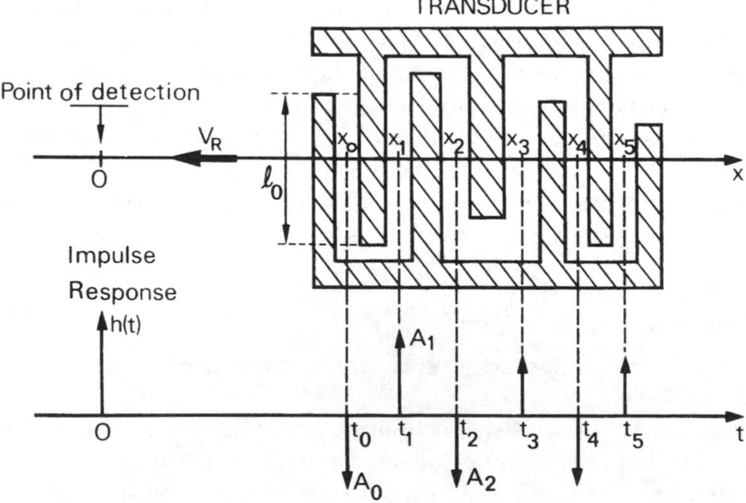

Fig. 2.10. Discrete source method. The transducer is modelled by an array of discrete sources at the centres of interdigital intervals, with amplitudes proportional to finger overlap lengths

In order to check the validity of this method, let us apply it to a simple transducer with N_s interdigital intervals (sources):

$$s_n = (-1)^n\,, \quad A_n = A_0 = \mathrm{Const.}\,, \quad t_n = t_0 + n\frac{d}{V_\mathrm{R}} = t_0 + \frac{n\pi}{\omega_0}\,.$$

The frequency response is

$$H(\omega) = A_0 \mathrm{e}^{-\mathrm{i}\omega t_0} \sum_{n=0}^{N_\mathrm{s}-1} (-1)^n \mathrm{e}^{-\mathrm{i}n\pi\omega/\omega_0}\,.$$

Putting $(-1)^n = \mathrm{e}^{\mathrm{i}n\pi}$,

$$H(\omega) = A_0 \mathrm{e}^{-\mathrm{i}\omega t_0} \sum_{n=0}^{N_\mathrm{s}-1} \left(\mathrm{e}^{-2\mathrm{i}\Delta\phi}\right)^n\,, \quad \Delta\phi = \frac{\pi}{2}\frac{\omega - \omega_0}{\omega_0}\,. \tag{2.18}$$

Carrying out the sum,

$$\begin{aligned} H(\omega) &= A_0 \mathrm{e}^{-\mathrm{i}\omega t_0} \frac{1 - \exp(-2\mathrm{i}N_\mathrm{s}\Delta\phi)}{1 - \exp(-2\mathrm{i}\Delta\phi)} \\ &= A_0 \frac{\sin(N_\mathrm{s}\Delta\phi)}{\sin\Delta\phi} \exp -\mathrm{i}[\omega t_0 + (N_\mathrm{s} - 1)\Delta\phi]\,, \end{aligned}$$

or alternatively, using $\Delta\phi = (\pi/2)(\omega/\omega_0) - \pi/2$,

$$H(\omega) = A_0 \mathrm{i}^{N_\mathrm{s}-1} \frac{\sin(N_\mathrm{s}\Delta\phi)}{\sin\Delta\phi} \mathrm{e}^{-\mathrm{i}\omega\tau}\,, \tag{2.19}$$

where the mean delay τ is defined by

$$\tau = t_0 + \frac{N_\mathrm{s} - 1}{4f_0} = t_0 + \frac{N_\mathrm{s} - 1}{2V_\mathrm{R}} d\,.$$

Near the frequencies $f_m = (2m+1)f_0$, at which both numerator and denominator are zero,

$$\Delta\phi = \frac{\pi}{2}\frac{f - f_m}{f_0} + m\pi\,,$$

and the frequency response has form $\sin X/X$:

$$H_m(f) \approx A_0 N_\mathrm{s} \frac{\sin X_m}{X_m}\,, \quad X_m = N_\mathrm{s}\frac{\pi}{2}\frac{f - f_m}{f_0}\,. \tag{2.20}$$

In the discrete source model, the transducer response is made up of bands of equal height, centred on odd harmonic frequencies of the synchronism frequency f_0, as shown in Fig. 2.11. These bands have 3 dB width ($X_m = \pm 0.885\pi/2$)

$$\frac{\Delta f}{f_0} = \frac{1.77}{N_\mathrm{s}} \approx \frac{0.885}{N}\,, \quad \text{when} \quad N_\mathrm{s} = 2N - 1 \gg 1\,. \tag{2.21}$$

The relative frequency difference between two consecutive zeros is $2/N_\mathrm{s}$. The spectrum is periodic because of sampling (Shannon's theorem): even harmonic bands disappear because the sampling function consists of two Dirac impulses of opposite signs, separated by $T_0/2$ (see Problem 2.3).

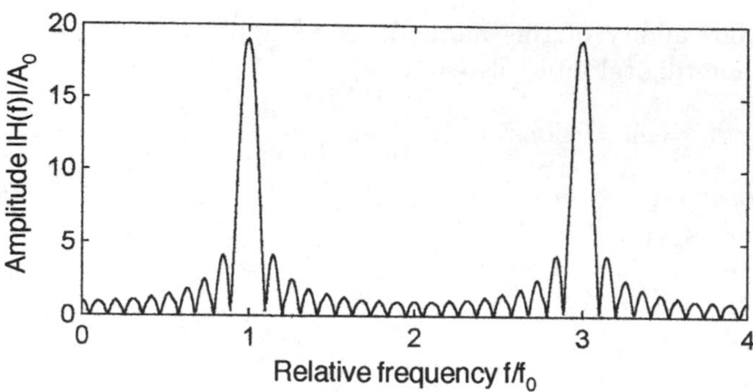

Fig. 2.11. Frequency response amplitude for a transducer with $N = 10$ pairs of fingers, calculated using the discrete source method

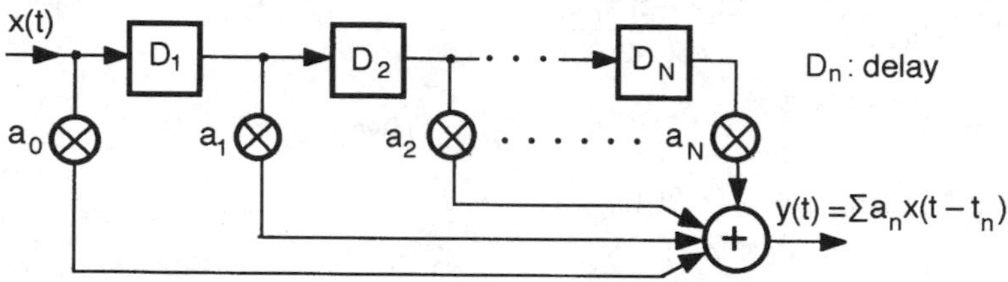

Fig. 2.12. Structure of a transversal filter. The input signal propagates at constant speed along the line. Samples are taken at fixed time intervals and multiplied by an appropriate factor before being added together. The weighted sum of these samples is the output signal. The time interval between samples is not necessarily constant

Viewed in this way, the interdigital transducer behaves like an ideal transversal filter. This type of filter [2.12] in fact comprises a delay line along which signals are sampled, as shown in Fig. 2.12. The principle of the transversal filter was first put forward in 1940 by Kallman [2.13]. However, devices available at the time to provide suitable delays were too cumbersome. The idea was not developed until the advent of surface elastic waves or digital processors, depending on the frequency.

The simple delta method only explains the shape of the transducer frequency response. In order to go further and calculate the response amplitude and radiation impedance, we must consider the energy, with the help of other physical characteristics, viz., the capacitance and electromechanical coupling coefficient of the material.

2.2 Impulse Response Model

The advantage of this method, based upon energy considerations, is that it predicts the absolute amplitude, and hence the power, of Rayleigh waves emitted by a transducer, assumed uniform in the x_3 direction and with constant finger overlap width w (see Fig. 2.2a). In order to define the response to electrical excitation $U(t)$, we introduce a scalar quantity $a(t)$ whose square represents the elastic power density per unit width of the Rayleigh wave beam emitted in both $\pm x_1$ directions (see Fig. 2.13a). The square $a^2(t)$ is given in terms of component P_1 of the Poynting vector and the mean transported power density P_R:

$$a^2(t) = \int_{-\infty}^{\infty} P_1(x_2, t)\,\mathrm{d}x_2 , \qquad P_\mathrm{R} = \langle a^2(t) \rangle . \tag{2.22}$$

Equations (5.46) and (5.86) in Vol. I show that in the harmonic regime, a is proportional to mechanical displacement or electric potential amplitude of the Rayleigh wave. In the impulsive regime, $a(t)$ represents instantaneous amplitude [2.14]. Neglecting any bulk waves which may be generated, Poynting's theorem applied to a closed volume V containing the transducer, of width w, implies [see (3.35), Vol. I]

$$\begin{aligned} \frac{\mathrm{d}W_\mathrm{s}}{\mathrm{d}t} &= \frac{\mathrm{d}W}{\mathrm{d}t} + 2\int_{-w/2}^{w/2}\int_{-\infty}^{\infty} P_1(x_2, t)\,\mathrm{d}x_2\,\mathrm{d}\,x_3 \\ &= \frac{\mathrm{d}W}{\mathrm{d}t} + 2wa^2(t) , \end{aligned} \tag{2.23}$$

where W_s and W are energy supplied by electrical sources and total energy stored in volume V, respectively. Referring to the arguments and notations of Sect. 5.3.4.3, Vol. I, assume that a constant voltage of 1 V has been applied across the transducer electrodes. These electrodes constitute a capacitance $C_\mathrm{T} = NC_1 w$, in which the stored electrical energy is

$$W_\mathrm{e} = \frac{1}{2} N C_1 w . \tag{2.24}$$

$C_1 = \varepsilon_\mathrm{p} + \varepsilon_0$ is the capacitance per pair of fingers and per unit length (see Appendix A).

2.2.1 Time Responses

If the electrodes are short-circuited at $t = 0$ (see Fig. 2.13b), the response $a_\mathrm{u}(t)$ to this unit voltage step $U(t) = 1 - \Gamma(t)$ is, by hypothesis, a sine curve of frequency $f_0 = V_\mathrm{R}/2d$ and duration N/f_0:

$$a_\mathrm{u}(t) = \begin{cases} -a_0 \sin(2\pi f_0 t) & \text{for } 0 < t < \Theta = N/f_0 , \\ a(t) = 0 & \text{for other } t \text{ values} . \end{cases} \tag{2.25}$$

A cosine function would not be possible because energy must be continuous at the origin.

Integrating (2.23) and using $W_\mathrm{s} = 0$ for $t > 0$, the generated acoustic energy is

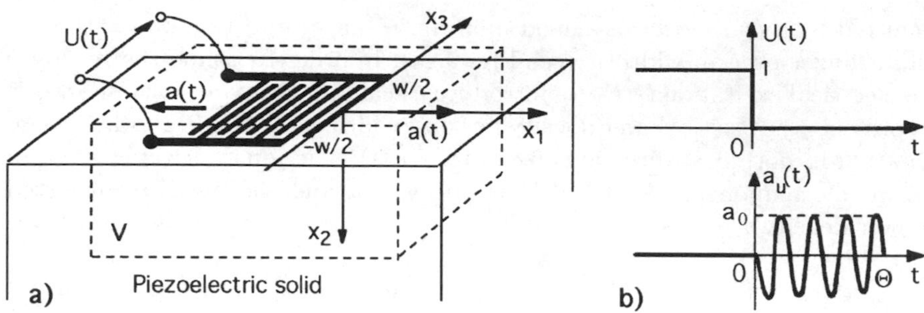

Fig. 2.13. (**a**) Interdigital transducer. (**b**) Response to a voltage step

$$W_{\mathrm{ac}} = W(0) - W(\infty) = 2w\int_0^\infty a_{\mathrm{u}}^2(t)\,\mathrm{d}t = \frac{Na_0^2 w}{f_0}\,.$$

From the definition of the electromechanical coupling coefficient K_{R} of the Rayleigh wave, $W_{\mathrm{ac}} = K_{\mathrm{R}}^2 W_{\mathrm{e}}$, and the amplitude of the step response is

$$a_0 = K_{\mathrm{R}}(f_0 C_1/2)^{1/2}\,. \tag{2.26}$$

The *impulse response* [for $U(t) = \delta(t)$] is the derivative of the step response:

$$h(t) = -\frac{\mathrm{d}a_{\mathrm{u}}}{\mathrm{d}t} = \pi\sqrt{2}K_{\mathrm{R}}C_1^{1/2}f_0^{3/2}\cos(2\pi f_0 t)\,, \quad 0 < t < N/f_0\,. \tag{2.27}$$

The *frequency response* is the Fourier transform of $h(t)$, which has the form $\sin X/X$ [see equation (2.37)]. If the finger overlap length w and interdigital distance d vary with the abscissa x_1, i.e., as a function of time $t = x_1/V_{\mathrm{R}}$, it is given by

$$\boxed{h(t) = \pi\sqrt{2}K_{\mathrm{R}}C_1^{1/2}f^{3/2}(t)\frac{w(t)}{w_0}\cos\phi(t)}\,, \tag{2.28}$$

where

$$\omega = \frac{\mathrm{d}\phi}{\mathrm{d}t} \quad \Rightarrow \quad \phi(t) = 2\pi\int_0^t f(t')\,\mathrm{d}t'\,. \tag{2.29}$$

Conversely, a given impulse response of form $e(t)\cos\phi(t)$, and therefore a given frequency response, can be obtained by making the right geometrical choice for the transducer. The fingers are arranged on either side of abscissa lines $x_n = V_{\mathrm{R}}t_n$, with times t_n defined by

$$\phi(t_n) = n\pi\,. \tag{2.30}$$

The overlap length $w_n = w(t_n)$ of two adjacent fingers satisfies:

$$w_n \propto e(t_n)f^{-3/2}(t_n)\,. \tag{2.31}$$

As we have seen in Sect. 2.1, detecting this elastic wave train by the inverse piezoelectric effect produces a voltage output similar to $h(t)$, when

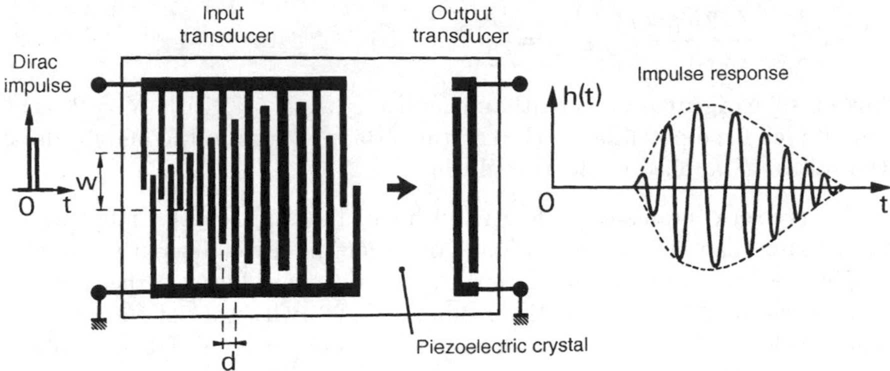

Fig. 2.14. Impulse response of a transducer. Variations in finger overlap length determine the shape of the envelope, whilst interdigital distance defines the instantaneous frequency

the receiver transducer is made of two fingers of length at least equal to the maximum value of $w(t)$ (see Fig. 2.14).

The impulse response gives the response to a general excitation by convolution. For example, if a unit amplitude sinusoidal voltage is applied to the transducer at the synchronism frequency, so that $U(t) = \cos(2\pi f_0 t)$, then we can use $h(t) = 2\pi a_0 f_0 \cos(2\pi f_0 t)$ to obtain

$$a_\mathrm{s}(t) = U(t) \otimes h(t) = 2\pi a_0 f_0 \int_0^{\Theta} \cos[2\pi f_0(t-\tau)] \cos(2\pi f_0 \tau)\,\mathrm{d}\tau \ ,$$

and hence

$$a_\mathrm{s}(t) = \pi a_0 f_0 \int_0^{\Theta} [\cos(2\pi f_0 t) + \cos 2\pi f_0(t - 2\tau)]\,\mathrm{d}\tau \ ,$$

where $\Theta = N/f_0 = NT_0$. The second term integrates to zero, being taken over a multiple of the period T_0, so that

$$a_\mathrm{s}(t) = \pi N a_0 \cos(2\pi f_0 t) \ . \tag{2.32}$$

Compared with the step excitation, the effect of synchronisation is to multiply the transducer response by $N\pi$. This phenomenon decreases as the frequency shifts away from f_0.

We now calculate the amplitude of the electric potential accompanying a Rayleigh wave on a free surface. From (5.86), Vol. I, this is related to the mean power density transported by the wave:

$$P_\mathrm{R} = \langle a_\mathrm{s}^2(t) \rangle = \frac{\pi^2 N^2 a_0^2}{2} |U|^2 = \frac{\pi^2}{4} N^2 K_\mathrm{R}^2 f_0 C_1 |U|^2 \ ,$$

where $|U|$ is the amplitude of the applied sinusoidal voltage. Using the expression for the capacitance per unit length $C_1 = \varepsilon_\mathrm{p} + \varepsilon_0$, it follows that

$$|\Phi| = \left(\frac{2K_R^2}{\varepsilon_p + \varepsilon_0} \frac{P_R}{\omega_0} \right)^{1/2} = \frac{\sqrt{\pi}}{2} K_R^2 N |U| \,. \tag{2.33}$$

Order of Magnitude. For lithium niobate, $K_R^2 \approx 5\%$, with $N = 20$, the amplitude of the potential at the crystal surface is close to the amplitude of the voltage U applied to the transducer.

If the receiver contains many fingers with equal overlap lengths, all of which are greater than the longest overlaps of fingers in the emitter, the impulse response for the configuration formed by the two transducers is the convolution of the impulse response of the emitter (2.28) with that of the receiver. The envelope of the impulse response for a configuration of two identical simple transducers (w = Const., d = Const.) is a triangle (autocorrelation of two rectangles) of width 2Θ, as we have already observed in Sect. 2.1. Its frequency response $H(f)$ is a curve of form $[\sin X/X]^2$. The response of a two-transducer configuration in which finger lengths are allowed to vary is more difficult to predict. An approximate solution involves mentally dividing the system into parallel bands, analysing the response of each band, and then summing [2.15]. The problem is simplified if a multistrip coupler is inserted between the two transducers (see Sect. 4.1). In fact, there is another method for weighting the response of a transducer. This consists in locally varying the density of the fingers, rather than their lengths, and even removing some of them. The amplitude of waves emitted by a group is proportional to the number of fingers.

2.2.2 Radiation Conductance

The electrical impedance of an interdigital transducer is made up of the static capacitance $C_T = NC_1 w$ and a radiation impedance arising from emission of surface waves. The radiation conductance $G_a(f)$ varies with frequency and is always accompanied by an imaginary part $B_a(f)$, called the motional susceptance (see Fig. 2.15). The electrical admittance of the transducer is

$$Y_T(f) = 2\pi \mathrm{i} f C_T + \mathrm{i} B_a(f) + G_a(f) \,.$$

The following calculation refers first to a simple transducer (w = Const., d = Const.), and then to a dispersive transducer (w = Const. but d varying linearly).

The *radiation conductance* $G_a(f)$ expresses the equality of mechanical energy generated by the transducer and electrical energy dissipated:

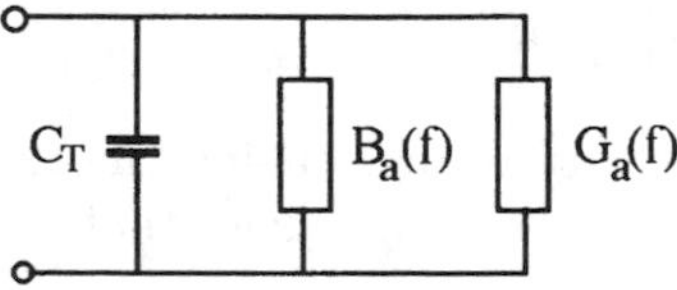

Fig. 2.15. Electrical admittance of a Rayleigh wave transducer

$$W_{\mathrm{ac}} = 2w \int_0^\infty a^2(t)\,\mathrm{d}t = \int_{-\infty}^\infty G_{\mathrm{a}}(f)|\overline{U}(f)|^2 \mathrm{d}f \,. \tag{2.34}$$

$\overline{U}(f)$ is the spectrum of the applied voltage $U(t)$, i.e., its Fourier transform. When $U(t) = \delta(t)$, we have $\overline{U}(f) = 1$ and the transducer response is $h(t)$. Given Parseval's relation between $h(t)$ and its Fourier transform $H(f)$, the frequency response of the transducer,

$$\int_0^\infty h^2(t)\,\mathrm{d}t = \int_{-\infty}^\infty |H(f)|^2 \mathrm{d}f \,, \tag{2.35}$$

it follows that

$$\boxed{G_{\mathrm{a}}(f) = 2w|H(f)|^2} \,. \tag{2.36}$$

For the simple transducer with impulse response given in (2.27),

$$H(f) = \frac{\pi}{\sqrt{2}} K_{\mathrm{R}} C_1^{1/2} f_0^{3/2} \int_0^{N/f_0} \left[\mathrm{e}^{-2\pi\mathrm{i}(f-f_0)t} + \mathrm{e}^{-2\pi\mathrm{i}(f+f_0)t} \right] \mathrm{d}t \,,$$

and the *frequency response* is the sum of two $\sin X/X$ curves:

$$H(f) = \frac{\pi}{\sqrt{2}} K_{\mathrm{R}} N (C_1 f_0)^{1/2} \left(\frac{\sin X_+}{X_+} \mathrm{e}^{-\mathrm{i}X_+} + \frac{\sin X_-}{X_-} \mathrm{e}^{-\mathrm{i}X_-} \right) \,, \tag{2.37}$$

where

$$X_\pm = N\pi \frac{f \pm f_0}{f_0} \,.$$

The 3 dB bandwidth is indeed inversely proportional to the number of finger pairs ($\Delta f/f_0 = 0.885/N$). In practice, it is sufficiently narrow and the part of the spectrum centred on f_0 does not overlap the part centred on $-f_0$. For $f > 0$, the *radiation conductance* is given by

$$G_{\mathrm{a}}(f) = \pi^2 K_{\mathrm{R}}^2 C_{\mathrm{T}} N f_0 \left(\frac{\sin X}{X} \right)^2 = G_0 \left(\frac{\sin X}{X} \right)^2 \,. \tag{2.38}$$

Its value G_0 at the centre frequency is proportional to the width w via the total capacitance of the transducer $C_{\mathrm{T}} = NC_1 w$.

This is the same formula up to a constant as the one found by Smith et al. [2.16] from a Mason equivalent circuit (see Sect. 2.6.3):

$$G_{\mathrm{a}}(f_0) = 8K_{\mathrm{R}}^2 C_{\mathrm{T}} N f_0 \,.$$

It is also the same up to a constant as the formula established by Auld [2.17] using normal mode theory:

$$G_{\mathrm{a}}(f_0) = 2.87\pi K_{\mathrm{R}}^2 C_{\mathrm{T}} N f_0 \,.$$

Order of Magnitude. A transducer on lithium niobate (Y-cut, Z propagation) operates at 100 MHz and comprises $N = 40$ finger pairs, with overlap width $w = 50\lambda = 1.75$ mm. Using the figures given in Fig. 5.27, Vol. I, the total capacitance of the transducer is $C_{\mathrm{T}} = 29$ pF and its radiation conductance is $G_0 = 0.055\ \Omega^{-1}$.

In any causal system, the imaginary part of the radiation admittance $B_{\mathrm{a}}(f)$ is the Hilbert transform of its real part (see Problem 2.4):

$$B_{\mathrm{a}}(f) = \frac{1}{\pi}\mathrm{PP}\int_{-\infty}^{\infty} \frac{G_{\mathrm{a}}(\nu)}{\nu - f}\,\mathrm{d}\nu = G_0 \frac{\sin 2X - 2X}{2X^2}\,, \tag{2.39}$$

where PP indicates principal part. Taking the static capacitance C_{T} into account, the transducer admittance is

$$Y(f) = G_0\left(\frac{\sin X}{X}\right)^2 + \mathrm{i}C_{\mathrm{T}}\omega_0\left(\frac{f}{f_0} + \pi K^2 N \frac{\sin 2X - 2X}{4X^2}\right)\,. \tag{2.40}$$

Figure 2.16 gives an example of how Y varies with frequency in a simple transducer with 10 pairs of fingers.

Regular dispersive structures are of particular interest because they were at the heart of developments of Rayleigh wave devices for radar. The frequency of the impulse response for a dispersive transducer varies linearly by $\Delta f/2$ on either side of a centre frequency f_0. Its conductance can be found easily if it is assumed that the impulse response spectrum is composed of two rectangles, centred on f_0 and $-f_0$, respectively, and of widths equal to the frequency variation Δf. This hypothesis is quite plausible if the product $\Theta\Delta f$ is greater than 50 (see Sect. 4.5.2.2). Each term in the Parseval relation (2.35) is then known:

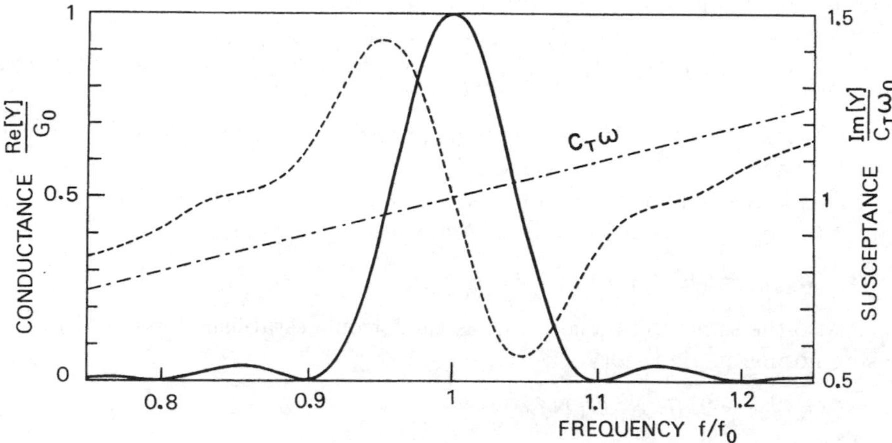

Fig. 2.16. Relative frequency (f/f_0) dependence of conductance (*full curve*) and susceptance (*dashed curve*) for a simple Rayleigh wave transducer with $N = 10$ pairs of fingers

$$\int_{-\infty}^{\infty} |H(f)|^2 \mathrm{d}f = 2|H(f_0)|^2 \Delta f \,, \tag{2.41}$$

and $[w(t)/w_0 = 1]$

$$\int_0^{\infty} h^2(t)\,\mathrm{d}t = \pi^2 K_\mathrm{R}^2 C_1 f_0^3 \Theta \,.$$

Hence,

$$|H(f_0)|^2 = \frac{\pi^2}{2} K_\mathrm{R}^2 C_1 \frac{f_0^3}{\Delta f} \Theta \,. \tag{2.42}$$

Then, from (2.36), the centre frequency conductance is

$$G_\mathrm{a}(f_0) = \pi^2 K_\mathrm{R}^2 C_1 w \frac{f_0^3}{\Delta f} \Theta = \pi^2 K_\mathrm{R}^2 C_1 w f_0 \left(\frac{f_0}{\Delta f}\right)^2 \Theta \Delta f \,. \tag{2.43}$$

Replacing N by $f_0/\Delta f$ in (2.38),

$$G_0 = \pi^2 K_\mathrm{R}^2 C_1 w N^2 f_0 = \pi^2 K_\mathrm{R}^2 C_1 w f_0 \left(\frac{f_0}{\Delta f}\right)^2 \,, \tag{2.44}$$

and comparing with (2.43), we find that the input conductance of a dispersive transducer is equal to that of a simple transducer with the same bandwidth, multiplied by a factor $\Theta\Delta f$ (compression factor, defined in Sect. 4.5.2.1).

Matching. Bandwidth. In reality, the interdigital transducer also emits bulk waves (see Sect. 2.1.1); the equivalent circuit in Fig. 2.15 contains several conductances. In order to ensure that the transducer is matched to the supply, its capacitance C_T is compensated by an inductance L, chosen in such a way that resonance occurs at the synchronism frequency f_0. For a simple transducer, the bandwidth is then maximal for an optimal value N_opt of the number of finger pairs. N_opt depends on the electromechanical coupling coefficient. Indeed, according to (2.3), the 'mechanical' bandwidth $(\Delta f)_\mathrm{m}$ due to the grating effect is inversely proportional to N. The quality factor of the electrical circuit containing the inductance L is given by

$$Q = \frac{C_\mathrm{T}\omega_0}{G_0} = \frac{2}{\pi K_\mathrm{R}^2 N} = \frac{f_0}{(\Delta f)_\mathrm{e}} \,, \tag{2.45}$$

where we have used (2.38) for the conductance G_0. The transducer bandwidth is maximal when $(\Delta f)_\mathrm{m}$ and $(\Delta f)_\mathrm{e}$ are equal:

$$\frac{(\Delta f)_\mathrm{m}}{f_0} = \frac{(\Delta f)_\mathrm{e}}{f_0} \quad \Rightarrow \quad \frac{0.885}{N} = \frac{\pi K_\mathrm{R}^2 N}{2} \,.$$

We therefore find

$$N_\mathrm{opt} = \frac{0.75}{K_\mathrm{R}} \quad \Rightarrow \quad \left(\frac{\Delta f}{f_0}\right)_\mathrm{opt} = 1.18 K_\mathrm{R} \,. \tag{2.46}$$

Application.

$$\mathrm{LiNbO_3}(Y\mathrm{cut},\ \vec{Z}),\ K_\mathrm{R}^2 = 4.8\% \qquad N_\mathrm{opt} = 3.5\ , \quad \left(\frac{\Delta f}{f_0}\right)_\mathrm{opt} = 26\%\ ,$$

$$\mathrm{SiO_2}(Y\mathrm{cut},\ \vec{X}),\ K_\mathrm{R}^2 = 0.22\% \qquad N_\mathrm{opt} = 16\ , \quad \left(\frac{\Delta f}{f_0}\right)_\mathrm{opt} = 5.5\%\ .$$

Given the order of magnitude of $1/G_\mathrm{a}(\omega) \ll 50\ \Omega$, it is often necessary to insert a transformer between supply and transducer. The behaviour of a receiver can be described using a current source (see Sect. 2.3.2).

2.3 Piezoelectric Permittivity Method

The discrete source and impulse response methods described above were established by idealising, in one case, the charge distribution (by a series of spikes), and, in the other, the potential distribution (by a sine curve). The first gave us the shape of the frequency response and the second, by bringing in the energy, gave us its amplitude and also the radiation impedance. It is possible, using the idea of surface piezoelectric permittivity, to express the electric potential associated with Rayleigh waves in terms of the actual charge density on the electrodes, in a completely general way. The characteristic features of the transducer can then be specified, with just a few extra hypotheses (the quasi-static approximation). This analysis includes reemission by the fingers, but not reflections from them.

The notion of surface piezoelectric permittivity was introduced in Vol I, Sect. 5.3.4.1. The method is analogous to that used in the bulk wave case, except that charges are distributed over the surface of the electrodes.

The relevant electrical quantities are charge density σ and potential Φ, taking values on the free surface $x_2 = 0$ of the piezoelectric solid. The only space variable is the abscissa $x_1 = x$. They are related in $(\omega,\ k)$ space by the piezoelectric permittivity $\overline{\varepsilon}$ defined in (5.72), Vol. I:

$$\overline{\sigma}(k,\ \omega) = [\overline{\varepsilon}(k,\ \omega) + \varepsilon_0]\,|k|\overline{\Phi}(k,\ \omega)\ , \tag{2.47}$$

or equivalently,

$$\overline{\Phi}(k,\ \omega) = \frac{\overline{\sigma}(k,\ \omega)}{|k|\,[\overline{\varepsilon}(k,\ \omega) + \varepsilon_0]}\ . \tag{2.48}$$

Quantities $\overline{\sigma}(k,\ \omega)$ and $\overline{\Phi}(k,\ \omega)$ are Fourier transforms, at a given angular frequency, of spatial distributions $\sigma(x,\ \omega)$ and $\Phi(x,\ \omega)$, found in the usual way:

$$\overline{g}(k,\ \omega) = \mathrm{FT}\,[g(x,\ \omega)] = \int_{-\infty}^{\infty} g(x,\ \omega)\mathrm{e}^{\mathrm{i}kx}\mathrm{d}x\ . \tag{2.49}$$

Recall that, in the case of a piezoelectric half-space, $\overline{\varepsilon}(k,\ \omega)$ is an even function of the phase speed $V = \omega/k$, giving an exact solution of the propagation equations in the medium for the mechanical boundary conditions on the

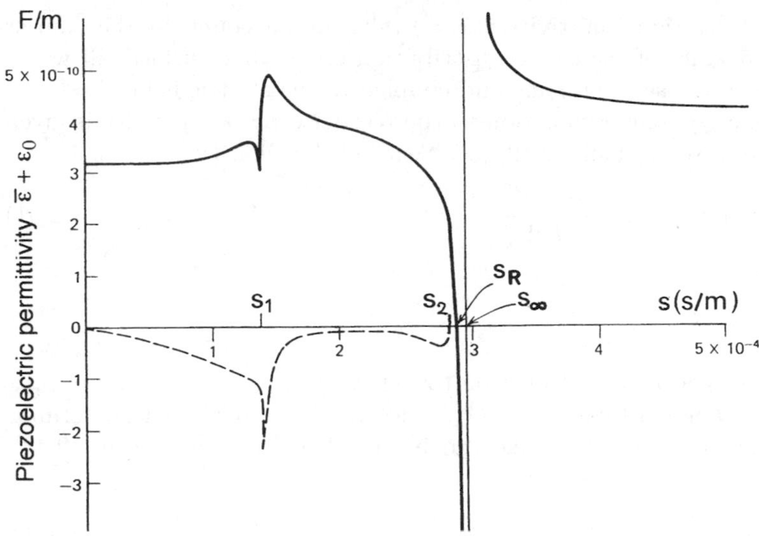

Fig. 2.17. Surface piezoelectric permittivity of a Y-cut lithium niobate crystal, with Z propagation, as a function of the inverse phase speed $s = k/\omega$. *Full curve*: real part. *Dashed curve*: imaginary part. $s_{\mathrm{R}} = 2.868 \times 10^{-4}$ s/m and $s_\infty = 2.932 \times 10^{-4}$ s/m. (From Fig. 2 in [2.18]. ©1977 IEEE)

free surface. As this solution does not cater for electrical boundary conditions, all phase speed values are possible (see Sect. 5.3.1, Vol. I). Fig. 2.17 shows how $\overline{\varepsilon}$, occurring in the denominator of (2.48), varies with slowness $s = k/\omega$ of the surface wave, in the case of Z-axis propagation in a Y-cut lithium niobate crystal [2.18]. The pole $s = s_\infty$ and zero $s = s_{\mathrm{R}}$ correspond to propagation on a metallised surface ($\Phi = 0$, speed $V_\infty = 1/s_\infty$), and on a non-metallised surface ($\sigma = 0$, speed $V_{\mathrm{R}} = 1/s_{\mathrm{R}}$), respectively. Discontinuities in the slope $\mathrm{d}\overline{\varepsilon}/\mathrm{d}s$ at s_1, s_2 occur at maximal slowness values for quasi-longitudinal and quasi-transverse waves, propagating and polarised in the sagittal plane YZ. Slowness curves in Fig. 4.33, Vol. I, show that $s_1 \approx 1.4 \times 10^{-4}$ s/m and $s_2 \approx 2.8 \times 10^{-4}$ s/m. The X polarised transverse wave, not piezoelectrically coupled (see Sect. 4.3.2b, Vol. I), does not contribute to surface permittivity. For slowness values $s < s_2$, surface piezoelectric permittivity becomes complex. Its imaginary part describes radiation of bulk waves towards the core of the solid. Indeed, from (5.73) in Vol. I, the real part $G(k, \omega)$ of the surface admittance,

$$Y(k, \omega) = \mathrm{i}w\omega|k|\,[\overline{\varepsilon}(k, \omega) + \varepsilon_0] = G(k, \omega) + \mathrm{i}B(k, \omega)\ ,$$

is proportional to the imaginary part of the piezoelectric permittivity:

$$G(k, \omega) = -w\omega|k|\,\mathrm{Im}\,[\overline{\varepsilon}(k, \omega) + \varepsilon_0]\ . \tag{2.50}$$

The elastic power generated in the form of bulk waves is therefore related to this imaginary part [which is negative, so that $G(k, \omega) > 0$].

The piezoelectric permittivity $\overline{\varepsilon}(k,\,\omega)$ takes into account, by the Fourier component $\overline{\sigma}(k,\,\omega)$ of the charge distribution on the free surface, all waves propagating in the sagittal plane and coupled to the electric field.

Near s_{R}, a good approximation to the piezoelectric permittivity is given by the Ingebrigtsen formula [2.19] (see Sect. 5.3.4.2, Vol. I):

$$\overline{\varepsilon}(k,\,\omega)+\varepsilon_0 \approx \varepsilon_{\mathrm{p}}\frac{\omega^2-k^2V_{\mathrm{R}}^2}{\omega^2-k^2V_{\infty}^2}\,, \tag{2.51}$$

where

$$\varepsilon_{\mathrm{p}}=\left[\varepsilon_{11}^{T}\varepsilon_{22}^{T}-\left(\varepsilon_{12}^{T}\right)^2\right]^{1/2}$$

is the effective permittivity [equation (5.71), Vol. I].

At a given angular frequency, $\overline{\Phi}(k,\,\omega)$ follows from $\overline{\sigma}(k,\,\omega)$. To determine the potential in x-space, we take the inverse Fourier transform of (2.48), which gives

$$\begin{aligned}\Phi(x,\,\omega)&=\frac{1}{2\pi}\int_{-\infty}^{\infty}\overline{\Phi}(k,\,\omega)\mathrm{e}^{-\mathrm{i}kx}\mathrm{d}k\\&=\frac{1}{2\pi}\int_{-\infty}^{\infty}\frac{\overline{\sigma}(k,\,\omega)}{|k|\left[\overline{\varepsilon}(k,\,\omega)+\varepsilon_0\right]}\mathrm{e}^{-\mathrm{i}kx}\mathrm{d}k\,.\end{aligned} \tag{2.52}$$

According to electrical boundary conditions, the potential $\Phi(x,\,\omega)$ equals the voltage across the electrodes and the charge density $\sigma(x,\,\omega)$ is zero between the electrodes. It comprises three terms: the quasi-static potential, localised under the transducer, and contributions from Rayleigh waves and bulk waves.

The *electrostatic potential* $\Phi_0(x)$ produced by charge distribution $\sigma_0(x)$ is found by replacing the surface piezoelectric permittivity $\overline{\varepsilon}(k,\,\omega)$ by ε_{p}, the dielectric constant used to calculate the spatial distribution of electrical quantities (see Appendix A). Let us now use (2.47) to write the electrostatic field,

$$E_0(x)=-\frac{\mathrm{d}\Phi_0}{\mathrm{d}x}\quad\Rightarrow\quad\overline{E}_0(x)=\mathrm{i}k\overline{\Phi}_0(k)\,,$$

in terms of the density $\overline{\sigma}_0(k)$:

$$\overline{E}_0(k)=\mathrm{i}\frac{k}{|k|}\frac{\overline{\sigma}_0(k)}{\varepsilon_{\mathrm{p}}+\varepsilon_0}=\mathrm{i}\,\mathrm{sign}(k)\frac{\overline{\sigma}_0(k)}{\varepsilon_{\mathrm{p}}+\varepsilon_0}\,.$$

The inverse Fourier transform, in the sense of distributions, of the function $\mathrm{sign}(k)$ is $-\mathrm{i}\pi/x$. The electrical field is then given by convolution as

$$E_0(x)=\frac{1}{\pi x}\otimes\frac{\sigma_0(x)}{\varepsilon_{\mathrm{p}}+\varepsilon_0}\,.$$

The electrostatic potential is obtained by integrating:

$$\Phi_0(x)=-\frac{1}{\pi}\frac{\log|x|}{\varepsilon_{\mathrm{p}}+\varepsilon_0}\otimes\sigma_0(x)\,. \tag{2.53}$$

2.3.1 Generation. Green Functions

The potential $\Phi_0(x)$ is localised in the region of the emitting transducer. Far from this region, the only potential at the surface of the piezoelectric solid is the potential accompanying Rayleigh waves emitted in $\pm x$ directions. This far-field contribution arises from the poles in the integrand of (2.52). Since the charge density $\sigma(x, \omega)$ is restricted to a bounded region of space, its Fourier transform $\overline{\sigma}(k, \omega)$ is finite for all values of k. The value $k = 0$ does not correspond to a pole because the total charge is zero, so that the numerator is also zero at $k = 0$:

$$\int_{-\infty}^{\infty} \sigma(x, \omega)\,\mathrm{d}x = \overline{\sigma}(k = 0, \omega) = 0\,. \tag{2.54}$$

The only poles in the integrand are the zeros of the quantity $\overline{\varepsilon}(k, \omega) + \varepsilon_0$. According to (5.77) in Vol. I, this vanishes for $k = \pm k_\mathrm{R}$, where $k_\mathrm{R} = \omega/V_\mathrm{R}$ is the Rayleigh wave number at angular frequency ω. These poles are located on the integration contour because we have neglected attenuation effects. In order to allow for these, a negative imaginary part must be associated with the positive pole, corresponding to propagation in the positive x direction. In other words, we replace $k_\mathrm{R} \to k_\mathrm{R} - \mathrm{i}\beta$, so that the propagation factor $\mathrm{e}^{-\mathrm{i}k_\mathrm{R}x}\mathrm{e}^{-\beta x}$ will tend to zero when $x \to +\infty$. Likewise, a positive imaginary part must be associated with the negative pole, corresponding to propagation in the negative x direction, and we replace $-k_\mathrm{R} \to -k_\mathrm{R} + \mathrm{i}\beta$. Equivalently, the integration contour can be modified to pass below the $-k_\mathrm{R}$ pole and above the $+k_\mathrm{R}$ pole, as shown in Fig. 2.18.

The integral in (2.52) is then calculated by the residue method, closing the integration contour by a semicircle at infinity in the lower (upper) half-plane when x is positive (negative), so that the factor $\mathrm{e}^{-\mathrm{i}kx}$ is zero on the semicircle. The contribution from the pole $+k_\mathrm{R}$ (integrating clockwise around C) is given by:

$$\begin{aligned}\Phi_\mathrm{R}^{+}(x, \omega) &= -2\pi\mathrm{i}\,\mathrm{Res}\left[\frac{1}{2\pi}\frac{\overline{\sigma}(k, \omega)\mathrm{e}^{-\mathrm{i}kx}}{|k|\,[\overline{\varepsilon}(k, \omega) + \varepsilon_0]}\right]_{k=k_\mathrm{R}}\\ &= -\mathrm{i}\frac{\overline{\sigma}(k_\mathrm{R}, \omega)}{k_\mathrm{R}(\mathrm{d}\overline{\varepsilon}/\mathrm{d}k)_{k_\mathrm{R}}}\,\mathrm{e}^{-\mathrm{i}k_\mathrm{R}x}\,, \quad x > 0\,.\end{aligned} \tag{2.55}$$

The contribution from the pole at $-k_\mathrm{R}$ (integrating anticlockwise around C') is

$$\Phi_\mathrm{R}^{-}(x, \omega) = \mathrm{i}\frac{\overline{\sigma}(-k_\mathrm{R}, \omega)}{k_\mathrm{R}(\mathrm{d}\overline{\varepsilon}/\mathrm{d}k)_{-k_\mathrm{R}}}\,\mathrm{e}^{\mathrm{i}k_\mathrm{R}x}\,, \quad x < 0\,. \tag{2.56}$$

As $\overline{\varepsilon}$ and $|\overline{\sigma}|$ are even functions of k, amplitudes of the two emitted surface waves are equal. In the Rayleigh wave case, the piezoelectric permittivity is only a function of phase speed $V = \omega/k$. The denominators of the two above expressions are independent of angular frequency ω. Using the approximate formula (2.51) for permittivity $\overline{\varepsilon}(k, \omega)$, the calculation in Sect. 5.3.4.4 , Vol. I, shows that

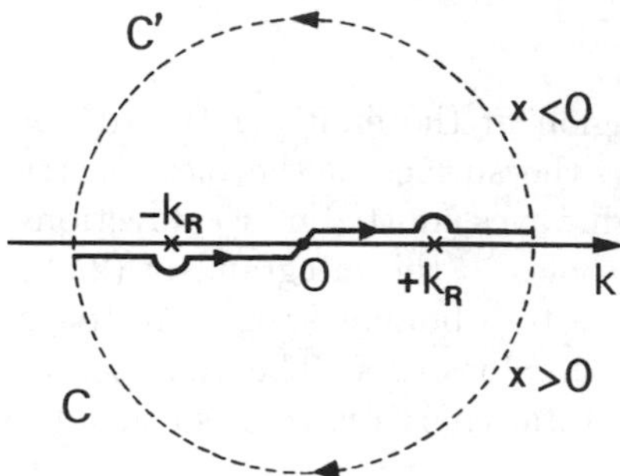

Fig. 2.18. Integration contours used to calculate contributions from poles in a medium without attenuation

$$k_{\mathrm{R}}\left(\frac{\mathrm{d}\overline{\varepsilon}}{\mathrm{d}k}\right)_{-k_{\mathrm{R}}} = -k_{\mathrm{R}}\left(\frac{\mathrm{d}\overline{\varepsilon}}{\mathrm{d}k}\right)_{k_{\mathrm{R}}} = V_{\mathrm{R}}\left(\frac{\mathrm{d}\overline{\varepsilon}}{\mathrm{d}V}\right)_{V_{\mathrm{R}}} = 2\frac{\varepsilon_{\mathrm{p}}+\varepsilon_0}{K_{\mathrm{R}}^2} . \tag{2.57}$$

This last equality can be taken as a more general definition of the electromechanical coupling coefficient K_{R} than the one given in Sect. 5.3.4.3, Vol. I. Setting

$$\Gamma_{\mathrm{R}} = -\frac{1}{k_{\mathrm{R}}(\mathrm{d}\overline{\varepsilon}/\mathrm{d}k)_{k_{\mathrm{R}}}} = \frac{K_{\mathrm{R}}^2/2}{\varepsilon_{\mathrm{p}}+\varepsilon_0} , \tag{2.58}$$

equations (2.55) and (2.56) can be summarised in a single formula:

$$\boxed{\Phi_{\mathrm{R}}^{\pm}(x,\,\omega) = \mathrm{i}\Gamma_{\mathrm{R}}\mathrm{e}^{-\mathrm{i}k_{\mathrm{R}}|x|}\overline{\sigma}(\pm k_{\mathrm{R}},\,\omega)} . \tag{2.59}$$

The electric potential is the sum of this term coming from Rayleigh waves generated by the charge distribution, the quasi-static potential $\Phi_0(x,\,\omega)$, and the potential $\Phi_{\mathrm{B}}(x,\,\omega)$ accompanying emitted bulk waves:

$$\Phi(x,\,\omega) = \Phi_0(x,\,\omega) + \Phi_{\mathrm{R}}(x,\,\omega) + \Phi_{\mathrm{B}}(x,\,\omega) . \tag{2.60}$$

These results can be interpreted by introducing a *Green function* $g(x,\,\omega)$ such that

$$\Phi(x,\,\omega) = g(x,\,\omega)\otimes\sigma(x,\,\omega) = \int_{-\infty}^{\infty} g(x-x',\,\omega)\sigma(x',\,\omega)\,\mathrm{d}x' . \tag{2.61}$$

Then the part of the Green function giving the quasi-static potential is

$$g_0(x) = -\frac{1}{\pi}\frac{\log|x|}{\varepsilon_{\mathrm{p}}+\varepsilon_0} . \tag{2.62}$$

From the identity

$$\mathrm{e}^{-ikx}\otimes\sigma(x,\,\omega) = \int_{-\infty}^{\infty}\mathrm{e}^{-\mathrm{i}k(x-x')}\sigma(x',\,\omega)\,\mathrm{d}x' = \mathrm{e}^{-ikx}\,\overline{\sigma}(k,\,\omega) , \tag{2.63}$$

the Green function $g_{\mathrm{R}}(x,\,\omega)$ associated with emitted Rayleigh waves is

$$g_{\mathrm{R}}(x,\,\omega) = \mathrm{i}\Gamma_{\mathrm{R}}\mathrm{e}^{-\mathrm{i}k_{\mathrm{R}}|x|} . \tag{2.64}$$

The potential due to bulk waves can be calculated exactly by numerical integration of (2.52) [2.20], or approximately by the stationary phase method, described in Appendix G. If this contribution can be neglected, i.e., if electrical charge on electrodes at the surface only excites Rayleigh waves, the electric potential contains just the two terms:

$$\Phi(x,\,\omega) = [g_0(x) + g_\mathrm{R}(x,\,\omega)] \otimes \sigma(x,\,\omega)\ . \tag{2.65}$$

Far from the electrodes, the potential accompanying emitted Rayleigh waves is no longer subject to the influence of their charges. It can then be found approximately without including the term $g_0(x)$, considered to be localised around the electrodes. Hence,

$$\Phi(x,\,\omega) \approx g_\mathrm{R}(x,\,\omega) \otimes \sigma(x,\,\omega)\ , \quad |x| > L/2\ ,$$

where L is the length of the transducer, centred on $x = 0$. Then by (2.63) and (2.64),

$$\Phi^{\pm}(x,\,\omega) \approx \mathrm{i}\Gamma_\mathrm{R}\mathrm{e}^{-\mathrm{i}k_\mathrm{R}|x|}\,\overline{\sigma}(\pm k_\mathrm{R},\,\omega)\ , \quad |x| > L/2\ . \tag{2.66}$$

This is an approximate form for the electric potential, only valid outside the electrode region.

Quasi-Static Approximation. Note that $\overline{\sigma}(k_\mathrm{R},\,\omega)$ is not the Fourier transform of $\sigma(x,\,\omega)$ in the usual sense, because the wave number variable $k_\mathrm{R} = \omega/V_\mathrm{R}$ is not independent of ω. When ω varies, $\sigma(x,\,\omega)$ varies and the Fourier integral must be recalculated for each value of the angular frequency. This fact enormously complicates the analysis. However, it ceases to be a problem if the charge density is determined by means of the quasi-static approximation [2.21]. The charge density contains two terms:

$$\sigma(x,\,\omega) \approx \sigma_0(x,\,\omega) + \sigma_\mathrm{R}(x,\,\omega)\ . \tag{2.67}$$

The first of these,

$$\sigma_0(x,\,t) = \sigma_1(x)U(t) \Rightarrow \overline{\sigma}_0(k,\,\omega) = \overline{\sigma}_1(k)\overline{U}(\omega)\ , \tag{2.68}$$

has spatial distribution $\sigma_1(x)$ independent of ω and can be determined in the absence of any waves for unit applied voltage $U = 1$. The second contribution σ_R is smaller and represents the induced charge density, when $U = 0$, due to piezoelectric waves propagating under the short-circuited electrode array.

The applied potential Φ_0 at the electrodes corresponds to the density σ_0, through relation (2.53). If the term $g_\mathrm{R}\sigma_\mathrm{R}$ can be neglected, replacing σ in (2.65) by the sum (2.67) leads to a potential at any point of the surface given by

$$\Phi(x,\,\omega) = [g_0(x) + g_\mathrm{R}(x,\,\omega)] \otimes \sigma_0(x,\,\omega) + g_0(x) \otimes \sigma_\mathrm{R}(x,\,\omega)\ . \tag{2.69}$$

The term $g_\mathrm{R} \otimes \sigma_0$ represents the potential on either side of the transducer. An expression for it can be deduced from (2.66), replacing $\overline{\sigma}$ by $\overline{\sigma}_0 = \overline{\sigma}_1\overline{U}$ to give

$$\boxed{\Phi^{\pm}(x,\,\omega) \approx \mathrm{i}\Gamma_\mathrm{R}\overline{\sigma}_1(\pm k_\mathrm{R})\overline{U}(\omega)\mathrm{e}^{-\mathrm{i}k_\mathrm{R}|x|}} \quad \text{for} \quad |x| > L/2\ . \tag{2.70}$$

From the symmetry properties of the Fourier transform of a real function, such as the charge distribution, we have $\overline{\sigma}_1(-k_R) = \overline{\sigma}_1^*(k_R)$ and the transducer is perfectly bidirectional, i.e., $|\Phi^+| = |\Phi^-|$. This is no longer true if we introduce a phase difference between applied potentials at the electrodes, which render the charge distribution complex. Then emission can be favoured in some particular direction (see Sect. 2.4.1).

It is essential to know σ_1. It can be obtained theoretically for transducers with simple structure, made from two comb-shaped electrodes in which fingers of the same size are separated by a constant interval. In general, however, we must adopt numerical techniques or an approximation using Chebyshev polynomials (see Appendix A). If the electrode structure is periodic, the calculation reduces to solving for a single-element charge distribution (the element factor).

Periodic Transducer. Many transducers consist of a periodic electrode array. Each electrode has width a, and they are brought to arbitrary potentials, U_n for the nth electrode. Since the potentials no longer alternate, the transducer is characterised by the period p of the electrode array. Electrodes are centred on abscissa points $x_n = np$ as shown in Fig. 2.19a.

By the superposition principle, the potential of the emitted wave Φ^+ or Φ^- is, outside the transducer, equal to the sum of the potentials produced by each electrode when it is raised to a potential U_n, the other electrodes being earthed (see Fig. 2.19b). The factor Γ_R is a characteristic constant for the material, so the total potential becomes

$$\Phi^+(\omega) \approx \mathrm{i}\Gamma_R \sum_n \overline{\sigma}_{1n}(k_R) U_n \mathrm{e}^{-\mathrm{i}k_R x_n} , \quad \Gamma_R = \frac{K_R^2}{2(\varepsilon_p + \varepsilon_0)} . \tag{2.71}$$

For an electrode situated far from the edges, the induced charge distribution $\overline{\sigma}_{1n}$ does not depend on the position of the electrode in the array. This *element factor* $\overline{\sigma}_{1e}(k)$ is a function of the surface metallisation ratio $\eta = a/p$. The frequency response of the transducer is:

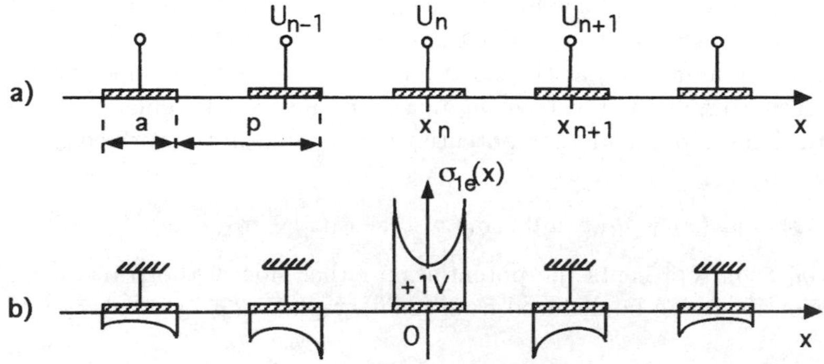

Fig. 2.19. (**a**) Periodic electrode array. (**b**) Single-element charge distribution produced by one electrode at 1 V, whilst the others are earthed

$$\Phi^+(\omega) \approx \mathrm{i}\Gamma_\mathrm{R}\overline{\sigma}_{1\mathrm{e}}(k)\sum_n U_n \mathrm{e}^{-\mathrm{i}kx_n}\,, \quad k = \frac{\omega}{V_\mathrm{R}}\,. \tag{2.72}$$

This involves the *array factor* $\sum U_n \mathrm{e}^{-\mathrm{i}kx_n}$, which depends only on the period and the sequence of voltages applied to the electrodes. This form is close to the one obtained by the discrete source method in Sect. 2.1.3.

The approach here is all the more interesting in that Datta and Hunsinger [2.22] have found an analytic expression for the element factor $\sigma_{1\mathrm{e}}(x)$. Its shape is shown in Fig. 2.19b. As $\sigma_{1\mathrm{e}}(x)$ is a real and even function of x, the same is true of its Fourier transform: $\overline{\sigma}_{1\mathrm{e}}(-k) = \overline{\sigma}_{1\mathrm{e}}(k)$. The formula is (see Appendix A.2)

$$\overline{\sigma}_{1\mathrm{e}}(k) = 2(\varepsilon_\mathrm{p}+\varepsilon_0)\frac{\sin \pi s}{P_{-s}(-\cos\eta\pi)}P_m(\cos\eta\pi)\,, \tag{2.73}$$

where $P_\nu(x)$ are Legendre functions, called Legendre polynomials when ν is an integer (see Appendix C). m is the integer part (IP) of the wave number k normalised to the wave number $G = 2\pi/p$ of the array, and s the decimal remainder:

$$s = \frac{kp}{2\pi} - m = \frac{f}{2f_0} - m \quad \rightarrow \quad 0 \le s \le 1\,, \quad m = \mathrm{IP}\left(\frac{f}{2f_0}\right)\,.$$

Hence,

$$\begin{array}{lll} 0 \le f < 2f_0 & m=0 & \rightarrow \quad P_0(\cos\eta\pi) = 1 \\ 2f_0 \le f < 4f_0 & m=1 & \rightarrow \quad P_1(\cos\eta\pi) = \cos\eta\pi \\ 4f_0 \le f < 6f_0 & m=2 & \rightarrow \quad P_2(\cos\eta\pi) = (3\cos^2\eta\pi - 1)/2\,. \end{array}$$

Further terms follow from the recurrence relation for Legendre polynomials, given in Appendix C.

For fixed metallisation ratio η, i.e., for given $x = -\cos\eta\pi$, the Fourier transform of the element factor depends on k and hence on the frequency, through the term $\sin\pi s$ and through the degree $\nu = -s$ of the Legendre polynomial $P_{-s}(x)$. In Fig. C.1, $P_\nu(x)$ is plotted in the relevant domain ($-1 \le \nu \le 0$) for various values of x. For each x, it is symmetric in $\nu = -0.5$, at which it reaches a maximum. Representative curves of $\overline{\sigma}_{1\mathrm{e}}(f)$ are plotted in Fig. 2.20 for three values of η. Extrema are located in the middle of frequency bands, when $s = 1/2$, i.e., when $f = (2m+1)f_0$. Given the relation (C.5) between the Legendre function of order $-1/2$ and the elliptic integral of the first kind $K(x)$, extremal values

$$A_m = (\varepsilon_\mathrm{p}+\varepsilon_0)\frac{\pi P_m(\cos\eta\pi)}{K[\cos(\eta\pi/2)]} = (\varepsilon_\mathrm{p}+\varepsilon_0)F_m(\eta)$$

are proportional to the harmonic coefficients $F_m(\eta)$ given in (2.13) and illustrated in Fig. 2.9. Each curve in Fig. 2.20 is made up of arcs whose peak value in the frequency band defined by integer m is equal to the coefficient $F_m(\eta)$ corresponding to the value of η. For example, when $m = 1$, i.e., $2f_0 < f < 4f_0$, $\overline{\sigma}_{1\mathrm{e}}(f) = 0$ if $\eta = 1/2$.

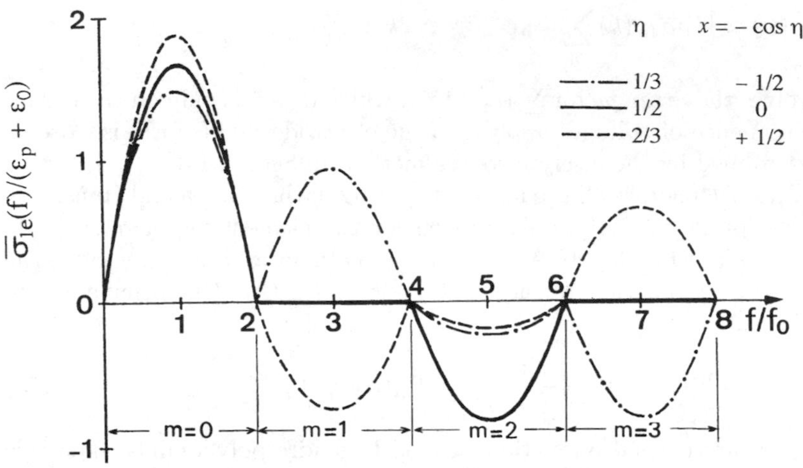

Fig. 2.20. Fourier transform $\overline{\sigma}_{1e}(f)$ of the element factor for three values of metallisation ratio: $\eta = 1/3,\ 1/2,\ 2/3$. (Figure 13 in [2.23]. ©1981 IEEE)

Given the formula for Γ_R, the complex amplitude of the potential generated by an element driven with a sinusoidal voltage of amplitude 1 V only depends on the electromechanical coupling coefficient of the material:

$$\Phi_{1e}^{+}(f) = \mathrm{i}\Gamma_R\overline{\sigma}_{1e}(k) \approx \mathrm{i}\frac{K_R^2}{2}F_m(\eta)\sin\left(\frac{\pi}{2}\frac{f}{f_0}\right), \quad m = \mathrm{IP}\left(\frac{f}{2f_0}\right). \tag{2.74}$$

Order of Magnitude. In the case of Y-cut, Z propagating lithium niobate, for $f = f_0 \to m = 0$ and $\eta = 0.5$, we have $F_0(0.5) = 1.7$. As $K_R^2 = 4.8\%$, the element potential $\Phi_{1e}(f_0)$ is equal to i 0.041 V.

Simple Transducer. Consider an emitter transducer of length L made from two combs with constant finger overlap width w (see Fig. 2.1). In order to calculate $\overline{\sigma}_1(k)$, we assume that the density $\sigma_1(x)$, which is zero outside the transducer ($|x| > L/2$), is equal to that of an infinite transducer of the same geometry. This assumption amounts to neglecting edge effects (number of finger pairs $N > 10$). The Fourier transform of the charge density $\sigma_1(x)$ [formula (A.7) with $U = 1$ V],

$$\overline{\sigma}_1(k) = (\varepsilon_p + \varepsilon_0)\sum_{m=0}^{\infty}\frac{F_m}{d}\int_{-L/2}^{+L/2}\cos(\chi_m x)\mathrm{e}^{\mathrm{i}kx}\mathrm{d}x\,,$$

is given by

$$\overline{\sigma}_1(k) = (\varepsilon_p + \varepsilon_0)\sum_{m=0}^{\infty}\frac{F_m}{d}\left[\frac{\sin(k-\chi_m)L/2}{k-\chi_m} + \frac{\sin(k+\chi_m)L/2}{k+\chi_m}\right].$$

Since $L = 2Nd$ and $\chi_m = (2m+1)2\pi f_0/V_R$,

$$k - \chi_m = \frac{2\pi}{V_\mathrm{R}}(f - f_m) , \quad f_m = (2m+1)f_0 ,$$

and it follows that

$$\overline{\sigma}_1(k_\mathrm{R}) = (\varepsilon_\mathrm{p} + \varepsilon_0)N \sum_{m=0}^{\infty} F_m \left[\frac{\sin X_m^-}{X_m^-} + \frac{\sin X_m^+}{X_m^+} \right] , \tag{2.75}$$

where $X_m^\pm = \pi N(f \pm f_m)/f_0$. From (2.70) and (2.58), the *frequency response* at the $x = L/2$ output of the transducer is, in the positive frequency domain,

$$H(\omega) = \frac{\Phi^+(L/2,\,\omega)}{\overline{U}(\omega)} = \mathrm{i}\frac{K_\mathrm{R}^2}{2} N \exp\left(-\mathrm{i}\frac{\omega L}{2V_\mathrm{R}}\right) \sum_{m=0}^{\infty} F_m \frac{\sin X_m^-}{X_m^-} . \tag{2.76}$$

The transducer frequency response is made up of $\sin X/X$ curves centred on odd harmonics of the synchronism frequency f_0, of constant bandwidth and amplitude proportional to coefficients F_m.

In the case of a transducer with metallisation ratio $\eta = 1/2$ (so that $\cos \eta\pi = 0$), the fact that $P_{2p+1}(0) = 0$ implies that there are no 3, 7, ... harmonics (coefficients F_m in Fig. 2.9 are zero). Experiment confirms this result for materials with low electromechanical coupling coefficient, as can be seen from the impulse response of a quartz transducer in Fig. 2.21. But it is no longer true for strongly piezoelectric crystals such as lithium niobate.

It is instructive to compare the response found here with that deduced via the discrete source model in Fig. 2.11, where all bands have the same amplitude, and also with that of the impulse response model in (2.37), which only retains the fundamental.

Impedance. The piezoelectric permittivity method provides the absolute amplitude at the solid surface of the potential Φ accompanying a Rayleigh wave. It is thus possible, using (5.85), Vol. I, and (2.58), to determine the transported power:

$$P = -\frac{\omega w}{4} k_\mathrm{R} \left(\frac{\mathrm{d}\overline{\varepsilon}}{\mathrm{d}k}\right)_{k_\mathrm{R}} |\Phi|^2 = \frac{\omega w}{4\Gamma_\mathrm{R}} |\Phi|^2 . \tag{2.77}$$

A similar wave of the same power is, of course, launched in the negative x direction.

Consider the admittance $Y(\omega)$ of the transducer, ratio of current intensity extracted from the external supply to voltage U. We expect to find two terms:

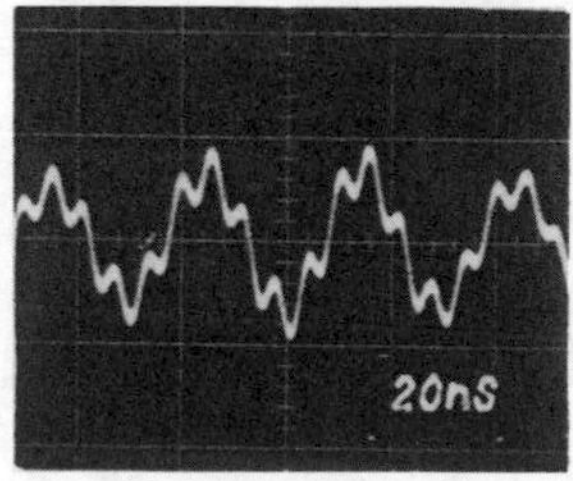

Fig. 2.21. Impulse response for an interdigital transducer with $\eta = 1/2$ (Y-cut α-quartz, propagation in the X direction). The 5 harmonic exists, but not the 3 harmonic

one arising from the charge density $\overline{\sigma}_0(k)$, and the other from the charge density $\overline{\sigma}_R(k)$. In fact, the second term separates into a real part G_a and an imaginary part B_a,

$$Y(\omega) = \overline{I}/\overline{U} = \mathrm{i}C_T\omega + G_a(\omega) + \mathrm{i}B_a(\omega) . \tag{2.78}$$

The admittance of the transducer is represented by the equivalent circuit in Fig. 2.15. Capacitance C_T is given by (A.27) in Appendix A, with $P_n = 1$ for one electrode and 0 for the other. Terms G_a, B_a arise from excitation of waves. Radiation conductance $G_a(\omega)$ is found immediately by equating externally supplied power with power carried in the two waves:

$$P = \frac{1}{2}G_a|U|^2 = \frac{1}{2}\frac{\omega w}{\Gamma_R}|\Phi|^2 . \tag{2.79}$$

Then, given (2.70),

$$G_a(\omega) = \omega w \Gamma_R |\overline{\sigma}_1(k_R)|^2 . \tag{2.80}$$

Susceptance $B_a(\omega)$ is the Hilbert transform of $G_a(\omega)$ (2.39), as already explained in Sect. 2.2.2. For a *simple transducer*, (2.75) implies, keeping only the fundamental term $F_0 = \pi/K(\sqrt{2}/2)$,

$$\overline{\sigma}_1(k_R) = \frac{\varepsilon_p + \varepsilon_0}{K(\sqrt{2}/2)} N\pi \frac{\sin X}{X} , \quad X = \pi N \frac{f - f_0}{f_0} .$$

Introducing the total capacitance $C_T = N(\varepsilon_p + \varepsilon_0)w$, the radiation conductance is given by

$$G_a(f) = \frac{\pi^3}{K^2} K_R^2 C_T N f \left(\frac{\sin X}{X}\right)^2 , \quad K = K(\sqrt{2}/2) = 1.854 . \tag{2.81}$$

This is similar to (2.38). The maximal value $G_a(f_0) = 2.87\pi K_R^2 C_T N f_0$ is slightly smaller than the value obtained by the impulse response method. The ratio π/K^2 of these two values corresponds to the proportion (91.4%) of the energy contained in the fundamental term (see Appendix A).

Impulse Response. We now return to the time domain, by inverting the Fourier transform of the potential, given on the positive x-axis by

$$\Phi^+(x, \omega) \approx \mathrm{i}\Gamma_R \overline{\sigma}_1\left(\frac{\omega}{V_R}\right) \exp\left(-\mathrm{i}\frac{\omega}{V_R}x\right) \overline{U}(\omega) . \tag{2.82}$$

From the convolution theorem,

$$\Phi^+(x, \omega) \approx \mathrm{i}\Gamma_R q(t) \otimes U(t) , \tag{2.83}$$

where the function $q(t)$ is obtained by inverting the Fourier transform:

$$\begin{aligned} q(t) &\approx \frac{1}{2\pi}\int_{-\infty}^{\infty} \overline{\sigma}_1\left(\frac{\omega}{V_R}\right) \exp \mathrm{i}\omega\left(t - \frac{x}{V_R}\right) \mathrm{d}\omega \\ &= \frac{V_R}{2\pi}\int_{-\infty}^{\infty} \overline{\sigma}_1(k) \mathrm{e}^{-\mathrm{i}k(x - V_R t)} \mathrm{d}k . \end{aligned}$$

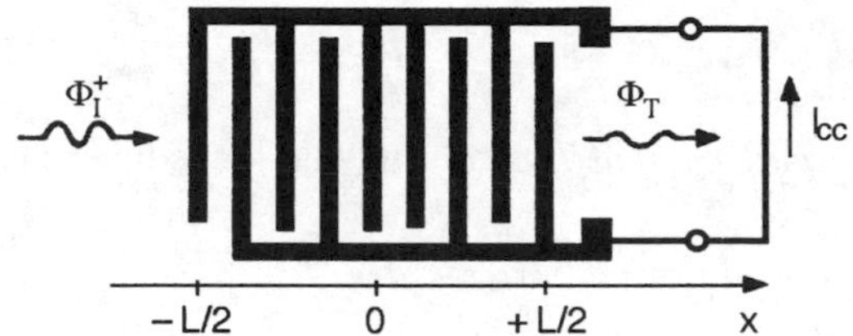

Fig. 2.22. Reception by a short-circuited transducer

Hence, $q(t) = V_R\sigma_1(x - V_Rt)$. Relation (2.83) gives the *impulse response*,

$$h(t) = \mathrm{i}\Gamma_R V_R \sigma_1(x - V_R t) \quad \text{for} \quad x > \frac{L}{2} , \tag{2.84}$$

where $\Gamma_R = K_R^2/2(\varepsilon_p + \varepsilon_0)$. This reproduces the quasi-static charge distribution. The amplitude of Rayleigh waves generated by applied voltage $U(t)$ across the transducer is proportional to the square of the electromechanical coupling coefficient.

2.3.2 Reception and Reemission

Consider a transducer with the same structure as the one above, operating as a receiver (see Fig. 2.22). In order to calculate the short-circuit current produced by an incident Rayleigh wave, let us assume that its electrodes are at zero potential. The incident wave is accompanied at the crystal surface by an electric potential

$$\Phi_I^+(x, \omega) = \Phi_I^+(\omega)\mathrm{e}^{-\mathrm{i}k_R x}\mathrm{e}^{-\mathrm{i}k_R L/2} , \tag{2.85}$$

where $\Phi_I(\omega)$ is the value of the potential at the transducer input $x = -L/2$. This wave generates charges on the electrodes with a density σ_R, in such a way that they remain at zero potential U. Now to any charge density, there corresponds an electric potential defined by (2.69) (in which $\sigma_0 = 0$ for the present case, since $U = 0$). At any point x_i on the electrode, the total electric potential is zero:

$$[g_0(x) \otimes \sigma_R(x, \omega)]_{x_i} + \Phi_I^+(x_i, \omega) = 0 . \tag{2.86}$$

With the notation of Appendix A, and using the correspondence (A.23) between charge density and potential, it then follows that [2.24]

$$\sigma_R(x_j, \omega) = -\sum_{i=1}^{P} B_{ij}\Phi_I^+(x_i, \omega) .$$

The current intensity entering the nth finger is given by (A.29):

$$I_{Rn} = -\mathrm{i}\omega w \sum_{i=1}^{P} \Phi_I^+(x_i, \omega) \sum_{j=1}^{P} B_{ij} p_n(x_j)\Delta x .$$

According to (A.24), the last sum is equal to $c_n(x_i)$ and converting $\sum \Delta x$ to an integral $\int \mathrm{d}x$,

$$I_{\mathrm{R}n} = -\mathrm{i}\omega w \int_{-\infty}^{\infty} \Phi_{\mathrm{R}}(x,\,\omega) c_n(x)\,\mathrm{d}x\,. \tag{2.87}$$

The total intensity of the *short-circuit current* I_{sc} entering the transducer is

$$I_{\mathrm{sc}} = \sum_{n=1}^{M} P_n I_{\mathrm{R}n} = -\mathrm{i}\omega w \int_{-\infty}^{\infty} \Phi_{\mathrm{I}}^{+}(x,\,\omega)\sigma_1(x)\,\mathrm{d}x\,,$$

where $\sigma_1(x) = \sum_{n=1}^{M} P_n c_n(x)$ is the charge distribution corresponding to applied voltage $U = 1$ V, introduced in the quasi-static approximation, Appendix A. Replacing the incident wave potential $\Phi_{\mathrm{I}}(x,\,\omega)$ by the expression in (2.85), and defining the Fourier transform

$$\overline{\sigma}_1(-k_{\mathrm{R}}) = \int_{-\infty}^{\infty} \sigma_1(x)\mathrm{e}^{-\mathrm{i}k_{\mathrm{R}}x}\mathrm{d}x = \overline{\sigma}_1^{*}(k_{\mathrm{R}})\,, \tag{2.88}$$

we find

$$I_{\mathrm{sc}}(\omega) = -\mathrm{i}\omega w\,\Phi_{\mathrm{I}}^{+}(\omega)\overline{\sigma}_1(-k_{\mathrm{R}})\mathrm{e}^{-\mathrm{i}k_{\mathrm{R}}L/2}\,, \quad k_{\mathrm{R}} = \frac{\omega}{V_{\mathrm{R}}}\,. \tag{2.89}$$

When operating as a receiver, the interdigital transducer is equivalent to a current generator of intensity I_{sc}, as shown in Fig. 2.23, with internal admittance Y_{T} made up of the three elements in Fig. 2.15.

Comparing with (2.70) reveals a reciprocity relation between the amplitude Φ_{I}^{+} of the incoming (incident) wave and the amplitude Φ_{S}^{-} of the outgoing (emitted) wave at the $x = -L/2$ port of the transducer [2.24]:

$$\left[\frac{I_{\mathrm{sc}}(\omega)}{\Phi_{\mathrm{I}}^{+}(\omega)}\right]_{\mathrm{reception}} = -\frac{\omega w}{\Gamma_{\mathrm{R}}}\left[\frac{\Phi_{\mathrm{S}}^{-}(\omega)}{U(\omega)}\right]_{\mathrm{emission}}\,. \tag{2.90}$$

The same relation holds between Φ_{I}^{-} and Φ_{S}^{+} at $x = L/2$.

If the receiver transducer is loaded by admittance Y_{L}, a potential difference U appears across the electrodes (Fig. 2.24):

$$U(\omega) = \frac{-I_{\mathrm{sc}}}{Y_{\mathrm{T}} + Y_{\mathrm{L}}} = \frac{\mathrm{i}\omega w}{Y_{\mathrm{T}} + Y_{\mathrm{L}}}\overline{\sigma}_1(-k_{\mathrm{R}})\mathrm{e}^{-\mathrm{i}k_{\mathrm{R}}L/2}\,\Phi_{\mathrm{I}}^{+}(\omega)\,. \tag{2.91}$$

The transducer is partially emitting because of this potential difference. It generates two waves: one is Φ_{S}^{+} in the direction of the incident wave, and the

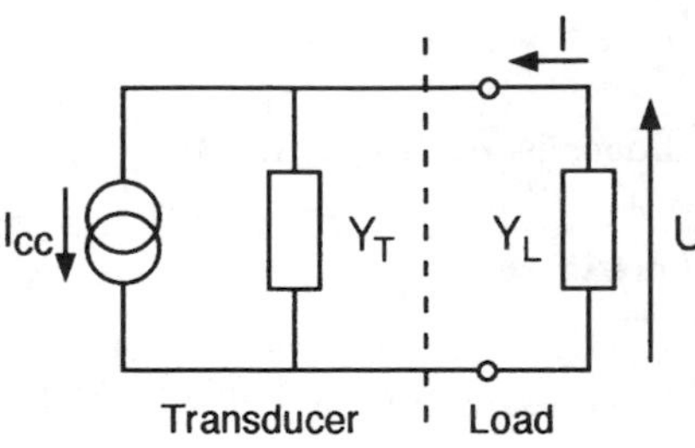

Fig. 2.23. Equivalent circuit for an interdigital receiver. Admittance Y_{T} is given by $Y_{\mathrm{T}}(\omega) = \mathrm{i}C_{\mathrm{T}}\omega + \mathrm{i}B_{\mathrm{a}}(\omega) + G_{\mathrm{a}}(\omega)$

Fig. 2.24. Reception by a transducer loaded by an admittance $Y_{\mathrm{L}} = G_{\mathrm{L}} + \mathrm{i}B_{\mathrm{L}}$

other Φ_{S}^{-} in the opposite direction. The potentials are given by (2.70). Since $\overline{\sigma}_1(-k_{\mathrm{R}}) = \overline{\sigma}_1^*(k_{\mathrm{R}})$,

$$\begin{aligned}\Phi_{\mathrm{S}}^{-}(-L/2,\,\omega) &\approx \mathrm{i}\Gamma_{\mathrm{R}}\overline{\sigma}_1(-k_{\mathrm{R}})U(\omega)\mathrm{e}^{-\mathrm{i}k_{\mathrm{R}}L/2}\\ &= -\frac{\omega w\Gamma_{\mathrm{R}}}{Y_{\mathrm{T}}+Y_{\mathrm{L}}}\left[\overline{\sigma}_1^*(k_{\mathrm{R}})\right]^2\mathrm{e}^{-\mathrm{i}k_{\mathrm{R}}L}\,\Phi_{\mathrm{I}}^{+}(\omega)\,,\end{aligned}$$

and since $\overline{\sigma}_1(k_{\mathrm{R}})\overline{\sigma}_1(-k_{\mathrm{R}}) = |\overline{\sigma}_1(k_{\mathrm{R}})|^2$,

$$\begin{aligned}\Phi_{\mathrm{S}}^{+}(L/2,\,\omega) &\approx \mathrm{i}\Gamma_{\mathrm{R}}\overline{\sigma}_1(k_{\mathrm{R}})U(\omega)\mathrm{e}^{-\mathrm{i}k_{\mathrm{R}}L/2}\\ &= -\frac{\omega w\Gamma_{\mathrm{R}}}{Y_{\mathrm{T}}+Y_{\mathrm{L}}}\left|\overline{\sigma}_1(k_{\mathrm{R}})\right|^2\mathrm{e}^{-\mathrm{i}k_{\mathrm{R}}L}\Phi_{\mathrm{I}}^{+}(\omega)\,.\end{aligned}$$

Using (2.80) for the radiation conductance, the surface potential accompanying the reflected wave is

$$\Phi_{\mathrm{R}} = \Phi_{\mathrm{S}}^{-}(-L/2,\,\omega) = \pm\frac{G_{\mathrm{a}}(\omega)}{Y_{\mathrm{T}}+Y_{\mathrm{L}}}\mathrm{e}^{-\mathrm{i}k_{\mathrm{R}}L}\Phi_{\mathrm{I}}^{+}(\omega)\,, \tag{2.92}$$

depending on whether the transducer is symmetric [$\overline{\sigma}_1(k_{\mathrm{R}})$ real, minus sign] or antisymmetric [$\overline{\sigma}_1(k_{\mathrm{R}})$ imaginary, plus sign].

The wave reemitted towards the right adds to the incident wave to give a transmitted wave. The potential at the $x = L/2$ end of the transducer is

$$\begin{aligned}\Phi_{\mathrm{T}} &= \Phi_{\mathrm{I}}^{+}(L/2,\,\omega) + \Phi_{\mathrm{S}}^{+}(L/2,\,\omega)\\ &= \left(1-\frac{\omega w\Gamma_{\mathrm{R}}}{Y_{\mathrm{T}}+Y_{\mathrm{L}}}\left|\overline{\sigma}_1(k_{\mathrm{R}})\right|^2\right)\mathrm{e}^{-\mathrm{i}k_{\mathrm{R}}L}\Phi_{\mathrm{I}}^{+}(\omega)\,.\end{aligned}$$

In terms of electrical admittances,

$$\Phi_{\mathrm{T}}(L/2,\,\omega) = \left(1-\frac{G_{\mathrm{a}}(\omega)}{Y_{\mathrm{T}}+Y_{\mathrm{L}}}\right)\mathrm{e}^{-\mathrm{i}k_{\mathrm{R}}L}\Phi_{\mathrm{I}}^{+}(\omega)\,. \tag{2.93}$$

These transmitted and reflected waves transport power which can be found using (2.77). Only a part of the incident power is converted into electrical power in the load conductance G_{L}:

$$P_{\mathrm{L}} = \frac{1}{2}G_{\mathrm{L}}|U|^2 = \frac{1}{2}G_{\mathrm{L}}\frac{\omega^2w^2}{|Y_{\mathrm{T}}+Y_{\mathrm{L}}|^2}\left|\overline{\sigma}_1(k_{\mathrm{R}})\right|^2\left|\Phi_{\mathrm{I}}^{+}\right|^2\,. \tag{2.94}$$

In terms of the radiation conductance (2.80) and incident power, given by $P_{\mathrm{I}} = \omega w\left|\Phi_{\mathrm{I}}^{+}\right|^2/4\Gamma_{\mathrm{R}}$, we have

$$P_{\mathrm{L}} = \frac{2G_{\mathrm{a}}G_{\mathrm{L}}}{|Y_{\mathrm{T}} + Y_{\mathrm{L}}|^2} P_{\mathrm{I}} \,. \tag{2.95}$$

The sum of the power transported by transmitted and reflected waves and the electrical power is, of course, equal to the incident power (Problem 2.5). There is therefore a direct relation between reflected power P_{R} and power converted into electrical form:

$$P_{\mathrm{R}} = \frac{|G_{\mathrm{a}}|^2}{|Y_{\mathrm{T}} + Y_{\mathrm{L}}|^2} P_{\mathrm{I}} = \frac{G_{\mathrm{a}}}{2G_{\mathrm{L}}} P_{\mathrm{L}} \,. \tag{2.96}$$

Let us investigate results for some particular values of terminal impedance:

- If the transducer is short-circuited, $Y_{\mathrm{L}} = \infty \Rightarrow P_{\mathrm{R}} = P_{\mathrm{L}} = 0$ and the incident wave is entirely transmitted.
- If the transducer is perfectly matched, $Y_{\mathrm{L}} = Y_{\mathrm{T}}^* \Rightarrow Y_{\mathrm{L}} + Y_{\mathrm{T}} = 2G_{\mathrm{a}}$ and one quarter of the incident power is transmitted, one quarter reflected and half converted.
- If the transducer susceptance is balanced by a lossless admittance, $G_{\mathrm{L}} = 0$, $Y_{\mathrm{L}} + Y_{\mathrm{T}} = G_{\mathrm{a}}$ and the incident wave is totally reflected.

These reemission phenomena suggest representing the transducer by a matrix giving, for example, potentials of outgoing waves and current intensity in electrodes in terms of potentials of incoming waves and voltage across the electrodes. More generally, the transducer can be represented by a hexapole and a scattering matrix or mixed matrix. Such a matrix representation works just as well for the emitter transducer as for the receiver transducer. It shows the effects of matching them to supply and load, and it also explains why there is a triple-transit echo due to reemission from the receiver.

2.4 Three-Port Circuit (Hexapole). Matrices

The interdigital transducer is basically a three-port device, with two acoustic ports and one electrical port (i.e., a hexapole). In the linear regime, it can be described by a 3×3 matrix, as can the bulk wave transducer (Sect. 1.2, Fig. 1.3). The matrix itself depends on the choice of variables. The relevant quantities at acoustic ports are clearly the waves. Adopting the same viewpoint for the electrical port, we obtain the scattering matrix (s) (also called the distribution matrix), which is homogeneous [2.25]. Consequences of the reciprocity principle are easily established (Appendix D). In practice, variables used for the electrical port are voltage U across the transducer terminals and current intensity I. The mixed matrix, also known as the P-matrix, resulting from this choice was introduced by Tobolka. Section 2.6 deals with equivalent circuits and the admittance matrix.

2.4.1 Scattering Matrix

Consider a wave with complex amplitude a_k. We need not specify its exact nature, mechanical or electrical, for the time being, provided that $a_k a_k^*/2$ is a power in the harmonic case. The wave enters by port k of the 3-port circuit and gives rise to an output wave of amplitude b_j at port j, such that

$$b_j = \sum_{k=1}^{3} s_{jk} a_k \,, \qquad s_{jk} = \sigma_{jk} \mathrm{e}^{\mathrm{i}\theta_{jk}} \,. \tag{2.97}$$

The symbol σ has also been used to denote electric charge, but carries indices here, which should avoid confusion in the context. The nine dimensionless coefficients s_{ij} are generally complex. They make up a matrix which gives the three output quantities in terms of the three input quantities. Relation (2.97) can be written out explicitly:

$$\begin{pmatrix} b_1 \\ b_2 \\ b_3 \end{pmatrix} = \begin{pmatrix} s_{11} & s_{12} & s_{13} \\ s_{21} & s_{22} & s_{23} \\ s_{31} & s_{32} & s_{33} \end{pmatrix} \begin{pmatrix} a_1 \\ a_2 \\ a_3 \end{pmatrix} \,. \tag{2.98}$$

Their physical meaning is illustrated in Fig. 2.25. s_{11}, s_{22} and s_{33} are reflection coefficients, whilst s_{jk} with $j \neq k$ are transmission coefficients between output j and input k. A term like σ_{jk}^2 equals the ratio of output power at j to input power at k. The system is passive and hence $\sigma_{jk} \leq 1$.

The reciprocity theorem, which implies that we may swap input and output quantities, reduces the number of independent coefficients to 6, since $s_{jk} = s_{kj}$, i.e., the matrix is symmetric [2.4]. If we can neglect electrical losses (Joule effect) and acoustic losses (attenuation, bulk wave generation), energy conservation also requires the matrix to be unitary:

$$\frac{1}{2}\sum_{j=1}^{3} b_j b_j^* = \frac{1}{2}(s_{jm}a_m)(s_{jn}a_n)^* = \frac{1}{2}a_m a_m^* \Rightarrow \sum_{j=1}^{3} s_{jm}s_{jn}^* = \delta_{mn} \,,$$

and hence

$$s_{1m}s_{1n}^* + s_{2m}s_{2n}^* + s_{3m}s_{3n}^* = \delta_{mn} \,, \quad m, n = 1, 2, 3 \,.$$

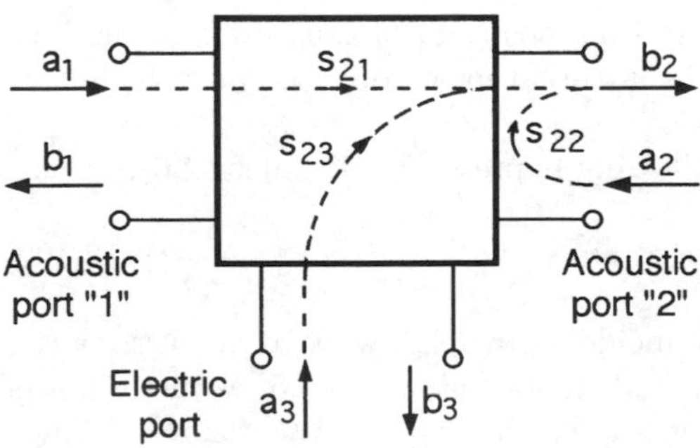

Fig. 2.25. The interdigital transducer is a device with one electrical and two acoustic ports. It is characterised by the scattering matrix s. *Dashed lines* symbolise generation of outgoing wave b_2 by the three incoming waves a_1, a_2 and a_3

Expanding this out,

$$1 = |s_{11}|^2 + |s_{12}|^2 + |s_{13}|^2 \tag{2.99a}$$

$$1 = |s_{12}|^2 + |s_{22}|^2 + |s_{23}|^2 \tag{2.99b}$$

$$1 = |s_{13}|^2 + |s_{23}|^2 + |s_{33}|^2 \tag{2.99c}$$

$$0 = s_{11}s_{12}^* + s_{12}s_{22}^* + s_{13}s_{23}^* \tag{2.99d}$$

$$0 = s_{11}s_{13}^* + s_{12}s_{23}^* + s_{13}s_{33}^* \tag{2.99e}$$

$$0 = s_{12}s_{13}^* + s_{22}s_{23}^* + s_{23}s_{33}^* \tag{2.99f}$$

yields nine relations, since the last three each have a real and an imaginary part.

Let us now proceed for the simple case of a *symmetric* or *antisymmetric transducer*. If the median line of the transducer is taken as the origin, the two acoustic ports are indistinguishable. Then $s_{22} = s_{11}$ and a purely electrical excitation $a_i = \delta_{i3}$ generates two acoustic waves b_1 and b_2 which are equal or opposite $s_{23} = \pm s_{13}$. There are four independent coefficients, viz., s_{11}, s_{12}, s_{13} and s_{33}, and there are still 5 independent relations among (2.99a–f). If we consider that only phase differences have any meaning, we can then express 6 of the 8 coefficients σ and θ in terms of the other two (Problem 2.6). For example, choosing the efficiency $\eta = \sigma_{13}^2 = \sigma_{23}^2$ and the angle $\phi = \theta_{11} - \theta_{12} \pm \pi$, we find successively

$$\sigma_{33}^2 = 1 - 2\eta \quad \Rightarrow \quad \eta \leq 0.5\,, \tag{2.100}$$

$$2\sigma_{11}\sigma_{12}\cos\phi = \eta\,, \quad \sigma_{11}^2 + \sigma_{12}^2 = 1 - \eta\,. \tag{2.101}$$

Operation as a Receiver. The efficiency of the interdigital receiver does not exceed 50%. The upper limit, reached for perfect matching to the load impedance, is a consequence of bidirectionality, as can be seen in Fig. 2.26. A unit power electrical excitation generates two symmetric waves, each carrying power 1/2. Conversely, two incident waves which are symmetric with respect to the transducer are coupled without loss if it is perfectly matched (Fig. 2.26a). A single wave arriving from the left is equivalent to the superposition of a symmetric and an antisymmetric mode (Fig. 2.26b). Only the symmetric pair produces a voltage across the transducer. The antisymmetric pair passes under the transducer without modification. Consequently, the interdigital-electrode receiver extracts at most half of the power of the incident Rayleigh wave. A fraction of the latter, at most equal to 1/4, is reemitted (Fig. 2.26c).

Solving (2.101), leads to the following expressions (Problem 2.6):

$$\left.\begin{matrix} 2\sigma_{12}^2 \\ 2\sigma_{11}^2 \end{matrix}\right\} = 1 - \eta \pm \sqrt{1 - 2\eta - \eta^2 \tan^2\phi}\,, \quad |\phi| < \arccos\left(\frac{\eta}{1-\eta}\right). \tag{2.102}$$

The reflection coefficient σ_{11} of the incident Rayleigh wave at acoustic port 1 is minimum for $\phi = 0$. For maximal efficiency value $\eta = 0.5$, acoustic power reflection and transmission coefficients are equal: $\sigma_{11}^2 = \sigma_{12}^2 = 0.25$. It follows,

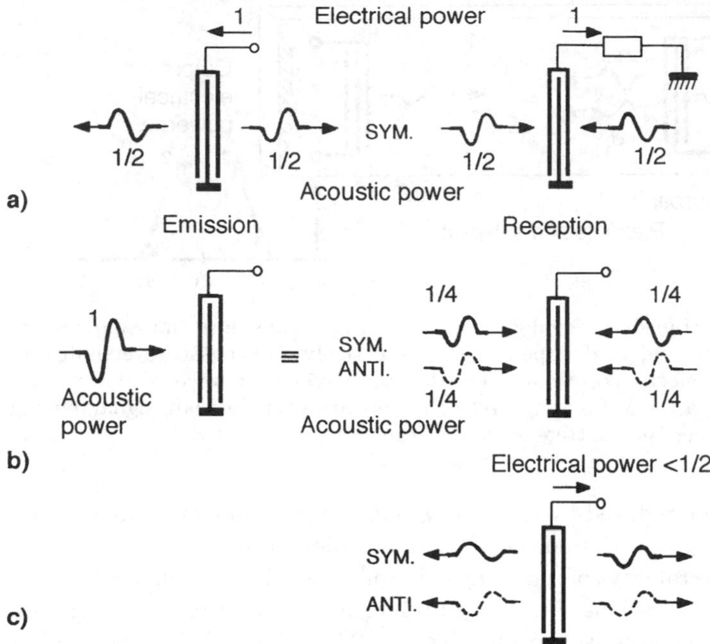

Fig. 2.26. (**a**)Two symmetric waves can be perfectly coupled in emission and reception at the electrical port of a symmetric transducer, with total power conversion. (**b**) Incident wave of unit power, approaching the receiver from the left, is decomposed into symmetric and antisymmetric modes. (**c**) Only the first mode is electrically coupled to the symmetric transducer, inducing an electrical signal of power at most 1/2, depending on the transducer efficiency. For imperfect matching, the symmetric mode is partially transmitted and the antisymmetric mode totally transmitted. Depending on their relative phase, transmitted modes can interfere constructively or destructively to produce reflected and transmitted waves of different amplitudes. Taken from [2.26]

for a two-transducer configuration, that an electrical signal applied to the emitter gives several signals at the receiver output (Fig. 2.27). Indeed, a unit power electrical signal generates a Rayleigh wave of power η at the receiver, assumed identical to the emitter. The receiver then converts a part η into an electrical signal of power η^2, reemitting the fraction $\eta\sigma_{11}^2$ back towards the emitter. The emitter in turn reemits the fraction σ_{11}^2 of this incident wave, i.e., $\eta\sigma_{11}^4$, which supplies an electrical signal of power $\eta^2\sigma_{11}^4$ after a further conversion at the receiver, and so on. The main signal is thus followed by a series of secondary signals appearing at times 3τ, 5τ, ... The first, which has travelled the emitter-receiver distance three times, is the most important. In the extreme case $\sigma_{11} = 0.5$, the level of this *triple-transit echo* is $10 \log \sigma_{11}^4$, or 12 dB below the primary signal. It arrives time 2τ after the primary, causing oscillations of frequency $1/2\tau$ in the frequency response of the system. Given the expression for σ_{11} in terms of efficiency η, the level of the triple-transit

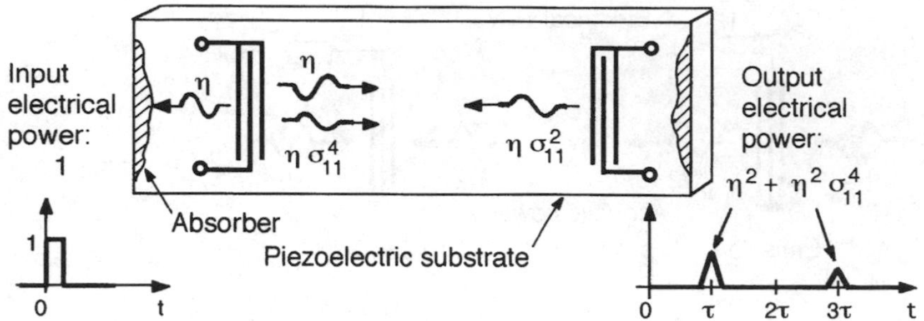

Fig. 2.27. Two-transducer configuration in which transducer impedances are matched to source and load impedances, respectively. Given the reversibility of the piezoelectric effect, receiver and emitter both reemit a wave of power $\eta\sigma_{11}^2$ when they receive a wave of power η. An echo appears after the main signal, having travelled three times the emitter-receiver distance

signal can only be reduced by reducing η, e.g., by mismatching impedances, thereby increasing insertion losses in the transmission line.

There are several ways of removing this spurious echo. One method, based on the above observations, involves building a transmission line with two emitters, placed on either side of the receiver [2.26]. Waves launched simultaneously by these symmetrical emitters are totally absorbed and converted by the central receiver, which is also symmetric. Another method is to make each transducer unidirectional. This technique also has the advantage of suppressing the intrinsic 3 dB loss in the bidirectional transducer. The first solution which comes to mind is to split the transducer into several elements, choosing the phases of applied voltages in such a way that emitted waves interfere positively in one direction and negatively in the other. Figure 2.28 shows a decomposition into two parts, separated by distance l, with phase shift ϕ between applied voltages. In one direction, the spatial phase difference ψ and electrical phase difference ϕ add together, whilst in the other, they subtract from one another. The transducer emits in just one direction provided that

$$\phi + \psi = \pi \quad \text{and} \quad \phi - \psi = 0 \quad \Rightarrow \quad \phi = \psi = \frac{\pi}{2} + 2n\pi \,.$$

This in turn implies

$$l = \frac{\lambda_0}{4} + n\lambda_0 \,.$$

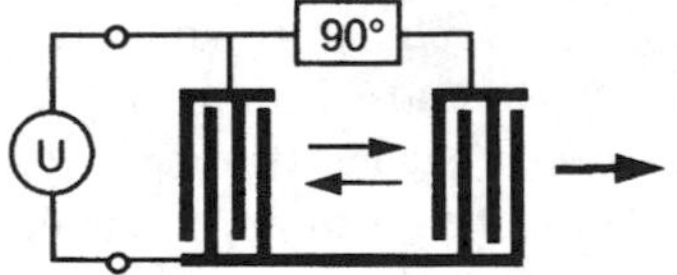

Fig. 2.28. Two-phase unidirectional transducer

However, having chosen the distance l, the operating frequency is predetermined. In principle, the phase condition is only satisfied within a narrow frequency band. It is not particularly convenient to add a phase shift circuit, although 3-phase transducers have been made in this way.

Present day techniques aim rather to take advantage of internal reflections. Before considering how reflections from the fingers could be exploited, e.g., to reduce losses in narrow bandwidth filters for mobile phones, a great deal of effort was expended in trying to eliminate them, treating them as unwanted secondary effects (Sect. 2.1.1). These reflections are taken into account locally in coupled mode analysis (Sect. 2.5) and globally in the mixed matrix.

2.4.2 Mixed Matrix

The mixed matrix, also called the P-matrix and denoted (P), is based on the fact that the practically useful variables at the electrical port are applied (or induced) voltage U and current intensity I [2.27]. As shown in Fig. 2.29, amplitudes of outgoing elastic waves and potential difference between electrodes are expressed in terms of incoming wave amplitudes and current intensity:

$$\begin{pmatrix} b_1 \\ b_2 \\ I \end{pmatrix} = \begin{pmatrix} P_{11} & P_{12} & P_{13} \\ P_{21} & P_{22} & P_{23} \\ P_{31} & P_{32} & P_{33} \end{pmatrix} \begin{pmatrix} a_1 \\ a_2 \\ U \end{pmatrix} . \tag{2.103}$$

Wave variables a_3 and b_3 at the electrical port are related to the voltage and current intensity by (see Appendix D)

$$a_3 = \frac{1}{2\sqrt{R_\mathrm{L}}}(U + R_\mathrm{L} I) , \quad b_3 = \frac{1}{2\sqrt{R_\mathrm{L}}}(U - R_\mathrm{L} I) , \tag{2.104}$$

where R_L ($G_\mathrm{L} = 1/R_\mathrm{L}$) is the reference resistance (conductance). Likewise, amplitudes a_i, b_i, $i = 1, 2$, of incoming and outgoing surface acoustic waves are related to particle velocities and forces by

$$a_i = \frac{1}{2\sqrt{Z_\mathrm{a}}}(F_i + Z_\mathrm{a} v_i) , \quad b_i = \frac{1}{2\sqrt{Z_\mathrm{a}}}(F_i - Z_\mathrm{a} v_i) , \tag{2.105}$$

or, equivalently,

$$F_i = \sqrt{Z_\mathrm{a}}(a_i + b_i) , \quad v_i = \frac{1}{\sqrt{Z_\mathrm{a}}}(a_i - b_i) , \tag{2.106}$$

where Z_a is the reference acoustic impedance.

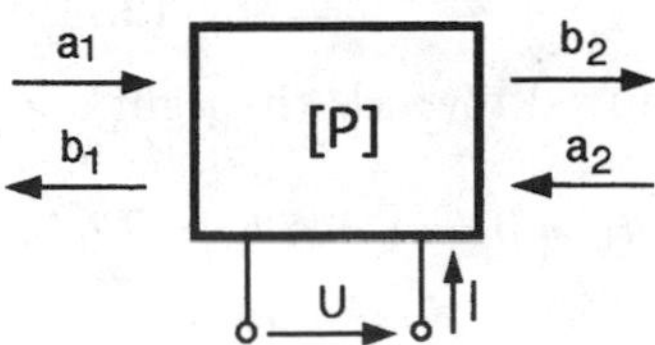

Fig. 2.29. P-matrix variables

The mean power carried by a Rayleigh wave in the x_1 direction is equal to the real part of the complex power, defined in (5.25), Vol. I. Using (5.20), Vol. I, and neglecting electrical power,

$$P = -\frac{1}{2}\mathrm{Re}\int_0^\infty \int_{-w/2}^{w/2} T_{j1} v_j^* \,\mathrm{d}x_2\,\mathrm{d}x_3 \,.$$

This can be put into the form $P = -\mathrm{Re}(Fv^*)/2$, where F has units of force (product of a stress and an area) and v has units of speed. The mean power radiated at port i is

$$P_i = -\frac{1}{2}\mathrm{Re}(F_i v_i^*) = \frac{1}{2}\left(|b_i|^2 - |a_i|^2\right) \,. \tag{2.107}$$

Amplitudes of incoming and outgoing waves are therefore given in units $[\mathrm{W}^{1/2}]$ or $[\mathrm{V}^{1/2}\mathrm{A}^{1/2}]$. The mean electrical power supplied is

$$P_{\mathrm{el}} = \frac{1}{2}\mathrm{Re}(IU^*) = \frac{1}{2}\mathrm{Re}(P_{33})\,|U|^2 \,. \tag{2.108}$$

This matrix contains one purely electrical term P_{33}, three mechanical terms P_{11}, $P_{12} = P_{21}$, P_{22}, and only two independent electromechanical terms P_{13} and P_{23}, since

$$P_{31} = -2P_{13} \,, \quad P_{32} = -2P_{23} \,. \tag{2.109}$$

They are given in units $[\mathrm{A}^{1/2}\mathrm{V}^{-1/2}]$. The last relation results from applying the reciprocity principle to the global admittance matrix, which relates the acoustic and electrical velocity vector $(v_1,\, v_2,\, I)$ to the mechanical and electrical force vector $(F_1,\, F_2,\, U)$ (see Appendix D).

The P-matrix entries have precise physical meaning. P_{33} is the input admittance $[\Omega^{-1}]$ of the transducer:

$$P_{33} = G_{\mathrm{a}}(\omega) + \mathrm{i}B_{\mathrm{a}}(\omega) + \mathrm{i}\omega C_{\mathrm{T}} \,. \tag{2.110}$$

When the electrical port is short-circuited ($U = 0$), the transducer becomes a simple reflector. P_{11} and P_{22} are reflection coefficients at the left and right acoustic ports (outputs) (at $x = 0$ and $x = L$, respectively) of the transducer:

$$P_{11} = r_0 \,, \quad P_{22} = r_L \,. \tag{2.111}$$

The term $P_{12} = P_{21}$ is the transmission coefficient between the two outputs:

$$P_{12} = P_{21} = t \,. \tag{2.112}$$

Electromechanical coupling terms P_{13} and P_{23} are transfer functions between electrical input and mechanical outputs 1 and 2, i.e., frequency responses:

$$P_{13} = H_{\mathrm{S}}(\omega) \,, \quad P_{23} = H_{\mathrm{R}}(\omega) \,, \tag{2.113}$$

where indices R and S denote Rayleigh waves emitted towards the right and left, respectively.

When there are no losses or incident waves ($a_1 = 0 = a_2$, so $b_1 = P_{13}U$, $b_2 = P_{23}U$), power conservation implies

$$|P_{13}|^2 + |P_{23}|^2 = \mathrm{Re}(P_{33}) = G_\mathrm{a}(\omega) , \tag{2.114}$$

and hence,

$$G_\mathrm{a}(\omega) = |H_\mathrm{R}(\omega)|^2 + |H_\mathrm{S}(\omega)|^2 . \tag{2.115}$$

This generalises (2.36) to the case where the transducer is not bidirectional, i.e., $|H_\mathrm{R}| \neq |H_\mathrm{S}|$.

Connection with Scattering Matrix. Components of matrix (s) are related to components of matrix (P) and to the load impedance on the electrical port. Let us now calculate acoustic coefficients s_{11}, s_{12} and s_{22}. When the electrical port of the transducer, operating as a receiver, is loaded with a conductance G_L such that $I = -G_\mathrm{L} U$, its behaviour is altered by reemission. A voltage U arises across the transducer terminals. U can be found from the last line of the matrix relation (2.103),

$$U = -\frac{P_{31}a_1 + P_{32}a_2}{G_\mathrm{L} + P_{33}} .$$

The first two lines of (2.103) yield the acoustic reflection coefficients,

$$s_{11} = P_{11} + 2\frac{P_{13}^2}{G_\mathrm{L} + P_{33}} , \qquad s_{22} = P_{22} + 2\frac{P_{23}^2}{G_\mathrm{L} + P_{33}} , \tag{2.116}$$

and acoustic transmission coefficients,

$$s_{21} = s_{12} = P_{12} + 2\frac{P_{13}P_{23}}{G_\mathrm{L} + P_{33}} . \tag{2.117}$$

In principle, the reflection coefficient s_{11} can be made equal to zero if the first term P_{11}, representing internal mechanical reflections ($U = 0$), exactly balances the second, which represents piezoelectric reemission [2.28]. Single-phase unidirectional transducers, or SPUDTs, are based upon this principle (Sect. 2.5.3).

Let us now consider the electrical coefficients s_{33} and electromechanical coefficients s_{13}, s_{23}. The electrical port of the transducer, operating as an emitter, is connected to a supply of e.m.f. U_0 and internal resistance R_L, as shown in Fig. 2.30. Since $a_1 = 0 = a_2$, we have $b_3 = s_{33}a_3$ and $I = P_{33}U$, so that (2.104) implies

$$s_{33} = \frac{U - R_\mathrm{L} I}{U + R_\mathrm{L} I} \quad \Rightarrow \quad s_{33} = \frac{G_\mathrm{L} - P_{33}}{G_\mathrm{L} + P_{33}} . \tag{2.118}$$

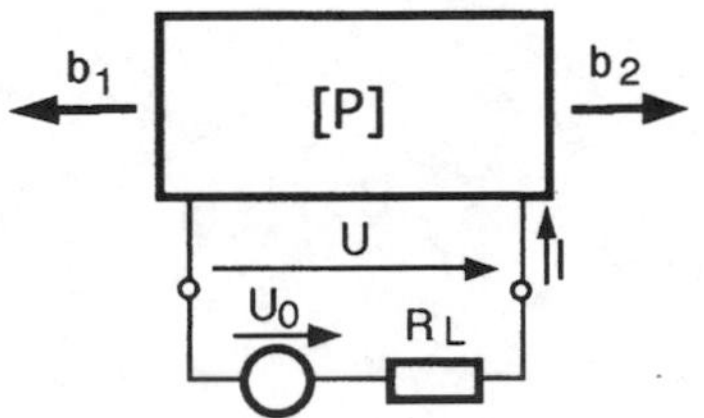

Fig. 2.30. Emitter transducer with supply U_0 and internal resistance $R_\mathrm{L} = 1/G_\mathrm{L}$

Amplitudes of emitted Rayleigh waves are

$$b_1 = s_{13}a_3 = P_{13}U \ , \quad b_2 = s_{23}a_3 = P_{23}U \ .$$

Since

$$a_3 = \frac{\sqrt{R_\mathrm{L}}}{2}(G_\mathrm{L}U + I) = \frac{1}{2\sqrt{G_\mathrm{L}}}(G_\mathrm{L} + P_{33})U \ ,$$

transfer coefficients between electrical input and acoustic outputs (frequency responses) are given by

$$s_{13} = 2\frac{\sqrt{G_\mathrm{L}}P_{13}}{G_\mathrm{L} + P_{33}} \ , \quad s_{23} = 2\frac{\sqrt{G_\mathrm{L}}P_{23}}{G_\mathrm{L} + P_{33}} \ . \tag{2.119}$$

Note that the matrix description also applies on a *local scale* whenever we need to split the transducer into several elements (Sect. 2.6).

Application to Piezoelectric Permittivity Model. The mean power carried by a Rayleigh wave beam of width w is given in terms of surface electric potential by (2.77):

$$P = \frac{1}{2}\alpha^2|\Phi|^2 \ , \quad \alpha = \left(\frac{\omega w}{2\Gamma_\mathrm{R}}\right)^{1/2} \ . \tag{2.120}$$

α is a potential/power conversion factor with units $[\mathrm{W}^{1/2}/\mathrm{V}]$. Using the notation of Sect. 2.3.2, amplitudes of incoming and outgoing waves are

$$a_1 = \alpha\Phi_\mathrm{I}^+ \ , \quad a_2 = \alpha\Phi_\mathrm{I}^- \ , \quad b_1 = \alpha\Phi_\mathrm{S}^- \ , \quad b_2 = \alpha\Phi_\mathrm{S}^+ \ . \tag{2.121}$$

Results of the piezoelectric permittivity model lead to expressions for P-matrix entries. There is no reflection when the transducer is short-circuited ($U = 0$), so P_{11} and P_{22} are zero. Total transmission occurs and allowing for the phase delay $k_\mathrm{R}L$, $P_{12} = P_{21} = \mathrm{e}^{-\mathrm{i}k_\mathrm{R}L}$. The P-matrix takes the form

$$\begin{pmatrix} \alpha\Phi_\mathrm{S}^- \\ \alpha\Phi_\mathrm{S}^+ \\ I \end{pmatrix} = \begin{pmatrix} 0 & \mathrm{e}^{-\mathrm{i}k_\mathrm{R}L} & P_{13} \\ \mathrm{e}^{-\mathrm{i}k_\mathrm{R}L} & 0 & P_{23} \\ P_{31} & P_{32} & P_{33} \end{pmatrix} \begin{pmatrix} \alpha\Phi_\mathrm{I}^+ \\ \alpha\Phi_\mathrm{I}^- \\ U \end{pmatrix} \ , \tag{2.122}$$

where $P_{33} = Y_\mathrm{T}$ is the transducer admittance.

From (2.70), when there is no incident wave $\Phi_\mathrm{I} = 0$, application of voltage U across the transducer terminals produces a Rayleigh wave propagating in the positive and negative directions, with potential

$$\Phi_\mathrm{S}^\pm = \Phi(\pm L/2, \omega) \approx \mathrm{i}\Gamma_\mathrm{R}\overline{\sigma}_1(\pm k_\mathrm{R})U(\omega)\mathrm{e}^{-\mathrm{i}k_\mathrm{R}L/2} \ .$$

Multiplying by α,

$$P_{13} = \mathrm{i}\left(\frac{\omega w}{2}\Gamma_\mathrm{R}\right)^{1/2}\overline{\sigma}_1(-k_\mathrm{R})\mathrm{e}^{-\mathrm{i}k_\mathrm{R}L/2} \ ,$$

$$P_{23} = \mathrm{i}\left(\frac{\omega w}{2}\Gamma_\mathrm{R}\right)^{1/2}\overline{\sigma}_1(k_\mathrm{R})\mathrm{e}^{-\mathrm{i}k_\mathrm{R}L/2} \ .$$

Conversely, when $U = 0$, the current induced by the incident wave of potential Φ_{I}^{+} is equal to the short-circuit current, the intensity of which is given in (2.89). Hence,

$$P_{31} = \frac{I_{\mathrm{sc}}}{\alpha \Phi_{\mathrm{I}}^{+}} = -\mathrm{i}(2\omega w \Gamma_{\mathrm{R}})^{1/2} \overline{\sigma}_1(-k_{\mathrm{R}}) \mathrm{e}^{-\mathrm{i}k_{\mathrm{R}}L/2} \,. \tag{2.123}$$

We can check that $P_{31} = -2P_{13}$ and $P_{32} = -2P_{23}$.

Values of scattering matrix entries depend on the load admittance G_{L}. From (2.116), the reflection coefficient at acoustic inputs,

$$s_{11} = s_{22} = \frac{2P_{13}^2}{G_{\mathrm{L}} + P_{33}} = -\frac{\omega w \Gamma_{\mathrm{R}} [\overline{\sigma}_1(-k_{\mathrm{R}})]^2}{G_{\mathrm{L}} + Y_{\mathrm{T}}} \mathrm{e}^{-\mathrm{i}k_{\mathrm{R}}L} \,,$$

agrees with (2.92), given expression (2.80) for $G_{\mathrm{a}}(\omega)$. Likewise, applying (2.117), the transmission coefficient,

$$s_{21} = s_{12} = P_{12} + \frac{2P_{13}P_{23}}{G_{\mathrm{L}} + P_{33}} = \left[1 - \frac{\omega w \Gamma_{\mathrm{R}} |\overline{\sigma}_1(-k_{\mathrm{R}})|^2}{G_{\mathrm{L}} + Y_{\mathrm{T}}}\right] \mathrm{e}^{-\mathrm{i}k_{\mathrm{R}}L} \,,$$

agrees with (2.93).

The above analysis based on the surface piezoelectric permittivity assumes that metallisation does not affect phase speed of waves or cause reflection. Present day techniques make use of internal reflections in the transducer. These reflections are accounted for in the coupled mode theory.

2.5 Coupled Mode Method

Before describing the method, let us review two facts. When a Rayleigh wave enters a simple, normally loaded receiving transducer, it undergoes a periodic speed variation and gives rise to the following two effects:

- It induces a voltage across each pair of fingers, which generates secondary waves. This reemission, giving rise to a triple-transit signal in the transmission line, was discussed in Sects. 2.1.1, 2.3.2 and 2.4.1. The amplitude of this spurious echo can be reduced if we accept to increase insertion losses by mismatching both emitter to supply and receiver to load. Otherwise we must build unidirectional transducers in which voltages across the various elements are out of phase.
- The wave is partially reflected at the two edges of each finger. These reflections are due to discontinuities in surface mechanical impedance, discussed in Sect. 2.1.1. Their effect is to lengthen the impulse response (infinitely, if there are no losses). One way of reducing them is to split each finger.

Of course, these two effects also occur for any wave produced by a finger pair and propagating within the emitter transducer.

As often happens, having expended considerable effort on the problem of eliminating these reflections, and despite some successes, scientists had the idea of actually using them (Sect. 2.5.3). The idea was all the more natural in that multistrip couplers (bulk wave suppressors, Sect. 4.1) had already been inserted in travelling wave filter lines for several years, and also that surface wave resonators were beginning to appear, containing two passive reflecting arrays. The first strategy was to compensate piezoelectric reemission in one direction by using internal reflections. The coupled mode analysis was worked out with this in mind. It will be applied here to the case of a short-circuited interdigital transducer, i.e., a simple reflective array, and then to an active transducer. The notions of reflection and generation centres are defined and used to make the transducer unidirectional.

The method is based on local analysis of the way two coupled waves evolve when propagating in opposite directions. We begin with the differential equation governing the amplitude of each emitted wave [2.29]. Recall first that when there is no transducer, and hence no array, the two waves propagate independently in opposite directions on the substrate (with zero attenuation, by hypothesis) and obey uncoupled equations:

$$\frac{\mathrm{d}a}{\mathrm{d}x} = -\mathrm{i}k_\mathrm{R}a \,, \quad \frac{\mathrm{d}b}{\mathrm{d}x} = \mathrm{i}k_\mathrm{R}b \,. \tag{2.124}$$

Harmonic solutions have form $A\mathrm{e}^{-ikx}$ and $B\mathrm{e}^{ikx}$, leaving out the factor $\mathrm{e}^{\mathrm{i}\omega t}$.

The transducer strip array, assumed passive (shorted electrodes), modifies the speed by ΔV, and hence changes the wave number k_R by Δk to the new value k'_R. There is an energy exchange between the two waves. By hypothesis, this exchange is small enough per array period to allow us to describe amplitudes by differential equations, rather than difference equations. Continuous coupling of the waves is expressed by two coefficients κ_{ab} and κ_{ba} (see Appendix B):

$$\frac{\mathrm{d}a}{\mathrm{d}x} = -\mathrm{i}k'_\mathrm{R}a - \mathrm{i}\kappa_{ab}b \,, \tag{2.125a}$$

$$\frac{\mathrm{d}b}{\mathrm{d}x} = \mathrm{i}k'_\mathrm{R}b - \mathrm{i}\kappa_{ba}a \,. \tag{2.125b}$$

By hypothesis, quantities a and b are such that the square of their modulus yields the power carried by the waves. Given orthogonality of the two modes, and the fact that they propagate in opposite directions, the power flow is proportional to the difference $|a|^2 - |b|^2$. If energy is exchanged between the two waves without loss, conservation of power implies

$$\frac{\mathrm{d}}{\mathrm{d}x}(|a|^2 - |b|^2) = 0 \quad \Rightarrow \quad \frac{\mathrm{d}}{\mathrm{d}x}(aa^*) = \frac{\mathrm{d}}{\mathrm{d}x}(bb^*) \,, \tag{2.126}$$

which in turn implies (Problem 2.7)

$$\kappa_{ba} = -\kappa^*_{ab} \,.$$

Coefficient κ_{ab} is a function of x in the electrode array of constant period p. It can be decomposed in Fourier series, putting $G = 2\pi/p$:

$$\kappa_{ab} = \sum_{n=-\infty}^{\infty} \kappa_{ab}(n)\mathrm{e}^{-\mathrm{i}nGx} . \tag{2.127}$$

It is useful to introduce new variables $R(x)$ and $S(x)$ by

$$a(x) = R(x)\mathrm{e}^{-\mathrm{i}Gx/2} , \tag{2.128a}$$

$$b(x) = S(x)\mathrm{e}^{+\mathrm{i}Gx/2} , \tag{2.128b}$$

which describe slow amplitude variations as the waves propagate. Equations (2.125a) and (2.125b) then become

$$\frac{\mathrm{d}R}{\mathrm{d}x} = -\mathrm{i}\left(k'_{\mathrm{R}} - \frac{G}{2}\right) R(x) - \mathrm{i}\sum_n \kappa_{ab}(n)S(x)\mathrm{e}^{-\mathrm{i}(n-1)Gx} ,$$

$$\frac{\mathrm{d}S}{\mathrm{d}x} = +\mathrm{i}\left(k'_{\mathrm{R}} - \frac{G}{2}\right) S(x) + \mathrm{i}\sum_n \kappa^*_{ab}(n)R(x)\mathrm{e}^{\mathrm{i}(n-1)Gx} .$$

Only the component $\kappa_{ab}(1) = \kappa$ gives an x-independent coupling coefficient. Defining a *detuning parameter* δ by

$$\delta = k'_{\mathrm{R}} - \frac{G}{2} , \tag{2.129}$$

the coupled equations become

$$\frac{\mathrm{d}R}{\mathrm{d}x} = -\mathrm{i}\delta R(x) - \mathrm{i}\kappa S(x) , \tag{2.130a}$$

$$\frac{\mathrm{d}S}{\mathrm{d}x} = \mathrm{i}\delta S(x) + \mathrm{i}\kappa^* R(x) . \tag{2.130b}$$

These are the same as (B.30) in Appendix B.

The parameter δ is given in terms of the difference between angular frequency ω and synchronism angular frequency ω_0 of the array, defined by

$$\frac{\omega_0}{V_{\mathrm{R}}} = \frac{G}{2} = \frac{\pi}{p} \quad \Rightarrow \quad f_0 = \frac{V_{\mathrm{R}}}{2p} . \tag{2.131}$$

Indeed, from (2.129),

$$\delta = \frac{\omega}{V_{\mathrm{R}} + \Delta V} - \frac{\omega_0}{V_{\mathrm{R}}} \approx \frac{\omega - \omega_0}{V_{\mathrm{R}}} , \quad \text{if} \quad \Delta V \ll V_{\mathrm{R}} . \tag{2.132}$$

2.5.1 Passive Array. Reflection and Transmission Coefficients

The above analysis encompasses reflection by any passive array, whether it be composed of metallic strips (shorted receiver transducer), or of grooves (see Fig. 2.31). Indeed, any periodic modification of surface impedance, either by affixing or removing material, will generate reflections (Sect. 2.1.1).

Such passive arrays are good reflectors if the number of elements they contain is large enough. Strips in metallic gratings are shorted in order to prevent piezoelectric reemission. These reflectors act as mirrors in surface wave resonators (see Sect. 4.3.4).

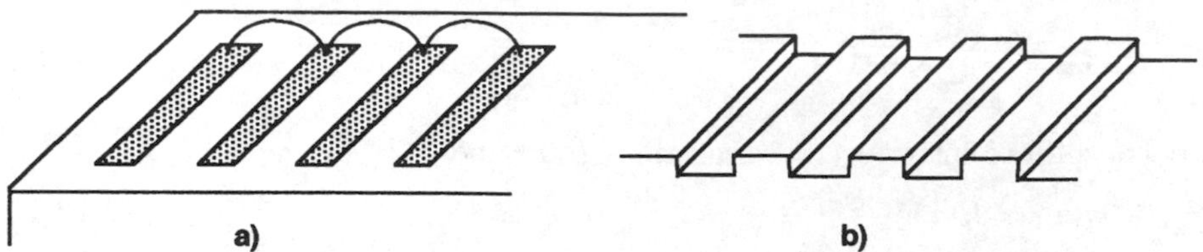

Fig. 2.31. Reflective array for surface waves. (**a**) Shorted metallic strips. (**b**) Grooves

The two coupled first order differential equations in (2.130a) and (2.130b) can be put into matrix form:

$$\frac{\mathrm{d}\boldsymbol{X}}{\mathrm{d}x} = \boldsymbol{F}\boldsymbol{X}(x)\,, \quad \boldsymbol{X} = \begin{pmatrix} R(x) \\ S(x) \end{pmatrix}\,, \quad \boldsymbol{F} = \begin{pmatrix} -\mathrm{i}\delta & -\mathrm{i}\kappa \\ \mathrm{i}\kappa^* & \mathrm{i}\delta \end{pmatrix}\,. \tag{2.133}$$

At a given frequency, the vector $\boldsymbol{X}$ describes the state of surface waves propagating in opposite directions, at any point of the surface.

The general solution is given in terms of the transition matrix $\boldsymbol{M}(x_1, x_0)$ of the array [equation (B.22), Appendix B]. This relates wave amplitude vectors at x_0 and x_1,

$$\boldsymbol{X}(x_1) = \boldsymbol{M}(x_1, x_0)\boldsymbol{X}(x_0)\,. \tag{2.134}$$

Uniform Array. When array parameters δ and κ depend only on frequency, the matrix $\boldsymbol{F}$ is constant and we have a *stationary system* of equations, i.e., x-invariant. The transition matrix $\boldsymbol{M}(x_1, x_0)$ is only a function of the difference $x = x_1 - x_0$. It is, in fact, the matrix exponential $\exp(\boldsymbol{F}x)$ (Appendix F). The square of $\boldsymbol{F}$ is

$$\boldsymbol{F}^2 = \begin{pmatrix} |\kappa|^2 - \delta^2 & 0 \\ 0 & |\kappa|^2 - \delta^2 \end{pmatrix} = -\gamma^2\boldsymbol{I}\,, \quad \gamma^2 = \delta^2 - |\kappa|^2\,. \tag{2.135}$$

This result facilitates calculation of higher powers, as in Appendix B.2, giving

$$\boldsymbol{M}(x) = \begin{pmatrix} \cos\gamma x - \mathrm{i}\dfrac{\delta}{\gamma}\sin\gamma x & -\mathrm{i}\dfrac{\kappa}{\gamma}\sin\gamma x \\ \mathrm{i}\dfrac{\kappa^*}{\gamma}\sin\gamma x & \cos\gamma x + \mathrm{i}\dfrac{\delta}{\gamma}\sin\gamma x \end{pmatrix}\,. \tag{2.136}$$

Hence the transition matrix is unitary, implying $||\boldsymbol{M}(x)|| = 1$, because of power conservation. The parameter γ can be real or imaginary, depending on whether $|\delta|$ is greater or less than $|\kappa|$.

In an array of length L (Fig. 2.32), amplitudes $R(x)$ and $S(x)$ of the two coupled waves $a(x)$ and $b(x)$, propagating in opposite directions, are given at the ends $x_1 = L$ and $x_0 = 0$ by (2.134), with $\boldsymbol{M}(L, 0) = \boldsymbol{M}(L)$,

$$R(L) = M_{11}(L)R(0) + M_{12}(L)S(0)\,, \tag{2.137a}$$

$$S(L) = M_{21}(L)R(0) + M_{22}(L)S(0)\,. \tag{2.137b}$$

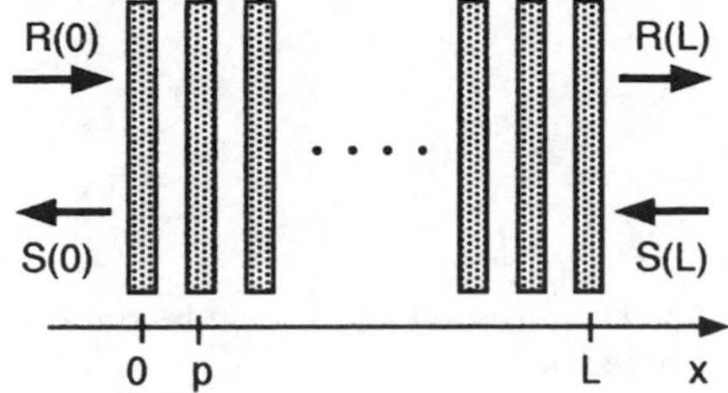

Fig. 2.32. Array of strips, grooves or other features which periodically modify the surface impedance of a (non-piezoelectric) substrate

In order to determine the reflection coefficient $r_0 = S(0)/R(0)$ at input $x = 0$, assume that an incident wave of amplitude $R(0)$ arrives at $x = 0$ and that $S(L) = 0$ at the other end of the array. Then (2.137b) yields

$$r_0 = -\frac{M_{21}(L)}{M_{22}(L)} = \frac{-\mathrm{i}\kappa^* \sin \gamma L}{\gamma \cos \gamma L + \mathrm{i}\delta \sin \gamma L} . \tag{2.138}$$

At $x = L$, the reflection coefficient is defined by $r_L = R(L)/S(L)$ when $R(0) = 0$:

$$r_L = \frac{M_{12}(L)}{M_{22}(L)} = \frac{-\mathrm{i}\kappa \sin \gamma L}{\gamma \cos \gamma L + \mathrm{i}\delta \sin \gamma L} . \tag{2.139}$$

Since the transition matrix is unitary, $||\boldsymbol{M}(L,\, 0)|| = 1$, the transmission coefficient t is given by

$$t = \frac{R(L)}{R(0)} \mathrm{e}^{-\mathrm{i}GL/2} = \frac{\mathrm{e}^{-\mathrm{i}GL/2}}{M_{22}(L)} = \frac{\gamma \mathrm{e}^{-\mathrm{i}GL/2}}{\gamma \cos \gamma L + \mathrm{i}\delta \sin \gamma L} . \tag{2.140}$$

The array in Fig. 2.32 can be represented by a scattering matrix relating output quantities $S(0)$ and $R(L)$ to input quantities $R(0)$ and $S(L)$:

$$\begin{pmatrix} S(0) \\ R(L) \end{pmatrix} = \begin{pmatrix} s_{11} = r_0 & s_{12} = t \\ s_{21} = t & s_{22} = r_L \end{pmatrix} \begin{pmatrix} R(0) \\ S(L) \end{pmatrix} .$$

Transmission and reflection coefficients can also be deduced from a transmission line model, introducing an *elementary transmission matrix.* This relates incident and reflected wave amplitudes a_{n+1} and b_{n+1} at the $(n+1)$th reflector to a_n and b_n at the nth reflector (Fig. 2.33),

$$\begin{pmatrix} a_{n+1} \\ b_{n+1} \end{pmatrix} = (T_1) \begin{pmatrix} a_n \\ b_n \end{pmatrix} . \tag{2.141}$$

The transmission matrix for an array composed of $N = L/p$ reflectors is the product of transmission matrices for each reflector (Problem 2.2). There is a relation between the reflection coefficient r_e for one (discrete) element in the transmission line model and the (continuous) parameter in the coupled mode model:

$$r_\mathrm{e} = \kappa p \quad \Rightarrow \quad |\kappa| L = N |r_\mathrm{e}| . \tag{2.142}$$

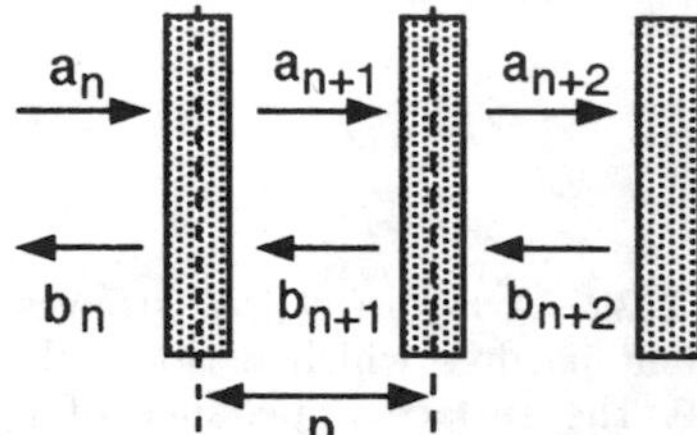

Fig. 2.33. Definition of the transmission matrix of a reflector

When $|\delta| < |\kappa|$, $\gamma = \mathrm{i}\chi$ is purely imaginary and the reflection coefficient of the array is given by

$$r_0 = -\mathrm{i}\frac{\kappa^* \sinh \chi L}{\chi \cosh \chi L + \mathrm{i}\delta \sinh \chi L}, \quad \chi^2 = |\kappa|^2 - \delta^2 . \tag{2.143}$$

The modulus of r goes through a maximum at the synchronism frequency f_0 of the array, such that

$$\delta = 0 \quad \Rightarrow \quad \chi = |\kappa| \quad \text{and} \quad |r(\delta = 0)| = \tanh |\kappa| L . \tag{2.144}$$

High reflectivity is achieved by satisfying $|\kappa|L > 1$.

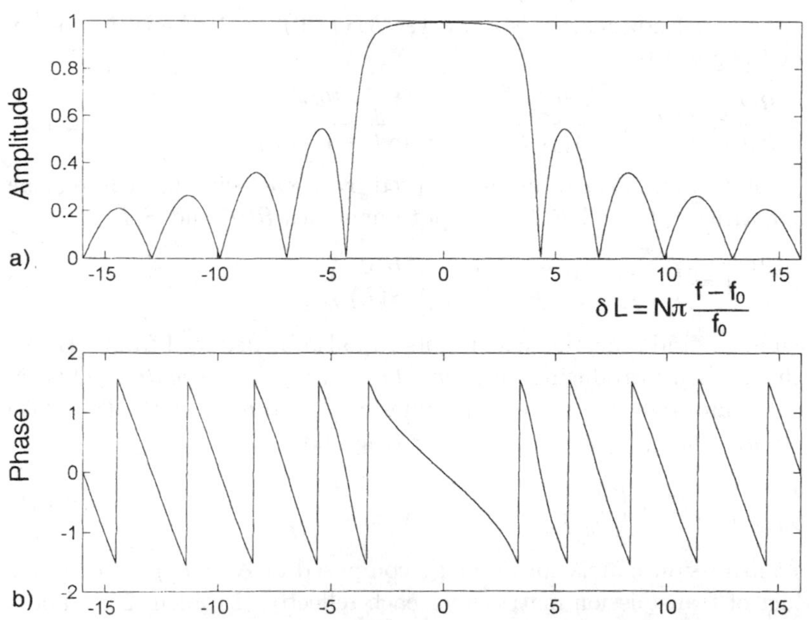

Fig. 2.34. Reflection coefficient of a periodic array, calculated using the coupled mode theory ($|\kappa|L = 3$). (**a**) Amplitude and (**b**) phase as functions of the parameter $L\delta \propto f - f_0$. Zero phase has been chosen at the centre of a finger so that the reflection coefficient κ is imaginary: $\kappa^* = \mathrm{i}|\kappa|$ [see equation (2.10)]

Figure 2.34 shows how the reflection coefficient varies with $L\delta$, which represents the deviation from frequency f_0 at which the Bragg condition is satisfied:

$$L\delta = \frac{2\pi}{V_R}(f - f_0)L\ , \quad f_0 = \frac{V_R}{2p}\ . \tag{2.145}$$

The curve has a central peak and sidelobes, separated by zeros at $\gamma L = n\pi$, i.e., where $L\delta = [(|\kappa|L)^2 + n^2\pi^2]^{1/2}$. The bandwidth, defined by the separation of the two zeros on either side of the central peak ($n = \pm 1$), is given by

$$\frac{\Delta f}{f_0} = \frac{p}{\pi}\Delta\delta = \frac{2p}{\pi L}\sqrt{(|\kappa|L)^2 + \pi^2}\ . \tag{2.146}$$

Depending on the value of $|\kappa|L$, the bandwidth is inversely proportional to the number of reflectors N, or depends only on the reflection coefficient r_e:

$$\begin{aligned} \frac{\Delta f}{f_0} &= \frac{2}{N}\ , \qquad |\kappa|L \ll \pi\ , \\ \frac{\Delta f}{f_0} &= 2\frac{|\kappa|p}{\pi} = 2\frac{|r_e|}{\pi}\ , \qquad |\kappa|L \gg \pi\ . \end{aligned} \tag{2.147}$$

Secondary peaks correspond approximately to

$$\gamma L = (n + 1/2)\,\pi \quad \Rightarrow \quad |r| = \frac{|\kappa|L}{\sqrt{(n + 1/2)^2\,\pi^2 + |\kappa|^2L^2}}\ .$$

Their relative amplitude increases with $|\kappa|L$.

Figure 2.35 shows that, within this frequency band, the amplitude of the mode R decreases inside the array. For example, at the centre frequency (Problem 2.9),

$$\begin{aligned} R(x) &= \frac{\cosh|\kappa|(L - x)}{\cosh|\kappa|L}R(0)\ , \\ S(x) &= -\mathrm{i}\frac{\kappa^*}{|\kappa|}\frac{\sinh|\kappa|(L - x)}{\cosh|\kappa|L}R(0)\ . \end{aligned} \tag{2.148}$$

This decrease is due to energy transfer, by reflection, towards the mode S, whose amplitude increases, given that it propagates in the opposite direction. Evanescence of the R and S modes is related to the fact that, within the stop band of the array, their wave numbers are complex (see Appendix B). A long enough array acts like a highly reflective mirror near the Bragg frequency.

Near the centre frequency $\delta^2 \ll |\kappa|^2$, the phase of the reflection coefficient varies linearly with frequency (Fig. 2.34b). From (2.143),

$$\phi = -\frac{\pi}{2} - \arg\kappa - \arctan\left(\frac{\delta}{\chi}\tanh\chi L\right) \approx -\frac{\pi}{2} - \arg\kappa - \frac{\delta}{|\kappa|}\tanh|\kappa|L\ .$$

Reflection is accompanied by a group delay, defined by

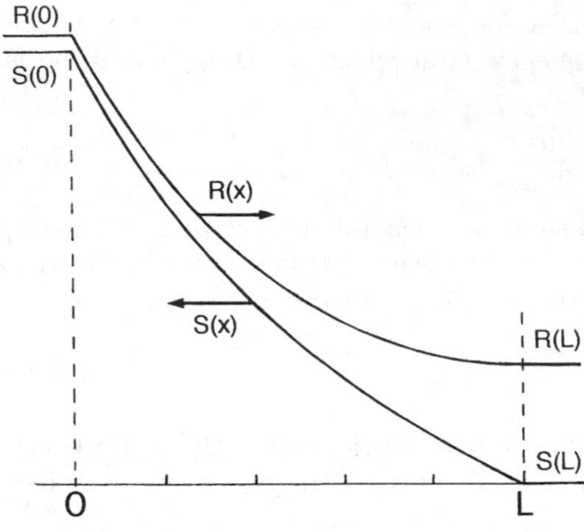

Fig. 2.35. Amplitude variation of incident and reflected modes inside a periodic array, when $\delta = 0$ ($|\kappa|L = 2$)

$$\tau_{\mathrm{g}} = -\frac{\mathrm{d}\phi}{\mathrm{d}\omega} = -\frac{1}{V_{\mathrm{R}}}\frac{\mathrm{d}\phi}{\mathrm{d}\delta} = \frac{1}{V_{\mathrm{R}}}\frac{\tanh|\kappa|L}{|\kappa|} \, .$$

This is constant. The wave is delayed as if it were reflecting from a plane located at an *effective distance* L_{e} from the edge of the array:

$$\tau_{\mathrm{g}} = \frac{2L_{\mathrm{e}}}{V_{\mathrm{R}}} \quad \Rightarrow \quad L_{\mathrm{e}} = \frac{\tanh|\kappa|L}{2|\kappa|} \approx \frac{1}{2|\kappa|} \, . \tag{2.149}$$

At the centre frequency, the penetration depth of the wave is $p/2|r_{\mathrm{e}}|$.

2.5.2 Transducer

If an x-independent voltage U is applied across the electrodes, the amplitude of each wave grows as it passes between each pair of fingers. This growth is relatively small and can be simply represented on the right-hand side of the above equations by a term proportional to the voltage U:

$$\frac{\mathrm{d}R}{\mathrm{d}x} = -\mathrm{i}\delta R(x) - \mathrm{i}\kappa S(x) + \mathrm{i}\zeta U \, , \tag{2.150a}$$

$$\frac{\mathrm{d}S}{\mathrm{d}x} = \mathrm{i}\delta S(x) + \mathrm{i}\kappa^* R(x) + \mathrm{i}\xi U \, . \tag{2.150b}$$

The generation parameters ζ and ξ of the two modes R and S are not independent. By hypothesis, the system is not dissipative, and so it evolves reversibly with respect to time. In the harmonic case, swapping t for $-t$ corresponds to conjugating complex variables. In this operation, the directions of propagation of the waves are reversed: $R^* \to S$, $S^* \to R$. These solutions must satisfy (2.150a) and (2.150b) with U^* as the excitation potential. This implies $\xi = -\zeta^*$.

A third equation must be added, relating the variation ΔI (assumed small) in the current intensity from one pair of fingers to the next to the voltage

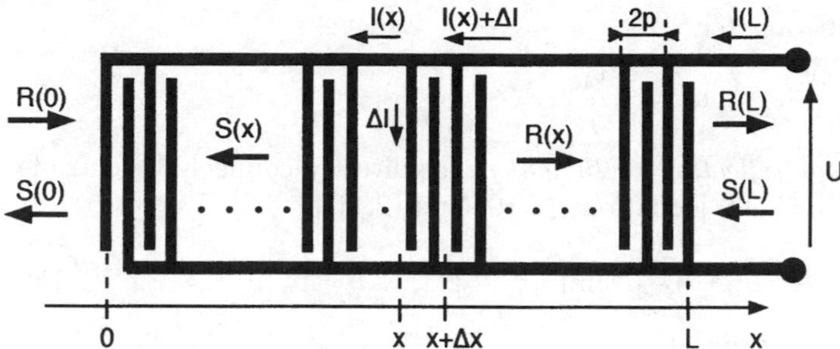

Fig. 2.36. Coupled mode representation of a transducer

U and the amplitude of the two waves propagating under the transducer (Fig. 2.36):

$$\frac{\mathrm{d}I}{\mathrm{d}x} = \mathrm{i}\alpha R(x) + \mathrm{i}\beta S(x) + \mathrm{i}C\omega U \, . \tag{2.151}$$

$C\Delta x$ is the capacitance of a pair of fingers ($\Delta x = p$). The total capacitance of the transducer is then $C_\mathrm{T} = \int_0^L C(x)\,\mathrm{d}x$.

The new coefficients α and β are not independent of the generation parameter. Let us use the complex Poynting theorem (Sect. 5.2.1, Vol. I) to express the fact that the supplied electrical power is completely transmitted to the two modes. In the steady harmonic case (ω real, $\omega'' = 0$), (5.24) of Vol. I becomes

$$\frac{\mathrm{d}\langle P\rangle}{\mathrm{d}x} = \mathrm{Re}\left(\frac{1}{2}\Phi j^*\right) ,$$

where

$$\langle P\rangle = \mathrm{Re}\left(\overline{P}\right) = \frac{1}{2}(|R|^2 - |S|^2) \quad \text{and} \quad j = \frac{\mathrm{d}I}{\mathrm{d}x}$$

are the mean transported power and the incoming current density per unit length in the x direction, respectively. By power conservation,

$$\frac{\mathrm{d}}{\mathrm{d}x}(RR^* - SS^*) = \frac{1}{2}\left(U\frac{\mathrm{d}I^*}{\mathrm{d}x} + U^*\frac{\mathrm{d}I}{\mathrm{d}x}\right) , \tag{2.152}$$

which implies (Problem 2.8)

$$\alpha = -2\zeta^* , \qquad \beta = -2\zeta \, . \tag{2.153}$$

Four parameters therefore characterise the transducer:

- The *capacitance* of one pair of fingers, given by (A.12), Appendix A:

$$2pC = C_1 w , \qquad C_1 = (\varepsilon_\mathrm{p} + \varepsilon_0)\frac{K(\sin\eta\pi/2)}{K(\cos\eta\pi/2)} \, . \tag{2.154}$$

- The *detuning parameter*,

$$\delta = \frac{\omega - \omega_0}{V_R}, \qquad \omega_0 = \pi \frac{V_R}{p}. \tag{2.155}$$

- The *single-finger reflection coefficient* r_e for directly connected electrodes and zero supply $U = 0$ [equations (2.10) and (2.11)],

$$\mathrm{i}\kappa p = r_e, \qquad r_e = \mathrm{i}\frac{\Delta Z}{Z} \sin\left(\eta\pi \frac{f}{f_0}\right). \tag{2.156}$$

- The *generation factor* ζ deduced from the piezoelectric permittivity model. The wave amplitude is $R = \alpha\Phi^+$, where constant $\alpha = (\omega w / 2\Gamma_R)^{1/2}$ is such that $|R|^2/2$ is the transported power (2.120). The increase in amplitude ΔR produced by an electrode at potential U, whilst the others are earthed (Fig. 2.37), is given by (2.74):

$$\Delta R = \alpha \Phi_{1e}^+(f) U = \mathrm{i}\left(\frac{\omega w \Gamma_R}{2}\right)^{1/2} \overline{\sigma}_{1e}(k) U.$$

The distance to the next active electrode is $2p$, so $\Delta R = 2\mathrm{i}p\zeta U$ and the generation factor per finger pair is given in terms of the single-element charge distribution $\overline{\sigma}_{1e}(k)$ by

$$2\mathrm{i}p\zeta = \mathrm{i}\left(\frac{\omega w \Gamma_R}{2}\right)^{1/2} \overline{\sigma}_{1e}(k), \quad \Gamma_R = \frac{K_R^2/2}{\varepsilon_p + \varepsilon_0}. \tag{2.157}$$

Reflection and generation coefficients are purely imaginary if the abscissa origin is chosen in the middle of a finger. This is the case, for example, if it is located in the middle of the first finger of a uniform transducer. Parameters κ and ζ in the coupled mode model are then real. Likewise for δ and C.

Finally, variations in the supplied current intensity are given by

$$\frac{\mathrm{d}I}{\mathrm{d}x} = -2\mathrm{i}\zeta^* R(x) - 2\mathrm{i}\zeta S(x) + \mathrm{i}C\omega U. \tag{2.158}$$

Amplitudes $R(x)$ and $S(x)$ of the two separating waves are given by the two coupled first order differential equations (2.150a,b). In matrix form, these are (using $\xi = -\zeta^*$)

$$\frac{\mathrm{d}\boldsymbol{X}}{\mathrm{d}x} = \boldsymbol{F}\boldsymbol{X}(x) + \boldsymbol{G}U, \quad \boldsymbol{X} = \begin{pmatrix} R(x) \\ S(x) \end{pmatrix}, \quad \boldsymbol{G} = \begin{pmatrix} \mathrm{i}\zeta \\ -\mathrm{i}\zeta^* \end{pmatrix}. \tag{2.159}$$

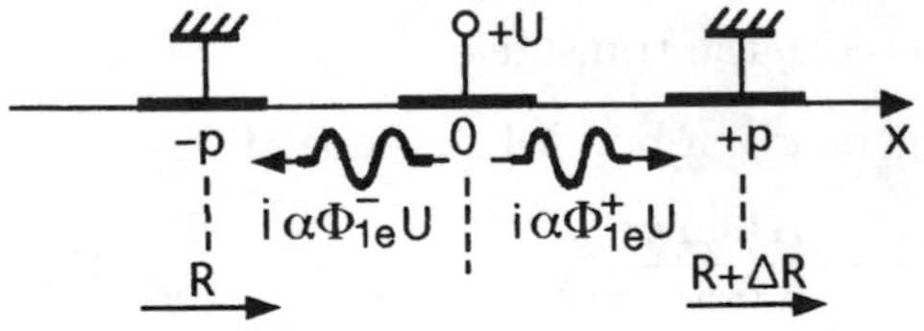

Fig. 2.37. Generation by a pair of fingers

Such a homogeneous presentation of the changes in amplitudes R, S and I allows us to attribute physical meaning to the various terms. It is analogous to the equation of state (F.1) of a linear system.

The general solution (see Appendix F) is given in terms of the transition matrix $\boldsymbol{M}(x_1, x_0)$, whose components M_{11}, M_{12}, M_{21} and M_{22} relate the wave amplitude vector at x_0 to the same at x_1:

$$\boldsymbol{X}(x_1) = \left[\int_{x_0}^{x_1} \boldsymbol{M}(x_1, x)\boldsymbol{G}(x)\,\mathrm{d}x\right] U + \boldsymbol{M}(x_1, x_0)\boldsymbol{X}(x_0)\,. \tag{2.160}$$

2.5.2.1 Frequency Response. Under an applied voltage $U(\omega)$ and when there are no incident waves at either end of the transducer, $R(0, \omega) = S(L, \omega) = 0$, frequency responses $H_\mathrm{R}(\omega)$, $H_\mathrm{S}(\omega)$ are defined by

$$H_\mathrm{R}(\omega) = \left.\frac{R(L, \omega)}{U(\omega)}\right|_{R(0)=0}, \qquad H_\mathrm{S}(\omega) = \left.\frac{S(0, \omega)}{U(\omega)}\right|_{S(L)=0}. \tag{2.161}$$

With $x_0 = L$ and $x_1 = 0$, and given the form of vector $\boldsymbol{G}$, the first line of (2.160) leads to

$$\begin{aligned} R(0, \omega) = \mathrm{i}U(\omega) \int_L^0 [M_{11}(0, x)\zeta(x) \; &- \; M_{12}(0,x)\zeta^*(x)]\,\mathrm{d}x \\ &+ M_{11}(0, L)R(L, \omega) = 0\,. \end{aligned}$$

Swapping integration limits yields

$$H_\mathrm{R}(\omega) = \frac{\mathrm{i}}{M_{11}(0, L)} \int_0^L [M_{11}(0, x)\zeta(x) - M_{12}(0,x)\zeta^*(x)]\,\mathrm{d}x\,. \tag{2.162}$$

With $x_0 = 0$ and $x_1 = L$, the second line of (2.160) gives, for the wave emitted in the other direction,

$$\begin{aligned} S(L, \omega) = \mathrm{i}U(\omega) \int_0^L [M_{21}(L, x)\zeta(x) \; &- \; M_{22}(L,x)\zeta^*(x)]\,\mathrm{d}x \\ &+ M_{22}(L, 0)S(0, \omega) = 0\,. \end{aligned}$$

Hence,

$$H_\mathrm{S}(\omega) = \frac{\mathrm{i}}{M_{22}(L, 0)} \int_0^L [M_{22}(L, x)\zeta^*(x) - M_{21}(L,x)\zeta(x)]\,\mathrm{d}x\,. \tag{2.163}$$

When parameters δ and κ only depend on frequency, the matrix $\boldsymbol{F}$ in (2.133) is constant. This is still true even if the generation factor ζ of the transducer depends on x, as happens when the finger overlap width varies. The transition matrix $\boldsymbol{M}(x_1, x_0)$, which is now only a function of the difference $x = x_1 - x_0$, is given by (2.136). Its components are:

$$\begin{aligned} M_{11}(0, x) &= M_{11}(-x) = \frac{\gamma+\delta}{2\gamma}\mathrm{e}^{\mathrm{i}\gamma x} + \frac{\gamma-\delta}{2\gamma}\mathrm{e}^{-\mathrm{i}\gamma x}\,, \\ M_{12}(0, x) &= \frac{\kappa}{2\gamma}(\mathrm{e}^{\mathrm{i}\gamma x} - \mathrm{e}^{-\mathrm{i}\gamma x})\,, \end{aligned}$$

and we also have

$$M_{11}(0,\, L) = \cos\gamma L + \mathrm{i}\frac{\delta}{\gamma}\sin\gamma L = M_{22}(L,\, 0)\,. \tag{2.164}$$

The *frequency response* of mode R can be written using Fourier transforms of the generation function and its complex conjugate,

$$I(\gamma) = \int_0^L \mathrm{e}^{\mathrm{i}\gamma x}\zeta(x)\,\mathrm{d}x\,, \quad J(\gamma) = \int_0^L \mathrm{e}^{\mathrm{i}\gamma x}\zeta^*(x)\,\mathrm{d}x = I^*(-\gamma^*)\,. \tag{2.165}$$

The result is

$$H_\mathrm{R}(\omega) = \frac{\mathrm{i}}{2}\frac{(\gamma+\delta)I(\gamma) + (\gamma-\delta)I(-\gamma) - \kappa[J(\gamma) - J(-\gamma)]}{\gamma\cos\gamma L + \mathrm{i}\delta\sin\gamma L}\,. \tag{2.166}$$

In the same way, relations

$$M_{21}(L-x) = \frac{\kappa^*}{2\gamma}\left[\mathrm{e}^{\mathrm{i}\gamma(L-x)} - \mathrm{e}^{\mathrm{i}\gamma(x-L)}\right]\,,$$

$$M_{22}(L-x) = \frac{\gamma+\delta}{2\gamma}\mathrm{e}^{\mathrm{i}\gamma(L-x)} + \frac{\gamma-\delta}{2\gamma}\mathrm{e}^{\mathrm{i}\gamma(x-L)}\,,$$

imply for mode S,

$$H_\mathrm{S}(\omega) = \frac{\mathrm{i}}{2}\frac{\mathrm{e}^{-\mathrm{i}\gamma L}[(\gamma-\delta)J(\gamma) + \kappa^* I(\gamma)] + \mathrm{e}^{\mathrm{i}\gamma L}[(\gamma+\delta)J(-\gamma) - \kappa^* I(-\gamma)]}{\gamma\cos\gamma L + \mathrm{i}\delta\sin\gamma L}\,. \tag{2.167}$$

The integral $I(\gamma)$ of the generation function $\zeta(x)$ can be calculated using the fast Fourier transform algorithm when γ is real. This is the case over the whole frequency range, except for a narrow band around the centre frequency such that $|\delta| < |\kappa|$, where γ is purely imaginary. In this frequency band, $|\omega - \omega_0| < |\kappa| V_\mathrm{R}$, known as the stop band, the modes no longer propagate, i.e., they are evanescent (Fig. 2.35). $I(\gamma)$ is then calculated numerically [2.30].

Uniform Transducer. Results are simpler when the generation parameter ζ is constant. Relation (2.162) gives directly

$$H_\mathrm{R}(\omega) = \frac{\mathrm{i}}{\gamma M_{22}(L)}\int_0^L [(\gamma\cos\gamma x + \mathrm{i}\delta\sin\gamma x)\zeta - \mathrm{i}\kappa\zeta^*\sin\gamma x]\,\mathrm{d}x\,, \tag{2.168}$$

and hence,

$$H_\mathrm{R}(\omega) = \frac{1}{\gamma M_{22}(L)}\int_0^L [(\kappa\zeta^* - \delta\zeta)\sin\gamma x + \mathrm{i}\gamma\zeta\cos\gamma x]\,\mathrm{d}x\,.$$

Another form is

$$H_\mathrm{R}(\omega) = \frac{(\kappa\zeta^* - \delta\zeta)(1-\cos\gamma L) + \mathrm{i}\gamma\zeta\sin\gamma L}{\gamma(\gamma\cos\gamma L + \mathrm{i}\delta\sin\gamma L)}\,. \tag{2.169}$$

For the wave emitted in the other direction, (2.163) implies

$$H_{\rm S}(\omega) = \frac{\rm i}{\gamma M_{22}(L)} \int_0^L \{[(\gamma \cos\gamma(L-x) \ + \ {\rm i}\delta \sin\gamma(L-x)]\,\zeta^* \\ - {\rm i}\kappa^* \zeta \sin\gamma(L-x)\}\ {\rm d}x\ .$$

Changing variable $L - x = u$, this gives

$$H_{\rm S}(\omega) = \frac{1}{\gamma M_{22}(L)} \int_0^L [(\kappa^*\zeta - \delta\zeta^*) \sin\gamma u + {\rm i}\gamma\zeta^* \cos\gamma u]\ {\rm d}u\ , \qquad (2.170)$$

or, equivalently,

$$H_{\rm S}(\omega) = \frac{(\kappa^*\zeta - \delta\zeta^*)(1-\cos\gamma L) + {\rm i}\gamma\zeta^* \sin\gamma L}{\gamma(\gamma\cos\gamma L + {\rm i}\delta \sin\gamma L)}\ . \qquad (2.171)$$

In general, attenuation of propagating waves can be characterised by means of a parameter α, replacing $-{\rm i}\delta$ by $-{\rm i}\delta - \alpha$ in (2.150a) and ${\rm i}\delta$ by ${\rm i}\delta + \alpha$ in (2.150b). The result is that, whatever the direction of propagation, δ is replaced by $\delta - {\rm i}\alpha$.

2.5.2.2 Generation and Reflection Centres. Directionality Factor. Expressions (2.169) and (2.171) show that amplitudes $R(L) = H_{\rm R}(\omega)U(\omega)$ and $S(0) = H_{\rm S}(\omega)U(\omega)$ of waves emitted in opposite directions are not equal. The transducer is more effective in one direction than the other. The directionality factor depends on the relative phase of coefficients ζ and κ, i.e., on the position of the *generation centre* with respect to the reflection centre. In order to make these ideas more precise, consider a path Δx corresponding to a finger pair. The amplitude of mode R grows by $\Delta R = G_1 \Delta x$ with $\Delta x = 2p$. For mode S, $\Delta S = G_2 \Delta x$ with $\Delta x = -2p$. G_1 and G_2 are the components of vector $\boldsymbol{G}$ in (2.159). The two emitted waves have the same amplitude $\Delta R = {\rm i}2p\zeta$ and $\Delta S = {\rm i}2p\zeta^*$. They are in phase if ζ is real. The latter is given by (2.157):

$$2p\zeta = \frac{K_{\rm R}}{2}\left(\frac{\omega w}{\varepsilon_{\rm p} + \varepsilon_0}\right)^{1/2} \overline{\sigma}_{\rm 1e}(k)\ ,\ \text{where}\ \overline{\sigma}_{\rm 1e}(k) = \int_{-\infty}^{\infty} \sigma_{\rm 1e}(x) {\rm e}^{{\rm i}kx} {\rm d}x\ ,$$

is the Fourier transform of the element factor $\sigma_{\rm 1e}(x)$. ζ is therefore real if the abscissa origin is such that $\sigma_{\rm 1e}(x)$ is an even function of x. This point (line), from which waves emitted in each direction are in phase, is called the generation centre. The Fourier transform represents a sum weighted by a phase factor ${\rm e}^{{\rm i}kx}$ which equals -1 if $x = \lambda/2$ and $+1$ if $x = \lambda$. The generation centre is therefore a weighted average position of the sources, with weight -1 or $+1$ depending on their distance $\lambda/2$ or λ. In the context of the discrete source model (Fig. 2.10), this rule provides an easy way of finding generation centres in structures with varying degrees of complexity. In the case of the uniform alternating transducer, the generation centre of a finger pair is located in the middle of the active finger (Fig. 2.38a).

A *reflection centre* is a point (line) from which waves are reflected with phase shift $\pm\pi/2$ [2.31]. From (2.156), the reflection centre of a finger is located at its centre. In simple transducers, reflection and generation centres

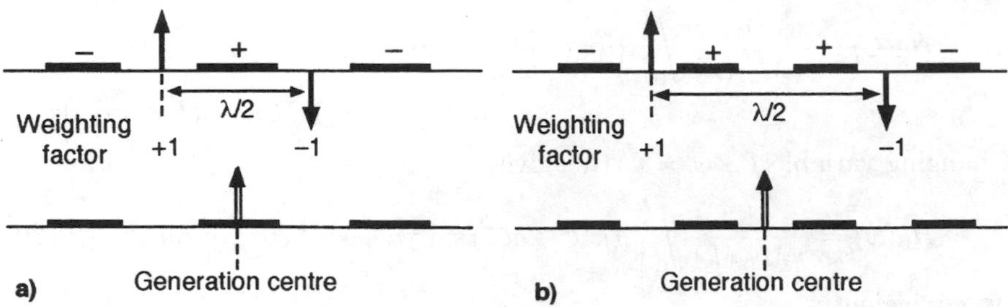

Fig. 2.38a,b. The generation centre is a weighted average position of the discrete sources, with weight -1 or $+1$ depending on their distance $\lambda/2$ or λ

coincide. Choosing the origin at the finger centre, parameters ζ and κ are real, so that $H_{\mathrm{R}}(\omega) = H_{\mathrm{S}}(\omega)$ and the transducer is bidirectional. If reflection and generation centres are separated, the transducer can be made unidirectional. Various procedures for doing so are described in Sect. 2.5.3.

Taking the phase of the generation factor as phase origin $\zeta = \zeta^*$, the frequency response (2.169) of a uniform transducer is

$$\begin{aligned} H_{\mathrm{R}}(\omega) &= 2\frac{\zeta}{\gamma}\sin(\gamma L/2)\frac{(\kappa-\delta)\sin(\gamma L/2) + \mathrm{i}\gamma\cos(\gamma L/2)}{\gamma\cos\gamma L + \mathrm{i}\delta\sin\gamma L} \\ &= \zeta L \frac{\sin(\gamma L/2)}{\gamma L/2}\frac{N_{\mathrm{R}}}{D} \ . \end{aligned} \tag{2.172}$$

The reflection coefficient is generally complex, $\kappa = \kappa' + \mathrm{i}\kappa''$. The frequency response of the other mode is found by replacing κ by κ^*. Formula (2.172) is valid for γ real or imaginary. If $|\delta| > |\kappa|$,

$$\frac{\gamma L}{2} = \sqrt{\delta^2 - |\kappa|^2}\frac{L}{2} = X \ ,$$

and if $|\delta| < |\kappa|$,

$$\gamma = \mathrm{i}\chi \Rightarrow \frac{\gamma L}{2} = \mathrm{i}\sqrt{|\kappa|^2 - \delta^2}\frac{L}{2} = \mathrm{i}X \ .$$

The frequency response contains a $\sin X/X$ or $\sinh X/X$ factor and a term due to internal reflections, whose modulus is 1 if $\kappa = 0$, i.e., if $\gamma = \delta$.

We define the *directionality factor* by

$$d(\omega) = \frac{|H_{\mathrm{R}}|^2 - |H_{\mathrm{S}}|^2}{|H_{\mathrm{R}}|^2 + |H_{\mathrm{S}}|^2} = \frac{|N_{\mathrm{R}}|^2 - |N_{\mathrm{S}}|^2}{|N_{\mathrm{R}}|^2 + |N_{\mathrm{S}}|^2} \ , \tag{2.173}$$

which is zero if the transducer is bidirectional and equal to ± 1 if it emits in just one direction. After some calculation (Problem 2.10), we find that for $|\delta| < |\kappa|$,

$$\begin{aligned} |N_{\mathrm{R,S}}|^2 &= |\kappa|^2\cosh 2X - \delta^2 - 2\delta\kappa'\sinh^2 X \pm \chi\kappa''\sinh 2X \ , \\ |D|^2 &= |\kappa|^2\cosh^2 2X - \delta^2 \ . \end{aligned} \tag{2.174}$$

Hence,

$$d(\omega) = \frac{\chi\kappa'' \sinh 2X}{|\kappa|^2 \cosh 2X - \delta^2 - 2\delta\kappa' \sinh^2 X}\,, \quad \chi = \sqrt{|\kappa|^2 - \delta^2}\,. \qquad (2.175)$$

The directionality factor is zero if $\kappa'' = 0$, i.e., if generation and reflection centres coincide. Otherwise, at the synchronism frequency ω_0, $\delta = 0 \Rightarrow X = |\kappa|L/2$, the factor

$$d(\omega_0) = \frac{\kappa''}{|\kappa|} \tanh |\kappa|L$$

is maximal when $\kappa'' = |\kappa|$, i.e., $\kappa = \mathrm{i}|\kappa|$. It is then equal to the reflection coefficient of the electrode array [see (2.144)]:

$$d_{\max} = \tanh |\kappa|L \quad \Rightarrow \quad \frac{|H_{\mathrm{R}}(\omega_0)|}{|H_{\mathrm{S}}(\omega_0)|} = \mathrm{e}^{|\kappa|L}\,. \qquad (2.176)$$

For example, if $|\kappa|L = 1$, the amplitude of waves launched in one direction is almost three times the amplitude for the opposite direction. Internal reflections within the transducer favour Rayleigh wave generation in one direction, at the expense of the opposite direction, provided that parameters ζ and κ are $\pi/2$ out of phase. This means that, taking the return trip into account, reflection centres must be $\lambda_0/8$ away from generation centres. Single-phase unidirectional transducers (SPUDTs) are based on this principle (see Sect. 2.5.3).

Using the results of Problem 2.10, the frequency response amplitude of a uniform SPUDT-type transducer ($\kappa' = 0$) with $|\delta| > |\kappa|$ is

$$|H_{\mathrm{R,S}}(\omega)|^2 = (\zeta L)^2 \left(\frac{\sin X}{X}\right)^2 \frac{\delta^2 - |\kappa|^2 \cos 2X \pm \gamma|\kappa| \sin 2X}{\delta^2 - |\kappa|^2 \cos^2 2X}\,, \qquad (2.177)$$

where $X = \gamma L/2$.

2.5.2.3 Admittance. The admittance is obtained by integrating (2.158) from 0 to L:

$$I(L, \omega) = \mathrm{i}\omega C_{\mathrm{T}} U - 2\mathrm{i} \int_0^L \zeta^*(x) R(x)\, \mathrm{d}x - 2\mathrm{i} \int_0^L \zeta(x) S(x)\, \mathrm{d}x\,,$$

where $C_{\mathrm{T}} = \int_0^L C(x)\, \mathrm{d}x$, and then dividing the incoming current intensity by the applied voltage $U(\omega)$. The admittance $Y(\omega) = I(L, \omega)/U(\omega)$ contains the total capacitance C_{T}, the susceptance $B_{\mathrm{a}}(\omega)$, and the radiation conductance $G_{\mathrm{a}}(\omega)$ [see (2.78)]. If there is no attenuation, the latter is given directly by (2.115). For a uniform transducer, and when $|\delta| > |\kappa|$, it is given by

$$|G_{\mathrm{a}}(\omega)|^2 = 2(\zeta L)^2 \left(\frac{\sin X}{X}\right)^2 \frac{\delta^2 - |\kappa|^2 \cos 2X - 2\delta\kappa' \sin^2 X}{\delta^2 - |\kappa|^2 \cos^2 2X}\,. \qquad (2.178)$$

Near the synchronism frequency ω_0, i.e., when $|\delta| < |\kappa|$,

$$|G_{\rm a}(\omega)|^2 = 2(\zeta L)^2 \left(\frac{\sinh X}{X}\right)^2 \frac{|\kappa|^2 \cosh 2X - \delta^2 - 2\delta\kappa' \sinh^2 X}{|\kappa|^2 \cosh^2 2X - \delta^2} \,. \tag{2.179}$$

Comparing (2.38) with the equation found by neglecting internal reflections ($\kappa = 0 \Rightarrow \gamma = \delta$),

$$G_{\rm a}(\omega) = 2(\zeta L)^2 \left(\frac{\sin X}{X}\right)^2 ,$$

shows that ζ is related to the conductance $G_0 = G_{\rm a}(\omega_0)$. Identifying $2(\zeta L)^2$ with G_0, the generation factor per finger pair $2p\zeta$ is given by

$$\zeta L = \frac{\pi}{\sqrt{2}} K_{\rm R} (C_{\rm T} N f_0)^{1/2} \Rightarrow 2p\zeta = \frac{\sqrt{\pi}}{2} K_{\rm R} (C_1 w \omega_0)^{1/2} \,. \tag{2.180}$$

When $\kappa \neq 0$, the conductance for $\omega = \omega_0$, i.e., $\delta = 0$ and $X = |\kappa| L/2$, depends on the modulus of the reflection coefficient:

$$G_{\rm a}(\omega) = \frac{G_0}{\cosh |\kappa| L} \left[\frac{\sinh(|\kappa| L/2)}{|\kappa| L/2}\right]^2 = \frac{2G_0}{(|\kappa| L/2)^2} \left(1 - \frac{1}{\cosh |\kappa| L}\right) .$$

Relations (2.178) and (2.179) can be used in two different ways:

- They bring out the effect of internal reflections in conventional transducers for which $\kappa' = |\kappa|$. Indeed, cancelling a factor $\delta + |\kappa| \cos \gamma L$,

$$G_{\rm a}(\omega) = G_0 \left(\frac{\sin X}{X}\right)^2 \frac{\delta - |\kappa|}{\delta - |\kappa| \cos 2X} , \qquad G_0 = 2(\zeta L)^2 \,.$$

  The function $G_{\rm a}(\omega)$ is no longer symmetric about $\omega = \omega_0$ (Fig. 2.39a). The conductance is G_0 for $\delta = -|\kappa| \Rightarrow \gamma = 0$. At the symmetric frequency $\omega = \omega_0 + |\kappa| V_{\rm R}$, it has value $G_0/(1 + |\kappa L|^2)$.
- In a single-phase unidirectional transducer ($\kappa' = 0$), the radiation conductance depends only on δ^2 :

$$G_{\rm a}(\omega) = G_0 \left(\frac{\sin X}{X}\right)^2 \frac{\delta^2 - |\kappa|^2 \cos 2X}{\delta^2 - |\kappa|^2 \cos^2 2X} ,$$

  and the curve regains its symmetry with respect to ω_0 (Fig. 2.39b).

The advantage of the coupled mode method is that it produces expressions for the main characteristics of a transducer (P-matrix elements, in the next section), showing how they vary with parameter changes. However, values of these parameters (capacitance, detuning parameter, reflection coefficient, generation factor) must be determined by other means, based on equations of state or experiment. For example, Chen and Haus [2.29] have shown that these coefficients can be found from the frequency variation caused by depositing the metallic grating on the substrate. They calculated this variation using equations of state and propagation equations, and applying the reciprocity theorem. The principle is simple but the calculation long. Assuming

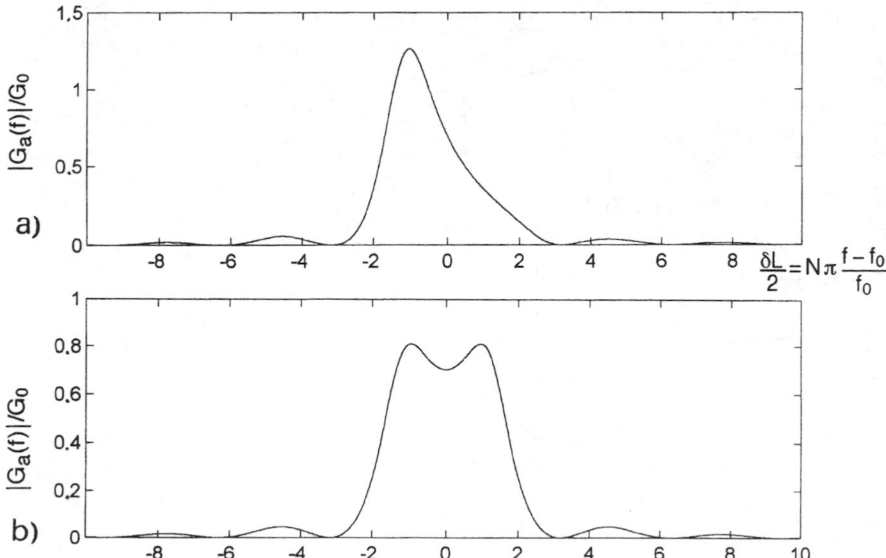

Fig. 2.39. Radiation conductance as a function of $L\delta/2$ for two types of 50 finger pair uniform transducer: (**a**) conventional, $\kappa L = 1$; (**b**) single-phase unidirectional, $\kappa L = \mathrm{i}$

partial isotropy of the substrate, formulas simplify and they were able to obtain results for lithium niobate and quartz in this way.

2.5.2.4 *P*-Matrix. Overall operation of the transducer is represented by a matrix (P), defined in Sect. 2.4.2. Its components are given by the above results. When the electrical port is shorted (U=0), the transducer becomes a passive array. P_{11} and P_{22} are reflection coefficients r_0 and r_L at left and right mechanical ports ($x = 0$ and $x = L$, respectively). $P_{12} = P_{21}$ is the transmission coefficient t between these two ports [see (2.111), (2.112)]. Electromechanical coupling terms P_{13} and P_{23} are the frequency responses $H_{\mathrm{S}}(\omega)$ and $H_{\mathrm{R}}(\omega)$ defined in (2.161). P_{33} is the transducer admittance Y_{T}, given by (2.110). The mixed matrix for the transducer takes the form

$$(P) = \begin{pmatrix} r_0 & t & H_{\mathrm{S}} \\ t & r_L & H_{\mathrm{R}} \\ -2H_{\mathrm{S}} & -2H_{\mathrm{R}} & Y_{\mathrm{T}} \end{pmatrix} . \tag{2.181}$$

When a source of conductance G_{L} is applied to the transducer, transfer functions s_{13} and s_{23} between the electrical port and the two mechanical ports are given by (2.119). The mismatching ratio $G_{\mathrm{L}}/G_{\mathrm{a}}$ therefore plays an important role with regard to losses and triple-transit level (as observed in Sect. 2.4.1). This role is illustrated in Fig. 2.40, due to Hartmann and Abbott [2.32].

These curves show, for three values of the coupling parameter $|\kappa|L$, the insertion losses and triple-transit level of a transmission line in which the gen-

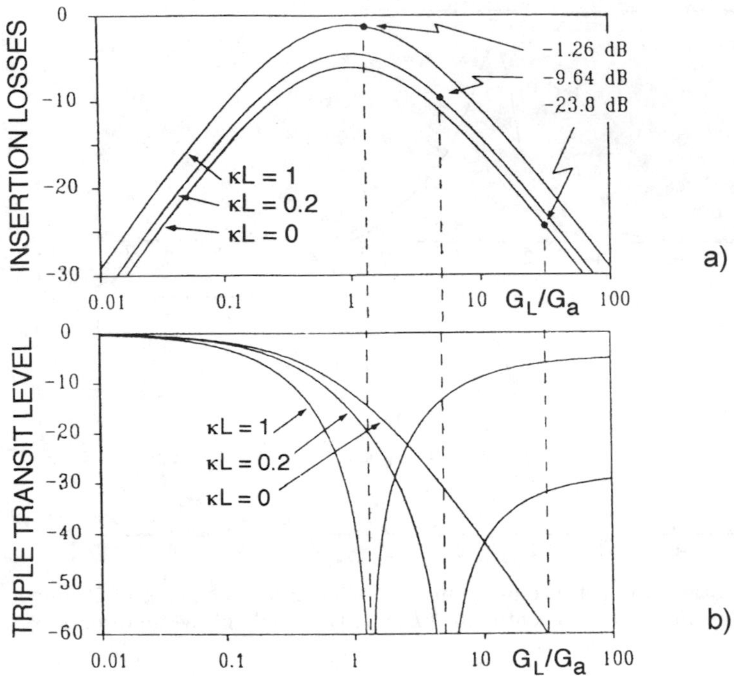

Fig. 2.40. Dependence of (**a**) insertion losses and (**b**) triple-transit echo level on G_L/G_a, for various values of the coupling coefficient $|\kappa|L$. From Hartmann and Abbot [2.32]. ©1989 IEEE

eration and reflection centres of the two transducers are shifted by $\lambda_0/8$, so as to achieve optimal directionality. The case $|\kappa|L = 0$ corresponds to a bidirectional transducer. Such a transducer must have 24 dB insertion losses if the triple-transit signal (TTS) is to be rejected at -60 dB (see Problem 2.6). In the opposite case, a transmission line comprising two SPUDT-type directional transducers has 1.3 dB losses for a TTS level ≥ 60 dB. In practice, it is therefore useless to require a value for $|\kappa|L$ greater than unity, if we wish to maintain very low levels of insertion losses and spurious echo.

2.5.3 Unidirectional Transducers

The fact that an ordinary transducer, with single or split fingers, emits waves in both directions is a priori a disadvantage, because an inherent 3 dB loss arises from the use of only one beam. It was natural to begin by seeking ways of using both beams. Various geometries were suggested, for example, placing a reflector on one side of the transducer. Then ways were found to favour one of the two beams, splitting the transducer into several elements, and applying suitably phase-shifted voltages (Fig. 2.28). Various devices (filters) were built, with emitter and receiver split into several elements and intercalated so as to insert one emitting element between two receiving elements. Quite satisfactory results were achieved (Sect. 4.3.1).

Whilst attempts were being made to remove the unwanted effects of reflections from the fingers (by splitting them and by choosing weakly piezoelectric substrates), efforts were also being made to assess them and understand their electrical and mechanical causes (Sect. 2.1.1). Hence the idea of using them to make the transducer unidirectional. This type of transducer has been given the acronym SPUDT, where 'single-phase' distinguishes them from the multiphase transducers mentioned above. Several techniques were devised for creating an asymmetry in the structure which would favour emission in one particular direction:

- in the case of fingers split into two elements, increasing the metallic thickness of one element, thereby increasing its reflection coefficient [2.28];
- introducing reflectors [2.33] or passive electrodes [2.34] between parts of the transducers;
- exploiting substrate anisotropy [2.35, 2.36], based on the fact that the phase difference between emitted waves and the voltage which generates them depends on the crystallographic orientation of the substrate. Hence the position of the effective generation centre relative to the electrodes depends on the cut and direction of propagation.

The following techniques have introduced a new point of view.

Distribution of Reflective Elements Inside Wide Electrodes. We shall now illustrate this idea, called DART (Distributed Acoustic Reflection Transducer) by its inventors [2.37]. Figure 2.41a shows that one period of the transducer contains two fingers of width $\lambda_0/8$, one finger of width $3\lambda_0/8$ and intervals of width $\lambda_0/8$. Lights of width $\lambda_0/8$ are produced in the wider finger. In the plane, the period is divided up into transverse strips A containing no lights and neighbouring strips B which contain one light. Any wave arriving from the left is partially reflected on the edges of the fingers or the lights. In an A strip, without lights, reflected wave phases appear as shown in Fig. 2.41b. Only waves reflected on the edges of the widest finger, with resultant r_{m}, will survive. A B strip, with lights, is one in which all fingers have width $\lambda_0/8$ and gives rise to no overall reflection (Sect. 2.1.1). Therefore, the wave R_{m} reflected by one period depends on the number of lights.

The second cause of reflection is piezoelectric reemission. Its effect depends on the transducer load. The equivalent source is located near active fingers and its intensity is a function of overlap length. Let R_{e} be the resulting reflected wave. The designer is free to choose the independent quantities R_{m}, R_{e} separately, thereby determining to what extent they compensate one another and rendering the transducer as directional as required.

Choice of Finger Width. The transducer based upon this technique has been called the EWC/SPUDT (Electrode Width Controlled SPUDT) by its inventors [2.32]. There are three structural differences with the previous de-

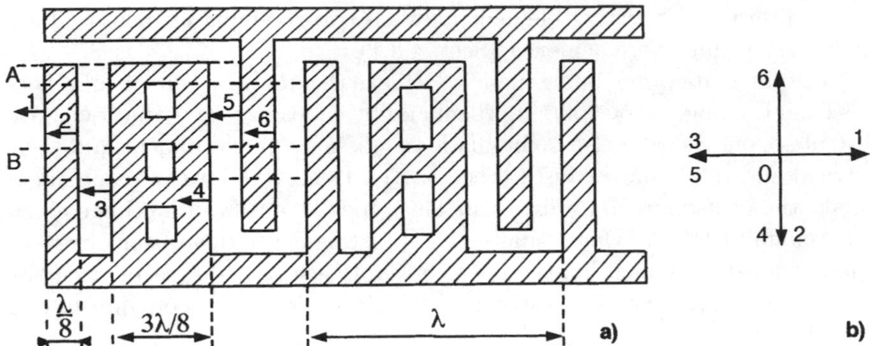

Fig. 2.41. DART transducer. (**a**) Construction. (**b**) Phase shifts of reflected waves in an *A*-type strip. From [2.37]

vice. The electrode of width $3\lambda_0/8$ is replaced by one of width $\lambda_0/4$, containing no lights (Fig. 2.42). The $\lambda_0/4$ electrode is itself replaced in certain places by a pair of $\lambda_0/8$ electrodes separated by distance $\lambda_0/8$. Mechanical reflection is controlled by choosing the number of $\lambda_0/4$ and $\lambda_0/8$ electrodes.

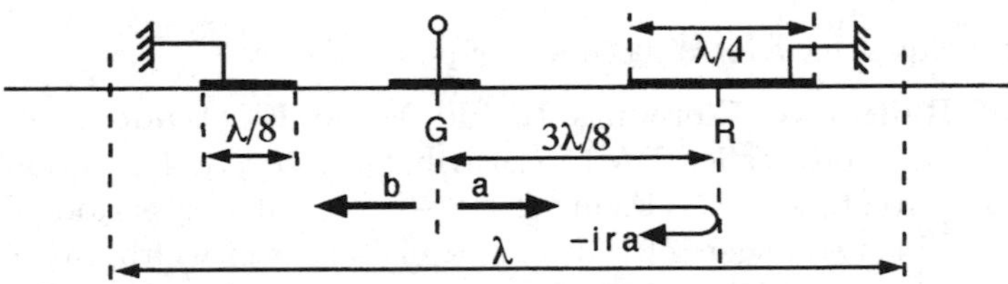

Fig. 2.42. Cell emitting preferentially towards the left. The reflection centre R located in the middle of the electrode of width $\lambda_0/4$ introduces a $-\pi/2$ phase shift. The generation centre G located in the middle of the active electrode generates waves towards both left and right. If the distance GR is $3\lambda_0/8$, waves moving rightward and reflected by R return in phase with leftward waves from G

Insertion of Resonators. This transducer is called the RSPUDT (Resonant SPUDT) by its inventors [2.38]. The basic idea is to create resonant cavities inside the transducer, e.g., by associating a cell which emits preferentially rightward with the cell in Fig. 2.42. This is shown in Fig. 2.43. The two cells are positioned so that their generation centres are distance λ_0 apart and emit waves in phase. The transducer is then composed of cells with positive, negative or even zero reflection coefficient. Unidirectionality is achieved when more cells emit in the useful direction than in the other.

In practice, the reflection coefficient of a finger depends not only on its width but also on the thickness of metal. Hence, the width of a reflecting element need not be $\lambda_0/4$. Moreover, the position of the reflection centre may

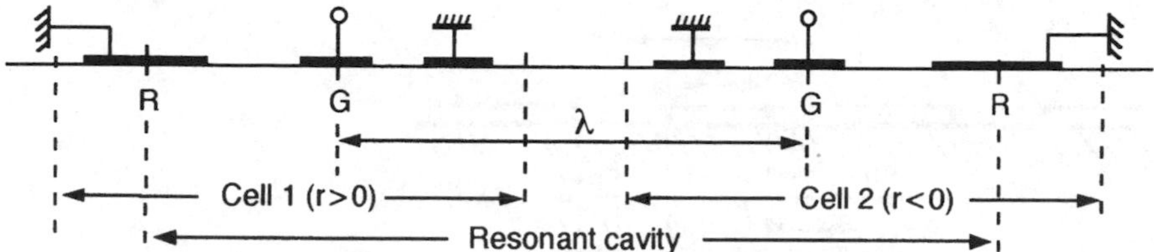

Fig. 2.43. Association of two cells emitting waves preferentially in opposite directions and in phase, thereby forming a resonant cavity. From [2.38]

be determined by several active fingers rather than just one. The principle illustrated by the EWC is clearly applicable to other structures. Optimisation techniques are required to design this type of transducer [2.39].

2.6 Adjacent Element Analysis

Transducer structures are discrete, being built up from successions of electrodes and intervals. The coupled mode model is based on slow and continuous variations of amplitudes R and S of the two waves propagating in opposite directions. It is therefore an approximation. These two quantities are governed by two coupled first order differential equations. As edge effects are neglected, the model is acceptable for long transducers with low internal reflection, i.e., narrow bandwidth transducers. However, it is difficult to relate model parameters to physical properties. More accurate studies have therefore been carried out, requiring more computation time.

There is a general procedure, applicable to any structure, whether it be reflector or transducer. This involves first choosing an elementary cell and then determining elements of a matrix, depending on the effects which are to be modelled. The matrix is transformed into a transfer matrix which is used to calculate the matrix appropriate to the complete reflector or transducer by simple multiplication. We shall present three models: s- or P-matrix, harmonic admittance and equivalent circuits.

2.6.1 P-Matrix Model

When the structure (grating or transducer) can be decomposed into identical cells, each one is described by an elementary matrix. It must be examined on a local scale, i.e., on the level of a single finger or finger pair.

In the *passive array* case, the elementary cell is, for example [2.40], made up of one finger and the two adjacent non-metallised surface zones (Fig. 2.44). It is represented, as in Sect. 2.4.1, by a scattering matrix (s) expressing outgoing wave amplitudes b_n and a_{n+1} in terms of incoming wave amplitudes a_n and b_{n+1}.

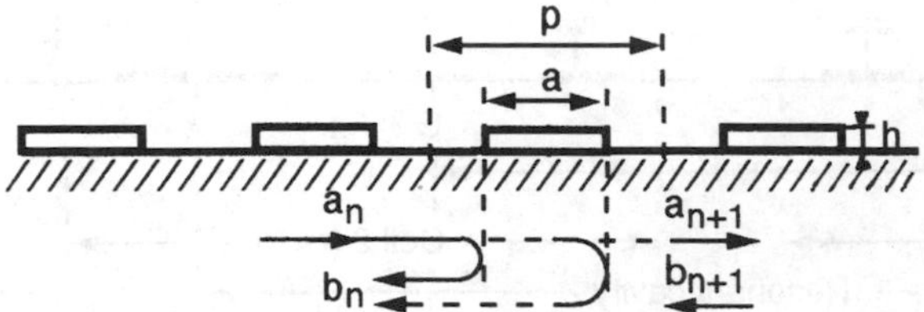

Fig. 2.44. Elementary cell c_i of an array, made up of one finger and two transmission line sections. A fraction a_{n+1} of incident wave a_n crosses the finger, and a small fraction b_n is mechanically or electrically reflected by the edges

The scattering matrix for the cell is given by

$$\begin{pmatrix} b_n \\ a_{n+1} \end{pmatrix} = \begin{pmatrix} s_{11} & s_{12} \\ s_{21} & s_{22} \end{pmatrix} \begin{pmatrix} a_n \\ b_{n+1} \end{pmatrix} .$$

For any incident wave a_n, coming from cells upstream of cell c_n, the relevant physical mechanisms are:

- mechanical reflection at upstream and downstream edges of the finger (neglecting multiple reflections);
- phase difference due to propagation at speed V_R over a free surface and propagation at speed V_∞ over a metallised surface;
- if the finger is inactive (transducers contain such fingers), reflection caused by disappearance of the tangential component of the electric field;
- piezoelectric reemission (which may be zero, if fingers are shorted);
- generation if the finger is connected to a voltage supply.

The elementary matrix (s) is therefore composed of three submatrices. An analytic expression for their terms can sometimes be found from the general equations (by a variational principle or the piezoelectric permittivity method). However, these expressions contain constants corresponding to the particular materials used. A rather accurate experimental determination of these parameters is therefore more useful.

Let us begin by investigating terms in the mechanical submatrix. Finger edges constitute a break in mechanical impedance and are sources of reflection for any external incident wave (internal reflections within the finger are neglected). Reflection from the upstream edge of the finger, taken as abscissa origin, are given by

$$(s_\mathrm{m}) = \mathrm{e}^{-\mathrm{i}\varphi_\mathrm{m}} \begin{pmatrix} \rho_\mathrm{m} & t_\mathrm{m} \\ t_\mathrm{m} & -\rho_\mathrm{m}^* \end{pmatrix} . \tag{2.182}$$

Phase difference φ_m and reflection coefficient ρ_m are given explicitly by

$$\rho_\mathrm{m} = C_1 \left(\frac{h}{\lambda}\right) + \mathrm{i}C_2 \left(\frac{h}{\lambda}\right)^2 \sin\eta\pi , \quad \varphi_\mathrm{m} = C_3 \left(\frac{h}{\lambda}\right)^2 \sin\eta\pi , \tag{2.183}$$

and the (real) transmission coefficient t_m is given by energy conservation, $t_\mathrm{m}^2 = 1 - |\rho_\mathrm{m}|^2$. λ is the free surface wavelength.

The first term of ρ_{m} describes effects of surface loading. It has been determined theoretically and checked experimentally for a certain number of metal–substrate associations [2.41]. The second term and phase shift angle φ_{m} express energy storage near the finger edge, in the form of evanescent modes [2.6, 2.9]. This phenomenon generates a phase difference between reflected and transmitted waves when there is a discontinuity. The sine only involves the metallisation ratio [2.42]. Coefficients C_1, C_2 and C_3 are determined by experiment and comparison with other models. The matrix for mechanical reflection at the downstream edge is obtained from (2.182) by permuting entries on the main diagonal.

The submatrix s_{e} for piezoelectric reflection (reemission), caused by the transducer load and significant for substrates with strong piezoelectric coupling, can be quite difficult to calculate, depending on the hypotheses made. For example, it can be calculated using the permittivity model. The reflection coefficient r_{e} is imaginary if the origin is located at the centre of the electrode. Its value depends on the metallisation ratio $\eta = a/p$ of the surface and electrical conditions. It is given in terms of Legendre functions $P_\nu(x)$ of variable $x = \cos\eta\pi$. For a *shorted* electrode,

$$r_{\mathrm{e}}^{\mathrm{sc}} = -\mathrm{i}kp\frac{K^2}{4}\left[P_{2s}(\cos\eta\pi) + \frac{P_s(-\cos\eta\pi)}{P_{-s}(-\cos\eta\pi)}P_{-2s}(\cos\eta\pi)\right] , \qquad (2.184)$$

where, as in Sect. 2.3.1, parameter s lies between 0 and 1:

$$s = \frac{kp}{2\pi} = \frac{f}{2f_0} , \qquad f_0 = \frac{V_{\mathrm{R}}}{2p} .$$

If the electrode is in *open* circuit,

$$r_{\mathrm{e}}^{\mathrm{oc}} = -\mathrm{i}kp\frac{K^2}{2}\left[P_{2s}(\cos\eta\pi) - \frac{P_s(\cos\eta\pi)}{P_{-s}(\cos\eta\pi)}P_{-2s}(\cos\eta\pi)\right] . \qquad (2.185)$$

At the synchronism frequency f_0, $s = 1/2$, $kp = \pi$ and using the results $P_1(x) = x$, $P_{-1}(x) = 1 = P_0(x)$ in Appendix C, equation (2.184) gives back (2.12). The transmission coefficient t_{e} follows from energy conservation: $t_{\mathrm{e}}^2 = 1 - |r_{\mathrm{e}}|^2$. The matrix (s_{e}) can be put into the form (D.12).

The matrix (s_{p}) represents propagation (wave number k) and attenuation (coefficient α) in each of the passive segments of length l on either side of the finger and is given simply by

$$(s_{\mathrm{p}}) = \mathrm{e}^{-(\mathrm{i}kl+\alpha l)}\begin{pmatrix} 0 & 1 \\ 1 & 0 \end{pmatrix} , \qquad l = (1-\eta)p/2 .$$

Once we have determined all elements of the matrix (s) for the unit cell, the transfer matrix (t) is given by (D.14), Appendix D. The transfer matrix expresses quantities to the right of the cell in terms of quantities to the left [see (2.141)]. For an array made up of N identical cells, the transfer matrix for the whole system is $(T) = (t)^n$ and can be found from (D.22), Appendix D.

A transducer is analysed in the same way, by dividing it up into sections. Each section is represented by a three-port hexapole (Sect. 2.4). The

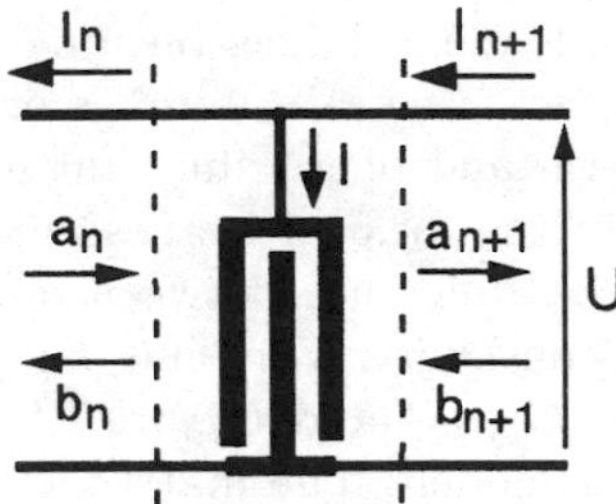

Fig. 2.45. Elementary cell for calculation of the matrix (p). Cell boundaries are arbitrary, provided that they coincide with those of adjacent cells

incoming current at each electrode is given, for example, by the quasi-static approximation (Problem 2.11). Figure 2.45 shows an elementary cell comprising a finger pair. The matrix (p) gives amplitudes of outgoing waves and current intensity I in terms of incoming wave amplitudes and voltage U. It is supposed identical for all cells:

$$\begin{pmatrix} b_n \\ a_{n+1} \\ I \end{pmatrix} = \begin{pmatrix} p_{11} & p_{12} & p_{13} \\ p_{21} & p_{22} & p_{23} \\ p_{31} & p_{32} & p_{33} \end{pmatrix} \begin{pmatrix} a_n \\ b_{n+1} \\ U \end{pmatrix} .$$

The matrix (p) can be rearranged to form a transmission matrix relating quantities to the right of the cell with those to the left, where $U_{n+1} = U_n = U$:

$$\begin{pmatrix} a_{n+1} \\ b_{n+1} \\ U \\ I_{n+1} \end{pmatrix} = \begin{pmatrix} t_{11} & t_{12} & t_{13} & 0 \\ t_{21} & t_{22} & t_{23} & 0 \\ 0 & 0 & 1 & 0 \\ t_{31} & t_{32} & t_{33} & 1 \end{pmatrix} \begin{pmatrix} a_n \\ b_n \\ U \\ I_n \end{pmatrix} . \tag{2.186}$$

Reciprocity relations imply

$$p_{21} = p_{12} , \quad t_{11}t_{22} - t_{12}t_{21} = 1 .$$

Comparing with relations for the matrix (p) and using $I_{n+1} = I_n + I$ yields the entries of the transmission matrix t_{ij}:

$$t_{11} = p_{21} - \frac{p_{11}p_{22}}{p_{12}} , \quad t_{12} = \frac{p_{22}}{p_{12}} , \quad t_{21} = -\frac{p_{11}}{p_{12}} ,$$

$$t_{22} = \frac{1}{p_{12}} , \quad t_{13} = p_{23} - \frac{p_{13}p_{22}}{p_{12}} , \quad t_{23} = -\frac{p_{13}}{p_{12}} ,$$

$$t_{31} = p_{31} - \frac{p_{32}p_{11}}{p_{12}} , \quad t_{32} = \frac{p_{32}}{p_{12}} , \quad t_{33} = p_{33} - \frac{p_{13}p_{32}}{p_{12}} .$$

These relations generalise results given in Appendix D. The first four relations are identical to (D.14). The transmission matrix (T) for a transducer containing N cascaded cells is the product of the N elementary matrices. Morgan has established a formula for the matrix (P) of a transducer comprising N identical cells in terms of entries of the elementary matrix (p) [2.43]. Characteristic responses of the transducer are then plotted and effects of the various parameters emphasised [2.44].

2.6.2 Harmonic Admittance. Numerical Model

Design of transducers with electrodes deposited on a piezoelectric substrate requires precise knowledge of several fundamental parameters: phase speed V_{sc} of surface waves propagating under the shorted electrode array; electromechanical coupling coefficient K_R; reflection coefficient r of a metallic strip; phase difference ψ between generation and reflection centres; and static capacitance C_1. In order to determine these parameters, the methods presented so far have appealed to experiment (for coupling coefficient and speed), a perturbation method [quasi-static approximation (Sect. 2.3), mechanical effects of a thin film], and an exact calculation [charge distribution, static capacitance (Appendix A)]. A model, coupled modes or P-matrix, was then worked out.

Hodé, Desbois and Ventura have provided a global analysis, valid for any periodic electrode array, based upon:

- The idea of *mutual admittance*, to describe electrical and acoustic coupling between two electrodes in the array. This is found by Fourier transforming the harmonic admittance of the array.
- Generalising the P-matrix formalism, to model the acoustic part of the harmonic admittance.
- Numerical modelling by finite element and boundary element analysis, to calculate the harmonic admittance, without making simplifying hypotheses concerning electrical and mechanical interactions in the elementary cell.
- Determining parameters V_{sc}, K_R, r, ψ, C_1 by comparing theoretical and numerical models at characteristic points of the harmonic admittance.

This powerful method is particularly useful for studying pseudo-surface waves. In this case, mechanical and electrical perturbations due to the array can no longer be taken separately. For example, the coupling coefficient for these modes depends on array parameters.

2.6.2.1 Harmonic and Mutual Admittance. Coupling of electrodes deposited on a piezoelectric substrate, due to elastic waves propagating under the periodic array, can be analysed using the notions of mutual admittance and harmonic admittance, introduced by Desbois [2.45]. Consider the metallic strip grating shown in Fig. 2.19a, assuming to begin with that electrode m is held at potential $U_m e^{i\omega t}$, whilst the other electrodes are earthed (impulsive spatial excitation). The current intensity $I_n e^{i\omega t}$ induced in electrode n is proportional to the voltage U_m. At a given frequency, the ratio $y_{mn} = I_n/U_m$ defines the *mutual admittance* between the two electrodes. In an infinite periodic array, translation invariance implies that it depends only on the difference $m - n$:

$$I_n = y_{m-n} U_m \ .$$

For a general excitation of all electrodes, the superposition principle implies that the incoming current at electrode n is the sum

$$I_n = \sum_{m=-\infty}^{\infty} y_{m-n} U_m \,, \tag{2.187}$$

which is a discrete convolution [2.12].

Now consider a harmonic spatial excitation defined by the dimensionless parameter s,

$$U_{\mathrm{m}}(s) = U_0 \exp(-2\pi \mathrm{i} m s) \,.$$

The current intensity at electrode n is then

$$\begin{aligned} I_n(s) &= U_0 \sum_{m=-\infty}^{\infty} y_{m-n} \exp(-2\pi \mathrm{i} m s) \\ &= U_n(s) \sum_{m=-\infty}^{\infty} y_{m-n} \exp[-2\pi \mathrm{i}(m-n)s] \,. \end{aligned}$$

The ratio of $I_n(s)$ to $U_n(s)$ is independent of the position of the electrode in the array. This is the *harmonic admittance* $Y(s)$ of the array. Putting $q = m - n$, we have

$$Y(s) = \frac{I_n(s)}{U_n(s)} = \sum_{q=-\infty}^{\infty} y_q \exp(-2\pi \mathrm{i} q s) \,. \tag{2.188}$$

$Y(s)$ is periodic, with unit period of spatial frequency s, and is analogous to the frequency response of a system sampled with unit period.

The mutual admittance between two electrodes a distance qp apart is the coefficient of order q in the Fourier series expansion of the harmonic admittance:

$$y_q = \int_0^1 Y(s) \exp(2\pi \mathrm{i} q s)\, \mathrm{d}s \,, \quad y_{-q} = y_q \,. \tag{2.189}$$

The last relation, arising from the reciprocity principle, shows that the harmonic admittance is an even function, symmetric about $s = 0.5$, corresponding to an alternating distribution of potentials $U_n = (-1)^n$. At a given frequency, the harmonic admittance of an array of period p is found by the numerical method presented in Sect. 2.6.2.3. The mutual admittance can be deduced from $Y(s)$ by a fast Fourier transform.

Figure 2.46 shows the harmonic admittance of an electrode array, deposited on ST-cut quartz, perpendicular to the X direction, with $fp = 1082.4$ m/s and metallisation ratio $\eta = 0.5$. $Y(s)$ contains a smooth part which is purely imaginary, corresponding to electrostatic coupling between electrodes, and several spikes, due to surface acoustic waves propagating under the array. The electrostatic contribution $Y_{\mathrm{e}}(s)$, determined in Appendix A, is very close to a function $\sin \pi s$. The exact form is

$$Y_{\mathrm{e}}(s) = 2\mathrm{i}\omega(\varepsilon_{\mathrm{p}} + \varepsilon_0) w |\sin \pi s| \,, \quad \text{when} \quad \eta = a/p = 0.5 \,.$$

The electrostatic mutual admittance is, according to (2.189),

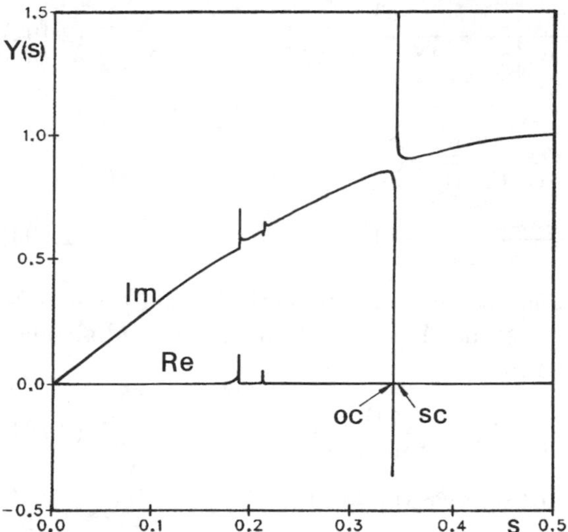

Fig. 2.46. Harmonic admittance $Y(s)$ of an electrode array ($fp = 1\,082$ MHz μm) on X propagating ST-cut quartz. $Y(s)$ is normalised to the maximal value of the electrostatic contribution, $2\omega(\varepsilon_\mathrm{p} + \varepsilon_0)w$. Figure 2 of [2.45]. ©1993 IEEE

$$y_\mathrm{em} = \frac{2C_1 f}{m^2 - 1/4}, \quad C_1 = (\varepsilon_\mathrm{p} + \varepsilon_0)w \,. \tag{2.190}$$

This decreases as $1/m^2$ in the far field $m > 5$. It dominates for $m < 10$.

The pole and zero are analogs of those arising for the surface piezoelectric permittivity (Sect. 2.3). They are due to the surface acoustic mode. To calculate its contribution to the harmonic admittance, assume that the intensity Y_sc and phase difference φ_sc of the acoustic interaction between two distant electrodes are constant. The mutual admittance can then be put into the form

$$\lim_{|q|\to\infty} y_q = Y_\mathrm{sc} \exp(-\mathrm{i}|q|\varphi_\mathrm{sc}) \,. \tag{2.191}$$

This is the case if the surface wave coupling the electrodes propagates without attenuation (pure Rayleigh mode), without diffraction (beam width w much larger than p), without reflection and without generation. The last two conditions are satisfied if the array is made up of massless, shorted electrodes and if the frequency is outside the stop band. For an array with given p, Y_sc and φ_sc are then two real characteristic parameters of the surface mode.

Extending (2.191) to all non-zero values of q, the harmonic admittance of an ideal array follows from (2.188):

$$Y(s) = y_0 + Y_\mathrm{sc} \sum_{q=1}^{\infty} \left(\mathrm{e}^{-\mathrm{i}\varphi_\mathrm{sc}} \mathrm{e}^{-2\pi\mathrm{i}s}\right)^q + Y_\mathrm{sc} \sum_{q=-1}^{-\infty} \left(\mathrm{e}^{-\mathrm{i}\varphi_\mathrm{sc}} \mathrm{e}^{2\pi\mathrm{i}s}\right)^{-q} \,.$$

Since $\sum_{q=1}^{\infty} z^q = z/(1-z)$,

$$Y(s) = y_0 + Y_{sc}\left[\frac{e^{-2\pi i s}}{e^{i\varphi_{sc}} - e^{-2\pi i s}} + \frac{e^{2\pi i s}}{e^{i\varphi_{sc}} - e^{2\pi i s}}\right], \tag{2.192}$$

and finally,

$$\begin{aligned} Y(s) &= y_0 + Y_{sc}\frac{\cos(2\pi s) - e^{-i\varphi_{sc}}}{\cos\varphi_{sc} - \cos(2\pi s)} \\ &= y_0 + Y_{sc}\left[\frac{i\sin\varphi_{sc}}{\cos\varphi_{sc} - \cos(2\pi s)} - 1\right]. \end{aligned} \tag{2.193}$$

When $s = 0$, all electrodes are at the same potential U and there is no current. Hence the harmonic admittance $Y(0) = I/U$ is zero. The admittance y_0 of each cell is therefore given by

$$y_0 = Y_{sc}\left[1 + \frac{i\sin\varphi_{sc}}{1 - \cos\varphi_{sc}}\right] = Y_{sc} + \frac{iY_{sc}}{\tan(\varphi_{sc}/2)}. \tag{2.194}$$

Substituting this into (2.193), $Y(s)$ takes the form

$$Y(s) = Y_{sc}\frac{i\sin\varphi_{sc}}{1 - \cos\varphi_{sc}}\left[\frac{1 - \cos\varphi_{sc}}{\cos\varphi_{sc} - \cos(2\pi s)} - 1\right].$$

The formula

$$Y(s) = \frac{iY_{sc}}{\tan(\varphi_{sc}/2)}\frac{1 - \cos(2\pi s)}{\cos\varphi_{sc} - \cos(2\pi s)} \tag{2.195}$$

shows that the harmonic impedance has a pole on the real axis of spatial frequencies s, at the point

$$s_{sc} = \frac{\varphi_{sc}}{2\pi}, \quad \text{with amplitude} \quad Y_p = \frac{iY_{sc}}{\tan(\varphi_{sc}/2)}. \tag{2.196}$$

This pole is related to the surface mode propagating at speed V_{sc} under the shorted infinite array. This mode induces a finite current intensity for a harmonic excitation of infinitesimal amplitude, provided that the phase shift φ_{sc} and harmonic excitation are synchronised. Numerical determination of s_{sc} gives the phase speed of this mode:

$$\varphi_{sc} = 2\pi fp/V_{sc} \quad \Rightarrow \quad V_{sc} = fp/s_{sc}. \tag{2.197}$$

Determination of Y_p gives the amplitude Y_{sc} of the mutual admittance, related to the electromechanical coupling coefficient.

2.6.2.2 Transducer and Reflective Array. In the general case, that is, in the stop band of a reflective array, analysis is carried out using the P-matrix formalism [2.39]. Figure 2.47 shows an elementary cell, of length equal to the period of the array. Its matrix (p) is given by

$$\begin{pmatrix} b_1 \\ a_2 \\ I \end{pmatrix} = \begin{pmatrix} r & t & \alpha_1 \\ t & r & \alpha_2 \\ \beta_1 & \beta_2 & y \end{pmatrix} \begin{pmatrix} a_1 \\ b_2 \\ U \end{pmatrix}.$$

The matrix entries, defined in Appendix D, are:

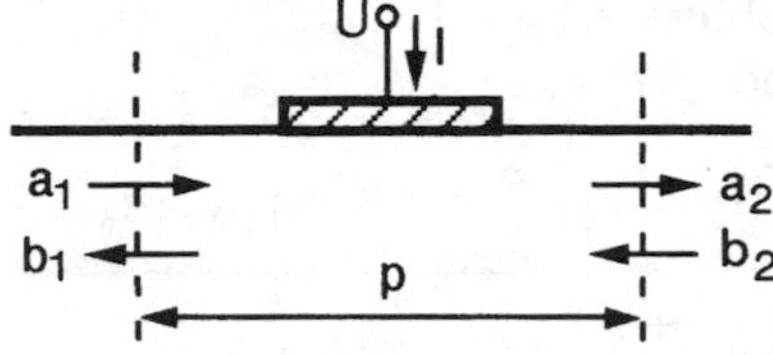

Fig. 2.47. However complex it may be, the elementary cell can be represented by a matrix (p)

- Reflection and transmission coefficients r and t of the shorted cell. r is imaginary if the origin is chosen at the reflection centre. Denoting its amplitude by $\sin\Delta$, the phase shift φ suffered by the surface wave over one period p satisfies

$$t = \cos\Delta\,\mathrm{e}^{-\mathrm{i}\varphi}\,, \quad r = -\mathrm{i}\sin\Delta\,\mathrm{e}^{-\mathrm{i}\varphi}\,, \quad t^2 - r^2 = \mathrm{e}^{-2\mathrm{i}\varphi}\,, \tag{2.198}$$

where Δ is a small angle ($t \approx 1$), whilst φ is close to π near the synchronism frequency.
- Electroacoustic coefficients α_1, α_2 and acoustoelectric coefficients β_1, β_2, equal and opposite to α_1, α_2.
- Electrical admittance $y = g + \mathrm{i}b$ of the cell.

Energy conservation implies (see Appendix D)

$$2g = |\alpha_1|^2 + |\alpha_2|^2\,. \tag{2.199}$$

Introducing an angle δ to characterise the electroacoustic asymmetry of the cell,

$$\begin{aligned} \alpha_1 + \alpha_2 &= 2\mathrm{i}\sqrt{g}\cos\delta\,\mathrm{e}^{-\mathrm{i}(\varphi+\Delta)/2}\,, \\ \alpha_1 - \alpha_2 &= 2\sqrt{g}\sin\delta\,\mathrm{e}^{-\mathrm{i}(\varphi-\Delta)/2}\,. \end{aligned} \tag{2.200}$$

If $\delta = 0$ $(\pm\pi/2)$, emission and reception are symmetric (antisymmetric).

The incoming current I at the centre electrode of the array in Fig. 2.48 contains three terms: the current of intensity yU corresponding to the cell itself, the current of intensity $\beta_1 R$ induced by the wave entering by the left-hand port (1), and the current of intensity $\beta_2 S$ induced by waves coming from the right. That is,

$$I = yU + \beta_1 R + \beta_2 S\,. \tag{2.201}$$

R and S are resultants at $x = -p/2$ and $x = +p/2$, respectively, of waves emitted by other cells.

In order to obtain a simple expression for propagation of waves across each cell, we seek transmission eigenmodes of the shorted circuit, i.e., the eigenvectors $\tilde{\boldsymbol{a}}$ and $\tilde{\boldsymbol{b}}$ of the transfer matrix (t). These constitute a basis in which cell traversal corresponds to multiplying by an eigenvalue Λ of (t), $\tilde{\boldsymbol{a}}_2 = \Lambda\tilde{\boldsymbol{a}}_1$, $\tilde{\boldsymbol{b}}_2 = \Lambda\tilde{\boldsymbol{b}}_1$. Eigenvalues are solutions of

$$\left\|\begin{matrix} t_{11} - \Lambda & t_{12} \\ t_{21} & t_{22} - \Lambda \end{matrix}\right\| = 0\,,$$

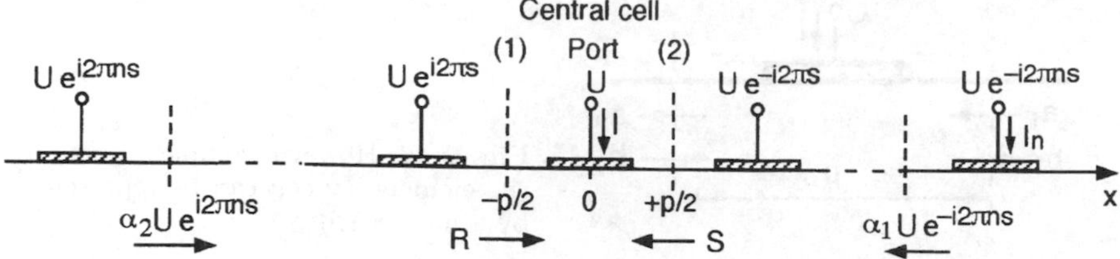

Fig. 2.48. Harmonic excitation of an electrode array. α_1, α_2 are generation coefficients of the elementary cell

where

$$t_{11} = \frac{\mathrm{e}^{-2\mathrm{i}\varphi}}{t}, \quad t_{12} = -t_{21} = \frac{r}{t}, \quad t_{22} = \frac{1}{t},$$

are components of the transfer matrix [see (D.15), Appendix D]. Since the latter is unitary, $||t|| = 1$, the characteristic equation

$$\Lambda^2 - \frac{\Lambda}{t}(\mathrm{e}^{-2\mathrm{i}\varphi} + 1) + ||t|| = 0$$

can be rewritten

$$\Lambda^2 - 2\Lambda\frac{\cos\varphi}{\cos\Delta} + 1 = 0, \quad \text{where} \quad \Lambda + \Lambda^{-1} = 2\frac{\cos\varphi}{\cos\Delta}. \tag{2.202}$$

This shows that, if Λ is a solution, then the other eigenvalue is Λ^{-1}. The relation

$$(\Lambda - \Lambda^{-1})^2 = (\Lambda + \Lambda^{-1})^2 - 4 = 4\frac{\cos^2\varphi - \cos^2\Delta}{\cos^2\Delta} \tag{2.203}$$

implies that eigenvalues are real in the stop band of the array:

$$\pi - \Delta < \varphi < \pi + \Delta \Rightarrow \Lambda - \Lambda^{-1} = \pm 2\frac{(\cos^2\varphi - \cos^2\Delta)^{1/2}}{\cos\Delta}. \tag{2.204}$$

These modes are evanescent. For other frequencies, in the passband, Λ is complex and modes propagate:

$$\Lambda = \mathrm{e}^{-\mathrm{i}\varphi_{\mathrm{sc}}}, \quad \cos\varphi_{\mathrm{sc}} = \frac{\cos\varphi}{\cos\Delta}. \tag{2.205}$$

Denote by $\tilde{\alpha}$ and $\tilde{\beta}$ emission and reception parameters of the mixed matrix in the transmission eigenmode basis (these are calculated in Appendix D). The wave emitted leftwards (rightwards) by the cell $n > 0$ (< 0), of amplitude $\tilde{\alpha}_1 U_n (\tilde{\alpha}_2 U_{-n})$, crosses $n-1$ cells before reaching acoustic port 2 (1) of the central cell (Fig. 2.48). Its complex amplitude is multiplied by Λ^{n-1}. For a harmonic excitation, $U_n = U\exp(-2\pi\mathrm{i}ns)$, resultants of eigenmode amplitudes are

$$\tilde{S}(p/2) = \tilde{\alpha}_1 U \sum_{n=1}^{\infty} \Lambda^{n-1}\mathrm{e}^{-2\pi\mathrm{i}ns},$$

$$\tilde{R}(-p/2) = \tilde{\alpha}_2 U \sum_{n=1}^{\infty} \Lambda^{n-1} \mathrm{e}^{2\pi i n s} .$$

The current intensity $I(s)$ entering the centre electrode is given by (2.201) and the harmonic admittance $Y(s) = I(s)/U(s)$ by

$$Y(s) = \tilde{y} + \tilde{\beta}_1 \tilde{\alpha}_2 \mathrm{e}^{2\pi \mathrm{i} s} \sum_{n=1}^{\infty} (\Lambda \mathrm{e}^{2\pi \mathrm{i} s})^{n-1} + \tilde{\beta}_2 \tilde{\alpha}_1 \mathrm{e}^{-2\pi \mathrm{i} s} \sum_{n=1}^{\infty} (\Lambda \mathrm{e}^{-2\pi \mathrm{i} s})^{n-1} .$$

Since $\tilde{\beta}_1 = -\tilde{\alpha}_1$ and $\tilde{\beta}_2 = -\tilde{\alpha}_2$, the formula

$$Y(s) = \tilde{y} - \tilde{\alpha}_1 \tilde{\alpha}_2 \left(\frac{\mathrm{e}^{2\pi \mathrm{i} s}}{1 - \Lambda \mathrm{e}^{2\pi \mathrm{i} s}} + \frac{\mathrm{e}^{-2\pi \mathrm{i} s}}{1 - \Lambda \mathrm{e}^{-2\pi \mathrm{i} s}} \right) \tag{2.206}$$

becomes

$$\begin{aligned} Y(s) &= \tilde{y} - \frac{\tilde{\alpha}_1 \tilde{\alpha}_2}{\Lambda} \frac{2\cos 2\pi s - 2\Lambda}{\Lambda^{-1} + \Lambda - 2\cos 2\pi s} \\ &= \tilde{y} + \frac{\tilde{\alpha}_1 \tilde{\alpha}_2}{\Lambda} \left[1 - \frac{\Lambda^{-1} - \Lambda}{\Lambda^{-1} + \Lambda - 2\cos 2\pi s} \right] . \end{aligned}$$

When $s = 0$, all electrodes are at the same potential U and there is no current. Hence, $Y(0) = 0$ and subtracting the expression for $Y(0)$,

$$Y(s) = \frac{\tilde{\alpha}_1 \tilde{\alpha}_2}{\Lambda} \frac{\Lambda^{-1} - \Lambda}{2} \left[\frac{1}{\frac{\Lambda^{-1} + \Lambda}{2} - 1} - \frac{1}{\frac{\Lambda^{-1} + \Lambda}{2} - \cos 2\pi s} \right] .$$

Putting

$$Y_{\mathrm{sc}} = -\frac{\tilde{\alpha}_1 \tilde{\alpha}_2}{\Lambda} , \quad \Lambda = \mathrm{e}^{-\mathrm{i}\varphi_{\mathrm{sc}}} \Rightarrow \begin{cases} \Lambda^{-1} + \Lambda = 2\cos\varphi_{\mathrm{sc}} \\ \Lambda^{-1} - \Lambda = 2\mathrm{i}\sin\varphi_{\mathrm{sc}} \end{cases} , \tag{2.207}$$

the harmonic admittance can be put into a form similar to (2.195)

$$\boxed{Y(s) = \frac{\mathrm{i} Y_{\mathrm{sc}}}{\tan(\varphi_{\mathrm{sc}}/2)} \frac{1 - \cos 2\pi s}{\cos\varphi_{\mathrm{sc}} - \cos 2\pi s}} . \tag{2.208}$$

If the frequency lies in the stop band, the pole at $s_{\mathrm{sc}} = \varphi_{\mathrm{sc}}/2\pi$ is no longer on the real axis, because the phase difference φ_{sc} is complex. Near the pole, an approximate expression is

$$Y(s) = Y_{\mathrm{r}} + Y_{\mathrm{p}} \frac{z_{\mathrm{p}}}{z - z_{\mathrm{p}}} , \quad z_{\mathrm{p}} = \sin^2 \frac{\varphi_{\mathrm{sc}}}{2} , \quad z = \sin^2 \pi s .$$

Y_{r} is a residual admittance, assumed constant, and Y_{p} is the pole amplitude, related to the electromechanical coupling coefficient.

For simplicity, let us now consider a symmetric cell, $\alpha_1 = \alpha_2 = \alpha_{\mathrm{S}}$. In the transmission eigenmode basis, $\tilde{r} = 0$ and $\tilde{t} = \Lambda$. Therefore, according to (D.48) in Appendix D, the amplitude Y_{sc} of the acoustic interaction is equal to the conductance G_{S} of the symmetric mode:

$$Y_{sc} = -\frac{\alpha_S^2}{\Lambda} = -\frac{\alpha_S^2}{\tilde{r} + \tilde{t}} = \tilde{G}_S \, .$$

Using (D.52) with $G_A = 0$ and $G_S = G$, it follows that

$$Y_{sc} = -\frac{\alpha_S^2}{\tilde{r} + \tilde{t}} = \tilde{G} = G \frac{\sin \varphi_{sc}}{\sin(\varphi + \Delta)} \, . \tag{2.209}$$

After some calculation,

$$Y_p = \frac{iY_{sc}}{\tan(\varphi_{sc}/2)} = \frac{iG}{\cos \Delta} \frac{\cos(\varphi - \Delta)/2}{\sin(\varphi + \Delta)/2} = \frac{iG}{\cos \Delta} A \, . \tag{2.210}$$

If Δ is positive, at the beginning of the stop band ($\varphi = \pi - \Delta$), $A = \sin \Delta$ is positive; at the end of the stop band ($\varphi = \pi + \Delta$), $A = 0$. The opposite result is obtained if $\Delta < 0$. We can therefore determine the sign of Δ, i.e., of the reflection coefficient, by locating the zero of Y_p. In the case of an asymmetric cell ($\delta \neq 0$), the general expression [2.46]

$$Y_p = \frac{iG}{\cos \Delta} \left[\frac{\cos(\varphi - \Delta)/2}{\sin(\varphi + \Delta)/2} \cos^2 \delta + \frac{\cos(\varphi + \Delta)/2}{\sin(\varphi - \Delta)/2} \sin^2 \delta \right] \tag{2.211}$$

shows that the zero of Y_p lies within the stop band. It is exactly in the middle ($\varphi = \pi$) if $\delta = \pi/4$, i.e., when directionality is maximal.

These results are illustrated in Fig. 2.49. For a transducer of period $p =$ 10 μm, curves show z_p and Y_p as functions of frequency. The first two graphs correspond to an electroacoustically symmetric configuration ($\delta = 0$). The crystal is ($Y + 128°$)-cut, X propagating lithium niobate and plots refer to two different values of finger thickness h. The curve $z_p(f)$ has a parabolic shape which bounds the stop band, for which φ_{sc} is complex and hence $z_p = \sin^2(\varphi_{sc}/2) \geq 1$. When $h/\lambda = 2\%$ (Fig. 2.49a), the amplitude Y_p of the pole vanishes at the end of the stop band (i.e., $r > 0$). Conversely, when $h/\lambda = 6\%$ (Fig. 2.49b), Y_p vanishes at the beginning of the stop band (i.e., $r < 0$). The reflection coefficient therefore changes sign for a certain electrode thickness. This phenomenon is related to the opposite signs of electrical and mechanical reflection coefficients. In the case of a quartz crystal ($Y + 42.5°$)-cut, ($X + 25°$)-propagating), the zero of $Y_p(f)$ is in the middle of the stop band (Fig. 2.49c). This natural directionality effect, corresponding to a separation of $\lambda/8$ between generation and reflection centres, was first observed by Wright [2.35].

2.6.2.3 Numerical Model. Given the dimensions of SAW devices – several hundred wavelengths long and several tens of wavelengths wide – an exact simulation of the way they work by Finite Element Methods (FEM) is not feasible. It would require too many nodes and hence too many variables. However, numerical methods can be developed to model all or part of the operation of SAW devices, up to a few approximations. The first of these, justified by the fact that acoustical apertures are sufficiently wide, is to treat the system as 2-dimensional, assuming electrodes to be infinitely wide.

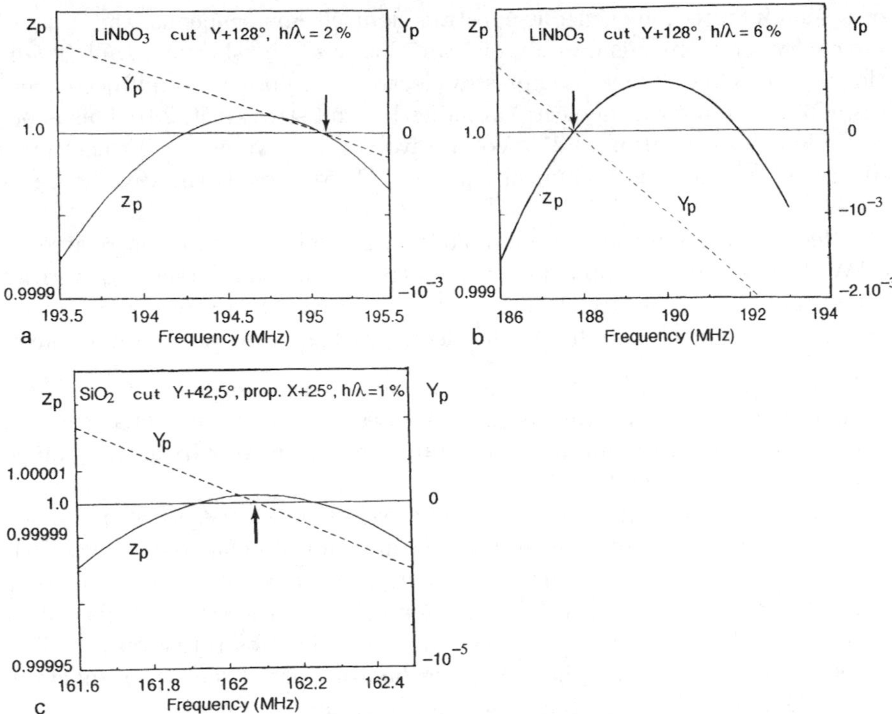

Fig. 2.49. $z_\mathrm{p}(f)$ and $Y_\mathrm{p}(f)$ for a $p = 10$ μm transducer, on a $(Y + 128°)$-cut, X-propagating lithium niobate crystal. (**a**) $h/\lambda = 2\%$. (**b**) $h/\lambda = 6\%$. (**c**) $(Y + 42.5°)$-cut, $(X + 25°)$-propagating quartz crystal, with $h/\lambda = 1\%$. Taken from [2.46]. ©1997 IEEE

Two types of model are then possible. One of these simulates the response of a complete filter, the other provides characteristic parameters for propagation and generation. We then model the infinite periodic electrode array in Fig. 2.48, excited by a harmonic potential, in order to calculate its harmonic admittance.

In both cases, the Boundary Element Method (BEM) can be applied. Even for quite large structures, this leads to computations of reasonable size. The method is based on the Green function for the electrical and mechanical response of all or part of the device (in this case, the piezoelectric substrate), to an electrical excitation (charge) and mechanical excitation (stress) at the surface (see Appendix E). In certain cases, particularly in electrostatics, an exact expression can be found for the Green function [see (2.62)]. However, most acoustic problems require numerical solution of an eighth order polynomial determinant (see Sect. 5.3.1, Vol. I). The Fourier transform of the Green function is calculated, replacing differential operators by multiplicative factors which are powers of the variables in the spectral domain. Given that the system of differential equations governing electroacoustic behaviour of a semi-infinite homogeneous substrate is itself homogeneous, the *spectral Green function* depends on only one variable, viz., the slowness k/ω. Traditionally, the Green function expresses electrostatic potential and mechanical displacements in terms of electrical induction and stresses (E.20). This is because Poisson's equation (3.21), Vol. I, involves the divergence of the electric displacement and the dynamical equation (3.15), Vol. I, the divergence of stress.

Mechanical quantities are eliminated when solving problems related to SAW devices. Indeed, applied quantities (potential) and detected quantities (current) are both electrical. They are eliminated differently depending on whether mechanical load due to the electrodes has been taken into account.

- If the electrode thickness is negligible, the surface can be treated as mechanically free, by putting normal stresses at the surface equal to zero. Quantities are then eliminated by simply reducing the Green function to its electrical term (Sect. 2.3).
- In the case where mechanical effects of the electrodes must be taken into account, we need first to understand their mechanical behaviour. Given their great diversity, the only reliable method, for general electrode shape, is to use FEM-type techniques. We can then calculate the stress-displacement relation induced by the electrode at the electrode–substrate interface. This relation, combined with the purely mechanical part of the Green function, can then be used to eliminate mechanical variables.

The difficulty with BEM lies in discretising the problem, particularly at the stage when charge and stress distributions must be given in terms of a theoretically infinite but discrete number of parameters which are to be determined numerically. These distributions are expressed as linear combinations of basis functions. It is essential to make a good choice of basis, since it will determine the number of parameters required. As the function $1/\sqrt{1-(2x/a)^2}$ provides a good description of singularities at electrode edges $x = \pm a/2$, expansion of the charge density,

$$\sigma_j(x') = \frac{2}{a_j}\sum_{n=0}^{\infty}\sigma_j^{(n)}\frac{T_n(2x'/a_j)}{\sqrt{1-(2x'/a_j)^2}}\,, \tag{2.212}$$

in a series of Chebyshev polynomials of the first kind $T_n(z)$ can be limited to an order R which is generally less than 10. Here, x' is the abscissa relative to the centre of the jth electrode, of width a_j. In order to calculate the parameters $\sigma_j^{(n)}$, the potential $\Phi_i(x)$ at electrode i is written as the convolution of charge distributions with the Green function $G(x)$ and identified with the imposed potential distribution. The latter is constant at each electrode. If N_e is the total number of electrodes,

$$\Phi_i(x) = \sum_{j=1}^{N_\mathrm{e}}\int_{-a_j/2}^{+a_j/2} G(x - x' + d_{ij})\sigma_j(x')\,\mathrm{d}x'\,,$$

where x is the abscissa relative to the centre of the ith electrode and d_{ij} the distance between ith and jth electrode centres. Substituting (2.212) into this relation,

$$\Phi_i(x) = \sum_{j=1}^{N_e} \sum_{n=0}^{R} \sigma_j^{(n)} \int_{-a_j/2}^{+a_j/2} G(x - x' + d_{ij}) \frac{T_n(2x'/a_j)}{\sqrt{1-(2x'/a_j)^2}} \frac{2}{a_j} \mathrm{d}x' \,. \tag{2.213}$$

In order to make the identification $\Phi_i(x) = U_i$, the Chebyshev polynomial basis proves itself useful once more, limiting to the first $(R+1)$ terms ($m = 0$ term being equal to 1):

$$\frac{1}{\pi} \int_{-a_i/2}^{+a_i/2} \Phi_i(x) \frac{T_m(2x/a_i)}{\sqrt{1-(2x/a_i)^2}} \frac{2}{a_i} \mathrm{d}x = U_i \delta_m \,, \tag{2.214}$$

where δ_m is the Kronecker delta. Since $T_m(\cos\theta) = \cos m\theta$, putting

$$x = \frac{a_i}{2} \cos\theta \,, \quad x' = \frac{a_j}{2} \cos\phi \,,$$

the required system of equations becomes

$$\sum_{j=1}^{N_e} \sum_{n=0}^{R} \frac{\sigma_j^{(n)}}{\pi} \int_0^{\pi} \int_0^{\pi} G\left(d_{ij} + \frac{a_i}{2}\cos\theta - \frac{a_j}{2}\cos\phi\right) \cos m\theta \cos n\phi \mathrm{d}\theta \mathrm{d}\phi = U_i \delta_m .$$

Introducing the spectral Green function $\overline{G}(k)$, defined by

$$G(x) = \frac{1}{2\pi} \int_{-\infty}^{\infty} \overline{G}(k) \mathrm{e}^{-\mathrm{i}kx} \mathrm{d}k \,,$$

into the previous relation, yields

$$\begin{aligned} U_i \delta_m = & \frac{1}{2\pi^2} \sum_{j=1}^{N_e} \sum_{n=0}^{R} \sigma_j^{(n)} \int_{-\infty}^{\infty} \overline{G}(k) \mathrm{e}^{-\mathrm{i}kd_{ij}} \mathrm{d}k \times \\ & \int_0^{\pi} \mathrm{e}^{-\mathrm{i}k(a_i/2)\cos\theta} \cos m\theta \, \mathrm{d}\theta \int_0^{\pi} \mathrm{e}^{\mathrm{i}k(a_j/2)\cos\phi} \cos n\phi \, \mathrm{d}\phi \,. \end{aligned}$$

Recalling the Bessel function integral equation,

$$\frac{1}{\pi} \int_0^{\pi} \mathrm{e}^{\mathrm{i}x\cos\phi} \cos n\phi \, \mathrm{d}\phi = \mathrm{i}^n J_n(x) \,, \tag{2.215}$$

we find, for $0 \le m \le R$,

$$\frac{1}{2} \sum_{j=1}^{N_e} \sum_{n=0}^{R} \mathrm{i}^{n-m} \sigma_j^{(n)} \int_{-\infty}^{\infty} \overline{G}(k) \mathrm{e}^{-\mathrm{i}kd_{ij}} J_m\left(\frac{ka_i}{2}\right) J_n\left(\frac{ka_j}{2}\right) \mathrm{d}k = U_i \delta_m \,. \tag{2.216}$$

In theory, the integral over wave number k required to calculate coefficients $\sigma_j^{(n)}$ must be numerical, since $\overline{G}(k)$ results from a numerical calculation.

However, in practice, this is not possible, since $G(k)$ varies extremely rapidly. The Green function must be decomposed into its electrostatic, Rayleigh wave and pseudo-surface wave contributions. Their specific variations, which are logarithmic, pole and pseudo-pole (see Sect. 2.3), are integrated analytically, whilst the residual part of $\overline{G}(k)$ is integrated numerically [2.47]. Once the system has been solved, we obtain the current intensity in each electrode by integrating the charge density. As an example, Fig. 2.50 shows the admittance of a wideband DART-SPUDT transducer, based on a $(Y+36°)$-cut lithium tantalate crystal, generating a pseudo-surface wave (Sect. 2.7). The agreement between calculated curve and measurement is all the more remarkable in that there is a Surface Skimming Bulk Wave (SSBW) with speed very close to that of the surface wave.

To calculate the harmonic admittance of a periodic transducer, the Green function $G(x)$ is replaced by a periodised function

$$G_{\mathrm{p}}(x,\, s) = \sum_{n=-\infty}^{+\infty} G(x+np)\mathrm{e}^{2\pi ins}\,, \tag{2.217}$$

which has Fourier transform

$$\overline{G}_{\mathrm{p}}(k,\, s) = \int_{-\infty}^{\infty} G_{\mathrm{p}}(x,\, s)\mathrm{e}^{\mathrm{i}kx}\mathrm{d}x = \overline{G}(k) \sum_{n=-\infty}^{+\infty} \exp -2\pi\mathrm{i}n\left(\frac{kp}{2\pi} - s\right)\,.$$

The standard relation

$$\sum_{n=-\infty}^{+\infty} \mathrm{e}^{-2\pi\mathrm{i}nx} = \sum_{q=-\infty}^{+\infty} \delta(x-q)\,,$$

applied with $x = kp/2\pi - s$, yields

$$\overline{G}_{\mathrm{p}}(k,\, s) = \overline{G}(k) \sum_{q=-\infty}^{+\infty} \delta\left(\frac{kp}{2\pi} - s - q\right)$$

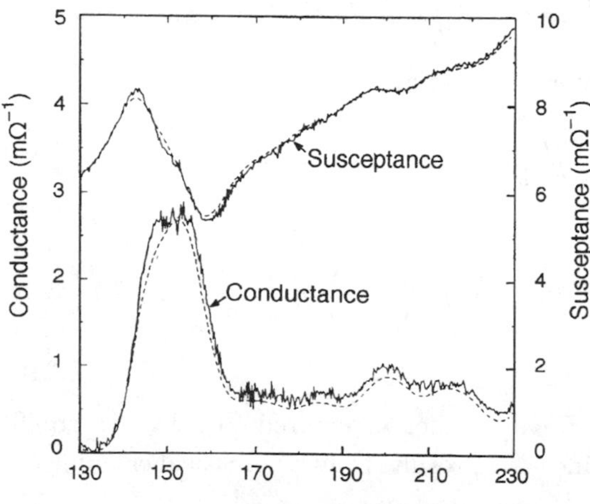

Fig. 2.50. Electrical admittance of a transducer with centre frequency 150 MHz, based on $(Y+36°)$-cut, X propagating lithium tantalate. *Dashed curve* simulation, *continuous curve* experiment. Figure 2 of [2.47]. ©1995 IEEE

$$= \frac{2\pi}{p}\overline{G}(k) \sum_{q=-\infty}^{+\infty} \delta\left[k - \frac{2\pi}{p}(s+q)\right], \tag{2.218}$$

which gives the spectral Green function. This can then be substituted into (2.216) to yield parameters $\sigma_j^{(n)}$.

2.6.3 Equivalent Circuits

It has been shown by Smith et al. [2.16] that a section of length $2d$ of an interdigital transducer can be represented by an equivalent electromechanical circuit. This representation is based on the assumption that one of the two electric field components plays a dominant role. If it is assumed that the normal component is constant and generates mainly Rayleigh waves, the period of length $2d$ is equivalent to two bulk wave resonators placed side by side, as shown in Fig. 2.51b, one being excited by a voltage U_3 and the other by $-U_3$. Substituting equivalent circuits (Redwood or Mason–Redwood, see Sect. 1.3.1) for each resonator, we obtain the equivalent circuit for the $2d$ section.

Resonators exciting Rayleigh waves in a direction perpendicular to the excitation field constitute the *crossed-field model.* If Rayleigh waves are generated rather by the tangential component of the electric field, the section of length $2d$ is equivalent to two resonators vibrating as in Fig. 2.51c, and hence to the circuit obtained by their association. In this case, the direction of wave propagation is parallel to the exciting electric field and this is referred to as the *in-line field model.*

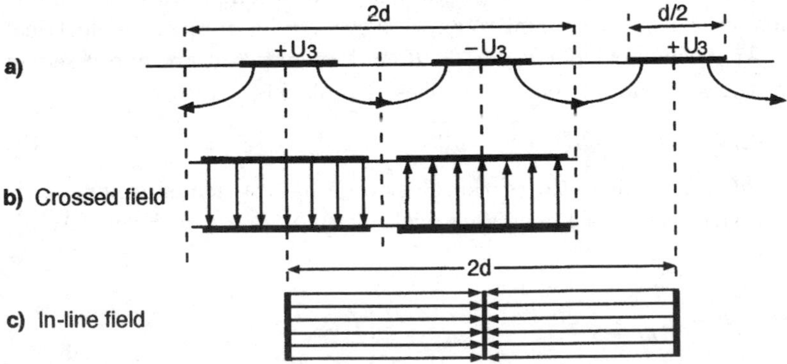

Fig. 2.51. Bulk wave resonator model for a surface wave transducer. (**a**) In a $2d$ period of the transducer, the normal component of the electric field changes sign, as does the tangential component. (**b**) If it is assumed that the normal component is constant and generates mainly Rayleigh waves, the section of length $2d$ can be replaced by a pair of bulk wave resonators, each of thickness d, vibrating as shown. This is the crossed-field model. (**c**) If Rayleigh waves are mainly excited by the tangential component, the two resonators vibrate as shown. This is the in-line field model. Taken from [2.16]

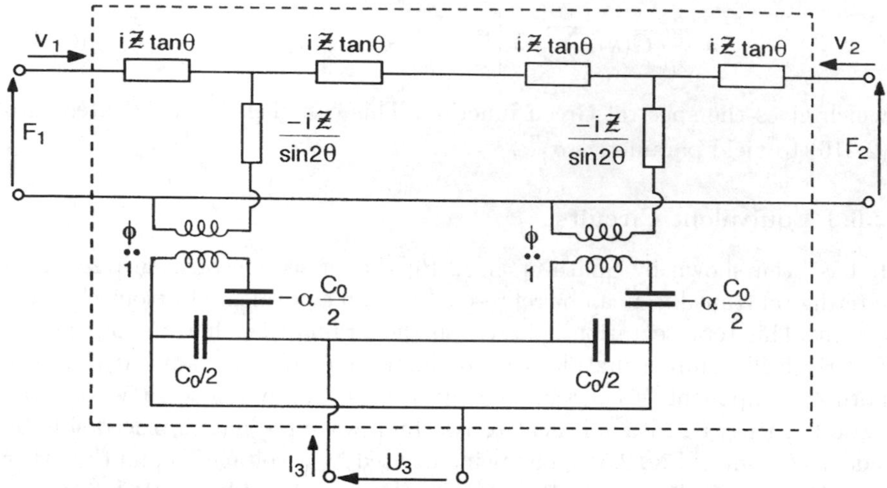

Fig. 2.52. Equivalent circuit for a length $2d$ section of an interdigital transducer, constructed by associating the circuits for two resonators of thickness d ($\theta = kd/2$), one excited by voltage U_3, the other by $-U_3$. Crossed-field model $\alpha = 0$. In-line field model $\alpha = 1$

The two equivalent electromechanical circuits for the $2d$ section, corresponding to the crossed-field and in-line field models, can be joined into a single circuit. Indeed, their circuit diagrams differ only by a capacitor with negative capacitance (Sect. 1.3.1). The diagram in Fig. 2.52 corresponds to the crossed-field model for $\alpha = 0$, and the in-line field model for $\alpha = 1$.

In order to characterise the elementary section of length $2d$ of the interdigital transducer, it is convenient to express everything in terms of electrical quantities. At mechanical ports, forces F and velocities v are expressed in terms of voltages U and current intensities I, replacing F_i, v_i by

$$U_i = F_i/\phi\,, \quad I_i = \phi v_i \quad (i = 1,\, 2)\,, \tag{2.219}$$

where $\phi = hC_0/2$ is the ratio of the electromechanical transformer. C_0 is the static capacitance of a finger pair with overlap length w. From (A.13), Appendix A,

$$C_0 = \varepsilon^S w = \frac{\varepsilon^S A}{d/2}\,,$$

where $A = wd/2$ is the area of one electrode. Mechanical impedance $\mathcal{Z}$ corresponds to electrical resistance $R = \mathcal{Z}/\phi^2$, i.e., to conductance $G = 1/R$ given by

$$G = \frac{\phi^2}{\mathcal{Z}} = \frac{h^2 C_0^2}{4\rho V A} = \frac{e^2 V_{\mathrm{R}}}{4(\varepsilon^S)^2 c^D}\,\frac{\varepsilon^S A}{d/2}\,\frac{C_0}{A} = K^2 C_0 \frac{V_{\mathrm{R}}}{2d} = K^2 C_0 f_0\,. \tag{2.220}$$

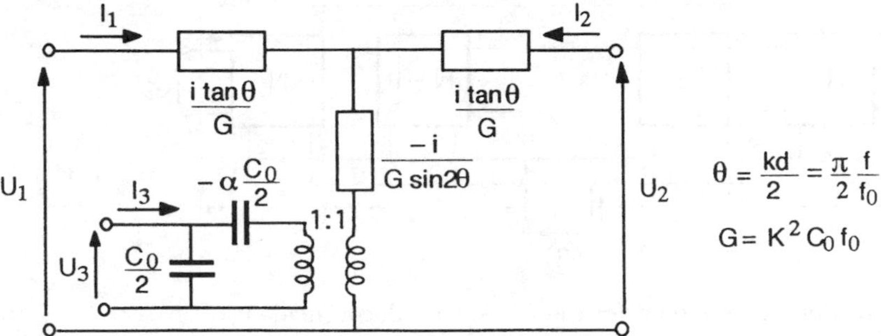

Fig. 2.53. Equivalent electrical circuit for a section of length d and static capacitance $C_0/2$. All quantities are electrical, unlike for the Mason circuit

The ratio of the electrical transformer in the equivalent circuit is then equal to 1 (see Fig. 2.53).

The unit shown in Fig. 2.53 is characterised by a matrix of electrical admittances y_{ij}, giving current intensities in terms of voltages:

$$I_i = y_{ij} U_j \qquad (i,\, j = 1,\, 2,\, 3)\ . \tag{2.221}$$

Symmetry of the circuit reduces the number of independent entries to four (y_{11}, y_{12}, y_{13}, y_{33}):

$$(y_{ij}) = \begin{pmatrix} y_{11} & y_{12} & y_{13} \\ y_{12} & y_{11} & -y_{13} \\ y_{13} & -y_{13} & y_{33} \end{pmatrix}\ . \tag{2.222}$$

For example, in the crossed-field model, admittances are given by (Problem 2.12)

$$\begin{aligned} y_{11} &= -\mathrm{i}G\cot 4\theta\ , \qquad & y_{12} &= \frac{\mathrm{i}G}{\sin 4\theta}\ , \\ y_{13} &= -\mathrm{i}G\tan\theta\ , \qquad & y_{33} &= \mathrm{i}(C_0\omega + 4G\tan 4\theta)\ . \end{aligned} \tag{2.223}$$

The diagram for the whole transducer is found by connecting the N sections with mechanical ports in cascade and electrical ports in shunt, as shown in Fig. 2.54. It is characterised by a matrix (Y) resulting from assembly of elementary units, obtained via a transfer matrix:

$$\overline{I}_i = Y_{ij}\overline{V}_j\ . \tag{2.224}$$

For the crossed-field model,

$$\begin{aligned} Y_{11} &= -\tfrac{\mathrm{i}}{Z_\mathrm{e}}\tfrac{1}{\tan 4N\theta}\ , \qquad & Y_{12} &= \frac{\mathrm{i}}{Z_\mathrm{e}}\frac{1}{\sin 4N\theta}\ , \\ Y_{13} &= -\tfrac{\mathrm{i}}{Z_\mathrm{e}}\tan\theta\ , \qquad & Y_{33} &= \mathrm{i}C_\mathrm{T}\omega + \frac{\mathrm{i}}{Z_\mathrm{e}}4N\tan\theta\ . \end{aligned} \tag{2.225}$$

At the synchronism frequency ($\theta = \pi/2$), the tan function becomes infinite and it must be expanded about this frequency in order to yield formulas for quantities such as input impedance and radiation conductance.

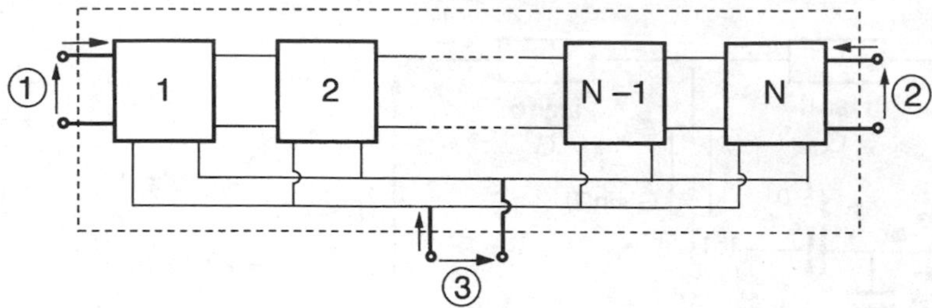

Fig. 2.54. Equivalent circuit for the whole transducer, made up of N sections with mechanical ports connected in cascade and electrical ports in shunt

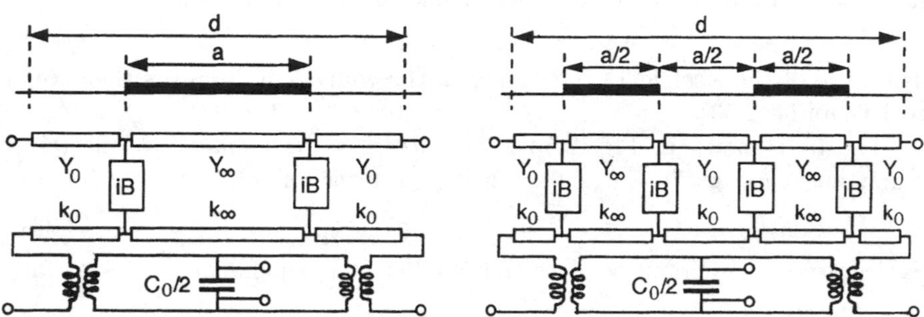

Fig. 2.55. Equivalent circuits for (*left*) a section of electrode finger and (*right*) a section of a split finger. Admittances and wave numbers of the bare surface (Y_0, k_0) and metallised surface (Y_∞, k_∞) are not the same. Susceptance iB represents energy localised near finger edges. Taken from Inagawa and Koshiba [2.51]. ©1994 IEEE

It is not obvious which model to choose. According to Smith et al. [2.48], the crossed-field model is preferable if more energy is stored by the normal electric field component than by the parallel component, and conversely. Experience shows that predictions of the crossed-field model correspond better to experimental results if the substrate has strong electromechanical coupling (e.g., lithium niobate) and that predictions of the in-line field model are preferable if coupling is weak (e.g., quartz).

The value to be taken for the electromechanical coupling coefficient of the material is not the one corresponding to the bulk wave resonator, but rather the value for surface waves. This is because the electric field is distributed differently. However, the value must be partly based on experience and corrected by a filling factor [2.49].

This type of equivalent circuit often yields quite satisfactory results for input impedance and radiation resistance, globally speaking. However, it has been modified on several occasions, e.g., by inserting inactive elements in series or reactive elements in parallel, with a view to modelling a range of different effects: reflections from impedance discontinuities, and energy stored under finger edges, partly due to evanescent bulk waves [2.50].

Figure 2.55 represents a more elaborate circuit [2.51] which is equivalent to a section of an ordinary finger and a section of a split finger.

One of the reasons for describing transducers by equivalent electromechanical circuits is that their operation and design can be studied (simulated) using algorithms developed for electrical circuits.

2.7 Materials and Technology

A surface wave component includes (at least) a piezoelectric substrate and an interdigital transducer. The two are inseparable. Piezoelectricity ensures direct and reverse electroacoustic conversion. The transducer structure determines the source distribution and hence the desired use. The monocrystalline substrate usually plays the double role of generating and propagating medium. Consequently, the Rayleigh wave materials described in Sect. 5.3.4.5, Vol. I, whose characteristics were summarised in Fig. 5.27, Vol. I, are of prime importance. However, the need for low-cost, general purpose filters has led to development of inhomogeneous substrates (isotropic solid + piezoelectric layer). The use of isotropic solids such as glasses is less justifiable nowadays, since the growing market has led to a fall in crystal prices.

Today, there is a great need for components operating at centre frequencies greater than 2 GHz, mainly due to the development of mobile phones. In order to contain the technological challenge ($f = 3$ GHz, $V = 3\,000$ m/s $\rightarrow \lambda/4 = 0.25$ μm), the benefits of higher propagation speeds are sought. Given the very small number of piezoelectric solids and the speed of Rayleigh waves, of order 3 to 4 000 m/s, two trends have been followed. One involves using a heterogeneous substrate. This is made up of a non-piezoelectric material chosen for its high speed (e.g., sapphire or diamond in the form of a membrane), carrying a piezoelectric overlay (e.g., ZnO, $LiTaO_3$, $LiNbO_3$). The other approach involves finding cuts in piezoelectric monocrystals liable to generate and propagate pseudo-surface waves with speeds higher than the Rayleigh wave speed. The dominant component in such waves is the horizontal transverse or even longitudinal component [2.52, 2.53] and they are also excited by interdigital transducers. Very strong coupling coefficients (20%) are associated with certain cuts and directions. These waves are sensitive to the thickness of the metal, which determines how close to the surface they are held, and they propagate in certain directions with very low losses. Table 2.1 is intended as a complement to Fig. 5.27, Vol. I.

Electrode Lithography. The finger width in an interdigital electrode is generally equal to a quarter wavelength ($\lambda_0/4$). When the operating frequency is less than 300 MHz, the wavelength is greater than 10 μm ($\lambda_0/4 > 2.5$ μm) for most materials. These transducers can be made by standard photolithographic techniques used in microelectronics.

Table 2.1. Characteristics of piezoelectric materials, generating and propagating media for surface waves and pseudo-surface waves

Crystal	Cut and direction		Type	Speed [m/s]	K^2 [%]	$\varepsilon_p/\varepsilon_0$	$\mathrm{d}\tau/\tau\mathrm{d}\theta$ [10^{-6} K^{-1}]
$LiNbO_3$	$Y + 41°$	X	Pseudo	4840	21	63	−90
$LiNbO_3$	$Y + 64°$	X	Pseudo	4695	11.6	52	−90
$LiNbO_3$	$Y + 128°$	X	SAW	3980	5.4	56	−75
$LiTaO_3$	$Y + 36°$	X	Pseudo	4240	4.7	50	−35
$LiTaO_3$	X	$Y + 112°$	SAW	3300	0.85	48	−18
Quartz	$Y + 36°$	X	SAW	3150	0.12	5	0

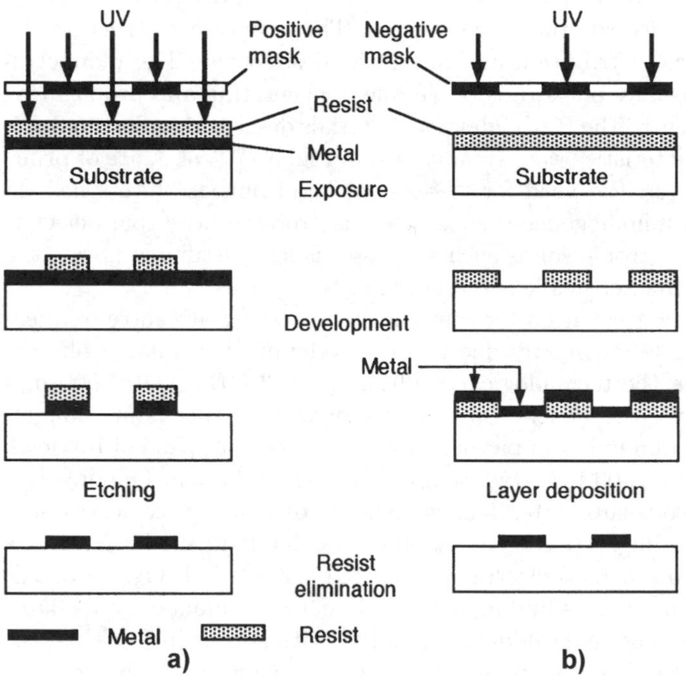

Fig. 2.56. Stages in the production of an interdigital transducer. (**a**) Classic approach. (**b**) Lift-off technique

A variation (known as lift-off) makes it possible to produce transducers operating up to 1 000 MHz. When the frequency is well above this value, the finger width $\lambda_0/4 < 0.8$ μm is of the same order as the wavelength of light. Diffraction effects then become prohibitive and comb patterns must be traced out using electron lithography or masking. Indeed, wavelengths associated with electrons are much shorter: $\lambda < 1$ Å for electrons of energy 10 keV.

Conventional technology (Fig. 2.56a) for integrated circuits involves variations of the following procedure. First, a homogeneous and uniform metallic film (e.g., a 5 000 Å aluminium film) is vacuum deposited onto a polished plane substrate (e.g., quartz or lithium niobate monocrystal, in the form of a small plate, several [cm^2] in area and one or two [mm] thick). The metal is then coated with a photosensitive resin, using a photoresist spinner. The plate carrying the photograph (chromium on glass) of the transducer is applied to the resin. When the latter is exposed to UV light, it is polymerised. Exposed parts of the resin, and then the metal which is no longer protected, are removed by etching. The photographic mask is produced from a large-scale drawing using a 1×1 m^2 coordinatograph. Several (even hundreds of) devices can be produced simultaneously on the same substrate.

Figure 2.56b shows the various stages of the lift-off technique which produces better defined finger edges [2.54]: resin is deposited directly on the substrate, the mask is applied and the resin exposed to UV can be removed by etching; the whole surface is then metallised and lift-off consists in dissolving the resin, which swells and takes any metal coating with it. This technique, used with the appropriate optical projection of the mask, leads to excellent results. An example is shown in Fig. 2.57.

Electron lithography requires a scanning microscope device [2.56, 2.57]. The resin is exposed to an electron pencil of diameter < 500 Å. With the lift-off technique, the procedure is as follows. First, a 200 Å aluminium film is deposited on an insulating substrate, in order to suppress electrostatic surface charge. Then a 4 000 Å layer of polymethyl methacrylate (PMM) resin is applied and exposed to a 20 keV electron beam which automatically reproduces the transducer pattern. Exposed resin is selectively etched and a second aluminium film (thickness $\sim$ 1 500 Å) is evaporated over both the resulting notches and remaining resin. Finally, the resin is dissolved, bringing

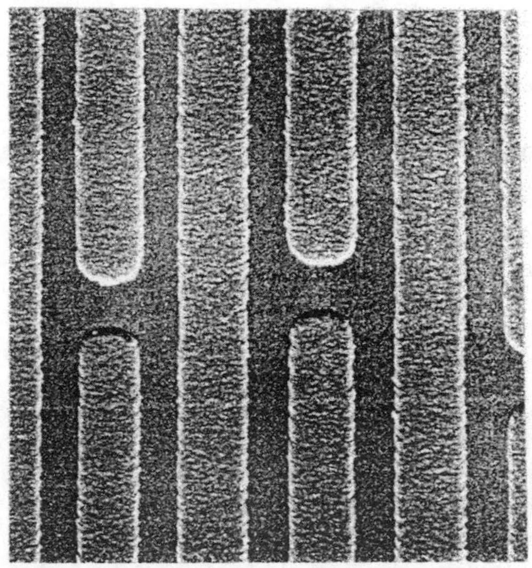

Fig. 2.57. Part of a transducer (f_0 = 2.45 GHz). Strip width 0.4 μm. Figure 3 in [2.55]. ©1996 IEEE

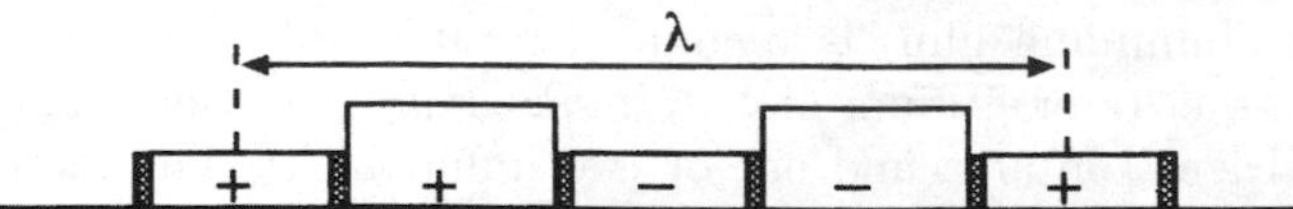

Fig. 2.58. Structure of a unidirectional transducer with centre frequency 10 GHz. The width of each aluminium finger is $\lambda_0/4 = 0.09$ μm, and its thickness is 350 or 500 Å. Each finger is insulated from its neighbours by a thin aluminium oxide film

the aluminium coating with it. Metallic strips produced in this way, future electrode fingers, are merely superimposed upon the original thin film of aluminium. Ionic bombardment removes film lying between the fingers, and also reduces finger height by an amount equal to the thickness of the original film. Transducers with fingers of width about 0.25 μm and operating at centre frequency $f_0 = 3\,500$ MHz, were first made by this process in 1970 [2.58]. It has since been perfected and transducers operating first at 5 GHz and then at 10 GHz have been produced [2.59]. Transducers built on a $LiNbO_3$ substrate [($Y+128°$)-cut, X propagating, speed 4 000 m/s] with centre frequency 10 GHz, require finger widths less than 0.1 μm [2.60]. Figure 2.58 shows the structure of $f_0 = 10$ GHz unidirectional transducers schematically.

In this structure, unidirectionality is produced by differing reflection coefficients at the fingers, due to their different heights (aluminium thickness). It could also be produced by using different metals. Each finger is insulated from its neighbour by an aluminium oxide film, formed by lateral oxidisation of the first deposited fingers. The second layer, in which the second series of fingers is cut, is evaporated at an oblique angle so as to leave the insulating regions at finger ends intact. Let us end by noting that the frequency response of transducers with $f_0 = 15$ GHz has been observed [2.60]. Given the demand for devices operating at higher and higher frequencies, techniques are being studied for reproducing smaller and smaller patterns (of size inversely proportional to the frequency). An example is dry process lithography using plasmas or ion or electron beams [2.61]. The term 'dry' distinguishes this process from etching which uses liquids.

Piezoelectric Layers. The above techniques are used to make transducers on a piezoelectric substrate which is generally monocrystalline. The substrate doubles as generating and propagating medium. In order to launch waves in a general non-piezoelectric substrate, a piezoelectric layer must be deposited on it, at least at its ends. Having produced monocrystal filters, it was natural at the beginning of the 1970s to seek a way of using a zinc oxide film. At the time, such films were being used to generate and detect high frequency ($f > 1$ GHz) bulk waves in a rod (see Sect. 1.5). These films were first deposited by the same methods, using diode and triode systems [2.62]. However, the problem posed was slightly different. Indeed, the active area required in a bulk wave transmission line is less than 1 mm^2. If the film is not uniform, it is simple enough to find a region with the right crystallographic orientation having these dimensions. Moreover, the number of such delay lines required per year was small at the time (and it still is, being just a few thousand). The time needed to deposit a film of thickness 1 μm was about one hour, of little relevance for the type of delay line used in most applications (e.g., radar) and manufactured as the need arose in batches of about ten.

The film area required for an interdigital transducer is, of course, higher, especially if it operates at frequencies less than 50 MHz. Likewise for its thickness (of order $\lambda_0/2$). It is then difficult to ensure geometric and crystallographic uniformity, when deposition times can exceed 24 hours. However, Japanese scientists had already succeeded in applying this technique to the fabrication of television filters, on a glass substrate [2.63]. Deposition times exceeded 50 hours! Because low manufacturing costs are related to fast, high-volume production, work was carried out to increase both speed and area of deposition. The solution adopted was to place magnets near the target in a diode system [2.64]. Electrons are then confined by the magnetron effect, following cycloidal paths around magnetic field lines parallel to the target surface (Fig. 2.59). The number of collisions increases thereby raising the plasma density. The deposition rate is multiplied by a factor of about 10 (> 10 μm/h). Using this kind of instrument, the Japanese company Murata has developed a very large number of television filters.

This setup has given rise to many variations in which, for example, cathode shapes are different, the target is no longer a zinc oxide disk but a zinc disk, the oxygen/argon plasma is maintained by a constant rather than a high frequency voltage and an electron gun is included [2.65, 2.66, 2.67]. Industrial systems are available commercially. ZnO film deposition is still a topical subject [2.68], with a view to increasing operational frequencies yet further. Devices operating at frequencies above 1 GHz have been built on glass substrates [2.69], and others on high-speed substrates such as sapphire and diamond films [2.70, 2.71].

Other piezoelectric materials can also be deposited in the form of thin films. Aluminium nitride AlN is a competitor for zinc oxide. Lithium niobate, first deposited by Foster in 1969 [2.72], and lithium tantalate are being

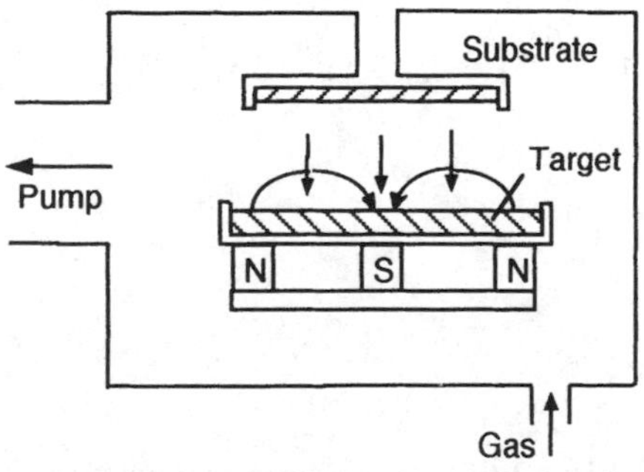

Fig. 2.59. Zinc oxide film deposition in a diode configuration with magnetron effect. The magnetic field produced by a magnet under the ZnO target increases the deposition rate. From [2.64]

investigated in the context of new techniques [2.73], viz., deposition based on laser pulse ablation of the target.

Problems

2.1 Consider an anisotropic material for which the slowness curve of the Rayleigh wave in the (x_1, x_3)-plane has a parabolic shape about the x_1 direction (Fig. 2.60): $s(\theta) = s_R(1 + \beta\theta^2)$, where $s_R = 1/V_R$ and β is the anisotropy parameter. The acoustic field is found at (x_1, x_3) by the angular spectrum method:

$$u(x_1, x_3) = \int_{-\infty}^{\infty} \overline{u}(k_3) \exp(-\mathrm{i}k_1 x_1 - \mathrm{i}k_3 x_3)\, \mathrm{d}k_3 \,, \qquad (2.226)$$

where $k_1^2 + k_3^2 = k^2(\theta)$.

(a) What does $\overline{u}(k_3)$ represent? Give an expression in the case of a uniform transducer of width w.

(b) Assuming θ to be small, find an expression for $u(x_1, x_3)$. What happens when $\beta = 1/2$?

Solution. $\overline{u}(k_3)$ is the Fourier transform of the profile of the source located at $x_1 = 0$:

$$\overline{u}(k_3) = \frac{1}{2\pi} \int_{-\infty}^{\infty} u(0, x_3) \exp(\mathrm{i}k_3 x_3)\, \mathrm{d}x_3 \,.$$

If this profile is uniform,

$$\overline{u}(k_3) = \frac{w}{2\pi} \sin\left(k_3 \frac{w}{2}\right) .$$

Since

$$k(\theta) = \omega s(\theta) = k_R(1 + \beta\theta^2) \,, \quad k_R = \omega s_R = \omega/V_R \,,$$

and $\beta\theta^2 \ll 1$ implies $k_3 \approx k\theta$, it follows that

$$k_1 \approx k(1 - \theta^2)^{1/2} = k_0(1 + \beta\theta^2)\left(1 - \frac{\theta^2}{2}\right) \approx k_0 - (1 - 2\beta)\frac{\theta^2}{2} k_0 \,.$$

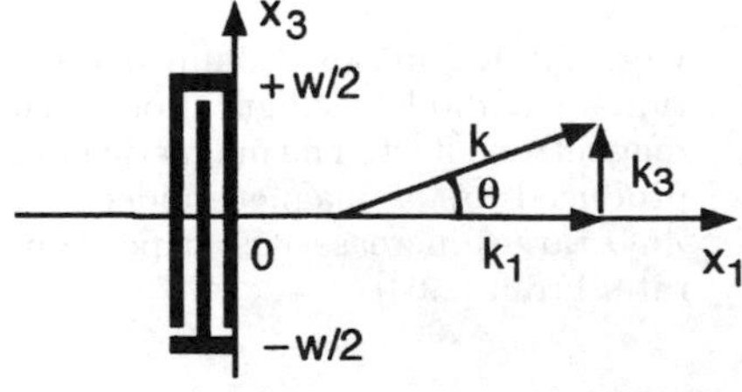

Fig. 2.60. Transducer of aperture w on an anisotropic medium

Substituting $k_1 \approx k_0 - (1-2\beta)k_3^2/2k_0$ into (2.226),

$$u(x_1, x_3) = \mathrm{e}^{-\mathrm{i}k_0 x_1} \int_{-\infty}^{\infty} \overline{u}(k_3) \exp\left[\mathrm{i}\frac{k_3^2}{2k_0}(1-2\beta)x_1 - \mathrm{i}k_3 x_3\right] \mathrm{d}k_3 .$$

If $\beta = 1/2$, $k_1 = k_0$ and the wave profile is x_1-independent. Diffraction effects are compensated by the curvature of the slowness surface. Generally, the beam profile at distance x_1 from the transducer is the profile it would have at distance $(1-2\beta)x_1$ in the isotropic medium.

2.2 Starting from the reflection and transmission coefficients r and t of an electrode or a groove, calculate reflection and transmission coefficients R and T for an array, made of N reflectors with period p. Assume that $r \ll 1$.

Solution. The incident wave a_n on a reflector gives rise to a reflected wave b_n and a transmitted wave a_{n+1} (Fig. 2.33). Measuring phases of a_n, b_n with respect to the centre $x_n = np$ of the nth reflector, we find

$$a_{n+1} = ta_n\mathrm{e}^{-\mathrm{i}kp} + rb_{n+1}\mathrm{e}^{-2\mathrm{i}kp} ,$$

$$b_n = ra_n + tb_{n+1}\mathrm{e}^{-\mathrm{i}kp} \quad \Rightarrow \quad tb_{n+1} = -ra_n\mathrm{e}^{\mathrm{i}\phi} + b_n\mathrm{e}^{\mathrm{i}\phi} ,$$

where $\phi = kp = \pi f/f_0$. Then, to first order ($t \approx 1$, $r^2 \ll 1$),

$$\begin{pmatrix} a_{n+1} \\ b_{n+1} \end{pmatrix} = \begin{pmatrix} \mathrm{e}^{-\mathrm{i}\phi} & r\mathrm{e}^{-\mathrm{i}\phi} \\ -r\mathrm{e}^{\mathrm{i}\phi} & \mathrm{e}^{\mathrm{i}\phi} \end{pmatrix} \begin{pmatrix} a_n \\ b_n \end{pmatrix} = (T_1) \begin{pmatrix} a_n \\ b_n \end{pmatrix} .$$

The transmission matrix (T_N) of an N-reflector array is found from T_1:

$$\begin{pmatrix} a_{N+1} \\ b_{N+1} \end{pmatrix} = (T_N) \begin{pmatrix} a_1 \\ b_1 \end{pmatrix} , \quad (T_N) = (T_1)^N = \begin{pmatrix} A_N & B_N \\ C_N & D_N \end{pmatrix} .$$

The condition of no reflection for the outgoing wave at the end of the array is

$$b_{N+1} = C_N a_1 + D_N b_1 = 0 , \quad a_{N+1} = A_N a_1 + B_N b_1 .$$

The reflection coefficient of the array is

$$R = \frac{b_1}{a_1} = -\frac{C_N}{D_N} \quad \Rightarrow \quad a_{N+1} = \frac{A_N D_N - B_N C_N}{D_N} a_1 ,$$

and the transmission coefficient is

$$T = \frac{a_{N+1}}{a_1} = \frac{||T_N||}{D_N} = \frac{||T_1||^N}{D_N} = \frac{1}{D_N} ,$$

because energy conservation implies $||T_1|| = 1$.

2.3 In the discrete source model (Fig. 2.10), the impulse response of a transducer results from sampling of the weighting function $w(x)$, reproducing variations in finger overlap length, by an alternating sequence of delta functions, separated by the interdigital distance d. Hence, $h(t) = w(x)e(x)$, where $x = V_\mathrm{R}t$ and $e(x)$ is the sampling function shown in Fig. 2.61b. What is the frequency response of the transducer?

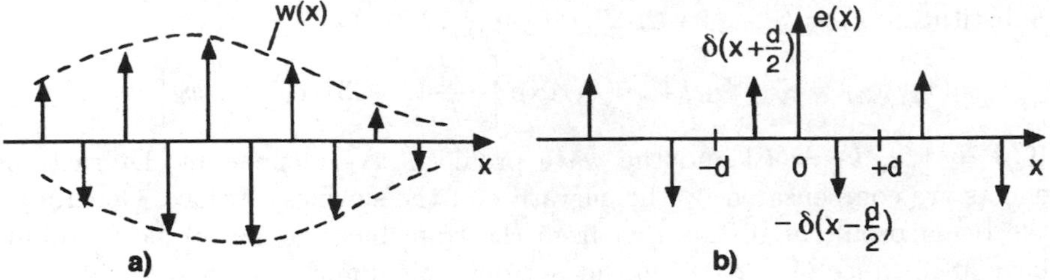

Fig. 2.61. Weighting and sampling functions

Solution. $e(x)$ is periodic with period $2d$ and its Fourier series expansion is

$$e(x) = \sum_{n=-\infty}^{\infty} E_n \exp\left(\mathrm{i}n\frac{2\pi x}{2d}\right) , \quad E_n = \frac{1}{2d}\int_{-d}^{+d} e(x) \exp\left(-\mathrm{i}n\frac{\pi x}{d}\right) \mathrm{d}x \, .$$

The frequency response

$$\begin{aligned} H(\omega) &= \int_{-\infty}^{\infty} h(t) \exp(-\mathrm{i}\omega t)\, \mathrm{d}t \\ &= \sum_{n=-\infty}^{\infty} E_n \int_{-\infty}^{\infty} w(V_{\mathrm{R}}t) \exp\left[-\mathrm{i}\left(\omega - n\frac{\pi}{d}V_{\mathrm{R}}\right)t\right] \mathrm{d}t \\ &= \sum_{n=-\infty}^{\infty} E_n W(\omega - n\omega_0) \end{aligned}$$

is given by the Fourier transform $W(\omega)$ of the weighting function $w(V_{\mathrm{R}}t)$, multiplying by coefficients E_0, E_1, E_2, ... and summing over its values relative to frequencies $0, \pm f_0, \pm 2f_0, \ldots$. Since $E_n = (\mathrm{i}/d)\sin(n\pi/2)$, i.e., $E_0 = E_2 = \ldots = 0$, only bands centred on odd multiples of the synchronism frequency f_0 survive.

2.4 Show that real and imaginary parts $G(f)$, $B(f)$ of a physical quantity $Y(f)$ satisfying the causality principle are related by Hilbert transform.

Solution. The Fourier transform $y(t)$ of $Y(f)$ is causal if $y(t) = \Gamma(t)y(t)$, where $\Gamma(t)$ is the unit step function. Applying the convolution theorem and using the Fourier transform of $\Gamma(t)$,

$$Y(f) = \left[\frac{1}{2}\delta(f) + \mathrm{PP}\left(\frac{1}{2\pi\mathrm{i}f}\right)\right] \otimes Y(f) \Rightarrow Y(f) = \mathrm{PP}\left(\frac{1}{\pi\mathrm{i}f}\right) \otimes Y(f) \, ,$$

where PP indicates principal part. Hence,

$$G(f) + \mathrm{i}B(f) = \frac{1}{\mathrm{i}\pi}\mathrm{PP}\int_{-\infty}^{\infty} \frac{G(\nu) + \mathrm{i}B(\nu)}{f - \nu}\mathrm{d}\nu \Rightarrow B(f) = \frac{1}{\pi}\mathrm{PP}\int_{-\infty}^{\infty} \frac{G(\nu)}{\nu - f}\mathrm{d}\nu \, .$$

2.5 For a receiver transducer, show that the sum of acoustic powers reflected, transmitted and converted into electrical signal equals the power P_{I} transported by the incident wave.

Solution. From (2.77), (2.92) and (2.93), reflected and transmitted powers are given by

$$P_{\mathrm{R}} = \frac{G_{\mathrm{a}}^2}{|Y_{\mathrm{T}} + Y_{\mathrm{L}}|^2} P_{\mathrm{I}} \,, \; P_{\mathrm{T}} = \left(1 + \frac{G_{\mathrm{a}}^2}{|Y_{\mathrm{T}} + Y_{\mathrm{L}}|^2} - 2G_{\mathrm{a}} \frac{\mathrm{Re}(Y_{\mathrm{T}} + Y_{\mathrm{L}})}{|Y_{\mathrm{T}} + Y_{\mathrm{L}}|^2}\right) P_{\mathrm{I}} \,,$$

where $G_{\mathrm{a}} = \mathrm{Re}(Y_{\mathrm{T}})$. Then formula (2.95) for the power supplied to the electrical load of conductance $G_{\mathrm{L}} = \mathrm{Re}(Y_{\mathrm{L}})$ yields

$$P_{\mathrm{R}} + P_{\mathrm{T}} + P_{\mathrm{L}} = \left(1 + \frac{2G_{\mathrm{a}}^2}{|Y_{\mathrm{T}} + Y_{\mathrm{L}}|^2} + \frac{2G_{\mathrm{a}} G_{\mathrm{L}}}{|Y_{\mathrm{T}} + Y_{\mathrm{L}}|^2} - 2G_{\mathrm{a}} \frac{G_{\mathrm{a}} + G_{\mathrm{L}}}{|Y_{\mathrm{T}} + Y_{\mathrm{L}}|^2}\right) P_{\mathrm{I}} \,,$$

which is just equal to P_{I}.

2.6 For a symmetric transducer, express σ_{11}^2 and σ_{12}^2 as functions of the efficiency $\eta = \sigma_{13}^2$ and the angle $\phi = \theta_{11} - \theta_{12} \pm \pi$. Show that the reflection coefficient σ_{11}^2 goes through a maximum when $\phi = 0$. Calculate the level of the triple-transit signal $\mathrm{TTS} = 10 \log(\sigma_{11}^4)_{\min}$ and insertion losses $\mathrm{IL} = 10 \log \eta^2$ for a filter comprising two identical transducers when $\eta = 0.5$, 0.4, 0.1 and $\eta \ll 1$.

Solution. Since $s_{13} = s_{23}$ and $s_{11} = s_{22}$:

$$\begin{aligned}
&(2.99\mathrm{a}, 2.99\mathrm{b}) && \Rightarrow \sigma_{11}^2 + \sigma_{12}^2 = 1 - \eta \,, \\
&(2.99\mathrm{c}) && \Rightarrow \sigma_{33}^2 = 1 - 2\eta \,, \\
&(2.99\mathrm{d}) && \Rightarrow 2\mathrm{Re}(s_{11} s_{12}^*) + \eta = 0 \quad \Rightarrow \quad 2\sigma_{11}\sigma_{12} \cos\phi = \eta \,.
\end{aligned}$$

Reflected and transmitted powers σ_{11}^2 and σ_{12}^2 are solutions of

$$\Sigma^2 + 2(1 - \eta)\Sigma + \eta^2(1 + \tan^2\phi) = 0 \,, \quad \Sigma = 2\sigma^2 \,.$$

Roots are

$$\left.\begin{matrix} 2\sigma_{12}^2 \\ 2\sigma_{11}^2 \end{matrix}\right\} = 1 - \eta \pm (1 - 2\eta - \eta^2 \tan^2\phi)^{1/2} \,, \quad |\phi| < \arccos\left(\frac{\eta}{1 - \eta}\right) .$$

The reflection coefficient σ_{11}^2 is minimum when $\phi = 0$:

$$(\sigma_{11}^2)_{\min} = \frac{1}{2}\left[1 - \eta - (1 - 2\eta)^{1/2}\right] = \frac{1}{4}\left[1 - (1 - 2\eta)^{1/2}\right]^2 .$$

The level of the triple-transit echo for a two-transducer filter depends on the efficiency η:

$$(\mathrm{TTS})_{\min} = 10 \log(\sigma_{11}^4)_{\min} = -12\,\mathrm{dB} + 40 \log\left[1 - (1 - 2\eta)^{1/2}\right] ,$$

i.e., it depends on insertion losses $\mathrm{IL} = 10 \log \eta^2 = 20 \log \eta$. Hence,

$\eta = 0.5 \quad \mathrm{IL} = -6\,\mathrm{dB} \quad \mathrm{TTS} = -12\,\mathrm{dB}$

$\eta = 0.4 \quad \mathrm{IL} = -8\,\mathrm{dB} \quad \mathrm{TTS} = -22.3\,\mathrm{dB}$

$\eta = 0.1 \quad \mathrm{IL} = -20\,\mathrm{dB} \quad \mathrm{TTS} = -51\,\mathrm{dB}$

$\eta \ll 1 \quad \rightarrow \quad \mathrm{TTS} = (-12 + 2\mathrm{IL})\,\mathrm{dB}.$

The level of the TTS in [dB] decreases twice as fast as the level of the direct signal.

2.7 Establish the power conservation relation $\kappa_{ba} = -\kappa_{ab}^*$ between coefficients κ_{ab} and κ_{ba}, defined in (2.125a, 2.125b).

Solution. Expand each side of (2.126),

$$\frac{\mathrm{d}}{\mathrm{d}x}(aa^*) = a^*\frac{\mathrm{d}a}{\mathrm{d}x} + \mathrm{cc.}\,,$$

where cc. denotes complex conjugate. Multiplying (2.125a) by a^*, it follows, since k'_R is real, that

$$\frac{\mathrm{d}}{\mathrm{d}x}(aa^*) = -\mathrm{i}k'_\mathrm{R}|a|^2 - \mathrm{i}\kappa_{ab}a^*b + \mathrm{cc.} = -\mathrm{i}(\kappa_{ab}a^*b - \kappa_{ab}^*ab^*)\,.$$

Likewise for (2.125b), the power conservation relation (2.126) implies

$$\mathrm{i}(\kappa_{ab}a^*b - \kappa_{ab}^*ab^*) = \mathrm{i}(\kappa_{ba}b^*a - \kappa_{ba}^*ba^*)\,,$$

and hence

$$(\kappa_{ba} + \kappa_{ab}^*)(a^*b - ab^*) = 0\,.$$

Equality is only possible for all a, b if $\kappa_{ba} = -\kappa_{ab}^*$.

2.8 Deduce relations (2.153) from the power conservation relation (2.152).

Solution. Since δ is real and $\xi = -\zeta^*$, (2.150a, 2.150b) lead to

$$\frac{\mathrm{d}}{\mathrm{d}x}(RR^*) = R^*\frac{\mathrm{d}R}{\mathrm{d}x} + \mathrm{cc.} = -\mathrm{i}\kappa R^*S + \mathrm{i}\zeta R^*U + \mathrm{i}\kappa^*RS^* - \mathrm{i}\zeta^*RU^*\,,$$

$$\frac{\mathrm{d}}{\mathrm{d}x}(SS^*) = S^*\frac{\mathrm{d}S}{\mathrm{d}x} + \mathrm{cc.} = \mathrm{i}\kappa^*S^*R + \mathrm{i}\xi S^*U - \mathrm{i}\kappa SR^* - \mathrm{i}\xi^*SU^*\,,$$

and to

$$\frac{\mathrm{d}}{\mathrm{d}x}(RR^* - SS^*) = \mathrm{i}\zeta R^*U + \mathrm{i}\zeta^*S^*U - \mathrm{i}\zeta^*RU^* - \mathrm{i}\zeta SU^*\,.$$

The right-hand side of (2.152) expands to give

$$\frac{1}{2}\left(U^*\frac{\mathrm{d}I}{\mathrm{d}x} + \mathrm{cc.}\right) = \frac{1}{2}(\mathrm{i}\alpha U^*R + \mathrm{i}\beta U^*S + \mathrm{i}C\omega|U|^2) + \mathrm{cc.}$$
$$= \frac{1}{2}(-\mathrm{i}\alpha^*R^*U - \mathrm{i}\beta^*S^*U) + \frac{1}{2}(\mathrm{i}\alpha RU^* + \mathrm{i}\beta SU^*)\,.$$

The equality in (2.152) can only be satisfied for all R, S and U if $\alpha = -2\zeta^*$ and $\beta = -2\zeta$.

2.9 Find expressions for amplitudes $R(x)$ and $S(x)$ of modes propagating under a passive array of length L as a function of $R(0)$, using $S(L) = 0$.

Solution. From (2.134) ($x_1 = x$, $x_0 = L$), with $S(L) = 0$,

$$R(x) = M_{11}(x-L)R(L) + M_{12}(x-L)S(L) \Rightarrow R(L) = \frac{R(0)}{M_{11}(-L)} ,$$

and it follows that

$$R(x) = \frac{M_{11}(x-L)}{M_{11}(-L)}R(0) \Rightarrow \frac{R(x)}{R(0)} = \frac{\gamma\cos\gamma(L-x) + \mathrm{i}\delta\sin\gamma(L-x)}{\gamma\cos\gamma L + \mathrm{i}\delta\sin\gamma L} .$$

At the centre frequency $\delta = 0 \Rightarrow \gamma = \mathrm{i}|\kappa|$:

$$R(x) = \frac{\cosh|\kappa|(L-x)}{\cosh|\kappa|L}R(0) \approx R(0)\mathrm{e}^{-|\kappa|x} ,$$

where the approximation is valid for $|\kappa|(L-x) \gg 1$.

In the same way, the amplitude of mode S,

$$S(x) = M_{21}(x-L)R(L) + M_{22}(x-L)S(L) ,$$

when $S(L) = 0$, is given by

$$S(x) = \frac{M_{21}(x-L)}{M_{11}(-L)}R(0) = -\frac{\mathrm{i}\kappa^*\sin\gamma(L-x)}{\gamma\cos\gamma L + \mathrm{i}\delta\sin\gamma L}R(0) .$$

Then for $\omega = \omega_0$,

$$S(x) = -\mathrm{i}\frac{\kappa^*}{|\kappa|}\frac{\sinh|\kappa|(L-x)}{\cosh|\kappa|L}R(0) .$$

2.10 Calculate the squares of the moduli of the numerator N_{R} and denominator D in (2.172).

Solution. First Case: $\delta^2 > |\kappa|^2 \Rightarrow \gamma$ real. Putting $\gamma L/2 = X$, with $\gamma^2 = \delta^2 - |\kappa|^2$, we find

$$N_{\mathrm{R}} = (\kappa - \delta)\sin X + \mathrm{i}\gamma\cos X = (\kappa' - \delta)\sin X + \mathrm{i}(\gamma\cos X + \kappa''\sin X) ,$$

and hence

$$|N_{\mathrm{R}}|^2 = (\kappa'^2 + \delta^2 - 2\kappa'\delta)\sin^2 X + \gamma^2\cos^2 X + \kappa''^2\sin^2 X + \gamma\kappa''\sin 2X$$

or

$$|N_{\mathrm{R}}|^2 = \delta^2 + |\kappa|^2(\sin^2 X - \cos^2 X) - 2\kappa'\delta\sin^2 X + \gamma\kappa''\sin 2X .$$

Results for mode S are found by changing $\kappa \to \kappa^*$, i.e., replacing κ'' by $-\kappa''$, whence

$$|N_{\mathrm{R,\,S}}|^2 = \delta^2 - |\kappa|^2\cos 2X - 2\kappa'\delta\sin^2 X \pm \gamma\kappa''\sin 2X .$$

Likewise, $|D|^2 = \gamma^2 \cos^2 2X + \delta^2 \sin^2 2X = \delta^2 - |\kappa|^2 \cos^2 2X$.

Second Case: $\delta^2 < |\kappa|^2 \Rightarrow \gamma = \mathrm{i}\chi$. Putting $\gamma L/2 = \mathrm{i}X$, with $\chi^2 = |\kappa|^2 - \delta^2$, we find

$$\begin{aligned} N_\mathrm{R} &= \mathrm{i}(\kappa - \delta) \sinh X - \chi \cosh X \\ &= \mathrm{i}(\kappa' - \delta) \sinh X - (\chi \cosh X + \kappa'' \sinh X) . \end{aligned}$$

This implies

$$|N_\mathrm{R}|^2 = (\kappa'^2 + \delta^2 - 2\kappa'\delta) \sinh^2 X + \chi^2 \cosh^2 X + \kappa''^2 \sinh^2 X + \chi\kappa'' \sinh 2X$$

or

$$|N_\mathrm{R}|^2 = -\delta^2 + |\kappa|^2 (\sinh^2 X + \cosh^2 X) - 2\kappa'\delta \sinh^2 X + \chi\kappa'' \sinh 2X .$$

Grouping expressions for the two modes,

$$|N_{\mathrm{R,\,S}}|^2 = -\delta^2 + |\kappa|^2 \cosh 2X - 2\kappa'\delta \sinh^2 X \pm \chi\kappa'' \sinh 2X .$$

For the denominator $D = \mathrm{i}\chi \cosh 2X - \delta \sinh 2X \Rightarrow |D|^2 = |\kappa|^2 \cosh^2 2X - \delta^2$.

2.11 Using (2.74) and the reciprocity relation (2.90), calculate the incoming current intensity I at a shorted electrode as a function of the electric potential Φ_I at its centre. Given equations (2.120) and (2.121), find the intensity I_n at the nth electrode in terms of incoming and outgoing wave amplitudes for the cell in Fig. 2.62.

Solution. The electric potential accompanying a wave emitted by an electrode at potential U is

$$\Phi_\mathrm{S}(f) = \mathrm{i}\Gamma_\mathrm{R} \overline{\sigma}_{1\mathrm{e}}(f) U .$$

The short-circuit current induced by an incident wave of potential Φ_I is therefore

$$I_\mathrm{sc} = -\frac{\omega w}{\Gamma_\mathrm{R}} (\mathrm{i}\Gamma_\mathrm{R} \overline{\sigma}_{1\mathrm{e}}) \Phi_\mathrm{I} = -\mathrm{i}\omega w \overline{\sigma}_{1\mathrm{e}} \Phi_\mathrm{I} , \quad \Phi_\mathrm{I} = \Phi_\mathrm{I}^+ + \Phi_\mathrm{I}^- .$$

At the electrode centre, the potential is the sum of the potentials produced by incoming and outgoing waves. Given the phase differences $\pm kp/2$ between

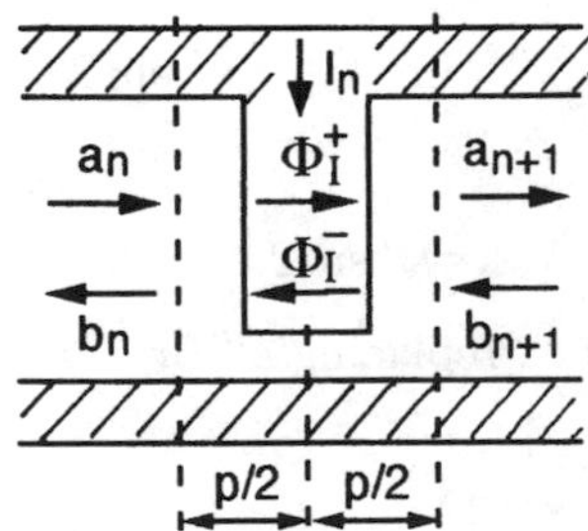

Fig. 2.62. Cell containing the nth electrode

the edges of the cell and its centre, and the coefficient of proportionality α, it follows that

$$\Phi_{\mathrm{I}}^{+} = \left(a_n \mathrm{e}^{-\mathrm{i}kp/2} + a_{n+1}\mathrm{e}^{\mathrm{i}kp/2}\right)/2\alpha \,,$$
$$\Phi_{\mathrm{I}}^{-} = \left(b_n \mathrm{e}^{\mathrm{i}kp/2} + b_{n+1}\mathrm{e}^{-\mathrm{i}kp/2}\right)/2\alpha \,, \quad \text{where} \quad \alpha = (\omega w/2\Gamma_{\mathrm{R}})^{1/2} \,.$$

The factor of 2 in the dominator ensures energy conservation (if $p \to 0$, $a_n = a_{n+1}$ and $\Phi_{\mathrm{I}}^{+} = a_n/\alpha$). Incoming current intensity at electrode n is

$$I_n = -\mathrm{i}\left(\frac{\omega w}{2}\Gamma_{\mathrm{R}}\right)^{1/2}\left[(a_n + b_{n+1})\mathrm{e}^{-\mathrm{i}kp/2} + (a_{n+1} + b_n)\mathrm{e}^{\mathrm{i}kp/2}\right]\overline{\sigma}_{1\mathrm{e}}(f) \,.$$

2.12 Using the equivalent circuit in Fig. 2.53, with $\alpha = 0$ (crossed-field model), draw the circuit representing a $2d$ length of an interdigital transducer. Calculate the entries of the admittance matrix (2.222).

Solution. The equivalent circuit for one pair of fingers is shown in Fig. 2.63. Kirchoff's rules imply for path EAM

$$U_1 = U_3 - \frac{\mathrm{i}}{G\sin 2\theta}(I_1 + I) + \frac{\mathrm{i}\tan\theta}{G}I_1 \,, \tag{2.227}$$

for path SBM,

$$U_2 = -U_3 - \frac{\mathrm{i}}{G\sin 2\theta}(I_2 - I) + \frac{\mathrm{i}\tan\theta}{G}I_2 \,, \tag{2.228}$$

and finally, for path ES,

$$U_1 - U_2 = \frac{\mathrm{i}\tan\theta}{G}(I_1 - I_2 - 2I) \,, \text{ or } 2I = I_1 - I_2 - \frac{G}{\mathrm{i}\tan\theta}(U_1 - U_2) \,. \tag{2.229}$$

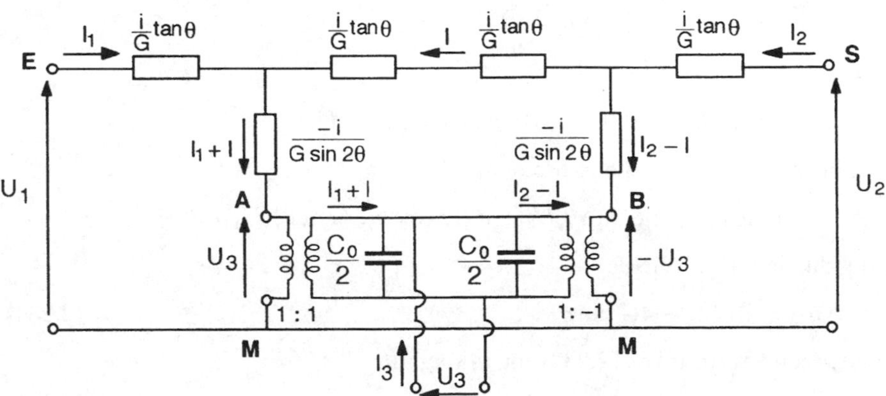

Fig. 2.63. Equivalent circuit for one pair of fingers in an interdigital transducer

Adding together (2.227) and (2.228) yields:

$$U_1 + U_2 = \frac{\mathrm{i}}{G}\left(\tan\theta - \frac{1}{\sin 2\theta}\right)(I_1 + I_2)$$

or

$$I_1 + I_2 = \mathrm{i}G\tan 2\theta(U_1 + U_2)\ ,$$

which implies

$$y_{22} = y_{11}\ , \quad y_{21} = y_{12}\ , \quad y_{23} = -y_{13}\ , \quad y_{11} + y_{21} = \mathrm{i}G\tan 2\theta\ . \tag{2.230}$$

Subtracting (2.228) from (2.227),

$$U_1 - U_2 = 2U_3 + \frac{\mathrm{i}}{G}\left(\tan\theta - \frac{1}{\sin 2\theta}\right)(I_1 - I_2) - \frac{\mathrm{i}}{G\sin 2\theta}2I\ .$$

Hence, by (2.229),

$$(U_1 - U_2)\left(1 - \frac{1}{2\sin^2\theta}\right) = 2U_3 + \frac{\mathrm{i}}{G}\left(\tan\theta - \frac{2}{\sin 2\theta}\right)(I_1 - I_2)\ .$$

Another variation, using $\tan\theta - 2/\sin 2\theta = -\cot\theta$, is

$$I_1 - I_2 = -\mathrm{i}G\cot 2\theta\,(U_1 - U_2) - 2\mathrm{i}G\tan\theta\, U_3\ . \tag{2.231}$$

Hence,

$$y_{11} - y_{12} = -\mathrm{i}G\cot 2\theta\ , \quad y_{13} = -\mathrm{i}G\tan\theta\ . \tag{2.232}$$

From (2.230), admittances y_{11} and y_{12} are given by

$$y_{11} = -\mathrm{i}G\cot 4\theta\ , \quad y_{12} = \frac{\mathrm{i}G}{\sin 4\theta}\ . \tag{2.233}$$

In order to determine the last admittance y_{33}, we express the current intensity I_3 in terms of the voltage U_3 in the primary circuit of the transformers (Fig. 2.63):

$$I_3 = \mathrm{i}C_0\omega U_3 - (I_1 - I_2 + 2I)\ .$$

Then by (2.229),

$$I_3 = \mathrm{i}C_0\omega U_3 - 2(I_1 - I_2) - \mathrm{i}G\cot\theta\,(U_1 - U_2)\ .$$

Replacing $I_1 - I_2$ by the expression in (2.231),

$$I_3 = (\mathrm{i}C_0\omega + 4\mathrm{i}G\tan\theta)U_3 + \mathrm{i}G(2\cot 2\theta - \cot\theta)(U_1 - U_2)\ ,$$

and the first term yields

$$y_{33} = \mathrm{i}(C_0\omega + 4G\tan\theta)\ . \tag{2.234}$$

The second term gives (2.232) for y_{13} again.

3. Elastic Waves and Light Waves

In this chapter, we shall present the relationships between elastic waves and light waves, under three headings: acousto-optic interactions, optical measurement of surface displacements, and generation of elastic waves by the photothermal effect.

The acousto-optic interaction between elastic waves in matter and light (electromagnetic) waves was first investigated in 1922 by Brillouin [3.1]. The first experiments were carried out in 1932 by Lucas and Biquard in France [3.2], and Debye and Sears in the United States [3.3]. Then, slightly later, further work was done by Raman and Nath in India [3.4]. This interaction arises because of changes in the dielectric permittivity of a medium due to the elastic wave, which then cause variations in the electric field of the light wave.

Diffraction of light by ultrasounds was observed during experiments on liquids. The effect was used to measure the propagation speed and attenuation of elastic waves in transparent materials. The maximum frequency of ultrasounds was around 30 MHz (in water, $\Lambda \approx 50$ μm and in ordinary glass, $\Lambda \approx 200$ μm).

New interest in elasto-optic interactions (often called acousto-optic interactions) was generated with the invention of the laser, a monochromatic light source of well defined shape and high energy density, together with development of piezoelectric transducers operating in the GHz range (high-frequency delay lines, Sect. 4.2.1; in a solid with $V = 5\,000$ m/s, $f = 1$ GHz $\Rightarrow \Lambda \approx 5$ μm). It was discovered that an elastic wave beam constituted a relatively simple means for rapidly changing the parameters of a light beam, e.g., intensity, direction, and frequency. Conversely, a light beam was an ideal probe for measuring certain characteristics of an elastic wave beam within a solid, e.g., homogeneity, attenuation and radiation pattern. At the time, studies were further justified by independent developments in both acoustics and optics: acoustoelectronic components were being developed for signal processing and ultrasonic imaging devices for non-destructive testing or obstetrics; optical fibres were under investigation for telecommunications and other purposes. This interaction, between laser beam and bulk elastic waves, gave rise to new components, such as modulators, deflectors, variable delay lines, spectrum analysers and tunable optical filters. The best known of all is the intensity modulator which has become an essential element in laser printers and optical interferometers. The acousto-optic interaction with surface waves has been less fruitful, possibly because integrated optics has not yet seen sufficient development.

The second part of the present chapter deals with probes used to measure mechanical displacements of a surface caused by various types of elastic wave. These displacements are of order a fraction of a nm, or several Å, for high-frequency waves. Devices based on deflection or diffraction of a light beam by surface ripples, which are used to study Rayleigh wave propagation, will be presented. Interferometers have proved themselves invaluable for investigating transducer vibration modes and radiation patterns. This is illustrated by examples of results obtained with a compact heterodyne interferometric probe. We shall also review the use of Doppler effect devices when the state of the moving surface is poor.

The third heading refers to generation of elastic wave beams by a laser. By emitting pulses lasting a few tens of ns and with energy of order 100 mJ, elastic waves are excited through the photoelastic effect. This generation technique is useful because no mechanical contact is required. The laser may be placed a good distance away from the medium in which waves are to be excited, for example, when the medium is at a high temperature. This technique is, of course, also well suited to optical detection.

3.1 Acousto-optic Interaction

The aim of this first section is to specify qualitatively, with the help of diagrams, the three main contexts in which this interaction occurs. Different names are often used: Brillouin effect, Debye–Sears effect, Lucas–Biquard effect, Raman–Nath effect or Bragg effect. We shall then review conditions of propagation for light waves in crystals, and define the elasto-optic tensor, whilst mentioning the electro-optic effect and describing the main results when interaction occurs at Bragg incidence, i.e., when there is a single diffracted beam. Surface waves also interact with light waves if the latter are concentrated close to the surface.

3.1.1 Three Cases of Interaction

The three cases are illustrated in Figs. 3.1, 3.2 and 3.3. A light beam passes through an isotropic solid in the form of a rod, in which an elastic wave beam is propagating.

First case. Figure 3.1. The width of the light beam is less than the wavelength Λ of the elastic waves. Slow variation in the optical refractive index of the medium only bends the light waves (mirage effect).

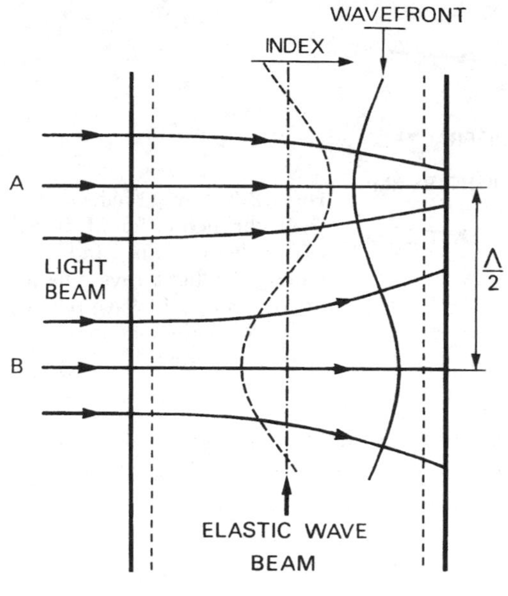

Fig. 3.1. Bending of light rays by an elastic wave. Rays in beam A converge as they cross a region of increasing index in the solid. Rays in beam B diverge, crossing a region of decreasing index. The elastic wave merely causes a periodic variation in the lateral dimensions of each light beam

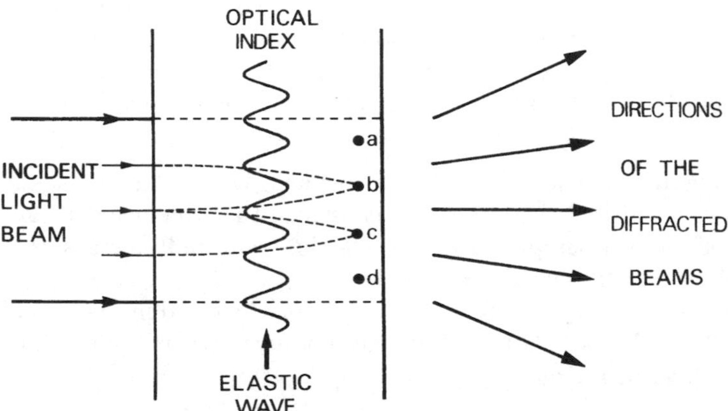

Fig. 3.2. Grating effect. When the light beam is wide compared with the elastic wavelength, many focal points a, b, ..., are formed by bending of light rays. These act as sources [3.2]. Interference of waves originating at these sources explains emission in the various directions. The same facts can also be interpreted in terms of the phase shift (of light waves) produced by the travelling wave of index variations (Sect. 3.1.4). A limiting value for the width of the elastic wave beam is given by the requirement that neighbouring diffracted beams should not overlap (the Raman–Nath condition)

Second Case. Figure 3.2. The width of the light beam is now greater than the elastic wavelength. Periodic variation of the index generates light beams of different intensities in different directions. This can be explained by assum-

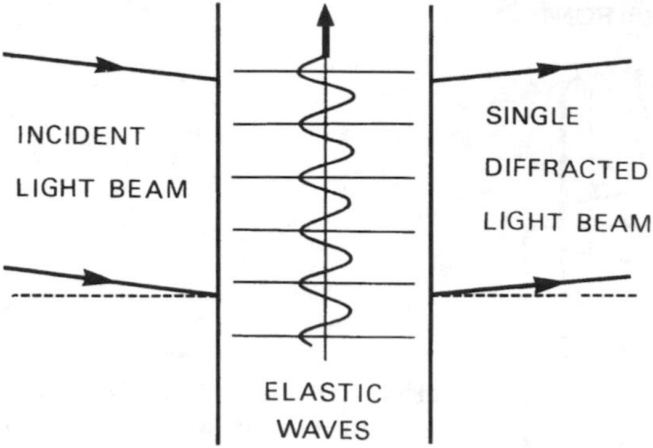

Fig. 3.3. Bragg effect. For the right inclination of the incident light beam, and provided that light rays cut homologous elastic wave planes, a single beam emerges from the interaction

ing that focal points are formed by bending of light rays and that these focal points then act as sources [3.2]. Interference of waves coming from the sources then explains how the beam divides into several distinct beams (grating effect). The phenomenon can also be analysed by considering the phase shift of light waves produced by the travelling wave of index variations [3.5]. This is the approach we shall adopt here. Whichever method is adopted, a limiting value for the width of the elastic wave beam is imposed by the requirement that neighbouring diffracted light beams should not overlap within the solid. Provided that the width of the elastic wave beam is just a fraction of this limit, which depends on both elastic and optical wavelengths (see Sect. 3.1.4), the interaction is said to occur under Raman–Nath conditions.

Third Case. Figure 3.3. The angle between incident light beam and elastic beam, and also the width of the latter, are chosen so that the incident light beam reflects off the various planes of the elastic beam and emerges as a single beam (Bragg conditions).

Interactions only partially satisfying the conditions involved in the second and third cases have also been studied but have not given rise to applications and will not be described here.

3.1.2 Propagation of Light Waves in Crystals

Light transmission through a crystal at rest gives rise to double refraction or birefringence. This is due to natural anisotropy of the crystalline medium. A slab of calcite ($CaCO_3$) splits incident light into two linearly polarised beams whose polarisations are perpendicular to one another. The anisotropy which causes double refraction is accentuated and even created in isotropic media or in cubic crystals, where it does not occur naturally, by applying external forces. An electric field thereby produces the electro-optic effect and mechanical strains lead to the elasto-optic effect studied here.

The behaviour of crystals with respect to light waves can be interpreted through the index ellipsoid, or index surface, predicted by Maxwell's equations for propagation of electromagnetic waves. We shall now summarise the main points.

3.1.2.1 Index Ellipsoid. In an isotropic medium, whichever direction is considered and whatever its polarisation may be, a light wave propagates at speed c:

$$c = \frac{1}{\sqrt{\varepsilon\mu}} = \frac{1}{\sqrt{\varepsilon_0\mu_0}}\frac{1}{\sqrt{\varepsilon_{\mathrm{r}}\mu_{\mathrm{r}}}} = \frac{c_0}{\sqrt{\varepsilon_{\mathrm{r}}\mu_{\mathrm{r}}}} = \frac{c_0}{n} ,$$

where ε_0, μ_0 are the dielectric constant (permittivity) and magnetic constant of the vacuum, whilst ε, μ, $\varepsilon_{\mathrm{r}} = \varepsilon/\varepsilon_0$, and $\mu_{\mathrm{r}} = \mu/\mu_0$ are absolute and relative values of the same for the medium. c_0 is the speed of light in vacuum and $n = \sqrt{\varepsilon_{\mathrm{r}}\mu_{\mathrm{r}}}$ is the refractive index of the medium. For a non-magnetic medium, $\mu_{\mathrm{r}} = 1$ and $n = \sqrt{\varepsilon_{\mathrm{r}}}$. Only relative permittivity ε_{r} will be relevant in this chapter. In order to simplify the notation, the index r will be dropped. The index ellipsoid, defined by

$$\frac{x_1^2 + x_2^2 + x_3^2}{\varepsilon} = \frac{x_i^2}{n^2} = 1 , \quad \text{or} \quad Bx_i^2 = 1 , \quad B = \frac{1}{\varepsilon} ,$$

is just a sphere.

Dielectric properties of an anisotropic medium can be described by a symmetric tensor ε_{ij}. Solving Maxwell's equations under the condition $D_i = \varepsilon_0\varepsilon_{ij}E_j$ implies the possible existence, for a direction defined by unit vector s_i, of two plane waves polarised perpendicularly to one another and propagating at different speeds [3.6]. Polarisations D_i and speeds $c = c_0/n$ of these two waves are given by the linear system (Problem 3.1)

$$(B_{ij} - B_{kj}s_is_k)\,D_j = \frac{D_i}{n^2} ,$$

where B_{ij} is the impermittivity tensor, inverse of the permittivity tensor:

$$\varepsilon_{ij}B_{jk} = \delta_{ik} . \tag{3.1}$$

Electric displacement vectors $\boldsymbol{D}^{(1)}$ and $\boldsymbol{D}^{(2)}$ for the two waves lie along axes of an ellipse found by intersecting the index ellipsoid, given by

$$B_{ij}x_ix_j = 1 , \tag{3.2}$$

with the diametral wave plane perpendicular to wave vector $\boldsymbol{k}$ (Fig. 3.4). Speeds are given by indices, which are equal to semi-axes of the ellipse.

The index ellipsoid clearly possesses the same symmetries as the crystal whose optical properties it describes. It follows that crystals can be divided into three categories. The same result can be obtained by studying the symmetric tensor B_{ij}, using crystal symmetries to reduce its components (Sect. 2.6.3, Vol. I).

Biaxial Crystals. These possess no axis of symmetry of order greater than 2 and therefore belong to the triclinic, monoclinic or orthorhombic systems. The three principal axes of a general ellipsoid are binary symmetry axes.

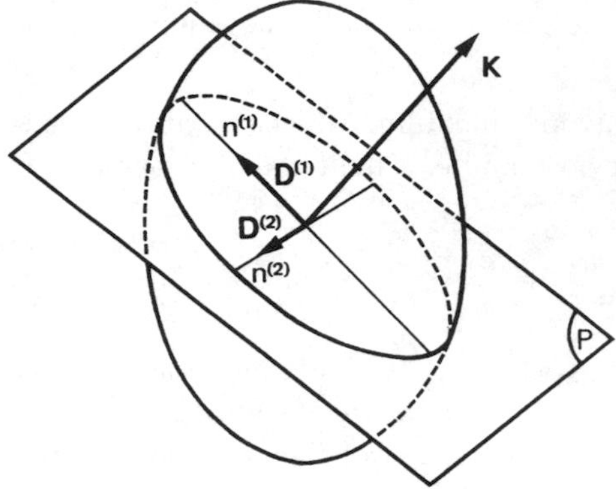

Fig. 3.4. Index ellipsoid of a crystal. Two waves with perpendicular polarisations $\boldsymbol{D}^{(1)}$ and $\boldsymbol{D}^{(2)}$ propagate in direction $\boldsymbol{k}$, at speeds $c_1 = c_0/n^{(1)}$, $c_2 = c_0/n^{(2)}$, respectively

If the crystal has a binary axis (monoclinic system) or three binary axes (orthorhombic system), they necessarily coincide with one or all three of the principal axes of the ellipsoid. Crystals in the triclinic system are not subject to any particular requirement from this point of view. In the principal axis system, the index ellipsoid equation is

$$\frac{x_1^2}{n_1^2} + \frac{x_2^2}{n_2^2} + \frac{x_3^2}{n_3^2} = 1 \,, \tag{3.3}$$

where n_1, n_2, n_3 are principal indices, lengths of the three semi-axes. A general ellipsoid has two circular diametral cross-sections which are symmetric with respect to two of the principal axes and contain the third. The normals to these particular cross-sections are called *optical axes*. There is only one index equal to the radius of the circle and no particular polarisation is imposed. Therefore, in each of these directions, all waves propagate at the same speed and there is no double refraction. Crystals behave as isotropic solids for these two special directions.

Uniaxial Crystals. These possess an axis of symmetry A_n of order n greater than 2. They belong to the trigonal, tetragonal and hexagonal systems. One principal axis of the index ellipsoid coincides with A_n and the index ellipsoid is an ellipsoid of revolution. The wave plane perpendicular to A_n intersects the ellipsoid in a circle, polarisation is indeterminate and the index, equal to the radius of the circle, is unique. The crystal behaves as an isotropic solid for waves propagating along axis A_n, known as the *optical axis*. The index ellipsoid of single-axis crystals is therefore defined by two numbers n_o and n_e. The first is the ordinary index, equal to the radius of the circular cross-section, whilst n_e is the extraordinary index, equal to the radius along the axis of revolution A_n. The index ellipsoid, with equation

$$\frac{x_1^2 + x_2^2}{n_\mathrm{o}^2} + \frac{x_3^2}{n_\mathrm{e}^2} = 1 \,, \tag{3.4}$$

is stretched along the optical axis if $n_\mathrm{e} > n_\mathrm{o}$. It is then said to be positive uniaxial (e.g., quartz). If $n_\mathrm{e} < n_\mathrm{o}$, the ellipsoid is flattened and the crystal is said to be negative uniaxial. The aim of Problem 3.2 is to show that the index for one of the vibrations is independent of the propagation direction (n_o), whilst the index of the other vibration varies between n_o and n_e.

Optically Isotropic Crystals. These possess several axes of symmetry of order greater than 2 and therefore belong to the cubic system ($4A_3$). The index ellipsoid must therefore have several axes of revolution and has to be a sphere, defined by a single index. Any light wave propagates in these crystals as it would through a glass or liquid, at the same speed and conserving its original polarisation, whatever the direction.

Let us end by making two observations.

- Differences between principal indices are always small, for all crystals, despite appearances in the figures, where they have been exaggerated for reasons of clarity. The index ellipsoid is thus very close to spherical. For quartz, $n_o = 1.5442$ and $n_e = 1.5533$.
- Permittivities vary with frequency and a particular ellipsoid corresponds to a limited frequency range. Hence, in a piezoelectric material, dielectric constants which determine the index ellipsoid are not the same as those featuring in equations for elastic wave speeds.

3.1.2.2 Index Surface. From the above, the index ellipsoid is found by marking off a length proportional to the index, in the direction of vibration. Another useful surface, called the index surface, is drawn by once again marking off a length equal to the index, but this time in the direction of the wave vector $\boldsymbol{k}$. Since the index is inversely proportional to the speed, this surface is similar to the *slowness surface* introduced in the study of dynamic elasticity. However, it only comprises two sheets because two orthogonal vibrations propagate in an arbitrary direction. The shape of the index surface is shown in Fig. 3.5 for a positive uniaxial crystal.

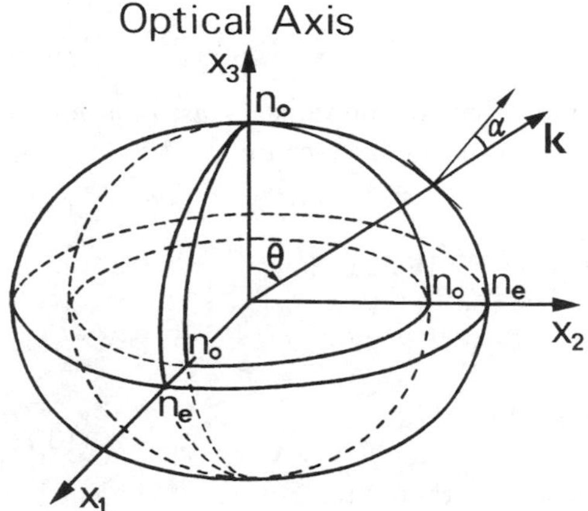

Fig. 3.5. Index surface for a positive uniaxial crystal ($n_e > n_o$). The two values of the indices $n^{(1)} = n_o = \text{Const.}$ and $n^{(2)}$, are marked off in the direction of the wave vector $\boldsymbol{k}$. The extraordinary ray makes angle α with $\boldsymbol{k}$, only zero when $\theta = 0$ and $\theta = \pi/2$

One of the sheets is spherical since one of the vibrations – the ordinary one – propagates at constant speed $c_{\mathrm{or}} = c_0/n_{\mathrm{o}}$, whatever the direction. The other sheet is an ellipsoid tangent to the sphere on the optical axis and with semi-major axis equal to n_{e}. Energy propagates along the normal which, for the elliptical sheet, makes a direction-dependent angle α with the wave vector. This angle is only zero along the principal axes. The index surface is useful in constructing the refracted plane wave from the incident wave (see the wave vector diagram in Sect. 3.1.5.3).

3.1.3 Elasto-optic Tensor

The optical properties of a crystal at rest are thus represented by the index ellipsoid, with equation

$$B_{ij}x_ix_j = 1\ .$$

Any deformation S_{kl} of the crystal will lead to modification of this ellipsoid, given by some variation ΔB_{ij} in the tensor B_{ij}. There are two cases. In the first, the crystal suffers a deformation involving no local rotation of the ellipsoid. In the second case, the deformation does produce a local rotation of the ellipsoid, that is, a rotation which varies from one point to another. Moreover, if the crystal is piezoelectric, the electro-optic interaction must be taken into account.

3.1.3.1 The Pockels Tensor. This theory applies when one of the following conditions is satisfied:

- crystal is optically isotropic and non-piezoelectric (cubic classes $m3m$, $m3$ and 432);
- deformation is homogeneous and static;
- deformation is caused by a longitudinal wave.

The variation ΔB_{ij} is then directly related to the deformation S_{kl}:

$$\boxed{\Delta B_{ij} = p_{ijkl}S_{kl}}\ . \tag{3.5}$$

Dimensionless quantities p_{ijkl} are components of the rank 4 *elasto-optic tensor*. The variation ΔB_{ij} is also given in terms of stresses by

$$\Delta B_{ij} = \pi_{ijkl}T_{kl}\ . \tag{3.6}$$

The tensor π_{ijkl} is the *piezo-optic tensor*. By Hooke's law,

$$T_{kl} = c_{klmn}S_{mn}\ ,$$

it follows that

$$p_{ijmn} = \pi_{ijkl}c_{klmn}\ . \tag{3.7}$$

Tensors B_{ij} and S_{kl} are symmetric and therefore, indices of the elasto-optic tensor can be contracted. Putting

$$B_{ij} = B_\alpha \,,$$

and, as in Sect. 3.2.1, Vol. I,

$$S_\beta = \begin{cases} S_{kl} & \beta \le 3 \,, \\ 2S_{kl} & \beta > 3 \,, \end{cases}$$

equation (3.5) can be written in matrix notation:

$$\Delta B_\alpha = p_{\alpha\beta} S_\beta \,, \quad \text{where } p_{\alpha\beta} = p_{ijkl} \,, \quad \forall \alpha,\, \beta = 1,\, 2,\, \dots,\, 6 \,. \tag{3.8}$$

However, $p_{\alpha\beta}$ is generally different from $p_{\beta\alpha}$, whereas for the elastic constants, $c_{\alpha\beta} = c_{\beta\alpha}$ follows from the existence of a thermodynamic potential (Sect. 3.2.2, Vol I). The elasto-optic tensor for triclinic crystals therefore has 36 independent components. The number of components is reduced by symmetries of the crystal and is calculated by the method described in Sect. 3.2.3, Vol. I. Results are gathered in Table 3.1. They apply equally to piezo-optic components π_{ijkl}. For constants $\pi_{\alpha\beta}$ defined by $\Delta B_\alpha = \pi_{\alpha\beta} T_\beta$, the following is used:

$$\pi_{\alpha\beta} = \begin{cases} \pi_{ijkl} & \beta \le 3 \,, \\ 2\pi_{ijkl} & \beta > 3 \,. \end{cases} \tag{3.9}$$

When there is a strain field, the index ellipsoid is thus given by

$$(B_{ij} + p_{ijkl} S_{kl})\, x_i x_j = 1 \,. \tag{3.10}$$

The Clausius–Mossotti relation [3.7] gives an idea of the order of magnitude of the p, although strictly only valid for liquids. It implies

$$\frac{n^2 - 1}{n^2 + 2} = \frac{N\alpha}{3\varepsilon_0} \quad \Rightarrow \quad \frac{1 - B}{1 + 2B} = \frac{N\alpha}{3\varepsilon_0} \,, \quad B = \frac{1}{\varepsilon} = \frac{1}{n^2} \,.$$

N is the dipole density per unit volume and α the polarisability, which is essentially of electronic origin in the optical frequency range. It follows by logarithmic differentiation that

$$\frac{1 + 2B}{1 - B} \Delta \left(\frac{1 - B}{1 + 2B} \right) = \frac{\Delta N}{N} + \frac{\Delta \alpha}{\alpha} \,.$$

Since N is proportional to the density ρ and α a function of ρ,

$$\frac{-3\Delta B}{(1 - B)(1 + 2B)} = \frac{\Delta \rho}{\rho} \left(1 + \frac{\rho}{\alpha} \frac{\mathrm{d}\alpha}{\mathrm{d}\rho} \right) \,.$$

Bringing in the strain $S = -\Delta\rho/\rho$ (longitudinal wave),

$$\Delta B = \frac{(1 - B)(1 + 2B)}{3} \left(1 + \frac{\rho}{\alpha} \frac{\mathrm{d}\alpha}{\mathrm{d}\rho} \right) S \,.$$

In liquids, the variation of polarisability with strain (compression) is small and the elasto-optic coefficient is given by

$$p \approx \frac{(1 - B)(1 + 2B)}{3} \,. \tag{3.11}$$

Table 3.1. Components $p_{\alpha\beta}$ of the Pockels elasto-optic tensor

• ∘ : non-zero component . : zero component
•—• : equal components •—∘ : equal and opposite components
× : component equal to $(p_{11} - p_{12}/2$

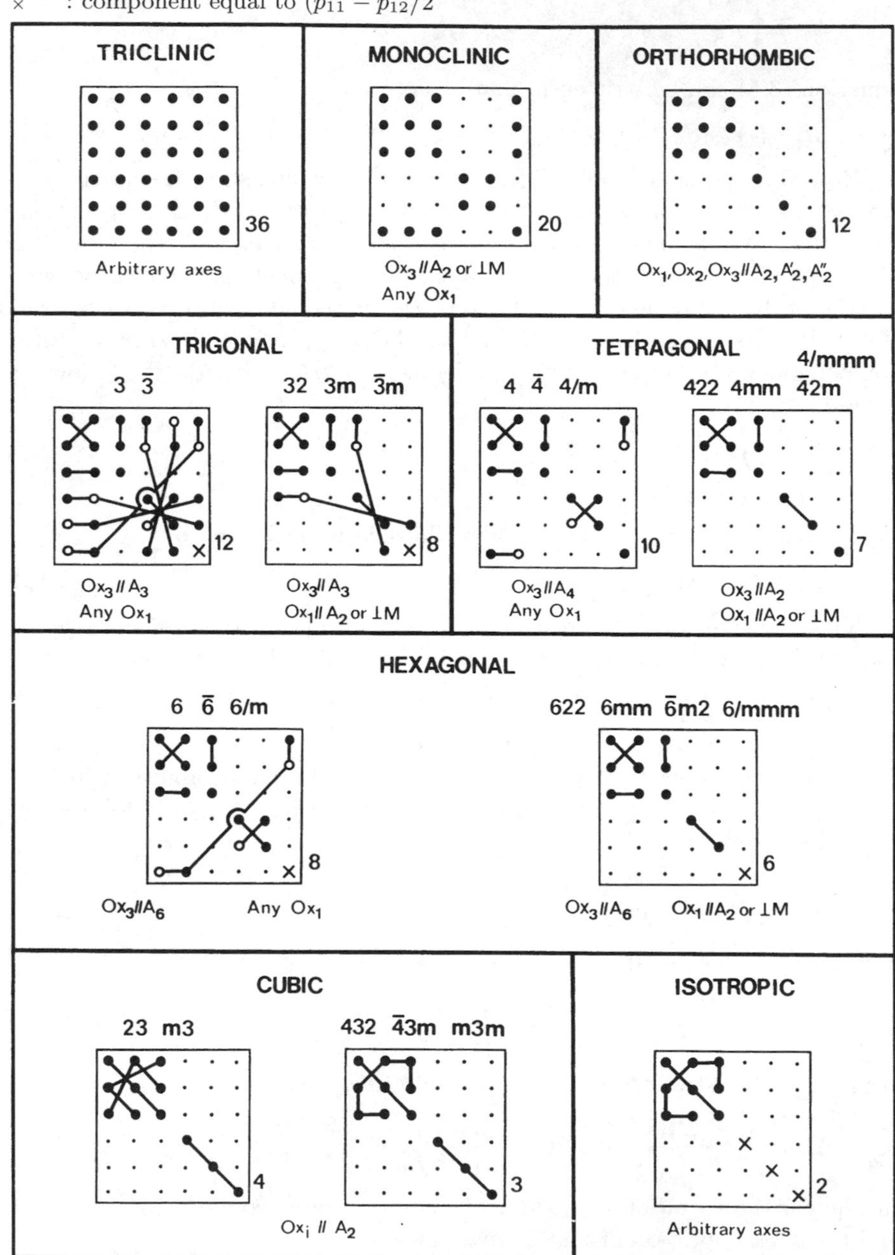

For example, in water $n = 1.33$ and hence $B = 0.57$, so that $p = 0.31$. In solids, the correction term is no longer negligible. Moreover, the variation in polarisability then depends on the type of deformation (dilation or shearing). In crystals, light polarisation, and propagation direction and polarisation of the elastic wave also come into play, so that a tensor description is required. However, relation (3.11) gives the right order of magnitude. Elasto-optic coefficients vary only slightly for the most commonly used crystals, and a value between 0.2 and 0.3 is obtained (see Table 3.2). Hence, variations ΔB_{ij} remain small, of order about 10^{-6} when $S = 10^{-5}$.

It is useful to relate variations ΔB_{jk} and $\Delta\varepsilon_{il}$ in components of the dielectric impermittivity and permittivity tensors. These variations are small and can be found by differentiating (3.1), which defined B_{jk}:

$$\Delta\varepsilon_{ij} B_{jk} + \varepsilon_{ij} \Delta B_{jk} = 0 \, .$$

Contracting with ε_{kl} and using $B_{jk}\varepsilon_{kl} = \delta_{jl}$,

$$\Delta\varepsilon_{il} = -\varepsilon_{ij} \Delta B_{jk} \varepsilon_{kl} \, .$$

In terms of strains,

$$\boxed{\Delta\varepsilon_{il} = -\varepsilon_{ij} p_{jkmn} \varepsilon_{kl} S_{mn}} \, . \tag{3.12}$$

As an example, we shall calculate variations $\Delta\varepsilon_{il}$ and Δn corresponding to strain S_{11} in the cases of an isotropic solid or cubic crystal of class $m3m$. In both cases, the dielectric tensor is diagonal:

$$\varepsilon_{il} = \varepsilon \delta_{il} \, .$$

Hence,

$$\Delta\varepsilon_{il} = -\varepsilon^2 p_{ilmn} S_{mn} \, ,$$

and for strain S_{11},

$$\Delta\varepsilon_{il} = -\varepsilon^2 p_{il11} S_{11} \, .$$

According to Table 3.1, the only non-zero elasto-optic components p_{il11} for crystals in class $m3m$ are

$$p_{1111} = p_{11} \, , \quad p_{2211} = p_{12} \, , \quad p_{3311} = p_{12} \, .$$

Permittivity variations are therefore given by:

$$\Delta\varepsilon_{11} = -\varepsilon^2 p_{11} S_{11} \, , \quad \Delta\varepsilon_{22} = \Delta\varepsilon_{33} = -\varepsilon^2 p_{12} S_{11} \, , \quad \Delta\varepsilon_{il} = 0 \text{ if } i \neq 1 \, .$$

The tensor ε_{ij} remains diagonal and reference axes are principal axes of the index ellipsoid of the deformed crystal, whose principal indices are now:

$$n_1 = \sqrt{\varepsilon + \Delta\varepsilon_{11}} \, , \quad n_2 = n_3 = \sqrt{\varepsilon + \Delta\varepsilon_{22}} \, .$$

Variations $\Delta\varepsilon_{ii}$ are small and $n = \sqrt{\varepsilon}$, so

$$n_i \approx n\left(1 + \frac{\Delta\varepsilon_{ii}}{2\varepsilon}\right) = n + \Delta n_i \, , \quad \text{where} \quad \Delta n_i = \frac{\Delta\varepsilon_{ii}}{2n} \, ,$$

Table 3.2. Elasto-optic constants for several materials, measured by diffraction of He-Ne laser light (0.6328 μm), except for quartz ($\Lambda_0 = 0.589$ μm)

Material	p_{11}	p_{12}	p_{21}	p_{22}	p_{13}	p_{31}	p_{33}	p_{23}	p_{32}	p_{14}	p_{41}	p_{44}	p_{45}	p_{55}	p_{16}	p_{61}	p_{66}	Ref.
Silica (isotropic)	0.121	0.270	p_{12}	p_{11}	p_{12}	p_{12}	p_{11}	p_{12}	p_{12}	0	0	[−0.075]	0	p_{44}	0	0	p_{44}	[3.8]
α-HIO_3 (222)	0.406	0.277	0.279	0.343	0.304	0.503	0.334	0.305	0.310	0	0	–	0	–	0	0	0.092	[3.9]
$PbMoO_4$ (4/m)	0.24	0.24	p_{12}	p_{11}	0.255	0.175	0.300	p_{13}	p_{31}	0	0	0.067	−0.01	p_{44}	0.017	0.013	0.05	[3.10]
TiO_2 (4/mmm)	0.011	0.172	p_{12}	p_{11}	0.168	0.096	0.058	p_{13}	p_{31}	0	0	–	0	p_{44}	0	0	–	[3.11]
TeO_2 (422)	0.007	0.187	p_{12}	p_{11}	0.340	0.090	0.240	p_{13}	p_{31}	0	0	−0.17	0	p_{44}	0	0	−0.046	[3.12]
$LiNbO_3$ (3m)	0.036	0.072	p_{12}	p_{11}	0.092	0.178	0.088	p_{13}	p_{31}	0.07[a]	0.155	–	0	p_{44}	0	0	[−0.018]	[3.11]
$LiTaO_3$ (3m)	0.080	0.080	p_{12}	p_{11}	0.094	0.086	0.150	p_{13}	p_{31}	0.031	0.024	0.022	0	p_{44}	0	0	[0.00]	[3.11]
α-quartz (32)	0.138	0.250	p_{12}	p_{11}	0.259	0.258	0.098	p_{13}	p_{31}	−0.029	−0.042	−0.068	0	p_{44}	0	0	[−0.056]	[3.13]

[a] Value determined by Reintjes, J., Schulz, M.B. (1968) J. Appl. Phys. **39**, 5254.
Values in square brackets are equal to $(p_{11} - p_{12})/2$.

and hence

$$\Delta n_1 = -\frac{n^3}{2} p_{11} S_{11} , \tag{3.13}$$

$$\Delta n_2 = \Delta n_3 = -\frac{n^3}{2} p_{12} S_{11} . \tag{3.14}$$

The crystal acquires an optical axis (Ox_1). There is birefringence for a generally polarised light wave entering the crystal along Ox_2 or Ox_3, but polarisation is nevertheless unchanged if it is parallel to one of the principal axes. The wave merely undergoes phase delay or advance if the index increases or decreases along this axis. The change in index is proportional to the photoelastic constant p_{11} or p_{12}, depending on whether the light is polarised parallel or normal to its direction of propagation. We shall study this simple case in Sect. 3.1.4.

3.1.3.2 The Nelson–Lax Tensor. A more general definition of the elasto-optic tensor is needed when there are local rotations. These may be generated, for example, by transverse elastic waves [3.14]. The 6-component symmetric tensor S_{kl} in (3.5) must be replaced by the gradient of the displacement $\partial u_k/\partial x_l$, which has 9 components (see Sect. 3.1.1.1, Vol. I):

$$\Delta B_{ij} = P_{ijkl} \frac{\partial u_k}{\partial x_l} . \tag{3.15}$$

The new elasto-optic tensor P_{ijkl} is no longer symmetric in the last two indices k, l. Let $P_{ij(kl)}$ and $P_{ij[kl]}$ be its symmetric and antisymmetric parts. From the identity

$$\frac{\partial u_k}{\partial x_l} = S_{kl} + \Omega_{kl} ,$$

in which the antisymmetric tensor

$$\Omega_{kl} = \frac{1}{2}\left(\frac{\partial u_k}{\partial x_l} - \frac{\partial u_l}{\partial x_k}\right) = -\Omega_{lk}$$

represents local rotations, it follows that

$$\Delta B_{ij} = \left(P_{ij(kl)} + P_{ij[kl]}\right)\left(S_{kl} + \Omega_{kl}\right) .$$

Finally,

$$\boxed{\Delta B_{ij} = P_{ij(kl)} S_{kl} + P_{ij[kl]} \Omega_{kl}} . \tag{3.16}$$

since the contracted product of a symmetric and an antisymmetric tensor is always zero (Problem 2.19, Vol. I).

Comparison of (3.5) and (3.16) shows that the symmetric part $P_{ij(kl)}$ corresponds to the tensor p_{ijkl}, whilst the antisymmetric part $P_{ij[kl]}$, referring only to local rotation of the index ellipsoid, is expressed in terms of components B_{ij}. In this local rotation ($S_{ij} = 0$), transforming coordinates x_i to $x'_i = x_i + u_i$, the new components B'_{ij} are given by

$$B'_{ij} = \frac{\partial x'_i}{\partial x_k} \frac{\partial x'_j}{\partial x_l} B_{kl} ,$$

where

$$\frac{\partial x'_i}{\partial x_j} = \delta_{ij} + \frac{\partial u_i}{\partial x_j} = \delta_{ij} + \Omega_{ij} ,$$

since $S_{ij} = 0$. The change $\Delta B_{ij} = B'_{ij} - B_{ij}$ is given by

$$\Delta B_{ij} = (\delta_{ik} + \Omega_{ik})(\delta_{jl} + \Omega_{jl}) B_{kl} - B_{ij} .$$

Neglecting second order terms,

$$\Delta B_{ij} = \Omega_{ik} B_{kj} + \Omega_{jl} B_{il} ,$$

or

$$\Delta B_{ij} = (B_{il}\delta_{jk} - B_{kj}\delta_{il}) \Omega_{kl} .$$

Comparison with (3.16), when $S_{kl} = 0$, shows that $P_{ij[kl]}$ is the antisymmetric part of the tensor in brackets here, viz.,

$$P_{ij[kl]} = \frac{1}{2} (B_{il}\delta_{jk} + B_{lj}\delta_{ik} - B_{ik}\delta_{jl} - B_{kj}\delta_{il}) . \tag{3.17}$$

Principal axes coincide with crystallographic axes except for crystals in triclinic and monoclinic systems. Then, in this reference system,

$$B_{il} = \frac{1}{n_i^2} \delta_{il} ,$$

where n_i, $i = 1, 2, 3$ are the principal optical indices. Equation (3.17) becomes

$$P_{ij[kl]} = \frac{1}{2} \left(\frac{1}{n_i^2} - \frac{1}{n_j^2} \right) (\delta_{il}\delta_{jk} - \delta_{ik}\delta_{jl}) , \tag{3.18}$$

and the only non-zero components are

$$\begin{aligned} P_{12[12]} &= -P_{12[21]} = -\frac{1}{2} \left(\frac{1}{n_1^2} - \frac{1}{n_2^2} \right) , \\ P_{23[23]} &= -P_{23[32]} = -\frac{1}{2} \left(\frac{1}{n_2^2} - \frac{1}{n_3^2} \right) , \\ P_{13[13]} &= -P_{13[31]} = -\frac{1}{2} \left(\frac{1}{n_1^2} - \frac{1}{n_3^2} \right) . \end{aligned} \tag{3.19}$$

The only differences with the Pockels tensor occur in the last three terms p_{44}, p_{55} and p_{66}, of the principal diagonal (see Table 3.1). For isotropic materials and crystals in the cubic system, there is no change, i.e., $P_{ijkl} = p_{ijkl}$, since $n_1 = n_2 = n_3$.

For uniaxial crystals in hexagonal, tetragonal and trigonal systems,

$$P_{12[12]} = 0 , \quad P_{23[23]} = P_{13[13]} ,$$

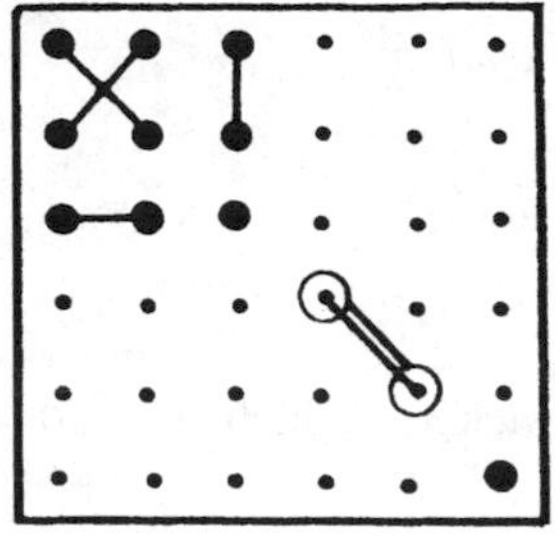

Fig. 3.6. Components of the elasto-optic tensor for crystals belonging to symmetry class 422. Circles represent extra components to be taken into account when local rotations of the index ellipsoid are involved. Lines joining circles, like those joining points, express equalities

since $n_1 = n_2$. Equality $p_{44} = p_{55}$ in the Pockels theory (see Table 3.1), i.e., $P_{23(23)} = P_{13(13)}$, splits into two equalities:

$$\begin{aligned} P_{2323} &= P_{23(23)} + P_{23[23]} = P_{1313} , \\ P_{2332} &= P_{23(23)} - P_{23[23]} = P_{1331} . \end{aligned} \tag{3.20}$$

For crystals in the orthorhombic system, the three antisymmetric components $P_{12[12]}$, $P_{23[23]}$ and $P_{13[13]}$ are non-zero. In each table, a circle is used to represent one of these extra components. For example, the table in Fig. 3.6 corresponds to class 422. Elasto-optic tensors of crystals in monoclinic and triclinic classes possess 7 and 18 extra components, respectively.

This theory has been confirmed in experiments carried out by Nelson et al. for strongly birefringent non-piezoelectric crystals: rutile ($4/mmm$) and calcite ($\overline{3}m$).

- For rutile [3.15]:

$$P_{2323} = P_{1313} = 9 \times 10^{-4} , \quad P_{2332} = P_{1331} = -255 \times 10^{-4} .$$

- For calcite [3.16]:

$$P_{2323} = P_{1313} = -11 \times 10^{-3} , \quad P_{2332} = P_{1331} = -105 \times 10^{-3} .$$

Equation (3.10), expressing variation in permittivity components, generalises to

$$\Delta\varepsilon_{il} = -\varepsilon_{ij} P_{jkmn} \varepsilon_{kl} \frac{\partial u_m}{\partial x_n} . \tag{3.21}$$

Piezoelectric Crystals. Electro-optic Tensor. The ellipsoid may also be deformed by an electric field applied to the crystal. This electro-optic effect is expressed through the relation

$$\Delta B_{ij} = r_{ijp} E_p .$$

The rank 3 electro-optic tensor r_{ijp} is zero, as also is the piezoelectric tensor, in crystals possessing a centre of symmetry.

In a piezoelectric crystal, plane elastic waves are generally accompanied by a longitudinal electric field. Along the direction parallel to unit vector $\boldsymbol{n}$, this field, deduced from (4.71–2), Vol. I, is given by:

$$E_p = -\frac{\partial \Phi}{\partial x_p} = \frac{n_p}{V}\Phi = \frac{n_p}{V}\frac{e_{qkl}n_q}{\varepsilon}n_k u_l \ , \qquad \varepsilon = \varepsilon_0 \varepsilon_{jk} n_j n_k \ .$$

This produces an indirect elasto-optic effect:

$$\Delta B_{ij} = \left(-r_{ijp} n_p \frac{n_q e_{qkl}}{\varepsilon_0 \varepsilon_{jk} n_j n_k} \right) \frac{\partial u_l}{\partial x_k} \ .$$

This effect is characterised by the tensor between brackets, which depends on the direction of propagation:

$$P^{\text{ind}}_{ijkl}(\boldsymbol{n}) = -\frac{r_{ijp} n_p n_q e_{qkl}}{\varepsilon_0 \varepsilon_{jk} n_j n_k} \ . \tag{3.22}$$

The contribution from this term is important, even predominant, in certain cases. For example, a wave propagating along [111] in an α-iodic crystal (class 222) generates an indirect elasto-optic effect with $P^{\text{ind}}_{3212} = 0.013$, whereas the direct component is zero [3.17].

3.1.4 Normal Incidence Diffraction

As we have seen, the dielectric tensor and optical index of a material are functions of the deformations it undergoes. Consequently, any elastic wave passing through a solid is accompanied by a wave of index variations travelling at the same speed. In order to analyse its effect on light waves, let us examine the simplest case, in which there is no change in polarisation: optically isotropic solid, longitudinal elastic wave, monochromatic light beam, polarised perpendicularly or parallel to the elastic wave vector and normally incident.

The succession of compressions and dilatations constituting the elastic wave propagate in direction Ox_2 and generate respectively an increase and a decrease in the index relative to its rest value n_0 :

$$n = n_0 + \Delta n \sin \Omega \left(t - \frac{x_2}{V} \right) \ , \tag{3.23}$$

where Ω and V are angular frequency and phase speed of the elastic wave.

Electromagnetic vibrations crossing a region of solid in which the index increases propagate more slowly than those crossing a region in which it decreases. The electric field of the light wave (which we refer to briefly as the light wave), propagating in direction x_1, is given by

$$a = A \cos \left(\omega t - \frac{2\pi}{\lambda_0} n x_1 \right) \ ,$$

taking zero phase at $x_1 = 0$ (see Fig. 3.7), where the elastic wave enters. λ_0 is the vacuum wavelength of the light.

Neglecting any attenuation, after crossing the elastic wave beam of thickness L, the light wave is given by

$$a = A \cos \left[\omega t + \Phi_0 + \Delta\Phi \sin \Omega \left(t - \frac{x_2}{V} \right) \right] \ , \tag{3.24}$$

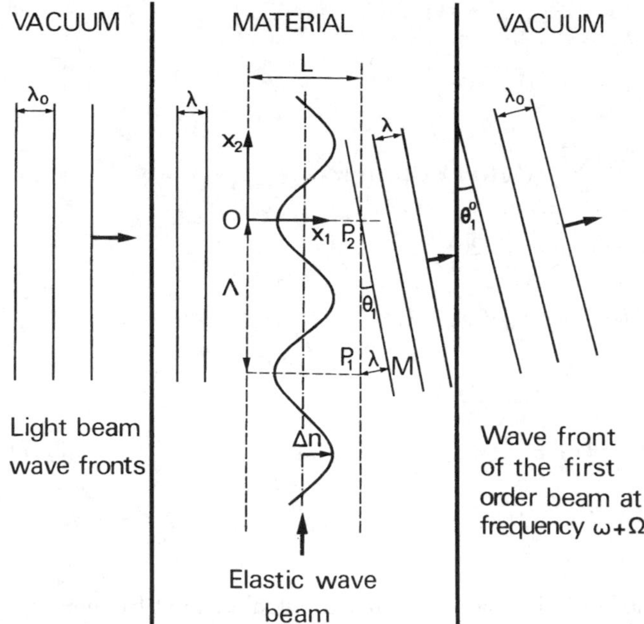

Fig. 3.7. The 2π phase shift (due to change Δn in the index) between points P_1 and P_2, separated by the elastic wavelength Λ, determines the angle θ_1 of the wave front of the first order beam, of frequency $\omega + \Omega$, viz., $\sin\theta_1 = \lambda/\Lambda$. In the vacuum, after passing through the solid, the deflection angle is θ_0^1, where $\sin\theta_0^1 = \lambda_0/\Lambda$

where $\Phi_0 = -2\pi n_0 L/\lambda_0 = -2\pi L/\lambda$ is the phase shift due to the mean index n_0. According to (3.13), the light wave undergoes a phase shift of

$$\Delta\Phi = -\frac{2\pi L}{\lambda_0}\Delta n = \frac{\pi L}{\lambda_0}n^3 pS \,. \tag{3.25}$$

This is often called the *Raman–Nath parameter* ν [3.4].

At each point of the elastic beam boundary ($x_1 = L$, x_2 fixed), the phase of the light wave is modulated, with a modulation which depends only upon time. Let us find the spectrum of this wave. Putting $\tau = t - x_2/V$, equation (3.24) becomes

$$a = A\mathrm{Re}\left[\exp \mathrm{i}(\omega t + \Phi_0)\exp(\mathrm{i}\Delta\Phi\sin\Omega\tau)\right] \,. \tag{3.26}$$

The term $\exp(\mathrm{i}\Delta\Phi\sin\Omega\tau)$ can be expanded in terms of Bessel functions $J_N(x)$. The generating function is

$$\exp\left[\frac{x}{2}\left(t - \frac{1}{t}\right)\right] = \sum_{N=-\infty}^{\infty} J_N(x)t^N \,,$$

and putting $t = \mathrm{e}^{\mathrm{i}\theta}$ implies

$$e^{ix \sin \theta} = \sum_{N=-\infty}^{\infty} J_N(x) e^{iN\theta} .$$

Hence,

$$\exp(i\Delta\Phi \sin \Omega\tau) = \sum_{N=-\infty}^{\infty} J_N(\Delta\Phi) \exp(iN\Omega\tau) .$$

Substituting this expansion into (3.26),

$$\frac{a}{A} = \mathrm{Re}\left[\sum_{N=-\infty}^{\infty} J_N(\Delta\Phi) \exp i(\omega t + \Phi_0 + N\Omega\tau)\right] .$$

Since $J_{-N}(x) = (-1)^N J_N(x)$, this yields

$$\frac{a}{A} = J_0(\Delta\Phi)\cos(\omega t + \Phi_0) + \tag{3.27}$$
$$\sum_{N=1}^{\infty} J_N(\Delta\Phi)\left[\cos(\omega t + N\Omega\tau + \Phi_0) + (-1)^N \cos(\omega t - N\Omega\tau + \Phi_0)\right] .$$

The spectrum of the light vibration contains a central peak at frequency ω and symmetrical sidelobes of frequency $\omega \pm N\Omega$, since $\tau = t - x_2/V$, where N is an integer. Amplitudes of carrier wave and side waves are Bessel functions J_0, J_1, J_2 of the phase variation $\Delta\Phi$ (see Fig. 3.8).

Order of Magnitude. $L = 5$ mm, $\Delta n = 10^{-6}$, $\lambda_0 = 0.5$ µm implies $\Delta\Phi = 2\pi \times 10^{-2} \approx 0.06$ rad. In this case, known as the Raman–Nath regime, we can apply the approximation

$$J_N(x) \approx \frac{x^N}{2^N N!} \quad \text{for } x \ll 1 , \tag{3.28}$$

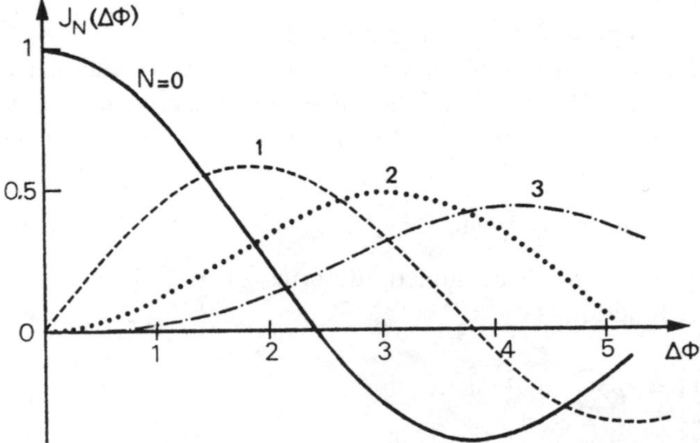

Fig. 3.8. Bessel functions J_0, J_1, J_2, ... representing amplitudes of carrier wave and spectral components as a function of phase variation $\Delta\Phi$

which implies $J_0(x) \approx 1$, $J_1(x) \approx x/2$, $J_2(x) \approx 0$.

At any given instant the phase of the carrier wave $\Phi = -2\pi L/\lambda$ does not depend on x_2. Its wave planes are therefore parallel to those of the incident wave. The same is not true for side spectral components, such as

$$J_N(\Delta\Phi)\cos\left[(\omega + N\Omega)t + \Phi_0 - \frac{N\Omega x_2}{V}\right],$$

for which the phase Φ_N is x_2-dependent, i.e., it depends on the emergence point on the boundary:

$$\Phi_N = \Phi_0 - \frac{N\Omega x_2}{V} = \Phi_0 - 2\pi\frac{N x_2}{\Lambda},$$

where Λ is the acoustic wavelength. Interference of waves having the same frequency, i.e., of the same order N, emitted from the boundary $x_1 = L$, is only constructive in one direction, at angle θ_N (Fig. 3.7). In this direction, for the wave emerging at P_1, the phase lag relative to the one emerging at P_2 is $-2\pi P_1M/\lambda$, due to the path difference P_1M. It compensates the initial phase advance $\Delta\Phi_N = 2\pi N P_2P_1/\Lambda$, so that

$$-\frac{2\pi}{\lambda}P_1M + \frac{2\pi N}{\Lambda}P_2P_1 = 0\,.$$

Hence, since $P_1M = P_2P_1\sin\theta_N$,

$$\sin\theta_N = N\frac{\lambda}{\Lambda}\,. \tag{3.29}$$

We thus conclude that the spectral component of order N and frequency $\omega + N\Omega$ is deflected through the angle θ_N given by (3.29). Spectral components of frequency $\omega - N\Omega$ are symmetrically deflected as shown in Fig. 3.9.

Summary. Sinusoidal index variations generated by an elastic wave have an analogous effect on light to a phased array. When light enters the crystal, parallel to elastic wave planes, it separates into several beams, symmetrically tilted relative to the incident beam, at angles θ_N given by

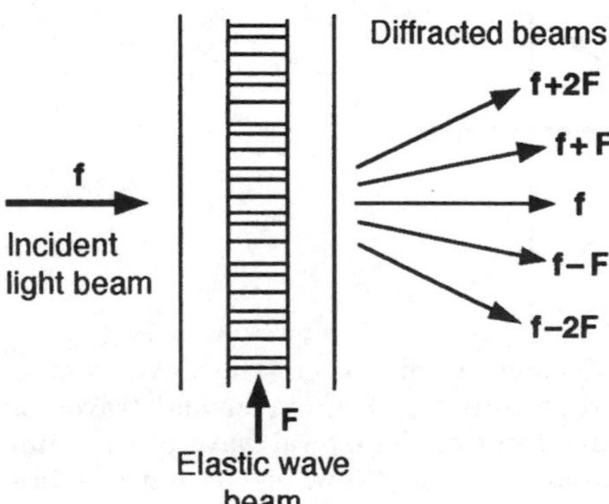

Fig. 3.9. Diffraction of a light beam normally incident on a narrow elastic wave beam. The light beam separates into several beams, symmetrically tilted relative to the incident beam

$$\boxed{\sin\theta_N = \pm N\frac{\lambda}{\Lambda}} . \tag{3.30}$$

In the case of silica with longitudinal elastic waves at 200 MHz and for red light from a He-Ne laser:

$$\begin{aligned}
&V_L = 5\,960 \text{ m/s} && \Rightarrow \Lambda \approx 30 \text{ µm} ,\\
&\lambda_0 = 0.633 \text{ µm and } n = 1.46 && \Rightarrow \lambda = 0.433 \text{ µm} ,\\
&\sin\theta_1 \approx \theta_1 = 1.45 \times 10^{-2} \text{ rad} = 0°50' .
\end{aligned}$$

However, the above reasoning only applies if the width L of the elastic beam is less than some critical value L_c.

Critical Thickness. Diffracted waves are generated right along the path of the carrier wave inside the ultrasonic beam and not only at the exit on its boundary. Analysing the elastic beam into thin slices parallel to the direction of propagation x_2, we can apply the above spectral analysis to each slice. Frequencies $\omega + N\Omega$ and direction of propagation θ_N of side waves are the same for slices at x_1 and $x_1 + l$ (see Fig. 3.10). Let us add together contributions for a given order from these two slices a distance l apart. The side wave front emitted at time $t - l/c$ from point x_1 and travelling in direction θ_N, arrives at Q_1 at time t, whilst the lateral wave front emitted at time t from $x_1 + l$ is at Q_2. There is a phase difference of

$$\phi(l) = \frac{2\pi l}{\lambda}(1 - \cos\theta_N) .$$

Since θ_N is small,

$$\cos\theta_N \approx 1 - \frac{1}{2}\left(\frac{N\lambda}{\Lambda}\right)^2 \quad \Rightarrow \quad \phi(l) \approx \frac{\pi l \lambda}{\Lambda^2} N^2 .$$

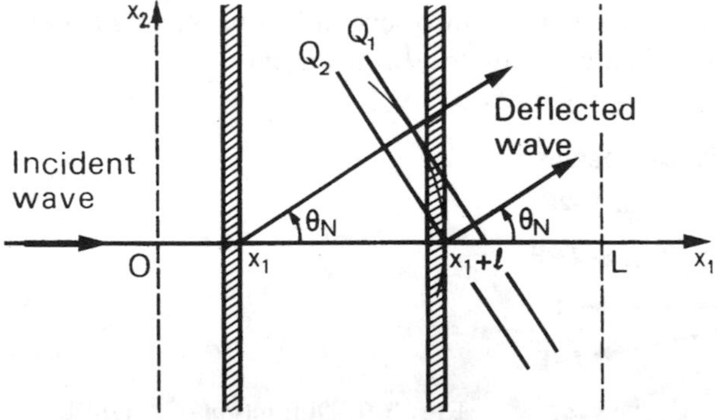

Fig. 3.10. Critical thickness for normal incidence diffraction. The wave front of the lateral light beam of order N, emitted at time $t - l/c$ from x_1 and travelling along θ_N, arrives at Q_1 at time t. The wave front of the lateral wave of the same order emitted from $x_1 + l$ at time t is then at Q_2. The two waves are out of phase by $\phi = 2\pi l(1 - \cos\theta_N)/\lambda \approx \pi l \lambda N^2/\Lambda^2$

Their phases are opposite when $\phi(l) = \pi$, i.e., for a distance $l_N = \Lambda^2/\lambda N^2$. Waves emitted from the two slices a distance l_N apart then interfere destructively. If the beam width is greater than l_N, the effect of one slice is cancelled by the slice distance l_N away. The thickness L of the elastic wave beam must not exceed the critical value of the first order:

$$L_c = l_1 = \frac{\Lambda^2}{\lambda} . \tag{3.31}$$

In terms of wave numbers, the condition on ϕ becomes

$$\phi = \frac{K^2}{2k} L < \pi ,$$

and putting

$$\boxed{Q = \frac{K^2}{k} L = 2\pi \frac{\lambda L}{\Lambda^2}} \quad \Rightarrow \quad Q = 2\pi \frac{L}{L_c} \ll 2\pi . \tag{3.32}$$

The parameter Q was introduced by Klein and Cook [3.18]. Physically, (3.32) implies that the diffraction into several beams which characterises the Naman–Rath regime is only possible if the spreading angle $2\delta \approx \Lambda/L$ of the acoustic beam is much larger than the diffraction angle θ_1 :

$$\delta \approx \frac{\Lambda}{2L} > \theta_1 \approx \frac{\lambda}{\Lambda} \quad \Leftrightarrow \quad Q \ll \pi .$$

In addition, the elastic power must be small enough to justify neglecting curvature of light rays as they cross the acoustic beam. The modulation depth $\Delta\Phi$ must satisfy (Problem 3.3)

$$Q\Delta\Phi < \pi .$$

Index variations then behave like a thin phased array.

The inequality $Q < \pi/2$ becomes more difficult to satisfy as the frequency increases. For example, for silica and $\lambda_0 = 0.633$ µm ($\lambda = 0.433$ µm):

100 MHz	$L_c = 8.4$ mm	$Q = 0.63 \Rightarrow L = 0.84$ mm ,
300 MHz	$L_c = 0.93$ mm	$Q = 0.63 \Rightarrow L = 0.09$ mm .

3.1.5 Bragg Incidence Diffraction

Interaction at the Bragg angle is the most important in practice, because it produces a single deflected beam. It is obtained by tilting the incident beam at an angle α so that constructive interference occurs for the first order diffracted beam, of angular frequency $\Omega + \omega$ (Fig. 3.11).

3.1.5.1 Isotropic Medium. Angle of Incidence. We seek the deflection angle θ for oblique incidence angle α. For two points P_1, P_2 in a slab parallel to Ox_2, the phase lag due to the path difference of the light rays balances the phase advance due to the elastic wave:

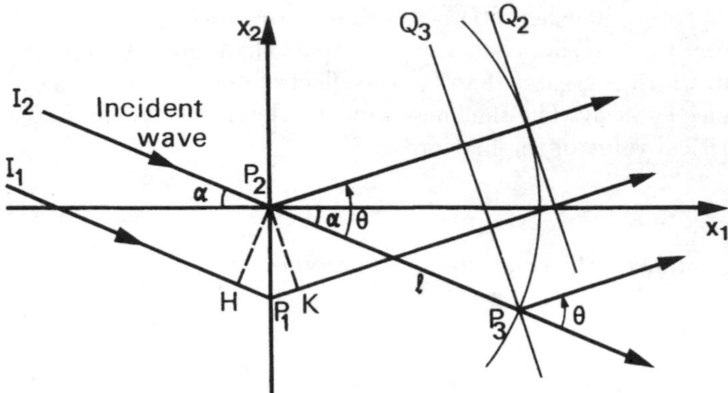

Fig. 3.11. Oblique incidence. Interference between first order side waves is constructive if effects produced by two incident rays I_1P_1 and I_2P_2, together with those produced by a single ray incident at two points P_2, P_3, are additive: $\sin\alpha + \sin(\theta - \alpha) = \lambda/\Lambda$ and $1 - \cos\theta = (\lambda/\Lambda)\sin\alpha \Rightarrow \alpha = \theta/2$

$$-\frac{2\pi}{\lambda}(HP_1 + P_1K) + \frac{2\pi}{\Lambda}P_1P_2 = 0 \, .$$

Hence,

$$\sin\alpha + \sin(\theta - \alpha) = \frac{\lambda}{\Lambda} \, . \tag{3.33}$$

We shall now establish the relation between α and θ which ensures that lateral waves emitted at two points P_2, P_3 of an incident light ray are in phase whatever their distance l. At time t, the phase difference φ between wave planes Q_2, Q_3 includes a term due to oblique propagation, $2\pi l(1 - \cos\theta)/\lambda$, and a term due to the elastic wave, $-(2\pi l/\Lambda)\sin\alpha$:

$$\varphi = 2\pi l\left(\frac{1 - \cos\theta}{\lambda} - \frac{\sin\alpha}{\Lambda}\right) \, .$$

The phase difference φ is zero for any l if

$$2\sin^2\frac{\theta}{2} = \frac{\lambda}{\Lambda}\sin\alpha \, ,$$

and hence, using (3.33), when

$$\sin\frac{\theta}{2} = \cos\left(\frac{\theta}{2} - \alpha\right)\sin\alpha \, .$$

This equation is satisfied when $\alpha = \theta/2$. The angle of incidence is given by

$$\boxed{\sin\alpha = \frac{\lambda}{2\Lambda} = \frac{\lambda F}{2V}} \, . \tag{3.34}$$

The light beam is deflected through $\theta = 2\alpha$. Elastic wave fronts appear to act as mirrors for the incident ray (see Fig. 3.12). Condition (3.34), which is

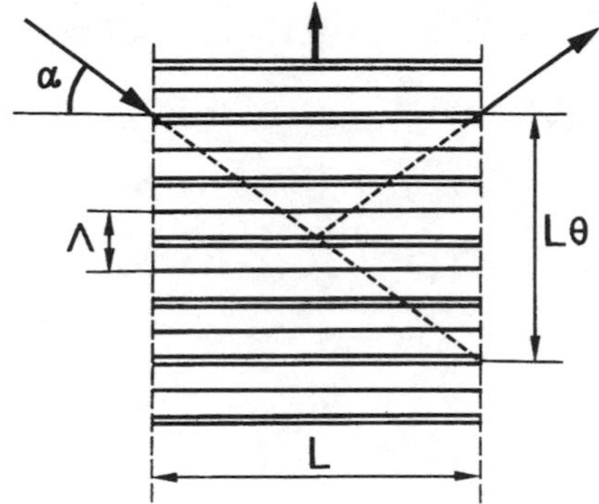

Fig. 3.12. Bragg condition. A light ray crosses several elastic wave fronts

reminiscent of selective X-ray reflection by crystal lattice planes, is known as the Bragg condition. Bragg incidence only favours the first order beam. However, the carrier wave generates side waves of all orders right along its path. The cumulative effect for the first order, and destructive for others, becomes all the more marked as the incident beam crosses further wave planes. This implies, for small α,

$$L\alpha \gg \Lambda \quad \Rightarrow \quad L \gg 2\frac{\Lambda^2}{\lambda} = 2L_{\mathrm{c}} \, ,$$

and hence, $Q > 4\pi$.

If the inclination of the incident light beam is symmetric to the one shown in Fig. 3.12 with respect to elastic wave fronts, it is the first order beam of angular frequency $\omega - \Omega$ which is favoured.

3.1.5.2 Coupled Mode Analysis. The travelling elastic wave produces a periodic index variation, i.e., a periodic variation in permittivity, moving at very low speed V relative to the light waves. Light propagation in the presence of a weak quasi-static perturbation can be treated using the coupled mode method (Appendix B). Two optical modes are strongly coupled by the acousto-optic interaction when the wave number tuning condition (B.15) holds:

$$k_2^{\mathrm{D}} - k_2^{\mathrm{I}} = q\frac{2\pi}{\Lambda} \, , \tag{3.35}$$

where $k_2^{\mathrm{I}} = -k\sin(\theta/2)$ is the component of the wave number of the incident light wave in the x_2 direction and $k_2^{\mathrm{D}} = k\sin(\theta/2)$ the same for the diffracted wave number. $2\pi/\Lambda$ is the array wave number G, i.e., the elastic wave number K. For sinusoidal modulation, the order q of the Fourier component is ± 1. The condition for diffraction at the Bragg angle is

$$2k\sin\frac{\theta}{2} = K \, . \tag{3.36}$$

It is equivalent to (3.34) and leads to the *wave vector diagram* in Fig. 3.13, relating incident and diffracted optical wave vectors $\boldsymbol{k}_{\mathrm{I}}$, $\boldsymbol{k}_{\mathrm{D}}$ to elastic wave vector $\boldsymbol{K}$:

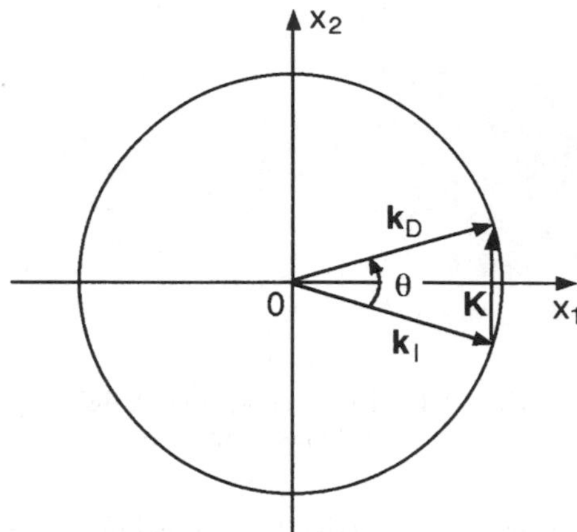

Fig. 3.13. Wave vector diagram. If the material is isotropic, the triangle is almost isosceles ($\Omega \ll \omega$)

$$\boxed{\boldsymbol{k}_{\mathrm{D}} = \boldsymbol{k}_{\mathrm{I}} \pm \boldsymbol{K}} \ . \tag{3.37}$$

This construction expresses conservation of momentum. In one case (sign +), the sum of the momenta of colliding incident photon and phonon equals the momentum of the diffracted phonon. In the other (sign −), the incident photon momentum equals the momentum sum of the diffracted photon and the phonon produced by interaction.

Motion of the index array leads by Doppler effect to a frequency change, proportional to the projection $V \sin(\theta/2)$ of the elastic wave speed V in the direction of the incident wave:

$$\frac{\Delta\omega}{\omega} = 2\frac{V}{c}\sin\frac{\theta}{2} \quad \Rightarrow \quad \Delta\omega = VK = \Omega \ll \omega \ .$$

This relation between angular frequencies,

$$\omega_{\mathrm{D}} = \omega_{\mathrm{I}} \pm \Omega \ , \tag{3.38}$$

expresses energy conservation.

Diffracted Beam Intensity. Figure of Merit. The electric field of the light wave is a solution of

$$\frac{\partial^2 E}{\partial x_3^2} + \frac{\partial^2 E}{\partial x_2^2} + \frac{\partial^2 E}{\partial x_1^2} + \omega^2 \mu\varepsilon_0(\varepsilon + \Delta\varepsilon)E = 0 \ .$$

This is similar to (B.1), Appendix B (with $a \to E$, $\rho \to \varepsilon_0\varepsilon$ and $c \to 1/\mu$). Incident and diffracted waves are coupled by the sinusoidal variation of the relative permittivity (3.12), given by

$$\Delta\varepsilon_{ij}(t) = 2\varepsilon_{ij}^{(1)}\cos(\Omega t - Kx_2) \ , \quad \text{where} \quad \varepsilon_{ij}^{(1)} = \frac{\Delta\varepsilon_{ij}}{2} \tag{3.39}$$

is the tensor of Fourier components at the fundamental frequency Ω of the perturbation.

The diffraction angle is usually very small ($\theta < 1°$). The two modes A and B then propagate in almost the same direction. Their amplitudes $A(x)$ and $B(x)$ obey (B.20):

$$\frac{\mathrm{d}A}{\mathrm{d}x} = -\mathrm{i}\delta\, A(x) - \mathrm{i}\kappa B(x)\,,$$
$$\frac{\mathrm{d}B}{\mathrm{d}x} = \mathrm{i}\delta\, B(x) - \mathrm{i}\kappa^* A(x)\,, \tag{3.40}$$

where δ is the detuning parameter in the x_1 direction of wave vector components k_1^A, k_1^B [equation (B.18) with $G = 0$]:

$$\delta = \frac{k_1^A - k_2^B}{2}\,, \quad k_1 = k\cos\frac{\theta}{2}\,. \tag{3.41}$$

The coupling coefficient of modes A and B, $\kappa = C_{AB}^{(1)}$, is obtained by transposing (B.14) to the optical case ($\rho \to \varepsilon_0\varepsilon$), where each mode N is represented by an electric field vector E_i^N:

$$\kappa = \frac{\omega}{4}\varepsilon_0 \int E_i^{A*}\varepsilon_{ij}^{(1)} E_j^B \mathrm{d}x_2\mathrm{d}x_3\,. \tag{3.42}$$

Orthogonality relation (3.120) (Problem 3.1) gives the electric field of normal modes (mean transported power density in the x_1 direction equal to $1\ \mathrm{W/m^2}$):

$$\int \boldsymbol{E}_N^* \cdot \boldsymbol{E}_M \mathrm{d}x_2\mathrm{d}x_3 = 2\frac{\mu\omega_N}{k_1^N}\delta_{NM} \Rightarrow \boldsymbol{E}_N = \left(2\frac{\mu\omega_N}{k_1^N}\right)^{1/2} \boldsymbol{e}_N\,,$$

where $\boldsymbol{e}_N$ is the unit vector defining polarisation of the Nth mode. The coupling coefficient between incident and diffracted light waves, I and D, due to Bragg angle acousto-optic interaction, is given by ($\omega_\mathrm{I} \approx \omega_\mathrm{D} = \omega$)

$$\boxed{\kappa = \frac{\mu\varepsilon_0\omega^2}{4\sqrt{k_1^\mathrm{I}k_1^\mathrm{D}}}\Delta\varepsilon\,, \quad \Delta\varepsilon = e_i^{\mathrm{I}*}\Delta\varepsilon_{il}e_l^\mathrm{D}}\,. \tag{3.43}$$

If diffraction angles are small, $k_1^\mathrm{I} = k_1^\mathrm{D} \approx k$, it follows for non-magnetic materials ($\mu = \mu_0$) that

$$\kappa = \frac{\omega^2}{4c_0^2 k}\Delta\varepsilon = \frac{k_0^2}{4k}\Delta\varepsilon = \frac{\pi}{2\lambda_0}\frac{\Delta\varepsilon}{n}\,, \tag{3.44}$$

since $\mu_0\varepsilon_0 = 1/c_0^2$.

Let us begin by applying this result to interactions without change of polarisation, i.e., the case of a longitudinal wave (strain $S = S_{22}$), propagating in an isotropic solid: $\Delta\varepsilon = -\varepsilon^2 pS$, with $\varepsilon = n^2$, so that

$$\kappa = -\frac{\pi}{2\lambda_0}n^3 pS\,. \tag{3.45}$$

Using (B.26) and (B.27), Appendix B, and considering the case of a single incident wave [$B(0) = 0$] at the Bragg angle ($\delta = 0 \Rightarrow \eta = |\kappa|$), amplitudes

$A(L)$ and $B(L)$ of incident and diffracted waves after covering distance L are given by

$$A(L) = A(0)\cos|\kappa|L\,,$$
$$B(L) = -\mathrm{i}\frac{\kappa^*}{|\kappa|}A(0)\sin|\kappa|L\,.$$

The fraction of incident power $I_0 = |A(0)|^2$ transferred to the diffracted beam $[I_\mathrm{D} = |B(L)|^2]$ is a function of the dimensionless parameter $|\kappa|L$ which characterises coupling over the whole interaction length:

$$R_\mathrm{max} = \frac{I_\mathrm{D}}{I_0} = \sin^2|\kappa|L\,. \tag{3.46}$$

Transfer is theoretically total when $|\kappa|L = \pi/2$. Energy is conserved, so that $|A(x)|^2 + |B(x)|^2 = |A(0)|^2$. The diffraction efficiency can also be deduced from this law (Problem 3.5).

Let us find the mean elastic power density per unit area, P. For a plane elastic wave, kinetic and potential densities are equal ($e_\mathrm{k} = e_\mathrm{p}$). If the propagation mode is pure, the energy flow speed equals the phase speed $V = \sqrt{c/\rho}$. From the results of Sect. 4.2.4, Vol. I, it follows that

$$P = \frac{1}{2}(e_\mathrm{k} + e_\mathrm{p})V = \frac{1}{2}cS^2V = \frac{1}{2}\rho V^3S^2\,. \tag{3.47}$$

Elasto-optic properties of materials are compared by introducing a *figure of merit* M which gauges the fraction of light deflected at Bragg incidence [3.19]:

$$\boxed{M = \frac{p^2n^6}{\rho V^3}}\,. \tag{3.48}$$

Replacing strain S by $\sqrt{2P/\rho V^3}$ in expression (3.45) for the coupling coefficient κ yields

$$|\kappa| = \frac{\pi}{2\lambda_0}\sqrt{2\frac{p^2n^6}{\rho V^3}P} = \frac{\pi}{\lambda_0}\sqrt{\frac{MP}{2}}\,. \tag{3.49}$$

The relative intensity of the diffracted beam, given by (3.46),

$$\frac{I_\mathrm{D}}{I_0} = \sin^2\left(\frac{\pi L}{\lambda_0}\sqrt{\frac{MP}{2}}\right) = \sin^2\left(\frac{\pi}{2}\sqrt{\frac{P}{P_1}}\right)\,, \quad P_1 = \frac{\lambda_0^2}{2ML^2}\,, \tag{3.50}$$

can theoretically attain a value of 1 (total deflection of the incident beam when $P = P_1$). The elastic power density required to deflect a given fraction of incident light is inversely proportional to the figure of merit M characterising the material. As already mentioned, the photoelastic coefficient p lies between 0.2 and 0.3 for the best crystal cuts. The important parameters, which are not in fact independent [3.20], are the index n and the speed V occurring in M to powers of 6 and 3, respectively. The refractive index rarely exceeds 2.5

for materials which are transparent in the visible. Longitudinal elastic wave speeds lie in the range 1 000 to 10 000 m/s. In general, low speed materials exhibit high attenuation with regard to elastic waves. In this respect, note that although liquids have high figures of merit, they can no longer be used at frequencies above 50 MHz. For example, at this frequency, attenuation in water is greater than 5 dB/cm and it grows as the square of the frequency. Table 3.3 gives attenuation coefficients at 500 MHz for various materials, together with figures of merit relative to that of silica ($n = 1.46$, $p = 0.27$, $\rho = 2.2 \times 10^3$ kg/m^3, $V = 5\,960$ m/s):

$$M_0 = 1.51 \times 10^{-15}\,\mathrm{s^3 kg^{-1}}\,.$$

α-iodic acid has a very high figure of merit but has the serious disadvantage of being very hygroscopic. The high value (794) in paratellurite (TeO_2) stems from the extremely low speed (616 m/s) of transverse waves polarised along $[1\bar{1}0]$ and propagating in direction [110]. However, its attenuation coefficient is too high beyond 100 MHz (2.8 dB/cm). Lead molybdate $PbMoO_4$ is an interesting material. TiO_2 and $LiNbO_3$ are useful at high frequencies because of their low attenuation. By (3.50), the elastic power $\mathcal{P} = PLH$ supplied by a rectangular transducer of cross-section LH, and required to deflect the whole light beam is

$$\mathcal{P}_1 = P_1 LH = \frac{\lambda_0^2}{2M}\frac{H}{L}\,. \tag{3.51}$$

In order to reduce this power, we must increase the interaction length L and reduce the height H of the elastic beam (limited by diffraction and light wave diameter). For example, for a transducer width $L = 15$ mm and $H = 1.5$ mm, and a He-Ne laser beam ($\lambda_0 = 0.633$ µm), $\mathcal{P}_1 = 13$ W in the case of silica, but only 0.56 W for lead molybdate.

Interaction Bandwidth. The phase synchronism condition $\delta = 0$ in Bragg angle interaction fixes the elastic wave frequency F_0 for a given angle of incidence α_0. However, natural spreading of the elastic wave beam leads to a certain bandwidth. At frequency $F \neq F_0$, the light beam, incident at the same angle with respect to the crystal, interacts with the elastic waves, whose direction of propagation $\boldsymbol{K'}$ is tilted at angle $\Delta\alpha = \Delta\theta/2$ (Fig. 3.14).

The detuning parameter is no longer zero. From (3.41),

$$2\delta = k\left[\cos\frac{\theta}{2} - \cos\left(\frac{\theta}{2} + \Delta\theta\right)\right] = 2k\sin\frac{\theta + \Delta\theta}{2}\,\sin\frac{\Delta\theta}{2}\,.$$

Then, using (3.36), and for small diffraction angles θ,

$$2\delta \approx K_0\frac{\Delta\theta}{2} \approx \Delta k_\mathrm{D} = k_\mathrm{D} - k'_\mathrm{D}\,,$$

where Δk_D is the change in diffracted wave number and $K_0 = 2\pi/\Lambda_0$.

The energy transfer between incident light wave, of amplitude $A(0)$, and diffracted wave, of amplitude $B(L)$, after an interaction length L, is given by (B.26) and (B.27) in Appendix B:

Table 3.3. Acousto-optic interaction conditions and figures of merit for several materials

Material	Elastic wave			Incident light wave ($\Lambda_0 = 0.6328$ µm)				Relevant	Figure of merit
	Propagation/ polarisation	Speed [m/s]	Attenuation (500 MHz) [dB/cm]	Useful range [µm]	Propagation/ polarisation[a]	Indices n_o	n_e	p_{ij}	M/M_0 $M = n^6 p^2/\rho V^3$ $[M_0 = 1.51 \times 10^{-15}]$
Silica	Any/L	5 960	3.0	0.2–2.5	Any/⊥	1.46		p_{12}	1
α-HIO_3	[001]/L	2 440	2.5[c]	0.4–1.3	[010]/⊥ ([100])	$n_1 = 1.98$[b]		p_{31}	55
$PbMoO_4$	[001]/L	3 630	2.5[d]		[010]/∥ or ⊥	2.38	2.26	p_{33} or p_{31}	23.7
TiO_2	[100]/L	8 015	0.5[e]	0.2–0.9	[001]/⊥ ([010])	2.58	2.9	p_{12}	2.6
TeO_2	[110]/T [1$\bar{1}$0]	616	70[f]	0.35–5	[001]/any	2.26	2.41	$(p_{11} - p_{12})/2$	794
TeO_2	[001]/L	4 200	3.8[f]	0.35–5	[010]/[100]	2.26	2.41	p_{13}	23
$LiNbO_3$	X/L	6 560	0.03[g]	0.5–4.5	Y/Z	2.29	2.20	p_{31}[i]	1.8
$LiTaO_3$	Z/L	6 180	0.02[g]		Any/Z	2.175	2.180	p_{33}	0.9
α-quartz	Z/L	6 363	0.6[h]		Y/Z	1.54	1.55	p_{13}	0.87

[a] Polarisation is defined parallel or perpendicular to the plane of elastic and optical wave vectors.
[b] The other two principal indices of (biaxial) α-HIO_3 are $n_2 = 1.96$ and $n_3 = 1.84$.
[c] Reference [3.9]. [d] Reference [3.10]. [e] Reference [3.27]. [f] Reference [3.22]. [g] Reference [3.28]. [h] Reference [3.29].
[i] For the beam undergoing no change in polarisation.

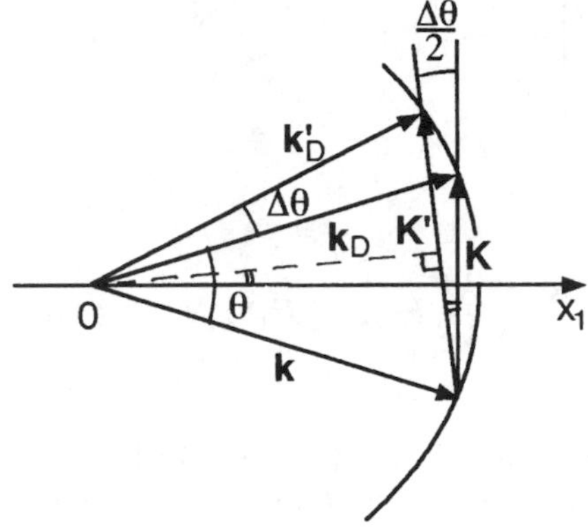

Fig. 3.14. Interaction bandwidth. Because the elastic wave beam spreads, at frequency $F \neq F_0$, interaction occurs with elastic waves whose wave vector satisfies the Bragg condition $\theta_0 + \Delta\theta = \lambda F/V$. Intensity of the deflected beam is reduced

$$B(L) = -\mathrm{i}\frac{\kappa^*}{\eta}A(0)\sin\eta L\ , \quad \eta^2 = |\kappa|^2 + \delta^2\ .$$

Efficiency R of energy transfer is thus reduced. The relative diffracted power (B.28) is a function of $|\kappa|L$ and the dimensionless parameter

$$\Delta\phi = 2\delta L \approx K_0 L\frac{\Delta\theta}{2} \approx \Delta k_{\mathrm{D}} L\ , \tag{3.52}$$

characterising *phase asynchronism* between incident and diffracted waves:

$$\begin{aligned} R &= \frac{|\kappa L|^2}{|\kappa L|^2 + \left(\frac{\Delta\phi}{2}\right)^2}\sin^2\sqrt{|\kappa L|^2 + \left(\frac{\Delta\phi}{2}\right)^2} \\ &= |\kappa L|^2 \mathrm{sinc}^2\sqrt{|\kappa L|^2 + \left(\frac{\Delta\phi}{2}\right)^2}\ . \end{aligned} \tag{3.53}$$

Phase asynchronism $\Delta\phi$ is due either to an elastic wave frequency change away from F_0 (case 1), or a variation $\Delta\alpha$ in the angle of incidence (case 2).

Case 1. From (3.34), it follows that

$$\Delta\phi = \pi\frac{\lambda}{\Lambda_0}\frac{L}{V}(F - F_0) = \pi\frac{\lambda L}{V^2}F_0(F - F_0)\ . \tag{3.54}$$

The bandwidth of the acousto-optic interaction depends on the elastic power density P, i.e., on the maximal diffraction efficiency. Since

$$|\kappa|L = \frac{\pi}{2}\sqrt{\frac{P}{P_1}} \quad \Rightarrow \quad \frac{|\kappa L|^2}{|\kappa L|^2 + \left(\frac{\Delta\phi}{2}\right)^2} = \frac{P/P_1}{P/P_1 + \left(\frac{\Delta\phi}{\pi}\right)^2}\ ,$$

the diffraction efficiency (3.53) is given in terms of the ratio $P/P_1 = \mathcal{P}/\mathcal{P}_1$ (3.51) by

$$R = \left[\frac{\pi}{2} \sqrt{\frac{P}{P_1}} \operatorname{sinc} \frac{\pi}{2} \sqrt{\frac{P}{P_1} + \left(\frac{\Delta\phi}{\pi}\right)^2} \right]^2 ,$$

or

$$R = \frac{P}{P_1} \frac{\sin^2 \frac{\pi}{2} \sqrt{\frac{P}{P_1} + \left(\frac{\Delta\phi}{\pi}\right)^2}}{\frac{P}{P_1} + \left(\frac{\Delta\phi}{\pi}\right)^2} , \quad \frac{\Delta\phi}{\pi} = \frac{\lambda L}{V^2} F_0 (F - F_0) . \tag{3.55}$$

When $\Delta\phi = 0$, we retrieve (3.50) for maximal efficiency. The -3 dB bandwidth is determined by the phase asynchronism corresponding to 50% reduction in deflected light intensity. As shown in Problem 3.7,

$$\left(\frac{\Delta\phi}{\pi}\right)_{\max} = 0.89,\ 0.87,\ 0.80, \quad \text{for} \quad R_{\max} = 0,\ 0.5,\ 1 ,$$

so that the bandwidth

$$\Delta F = \frac{2V^2}{\lambda L F_0} \left(\frac{\Delta\phi}{\pi}\right)_{\max} \approx 1.7 \frac{V^2}{\lambda L F_0} = 1.7 \frac{V}{\lambda} \frac{\Lambda_0}{L} \tag{3.56}$$

is limited by the divergence angle Λ_0/L of the elastic wave beam. Hence, $\Delta F = 100$ MHz for $V = 4\,000$ m/s, $L = 1.5$ mm, $\lambda = 0.3\,\mu$m and $F_0 = 600$ MHz.

Case 2. In this case, there is a variation $\Delta\alpha = \Delta\theta/2$ in the angle of incidence. The diffracted intensity is obtained by substituting the phase asynchronism parameter

$$\Delta\phi = K_0 L \Delta\alpha = 2\pi \frac{F_0}{V} L \Delta\alpha$$

into (3.53). Rotating the crystal for constant applied voltage across the transducer, and assuming a weak interaction $\mathcal{P} \ll \mathcal{P}_1$, variations in deflected light intensity with $\Delta\alpha = \alpha - \alpha_0$ (where α_0 is the Bragg angle) reproduce the far field radiation pattern for power emitted by the transducer:

$$\frac{I(\alpha_0 + \Delta\alpha)}{I(\alpha_0)} = \operatorname{sinc}^2 \left(\pi \frac{L}{\Lambda_0} \Delta\alpha \right) .$$

Number of Resolvable Directions. The angular deflection θ of the light beam due to elastic waves of frequency F equals twice the Bragg angle:

$$\theta = 2\alpha \approx \frac{\lambda}{\Lambda} = \frac{\lambda}{V} F .$$

A change ΔF induces a change $\Delta\theta = (\lambda/V)\Delta F$ in this deflection.

Assuming that the angular width of the light beam of diameter D is defined by natural diffraction angle $\Delta\beta = \lambda/D$, the number of distinct directions N_{B} within angle $\Delta\alpha$ is

$$N_{\mathrm{B}} = \frac{\Delta\theta}{\Delta\beta} = \frac{D}{V}\Delta F \ , \quad \text{or} \quad N_{\mathrm{B}} = T_{\mathrm{B}}\Delta F \ , \tag{3.57}$$

where T_{B} denotes the elastic wave transit time across the light beam. For example, the value $N_{\mathrm{B}} = 300$ is obtained when $\Delta F = 100$ MHz and $T_{\mathrm{B}} = 3$ µs. If the elastic wave speed is $V = 4\,000$ m/s, i.e., 4 mm/µs, the diameter of the light beam must be 12 mm. The laser beam must generally be broadened optically. Moreover, the crystal length, at least equal to $T_{\mathrm{B}}V$, must not cause excessive attenuation of elastic waves at the operating frequency, itself determined by the required bandwidth. Attenuation at 300 MHz in lead molybdate is greater than 1 dB/cm.

Let us summarise the three main points about Bragg incidence interaction:

- The intensity of the single diffracted light beam is a function of the intensity of the elastic wave beam.
- The angle through which the beam is deflected is proportional to the elastic wave frequency.
- The elastic wave frequency occurs in the diffracted beam frequency.

The first property is exploited in modulating light beams and the second in deflecting them (see Chap. 5). The third is used to change the frequency of a light wave in heterodyne devices (Sect. 3.2.2).

3.1.5.3 Anisotropic Media. Change of Polarisation. Let us now consider some cases in which the light wave polarisation is changed by the interaction. In particular, we shall discuss the collinear interaction, with no beam deflection.

We previously investigated the case of an optically isotropic solid, longitudinal waves and incident and diffracted beams having the same polarisation. The triangle in Fig. 3.13 then turns out to be almost isosceles ($F \ll f \Rightarrow k_{\mathrm{I}} \approx k_{\mathrm{D}}$). In a crystal, the refractive index varies with direction of propagation. Phase speeds of incident and diffracted waves are then no longer the same and wave vectors $\boldsymbol{k}_{\mathrm{I}}$, $\boldsymbol{k}_{\mathrm{D}}$ are not the same length. In some cases, incident and diffracted waves have different polarisations. The triangle is no longer isosceles and angles of incidence and diffraction α_{I}, α_{D} are no longer equal [3.23].

Consider the example of a uniaxial trigonal crystal belonging to one of the classes $3m$, 32 or $\bar{3}m$ characterised by the elasto-optic tensor in Table 3.1. By hypothesis, all beams are located in a plane perpendicular to the optical axis Ox_3 (Fig. 3.15). A transverse elastic wave propagating along crystallographic axis $X = Ox_1$ and polarised in the (x_2, x_3)-plane produces a variation $\Delta\varepsilon_{il}$ which can be found from (3.21):

$$\Delta\varepsilon_{il} = -\varepsilon_{ii}\varepsilon_{ll}\left(P_{il31}\frac{\partial u_3}{\partial x_1} + P_{il21}\frac{\partial u_2}{\partial x_1}\right) .$$

The only non-zero components of the Nelson–Lax tensor P_{il31} and P_{il21} are P_{1231}, P_{1331} and P_{1221}, P_{1321}. Using (3.22) with $n_1 = 1$, $n_2 = n_3 = 0$, the electro-optic effect

$$P^{\text{ind}}_{il31} = -r_{il1}e_{131}/\varepsilon_0\varepsilon_{11}\,, \quad P^{\text{ind}}_{il21} = -r_{il1}e_{121}/\varepsilon_0\varepsilon_{11}\,,$$

is zero for classes 32 and $\overline{3}m$, in which $e_{15} = 0 = e_{16}$. For class $3m$, components $P_{\alpha 31}$ and $P_{\alpha 21}$ are modified:

$$P^{\text{ind}}_{\alpha 31} = -r_{\alpha 1}e_{15}/\varepsilon_0\varepsilon_{11}\,, \quad P^{\text{ind}}_{\alpha 21} = -r_{\alpha 1}e_{16}/\varepsilon_0\varepsilon_{11}\,, \quad \alpha = 5, 6\,.$$

The tensor $\Delta\varepsilon_{il}$ takes the form

$$\Delta\varepsilon_{il} = \begin{pmatrix} 0 & \Delta\varepsilon_{12} & \Delta\varepsilon_{13} \\ \Delta\varepsilon_{12} & 0 & 0 \\ \Delta\varepsilon_{13} & 0 & 0 \end{pmatrix}, \tag{3.58}$$

where, for classes 32 and $\overline{3}m$,

$$\Delta\varepsilon_{12} = -\varepsilon_{11}\varepsilon_{22}\left(P_{1231}\frac{\partial u_3}{\partial x_1} + P_{1221}\frac{\partial u_2}{\partial x_1}\right),$$

$$\Delta\varepsilon_{13} = -\varepsilon_{11}\varepsilon_{33}\left(P_{1331}\frac{\partial u_3}{\partial x_1} + P_{1321}\frac{\partial u_2}{\partial x_1}\right).$$

The incident light wave, polarised in the x_3 direction, is accompanied by an electric field with components (0, 0, $E_3 = E\cos\omega t$), which produces electric displacement

$$D_i = \varepsilon_0(\varepsilon_{i3} + \Delta\varepsilon_{i3})E_3\,.$$

Since,

$$\Delta\varepsilon_{i3} = (\Delta\varepsilon_{i3})_0 \sin\Omega t\,,$$

the diffracted wave of frequency $\omega + \Omega$ arises from the product $\Delta D_i = \varepsilon_0\Delta\varepsilon_{i3}E_3$. It is polarised in the x_1 direction, because only $\Delta\varepsilon_{13}$ is non-zero. The elasto-optic interaction changes the polarisation of the light beam. Construction of the wave vector diagram thus involves two sheets of the index surface, i.e., two circles of radii n_e and n_o in the (x_1, x_2)-plane, since the index surface is a surface of revolution obtained by rotation about optical axis Ox_3. The end of the incident wave vector $\boldsymbol{k}_\text{I}$ is located on the circle of radius n_e corresponding to polarisation parallel to optical axis Ox_3. The end of the wave vector $\boldsymbol{k}_\text{D}$ for the diffracted wave, polarised perpendicular to Ox_3, is located on the circle of radius n_o (see Fig. 3.15).

Note that, for an extraordinary incident ray, there are generally two symmetrically diffracted rays corresponding to different acoustic frequencies. It is easy to show (Problem 3.8) that algebraic values of incident and diffracted angles are

$$\sin\alpha_\text{I} = -\frac{1}{2n_\text{e}}\left[\frac{\Lambda}{\lambda_0}(n_\text{e}^2 - n_\text{o}^2) + \frac{\lambda_0}{\Lambda}\right],$$

$$\sin\alpha_\text{D} = -\frac{1}{2n_\text{o}}\left[\frac{\Lambda}{\lambda_0}(n_\text{e}^2 - n_\text{o}^2) - \frac{\lambda_0}{\Lambda}\right]. \tag{3.59}$$

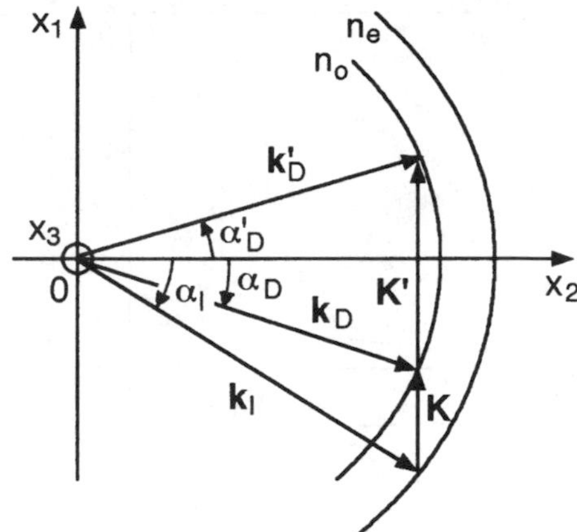

Fig. 3.15. Wave vector diagram. Positive uniaxial crystal ($n_e > n_o$). Interaction of an extraordinary incident ray (index n_e, wave vector $\boldsymbol{k}_I$) with transverse elastic wave (wave vector $\boldsymbol{K}$ or $\boldsymbol{K}'$) results in an ordinary diffracted wave (index n_o, wave vector $\boldsymbol{k}_D$ or $\boldsymbol{k}'_D$)

Figure 3.16 shows how incident and diffracted angles vary with wavelength ratio $\lambda_0/\Lambda = (\lambda_0/V)F$ in α-quartz ($n_o = 1.54$, $n_e = 1.55$). The minimal elastic wave frequency,

$$F_{\min} = \frac{V}{\lambda_0}(n_e - n_o) , \quad \text{reached when} \quad \frac{\lambda_0}{\Lambda} = n_e - n_o , \tag{3.60}$$

corresponds to a *collinear interaction* of light beams, in the same direction ($\alpha_I = \alpha_D = -90°$). The maximal elastic wave frequency,

$$F_{\max} = \frac{V}{\lambda_0}(n_e + n_o) , \quad \text{reached when} \quad \frac{\lambda_0}{\Lambda} = n_e + n_o , \tag{3.61}$$

corresponds to a collinear interaction between light beams propagating in opposite directions ($-\alpha_I = \alpha_D = +90°$). Given the speed $V = 3\,300$ m/s of the slow transverse wave in quartz and using a He-Ne laser,

$$F_{\min} = 52.1\,\text{MHz} , \quad F_{\max} = 16.1\,\text{GHz} .$$

Between these two extremes, there is a wide band of frequencies over which the angle of incidence varies little, whilst the angle of diffraction varies linearly with frequency. These conditions are obtained for a frequency F_0 such that

$$\frac{\lambda_0}{\Lambda} = (n_e^2 - n_o^2)^{1/2} \quad \Rightarrow \quad F_0 = \frac{V}{\lambda_0}(n_e^2 - n_o^2)^{1/2} = (F_{\min}F_{\max})^{1/2} . \tag{3.62}$$

For the corresponding angle of incidence $\sin\alpha_I = -(n_e^2 - n_o^2)^{1/2}/n_e$, the angle of diffraction is zero. The acoustic wave vector is then tangent to the index surface of the diffracted wave. These conditions for interaction with *tangential phase tuning* are used in practice to make light deflectors capable of operating over a broad bandwidth. However, if the interaction length is large, the diffracted beam, polarised in the XY-plane (index n_o) may interact under normal incidence to give a second diffracted beam, symmetrical with the incident beam. This interaction, in which roles of light beams are permuted, corresponds to the dashed lines in Fig. 3.16.

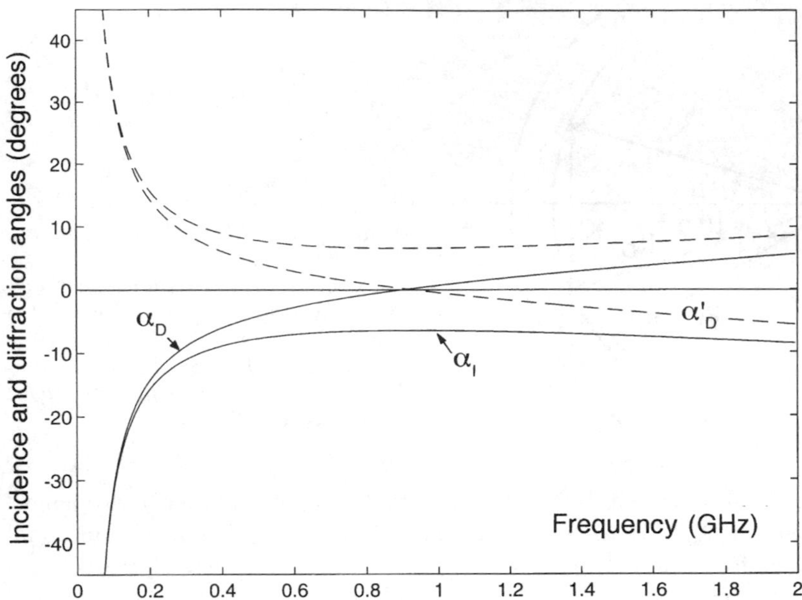

Fig. 3.16. Interaction with change of polarisation in a positive uniaxial crystal (α-quartz, $n_\mathrm{o} = 1.54$, $n_\mathrm{e} = 1.55$). Variation of angle of incidence α_I and angle of diffraction α_D with elastic wave frequency

Collinear Interaction. Let us investigate the effect of a longitudinal elastic wave propagating in the x_1 direction (strain S_{11}) on a light wave polarised in the x_3 direction and also propagating in the x_1 direction, still for the case of a trigonal crystal with 32 symmetry. By the Pockels theory, which is valid in such conditions, the change in the permittivity tensor,

$$\Delta\varepsilon_{il} = -\varepsilon_{ii}\varepsilon_{ll}p_{il11}S_{11} ,$$

reduces to

$$\Delta\varepsilon_{il} = \begin{pmatrix} \Delta\varepsilon_{11} & 0 & 0 \\ 0 & \Delta\varepsilon_{22} & \Delta\varepsilon_{23} \\ 0 & \Delta\varepsilon_{23} & \Delta\varepsilon_{33} \end{pmatrix} ,$$

where

$$\Delta\varepsilon_{11} = -(\varepsilon_{11})^2 p_{11} S_1 , \quad \Delta\varepsilon_{22} = -(\varepsilon_{11})^2 p_{12} S_1 ,$$
$$\Delta\varepsilon_{23} = -\varepsilon_{11}\varepsilon_{33} p_{41} S_1 , \quad \Delta\varepsilon_{33} = -(\varepsilon_{33})^2 p_{31} S_1 .$$

Here, we have used the table for the $p_{\alpha\beta}$. The electro-optic effect, which occurs only for class 32 ($P_{il11}^{\mathrm{ind}} = -r_{il1}e_{11}/\varepsilon_{11}$), modifies components p_{11}, p_{12} and p_{41}.

The wave of frequency $\omega + \Omega$ has polarisation $\Delta D_i = \Delta\varepsilon_{i3}E_3$, i.e.,

$$\Delta D_1 = 0 , \quad \Delta D_2 = \Delta\varepsilon_{23}E_3 , \quad \Delta D_3 = \Delta\varepsilon_{33}E_3 .$$

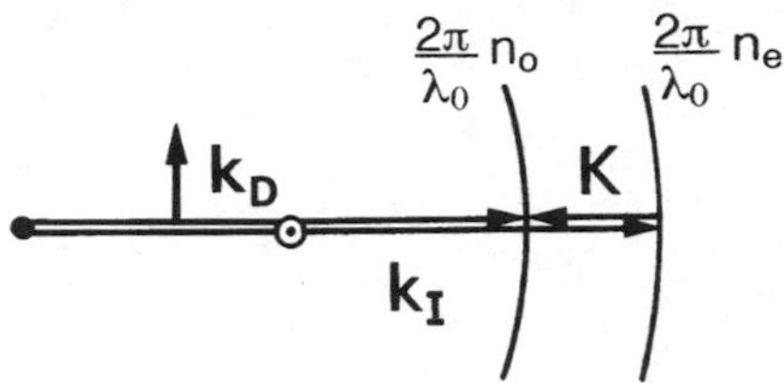

Fig. 3.17. Starting with an extraordinary ray $\boldsymbol{k}_\mathrm{I}$ of wavelength λ_0, collinear interaction produces an ordinary diffracted ray $\boldsymbol{k}_\mathrm{D}$ if the elastic wave frequency is $F_0 = V(n_\mathrm{e} - n_\mathrm{o})/\lambda_0$

In collinear interaction, the Bragg condition is only satisfied if there is a change in polarisation of the light beam (otherwise $k_\mathrm{D} = k_\mathrm{I} \Rightarrow K = 0$). The diffracted ray therefore has ordinary polarisation parallel to Ox_2, i.e., $(0, \Delta\varepsilon_{23}E_3, 0)$. The wave vector diagram shown in Fig. 3.17 for a positive uniaxial crystal ($n_\mathrm{e} > n_\mathrm{o}$) implies:

$$K = k_\mathrm{I} - k_\mathrm{D} = \frac{2\pi}{\lambda_0}(n_\mathrm{e} - n_\mathrm{o}) = \frac{2\pi}{V}F\,.$$

Hence, the elastic wave frequency is

$$F_0 = \frac{V}{\lambda_0}(n_\mathrm{e} - n_\mathrm{o})\,. \tag{3.63}$$

In contrast, for a given elastic wave frequency, only those light waves satisfying (3.63) will be diffracted. If their wavelength deviates from λ_0, the asynchronous phase condition is no longer satisfied and diffraction efficiency decreases. Using results of Problem 3.10, the −3 dB bandwidth of the acousto-optic filter is inversely proportional to interaction length L:

$$\Delta\lambda \approx 0.85\frac{\lambda_0^2}{L|\Delta n|}\,.$$

Tunable optical filters have been built according to this principle [3.24]. They are tuned by varying the frequency of stationary or travelling elastic waves. For example, varying the frequency of transverse elastic waves propagating in a calcium molybdate ($CaMoO_4$) crystal between 40 and 68 MHz, light waves can be selected in the range 6 700 to 5 100 Å with a bandwidth of 8 Å. Configurations which are non-collinear in terms of wave vectors but collinear for group velocities, i.e., for incident and diffracted light rays, have been used in tellurium oxide crystals.

Tangential Interaction. The relevant configuration is shown in Fig 3.18. Near the centre frequency of the interaction, the angle of diffraction varies linearly. Indeed, from (3.59) and (3.62),

$$\sin\alpha_\mathrm{D} = \frac{\lambda_0}{2Vn_0}\frac{F^2 - F_0^2}{F_0} \quad\Rightarrow\quad \alpha_\mathrm{D} \approx \frac{\lambda}{V}(F - F_0)\,.$$

The tangential interaction is of interest because a small change $\Delta\alpha = \Lambda/L$ in the direction of $\boldsymbol{K}$, due to spreading of the elastic wave beam, causes a large deflection of the diffracted beam:

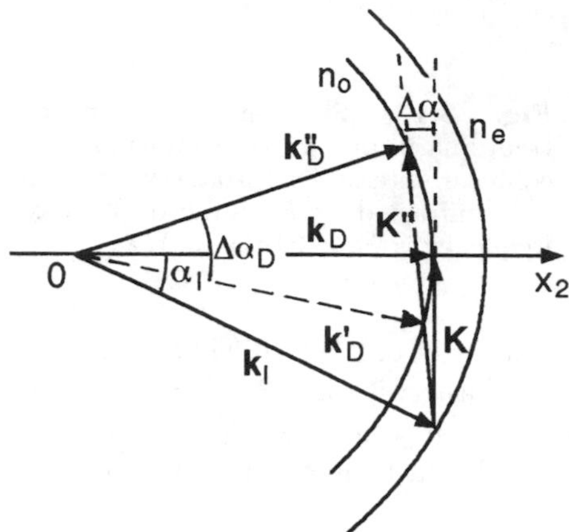

Fig. 3.18. Tangential interaction

$$\Delta\alpha = \frac{k_{\mathrm{D}} - k_{\mathrm{D}}\cos\dfrac{\Delta\alpha_{\mathrm{D}}}{2}}{K} \approx \frac{\Lambda}{\lambda_{\mathrm{D}}}\frac{(\Delta\alpha_{\mathrm{D}})^2}{8} = \frac{\Lambda}{L} \Rightarrow \Delta\alpha_{\mathrm{D}} = \left(\frac{8\lambda_{\mathrm{D}}}{L}\right)^{1/2} .$$

The change in wave number $\Delta K = k_{\mathrm{D}}\Delta\alpha_{\mathrm{D}}$ corresponds to a change in elastic wave frequency:

$$(\Delta F)_{\tan} = \frac{V}{2\pi}\Delta K = \frac{V}{\lambda_{\mathrm{D}}}\left(\frac{8\lambda_{\mathrm{D}}}{L}\right)^{1/2} = 2V\left(\frac{2}{\lambda_{\mathrm{D}}L}\right)^{1/2} . \tag{3.64}$$

This bandwidth is much larger than the one obtained for interaction in an isotropic medium. According to (3.56),

$$\frac{(\Delta F)_{\tan}}{(\Delta F)_{\mathrm{iso}}} \approx \frac{2\sqrt{2}}{1.7}\frac{(\lambda L)^{1/2}}{\Lambda} .$$

Since $\Lambda = \lambda_0/(n_{\mathrm{e}}^2 - n_{\mathrm{o}}^2)^{1/2}$ [equation (3.62)] and $\lambda_{\mathrm{D}}(n_{\mathrm{e}} + n_{\mathrm{o}}) \approx 2\lambda_0$, it follows that

$$\frac{(\Delta F)_{\tan}}{(\Delta F)_{\mathrm{iso}}} \approx \frac{2\sqrt{2}}{1.7}\frac{\left[L\lambda_{\mathrm{D}}(n_{\mathrm{e}}^2 - n_{\mathrm{o}}^2)\right]^{1/2}}{\lambda_0} \approx 2.3\left(\frac{L}{\lambda_0}\Delta n\right)^{1/2} . \tag{3.65}$$

Example. For interaction length $L = 8$ mm in a lithium niobate crystal ($\Delta n = 0.01$) and $\lambda_0 = 0.8$ μm, the multiplicative factor is 23.

In summary, the key parameter in determining the acousto-optic interaction is phase asynchronism, defined by

$$\Delta\phi = \Delta k\, L = 2\delta\, L ,$$

where L is interaction length, and Δk the difference between incident and diffracted wave numbers. Using coupled mode theory (Appendix B), coupling between modes is efficient if $\Delta\phi$ is smaller than a value of order π, which depends little on diffraction efficiency (see Problem 3.7). At normal incidence, phase synchronism cannot be perfect. For the Nth mode,

$$\Delta k_N = k(1 - \cos\theta_N) \approx \frac{k}{2}\theta_N^2 = \pi\frac{\lambda}{\Lambda^2}N^2 .$$

In the Raman–Nath regime, that is, when there is multiple diffraction ($N \geq 2$), the condition $\Delta\phi < \pi$ implies

$$\frac{\lambda L}{\Lambda^2} < \frac{1}{N^2} , \quad \text{hence} \quad Q \ll 2\pi ,$$

where $Q = K^2L/k$ is the Klein–Cook parameter. This is reached if the array thickness L is small, i.e., $L \ll \Lambda^2/\lambda$.

At oblique incidence, at the Bragg angle, phase synchronism is perfect for just one mode. The cumulated effect is strong when $Q > 4\pi$. In the intermediate region $1 < Q < 2\pi$, there is no exact theoretical solution and diffracted light intensity must be found numerically.

3.1.6 Interaction with Surface Elastic Waves

Up to now, we have been considering bulk waves, either longitudinal or transverse. However, light waves are also diffracted by surface waves, and in particular, by Rayleigh waves. Since these involve at least one component of longitudinal displacement and one component of transverse displacement, we must expect to find, in the general case, combinations of all the effects described above. For example, only part of the diffracted light may undergo a polarisation rotation [3.25]. Moreover, surface deformations due to Rayleigh waves diffract reflected light. This case, which does not involve modulation of the permittivity ε_{ij}, is treated in Sect. 3.2.

Modulation of dielectric permittivity is only relevant if light waves cross the elastic wave beam laterally. As the thickness of mechanically vibrating matter is of the order of the elastic wavelength, this interaction condition requires concentration of light energy close to the surface. One solution is then to guide light into a thin layer deposited on (or produced within) the substrate, which is generally piezoelectric and serves to carry the elastic waves (see Fig. 3.19). The guide thickness (≈ 1 μm) is less than the elastic wavelength (≈ 30 μm at 100 MHz) and its optical index greater than that of the substrate. As for bulk elastic waves, there is a critical width L_c for the surface wave beam, which separates two extreme situations. For $L \ll L_c$, the incident light beam, normal to the elastic beam, splits into beams with different frequencies and directions. For $L \gg L_c$, the light beam, encountering the elastic beam at the Bragg angle ($\sin\alpha = \lambda/2\Lambda$), is deflected through angle 2α. With regard to applications, Bragg incidence interaction is the most useful since elastic wave frequencies can be very high (> 1 GHz), making the passband very wide, and it provides a single deflected beam in a well defined direction. This is illustrated in Fig. 3.19. Elastic waves, excited by an interdigital transducer, generate periodic index variations in the optical guide. These have the same effect on light as a moving grating.

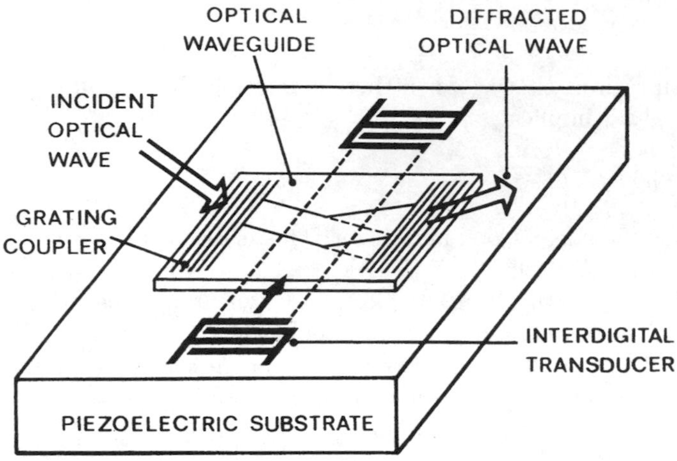

Fig. 3.19. Bragg incidence diffraction by an elastic wave beam of a light beam guided within a thin layer. Light waves are introduced into the optical guide by means of a strip grating (or a prism) and extracted in the same way

If the diffracted light propagates in the same mode, with the same polarisation as the incident light, interaction is characterised by the properties established above in the bulk wave case:

- The intensity of the diffracted light beam is proportional (under linear operating conditions) to the electric power applied to the transducer (and also the incident light intensity).
- The deflection angle increases in proportion to elastic wave frequency.
- The diffracted beam frequency is that of the incident beam plus or minus the elastic wave frequency, depending on whether the angle between optical and elastic wave vectors is greater than or less than 90°.

The plane guide is a layer of glass deposited on the substrate, or better, the surface of the substrate itself, whose index has been increased. The most widely used substrate is lithium niobate. The guide can be made in various ways: thermal exodiffusion of lithium, or diffusion of metals such as Ni, Au, Ag and Ti. Titanium diffusion gives good results. A film of thickness about 300 Å is deposited onto a Y-cut $LiNbO_3$ crystal, possibly through a mask if a precise guide width is required. The surface is then heated to about 1000°C in oxygen for around 80 hours. All the light is deflected when an electric power of 0.1 W is applied to the transducer. When equipped with an ordinary transducer, the bandwidth of this device is limited to a few tens of [MHz], but it can be increased using an appropriate configuration of transducers operating at different frequencies.

Light waves are introduced and extracted from the plane guide by means of an intermediary. If they enter or leave via the top of the guide, a grating

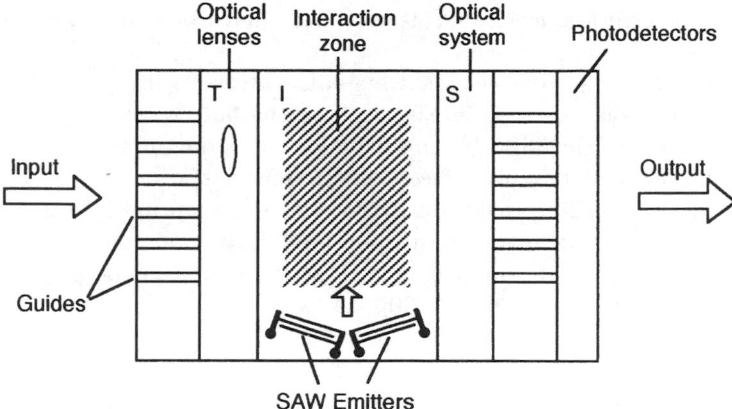

Fig. 3.20. General setup for the treatment of several optical beams by surface acousto-optic interaction. Main components: guides, shaping optics T, interaction zone I, output optics S, guides and detectors. Taken from [3.26]

or prism is used; if they enter or leave via the sides, a lens (e.g., a cylindrical lens) can be used.

In fact, this surface wave deflector or modulator of light does not aim to produce the same characteristics as bulk wave devices. It is employed rather as a better way of making integrated optical devices. Indeed, several guides and lenses of various shapes can be fabricated on the same crystal. For example, a lens can be made by depositing a Cr/Au protective film onto the diffusion-deposited titanium guide, and then etching out a window in the shape of the lens. Immersion in the appropriate bath increases the extraordinary index of the unprotected region. In this way devices suitable for simultaneous treatment of several optical beams can be investigated. Figure 3.20 shows how they work. Each input guide receives an optical beam. T represents some particular action (magnification by a lens placed at each guide output) or a collective action (concentration of beams or modification of their directions using a single lens). I represents the interaction with one or more surface elastic wave beams propagating in appropriate directions. The resulting light beam(s) are then transformed as they cross S and directed towards output channels with their associated detectors. Elementary components have been designed which can connect one input channel (among 4) to one output channel (among 8) and multiply a vector by a 4×4 matrix [3.26].

Of course, it would be interesting to construct optical sources (laser diodes) and detectors on the same substrate. This objective explains why attempts have been made to use gallium arsenide substrates. This is a semiconducting material very widely used in electronics, even in preference to silicon for applications requiring high mobility carriers. However, this material is only slightly piezoelectric and elastic waves must be generated using a zinc oxide (ZnO) film. Several Bragg cells, equipped with integrated lenses and operating at centre frequencies of order 1 GHz, have been designed in laboratory conditions.

Guided light waves and surface elastic waves also interact when propagating in the same direction, as shown in Fig. 3.17 for the bulk wave case. This *collinear interaction* can thus also be applied when designing optical filters which are tunable via variation of surface wave frequencies.

The article by Tsaï [3.26] provides a review of results up until 1992, in the area of integrated acousto-optics for lithium niobate and gallium arsenide substrates. Since then, several components such as tunable acousto-optic filters have been designed in laboratory [3.30].

3.2 Optical Measurement of Mechanical Displacements

Mechanical displacements considered here are those produced perpendicular to a surface by elastic waves of relatively high frequency (> 100 kHz). They are generally rather small, being of order a fraction of a [nm]. Optical devices capable of measuring such displacements fall into two categories. The first includes deflection and diffraction probes which use surface ripples, whilst the second contains interferometric probes which measure any displacement normal to the surface. All these optical probes are less sensitive than piezoelectric detectors but carry out local measurement without mechanical contact and have high bandwidth.

3.2.1 Deflection and Diffraction Probes

These are suitable for studying waves which cause surface ripple, such as the Rayleigh wave. This wave propagates at the surface of any solid, to a depth of the order of the wavelength (see Sect. 5.3, Vol. I).

There are two cases depending on whether the dimension d of the probe light beam (in the direction of propagation of the wave) is greater than or less than the elastic wavelength Λ:

- if $d \gg \Lambda$, the light beam is divided into several parts, and the result is a diffraction effect;
- if $d < \Lambda$, the direction of the light beam is modified, and the result is deflection.

3.2.1.1 Deflection. Figure 3.21 shows the operating principle for the probe. The light beam of wavelength λ, reflected by the surface, oscillates as the wave passes. It is partially masked by an edge (in practice, the edge of the photodetector), so that the photocurrent intensity is modulated at the wave frequency.

Let $u \sin[\Omega(t - x/V)] = u \sin(\Omega t + \phi)$ be the mechanical displacement. The greatest slope in the surface undulation is $\alpha = 2\pi u/\Lambda$ and the amplitude

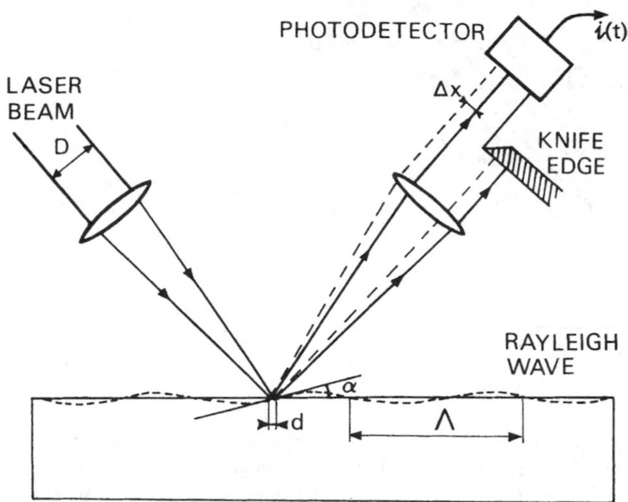

Fig. 3.21. Deflection of a light beam by a (Rayleigh) surface wave. The dimension d of the light beam in the direction of propagation of the elastic wave is small compared with the wavelength Λ of the latter ($V = 3\,000$ m/s, $f = 30$ MHz, implying $\Lambda = 100$ μm). The beam direction oscillates at the wave frequency. The width of the photodetector is less than the zone swept out by the light beam, so that the received flux varies

of the angular deflection is $2\alpha = 4\pi u/\Lambda$. After the second lens, with focal length L assumed equal to that of the first, the beam displacement is $\Delta x = 2\alpha L$. If half the beam of diameter D and power P_L is masked, the current intensity i delivered by the photodetector varies according to

$$i(t) = sP_\mathrm{L}\frac{\Delta x}{D} = \frac{4\pi u}{\Lambda}\frac{L}{D}sP_\mathrm{L}\sin(\Omega t + \phi) , \tag{3.66}$$

where s is its response factor in [A/W]. The diameter d of the focal spot on the surface is diffraction limited. If k is the optical wave number, we find

$$d = \lambda\frac{L}{D} = \frac{2\pi}{k}\frac{L}{D} .$$

Let $I_0 = sP_\mathrm{L}/2$ be the continuous part of the current intensity. Then,

$$i(t) = 4\frac{d}{\Lambda}kuI_0\sin(\Omega t + \phi) . \tag{3.67}$$

This remains true provided that $d < \Lambda$. If the focal spot covers several elastic wavelengths, the beam is diffracted (Sect. 3.2.1.2). The optimal value is $d \approx \Lambda/2$ [3.31]:

$$i_\mathrm{opt} = 2kuI_0\sin(\Omega t + \phi) . \tag{3.68}$$

This technique, originally developed by Adler et al. [3.32], is particularly convenient for wave front scanning. It is used in Scanning Laser Acoustic Microscopy SLAM [3.33] for non-destructive testing. However, it is not suitable

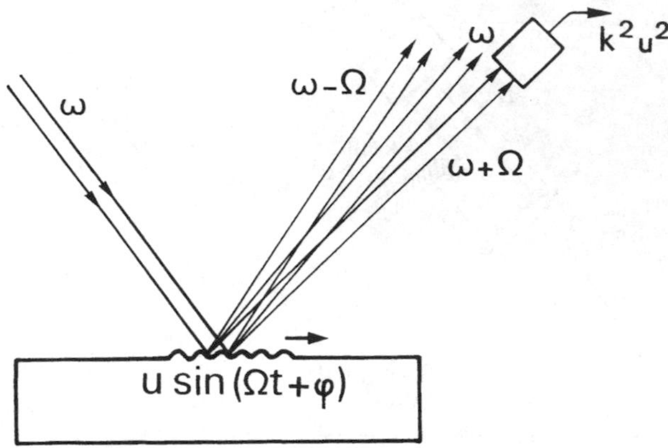

Fig. 3.22. Diffraction of a light beam by a Rayleigh wave. The surface perturbation acts like a phased array on the light beam which covers several wave fronts. The output signal of the photodiode sensing the first order beam is proportional to the square of the mechanical displacement amplitude

for poor surfaces and works only for a limited frequency range. In fact, the maximal value $F_{\max}$, defined by $d = \Lambda$, is equal to $DV/\lambda L$. For $L = 10$ cm, a He-Ne laser with $D = 1$ mm and an elastic wave phase speed $V = 3\,000$ m/s, the maximal frequency is 47 MHz.

Modifying this arrangement by using a lens of short focal length, a double photodiode and the stroboscope effect, Engan was able to observe displacements of amplitude 5 Å caused by 100 MHz Rayleigh waves, with a signal-to-noise ratio of 60 dB [3.34].

This way of measuring displacements by deflection of a light beam has been used in other areas, e.g., to detect oscillations of the tiny cantilever in an atomic force microscope.

3.2.1.2 Diffraction. When the light beam of power P_{L} covers several wave fronts ($d \gg \Lambda$), the surface corrugation acts as a phased array (see Fig. 3.22). For angle of incidence θ_0, the beam is diffracted into beams of frequencies $\omega \pm m\Omega$ and directions

$$\sin\theta_m = \sin\theta_0 \pm m\lambda/\Lambda\,, \quad m = 0,\, 1,\, \ldots\,. \tag{3.69}$$

For a perfectly reflecting surface, the power P_m of the order m beam is given by the square of the Bessel function $J_m(\Delta\Phi)$, whose argument is the phase difference $\Delta\Phi = 2ku\cos\theta_0$ produced by normal displacement of amplitude u at the surface:

$$P_m/P_{\mathrm{L}} = [J_m(\Delta\Phi)]^2\,.$$

When $ku \ll 1$, $J_1(\Delta\Phi) \approx \Delta\Phi/2$ and the intensity of the first order diffracted beam is proportional to the square of the mechanical displacement:

$$P_1/P_{\mathrm{L}} = (ku\cos\theta_0)^2\,. \tag{3.70}$$

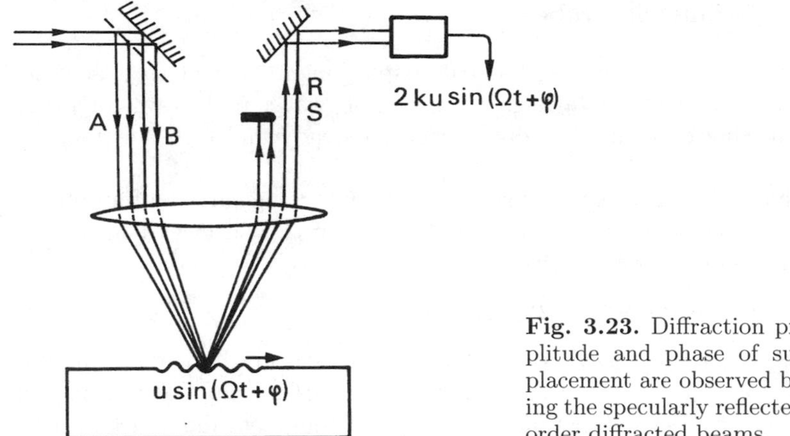

Fig. 3.23. Diffraction probe. Amplitude and phase of surface displacement are observed by interfering the specularly reflected and first order diffracted beams

Mechanical displacement can therefore be deduced by measuring the relative intensity of this beam. Since beams must separate, the method only applies to high frequencies ($F > 100$ MHz). Moreover, it is not sensitive to small displacements because the photocurrent intensity is proportional to u^2. For example, with a He-Ne laser, $ku \approx 10^{-3}$ when $u = 1$ Å. This simple technique has been used to measure the spatial distribution of power for a surface elastic wave beam in steady regime [3.35].

The above method does not give the phase of the elastic wave, even though it is contained in the phase of the diffracted light beam. It is found by interfering this beam with part of the incident beam. Figure 3.23 shows the setup due to Rouvaen et al. [3.36], transforming an earlier experiment for bulk waves (see Sect. 4.2.1).

The light beam is divided into two parallel beams A and B. The specularly reflected part R of beam A, taken as reference, interferes with the first order diffracted part S of beam B. This interference produces current intensity I at the photodiode (a quadratic detector), given by

$$I = s(E_\mathrm{R} + E_\mathrm{S})(E_\mathrm{R} + E_\mathrm{S})^* = I_0 + i(t) , \tag{3.71}$$

where

$$E_\mathrm{R} = E_0 \exp \mathrm{i}(\omega t + \Phi_\mathrm{R}) , \tag{3.72a}$$

$$E_\mathrm{S} = E_0 ku \exp \mathrm{i}(\omega t + \Omega t + \phi + \Phi_\mathrm{S}) , \tag{3.72b}$$

are electric fields of beams R and S, and Φ_R, Φ_S their phases. The term $i(t)$ contains the frequency and phase of the elastic wave. It is proportional to the amplitude u of the displacement:

$$i(t) = kuI_0 \cos(\Omega t + \phi + \Phi_\mathrm{S} - \Phi_\mathrm{R}) , \quad I_0 = 2s|E_0|^2 . \tag{3.73}$$

The method, applicable only in steady regime, has similar sensitivity to the deflection probe. It is fairly stable with regard to optical path fluctuations, the two beams A and B being close together.

3.2.2 Interferometric Probes

These probes measure any mechanical displacement normal to a surface, whatever its origin (e.g., Rayleigh wave, Lamb wave, bulk wave, or other vibration or shock). It applies equally well to steady and transient regimes [3.37].

The phase of a light beam reflected by the vibrating surface of an object is modulated by the displacement $u\sin(\Omega t+\phi)$ of this surface. The electric field E_S of the probe wave is given by

$$E_S = E_0 \exp \mathrm{i}[\omega t + \Phi_S + 2ku\sin(\Omega t+\phi)] \,. \tag{3.74}$$

The spectrum of this signal contains a central line at the carrier angular frequency ω and lateral lines at frequencies $\omega \pm m\Omega$. Amplitudes of lateral lines are given by Bessel functions J_1, J_2, ... of the phase difference ku (see Sect. 3.1.4). For small displacements compared with λ ($ku \ll 1$), only the $\omega \pm \Omega$ lines have a significant amplitude [$J_1(2ku) \approx ku$] and we can expand:

$$\begin{aligned} E_S = {} & E_0\{\exp \mathrm{i}(\omega t + \Phi_S) + \\ & ku \exp \mathrm{i}[(\omega+\Omega)t + \Phi_S + \phi] - ku \exp \mathrm{i}[(\omega-\Omega)t + \Phi_S - \phi]\} \,. \end{aligned} \tag{3.75}$$

At high frequencies, greater than a few [MHz], information about mechanical displacements can in principle be extracted directly by spectroscopy using a Fabry–Pérot cavity as frequency discriminator (Sect. 3.2.2.3).

However, the most common method involves mixing the probe beam E_S (3.74) with a reference beam E_R (3.72a) coming from the same source. The photodetector receiving the two beams, assumed to be of the same intensity, produces a current of intensity (3.71):

$$I_0 = s\left[|E_R|^2 + |E_S|^2 + 2\mathrm{Re}\,(E_S E_R^*)\right] \,,$$

and hence

$$I = I_0 + I_0 \cos\left[2ku\sin(\Omega t+\phi) + \Phi_S - \Phi_R\right] . \tag{3.76}$$

Comparing with (3.50) brings out the significance in interferometry of fluctuations in phases Φ_S, Φ_R, i.e., fluctuations in the corresponding optical paths L_S, L_R. Variations $\Phi_S - \Phi_R = 2\pi(L_S - L_R)/\lambda$ must be small compared with ku but not with the phase ϕ of the elastic signal.

Among the various ways put forward to reduce effects of such fluctuations, we may cite:

- displacing the mirror which returns the reference beam so as to maintain constant phase difference $\Phi_S - \Phi_R$ (stabilised Michelson interferometer);
- shifting the frequency of one or both light beams and then processing the photodetector output signal in consequence (heterodyne interferometer).

3.2.2.1 Stabilised Michelson Interferometer. This has a similar optical configuration to a classical interferometer (Fig. 3.24a).

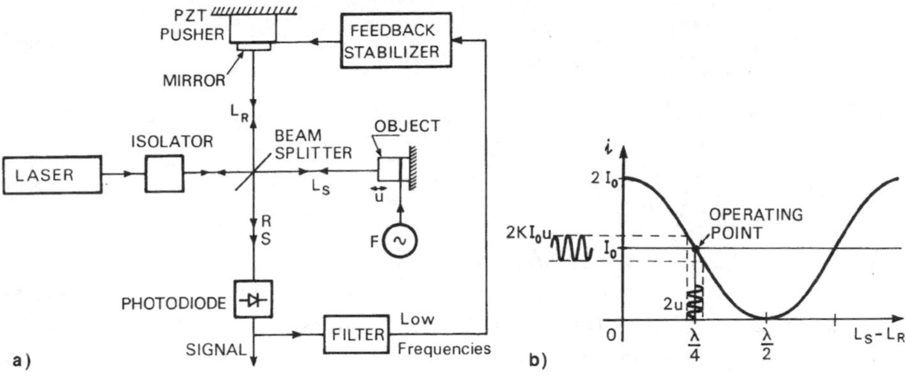

Fig. 3.24. Stabilised Michelson interferometer. (**a**) The position of the reference mirror is controlled by the low frequency part of the photodiode output signal in such a way that operating conditions are fixed, despite optical path fluctuations. (**b**) Dependence of photocurrent intensity on optical path difference. Sensitivity is maximal when this difference equals $\lambda/4$

The laser beam of power P_L is divided into two equal parts which reflect from the object (probe beam S) and the mirror (reference beam R), before mixing in the photodiode. An insulator prevents the half beam which returns towards the source from entering the laser cavity and causing instabilities. The photocurrent intensity resulting from interference between beams R and S is a sine function of optical path difference $L_S - L_R$, with period λ (Fig. 3.24b). Maximum sensitivity occurs when

$$L_S - L_R = (n + 1/4)\lambda \quad \Rightarrow \quad \Phi_S - \Phi_R = \pi/2 \,(\mathrm{mod}\ 2\pi) \,. \tag{3.77}$$

A feedback loop controls the position of the reference mirror [3.38] so that the system continues to operate in phase quadrature, despite random variations in optical path. The mirror is supported on a piezoelectric pusher controlled by the low-frequency ($F < 1$ kHz) part of the detected signal. This contains most of the thermal and mechanical perturbation spectrum. Periodic adjustment is needed to allow for the limited dynamic range of feedback.

In such phase quadrature operating conditions, and when displacements are small $ku \ll 1$, the photocurrent intensity is

$$I = I_0 + 2kuI_0 \sin(\Omega t + \phi) = I_0 + i_S(t) \,. \tag{3.78}$$

Sensitivity is limited by shot noise (or photon noise) originating in the direct current component I_0. This generates a current i_N with mean squared intensity proportional to the bandwidth B of the electronic detection circuit:

$$\left\langle i_N^2(t) \right\rangle = 2eI_0B \,, \tag{3.79}$$

where e is the electron charge. The smallest measurable displacement $u_{\min}$ is defined as the one giving unit signal-to-noise ratio:

$$\frac{\mathrm{S}}{\mathrm{N}} = \sqrt{\frac{\langle i_{\mathrm{S}}^2(t)\rangle}{\langle i_{\mathrm{N}}^2(t)\rangle}} = ku\sqrt{\frac{I_0}{eB}} \, . \tag{3.80}$$

This implies

$$\boxed{u_{\min} = \frac{\lambda}{2\pi}\sqrt{\frac{eB}{I_0}}} \, . \tag{3.81}$$

For a He-Ne laser ($\lambda = 6\,328$ Å), power $P_0 = 1$ mW received by the photodiode ($I_0 = 0.3$ mA) and bandwidth $B = 1$ Hz, the minimum value is given by $u_{\min} = 2.3 \times 10^{-5}$ Å $\mathrm{Hz}^{-1/2}$. This argument assumes that thermal noise $4kTB/R$ in the load resistance R of the photodiode is less than noise due to the continuous current I_0. This means

$$R \gg 2kT/eI_0 \, .$$

In practice, the condition is $R \gg 150\,\Omega$. As the frequency increases, it becomes more difficult to satisfy, as a result of spurious capacitances which reduce the bandwidth.

3.2.2.2 Heterodyne Probe. In a heterodyne interferometer, the frequency of one or both beams is shifted. The frequency change by $\pm f_{\mathrm{B}}$ distinguishes this type of interferometer from the homodyne interferometers described above. It is effected by an acousto-optic modulator in which the elastic wave frequency f_{B} is of order several tens of MHz. Electric fields E_{S} and E_{R} are given by (3.74) and (3.72a) with ω replaced by ω_{S} for the probe beam and by ω_{R} for the reference beam. In the alternating part of the current supplied by the photodetector [see (3.76)], a term of frequency $\omega_{\mathrm{S}} - \omega_{\mathrm{R}} = \omega_0$ occurs (equal to either ω_{B} or $2\omega_{\mathrm{B}}$, depending on whether the frequency of one or both beams has been shifted):

$$i_{\mathrm{S}}(t) = I_0 \cos\left[\omega_0 t + 2ku\sin(\Omega t + \phi) + \Phi_{\mathrm{S}} - \Phi_{\mathrm{R}}\right] . \tag{3.82}$$

Modulation of the probe beam phase by surface displacements is thereby transposed into the radio frequency domain. The spectrum of $i(t)$ [see (3.27)] contains a central line at ω_0 and lateral lines at $\omega_0 \pm m\Omega$ whose heights are given by Bessel functions $J_m(2ku)$. If displacement u is small compared with the optical wavelength ($ku \ll 1$), which means in practice that $u < 300$ Å, the spectrum reduces to the carrier frequency ω_0 and two lateral lines $\omega_0 \pm \Omega$ with amplitudes $J_1(2ku) \approx ku$ (Fig. 3.26b):

$$\begin{aligned} i_{\mathrm{S}}(t) = \; & I_0\{\cos(\omega_0 t + \Phi_{\mathrm{S}} - \Phi_{\mathrm{R}}) + ku\cos[(\omega_0 + \Omega)t + \phi + \Phi_{\mathrm{S}} - \Phi_{\mathrm{R}}] \\ & -ku\cos[(\omega_0 - \Omega)t - \phi + \Phi_{\mathrm{S}} - \Phi_{\mathrm{R}}]\} \, . \end{aligned} \tag{3.83}$$

The ratio r of the heights of central and lateral lines yields the absolute amplitude of mechanical vibrations in the steady regime, independently of the luminous power reflected by the surface. Using a He-Ne laser beam, u [Å] $\approx 1\,000/r$.

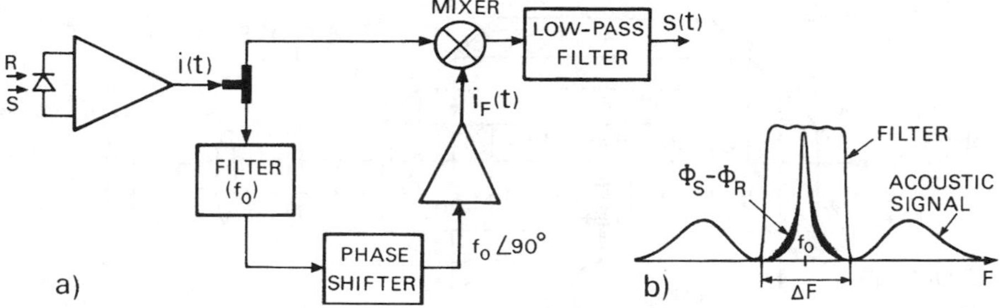

Fig. 3.25. Wideband coherent detection of the heterodyne probe. (**a**) Mixing the part of the photocurrent at carrier frequency f_0, phase shifted by $\pi/2$, with the other, unmodified part, and then filtering, yields the vibration of the object. (**b**) In order to eliminate fluctuations $\Phi_S - \Phi_R$, by taking differences, the filter passband, centred on carrier frequency f_0, must be wider than their spectrum

Coherent Electronic Detection. Formula (3.83) indicates that random phase fluctuations $\Phi_S - \Phi_R$ affect the central line and lateral lines of the spectrum in the same way. In principle, they can be eliminated by coherent detection [3.39], in which we measure the frequency difference between the central line and one of the lateral lines. Such detection can be carried out in various ways.

Figure 3.25 shows a wideband treatment suitable for detecting displacements in the transient regime. The signal remaining after eliminating acoustic phase modulation by means of a narrow passband filter, centred on frequency f_0, is phase shifted by $\pi/2$:

$$i_F(t) \propto \cos(\omega_0 t + \pi/2 + \Phi_S - \Phi_R) . \tag{3.84}$$

Having eliminated the term of frequency $2\omega_0$ with the help of a low pass filter, the signal which results by mixing with photocurrent $i_S(t)$, given by (3.82),

$$s(t) = i_S(t) i_F(t) \propto I_0 \cos\left[\frac{\pi}{2} - 2ku\sin(\Omega t + \phi)\right] ,$$

is proportional to the mechanical displacement when $ku \ll 1$:

$$s(t) \approx 2kuI_0 \sin(\Omega t + \phi) . \tag{3.85}$$

A phase locked loop circuit can be used to measure very low frequency surface displacements ($F < 1$ kHz) [3.41].

Compact Optical Arrangement. One arrangement consists of a modified Michelson interferometer in which the beam splitter is an acousto-optic modulator operating at Bragg incidence [3.31]. Probe and reference beams are not perpendicular. They leave the modulator separated by a small diffraction angle θ ($f_B \approx 70$ MHz implies $\theta \approx 10$ mrad). The various components (laser and photodetector on one side and mirror and sample on the other) are thus disposed relatively far from the modulator (> 50 cm). In addition,

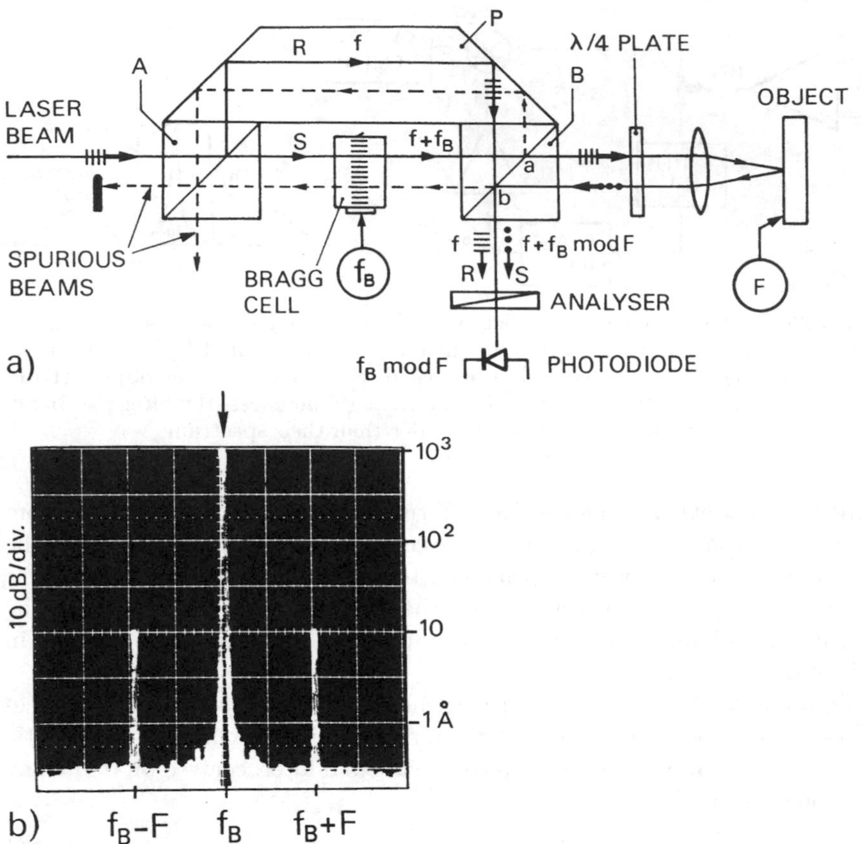

Fig. 3.26. Heterodyne interferometric probe. (**a**) Compact optical arrangement. (**b**) Spectrum of photodiode output signal. A sub-angström surface displacement amplitude (at $F = 100$ kHz) can be read off directly from the spectrum analyser (bandwidth 3 kHz)

beams cross the modulator twice, one of them returning towards the laser cavity where it generates instabilities which add to optical path fluctuations.

A Mach–Zehnder type structure is preferable, like the asymmetric arrangement shown in Fig. 3.26 [3.40].

The horizontally polarised source beam is split by a beam-splitter cube A into reference and probe beams R and S. The first is directed by a prism P towards the photodiode. The frequency of the second is shifted f_{B} (70 MHz) by the acousto-optic modulator, which has collinear input and output. It is then reflected by the sample, which is vibrating at frequency F. After twice crossing the quarter-wave plate, it returns vertically polarised, to be reflected at 90° by the polarisation splitter cube B, towards the photodiode. The two beams R and S, with frequencies f and $f + f_{\mathrm{B}}$, cross an analyser oriented at 45° to their respective polarisations before interfering at the photodiode. The

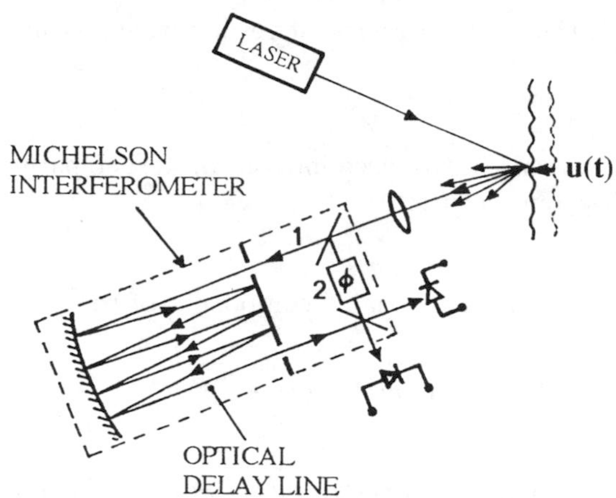

Fig. 3.27. Principle of time-delay differential interferometry

latter delivers a current with phase modulated by vibration F of the sample surface.

The lens focussing the probe beam on the object specifies the observation point, increases the amount of reflected light gathered in the case of a scattering surface and renders the probe less sensitive to sample inclination when it is displaced. The shift relative to the axis of the two cubes eliminates interference signals arising from reflection of the probe beam (at point a) and of the reference beam (at point b) at the interface of splitter cube B. This probe can be adjusted very rapidly (in a few minutes) because the modulator is only crossed once, and then only by the probe beam, and the two beams are either parallel or perpendicular. Since the optical part is compact, operation is stable. Fluctuations in output signal amplitude and phase are 0.2% and 0.2°, respectively [3.42].

3.2.2.3 Doppler Velocimetry. In the homodyne and heterodyne interferometers described above, the reflected wave, whose wave surface may be significantly perturbed, is mixed with a quasi-plane reference wave. For scattering surfaces, that is, when there is speckle, these probes can only gather a phase coherent light grain and their geometrical étendue is limited. In order to overcome this problem, we must appeal to a single wave, namely the wave coming from the surface whose displacement we wish to measure. Information about motion results from the beat between one part of this wave and another retarded by time τ. Wave front distortion during reflection is of little importance since, to a first approximation, it affects both waves in the same way. Interference can be implemented, for example, by a Michelson interferometer (Fig. 3.27), and retardation by a multiple-reflection device, an optical fibre or a Fabry–Pérot cavity [3.43].

Let $\Delta\Phi = \omega\tau$ be the phase shift due to the optical path difference between paths 1 and 2, where $\tau = L/c$. If $2ku(t)$ is the phase shift in beam 1 due to surface displacement at time t, then the phase shift in beam 2 at the same instant is $2ku(t-\tau)$. By (3.76), the photocurrent that results from beating the two beams is

$$I = I_0 + I_0 \cos\{2k[u(t) - u(t-\tau)] + \omega\tau + \phi\}\ , \tag{3.86}$$

where ϕ is an adjustable phase difference introduced into one of the two paths with a view to maintaining phase quadrature:

$$\omega\tau + \phi = \pm\pi/2\ (\mathrm{mod}\ 2\pi)\ . \tag{3.87}$$

For displacements of amplitude $u \ll \lambda$, the variable component of the photocurrent is given by

$$i(t) = I_0 \sin\{2k[u(t) - u(t-\tau)]\} \approx 2kI_0[u(t) - u(t-\tau)]\ . \tag{3.88}$$

This result can be used in several ways.

- If the retardation τ is great compared with the duration Θ of the signal, $u(t-\tau) = 0$ when $u(t) \neq 0$, and the current intensity is proportional to normal displacement of the surface.
- If on the other hand the retardation τ is small compared with Θ, then $u(t)-u(t-\tau) \approx \tau \mathrm{d}u/\mathrm{d}t$ and the signal is proportional to the normal component of velocity $v = \mathrm{d}u/\mathrm{d}t$. The device then operates as a Doppler velocimeter. After reflection by the moving surface, the incident wave $E_0 \exp(\mathrm{i}\omega t)$ becomes $E_1 \exp \mathrm{i}[\omega t + 2ku(t)]$. Its instantaneous angular frequency, i.e., the time derivative of the phase,

$$\Phi_1(t) = \omega t + 2ku(t) \quad \Rightarrow \quad \omega_1 = \frac{\mathrm{d}\Phi_1}{\mathrm{d}t} = \omega\left[1 + 2\frac{v(t)}{c}\right]\ , \tag{3.89}$$

  is shifted with respect to the angular frequency ω of the incident wave. This Doppler effect appears in the spectrum of the reflected wave in the form of side bands, as shown by (3.83), which is valid for sinusoidal displacements $u(t)$. In Monchalin's setup, this frequency variation is transformed into an amplitude variation by a confocal Fabry–Pérot cavity of length 50 cm [3.43]. The bandwidth of the cavity lies between 1.5 and 10 MHz, depending on reflectivity of the mirrors. Its length is controlled by piezoelectric pushers (Fig. 3.28a) so as to maintain the laser frequency half-way up the transmission curve (operating as a frequency discriminator, see Fig. 3.28b).

The optical setup can sense a light spot of diameter 1 mm at a distance of 1.5 m. Experiments have been devised to detect mechanical displacements generated by laser pulses in hot solids, or solids whose surfaces are of poor optical quality, such as composite materials [3.44]. They have led to construction of a Laser Ultrasonic Inspection System (LUIS) for components of aeronautic structures.

3.2.3 Applications

The results presented below demonstrate the possibilities of the compact heterodyne probe. Figure 3.29 gives and idea of the sensitivity of such an

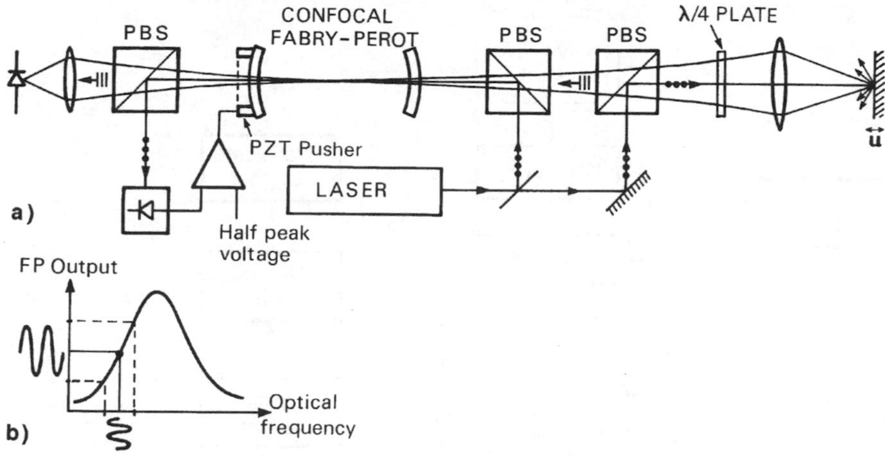

Fig. 3.28. (**a**) Confocal Fabry–Pérot cavity interferometer. After Monchalin [3.43]. PBS is a polarising beam splitter. (**b**) The position of one cavity mirror is controlled in such a way that the optical frequency is maintained on the characteristic slope

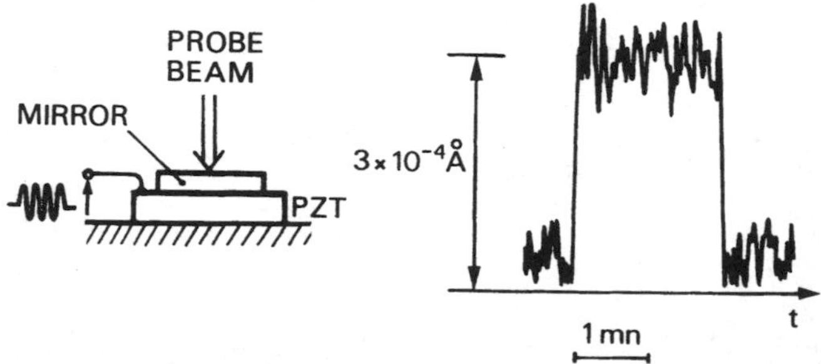

Fig. 3.29. Measurement of an amplitude of a few 10^{-4} Å in harmonic regime, using 1 Hz bandwidth electronic detection. The surface vibrating at 160 kHz is a mirror fixed on a piezoelectric ceramic (PZT) transducer

instrument. It represents the thickness vibration at frequency 160 kHz of a piezoelectric ceramic (PZT) transducer carrying a mirror. The jump by 3×10^{-4} Å shows that an amplitude of 10^{-4} Å is detectable. The theoretical limit corresponding to these experimental conditions (2 mW He-Ne laser, $B =$ 1 Hz) is almost reached [see (3.81)]. Equipped with a 100 mW YAG laser, with frequency doubler ($\lambda = 532$ nm), the probe has sensitivity 10^{-5} Å $\mathrm{Hz}^{-1/2}$. The following two experiments have been carried out using a He-Ne laser (SH 130 probe commercialised by BM Industries).

(a) In ultrasonic imaging, lateral resolution is improved by focussing the acoustic beam and increasing the centre frequency of the transducer. Standard calibration using a miniaturised hydrophone or ball-shaped reflector are limited both from the bandwidth point of view (15 MHz) and

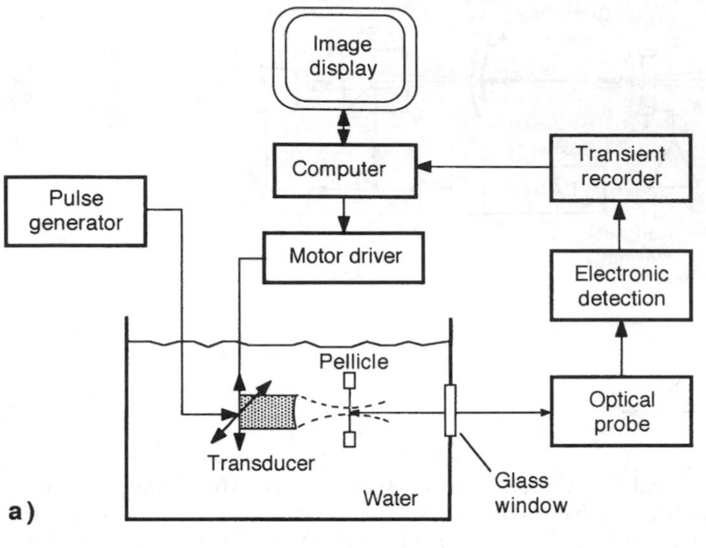

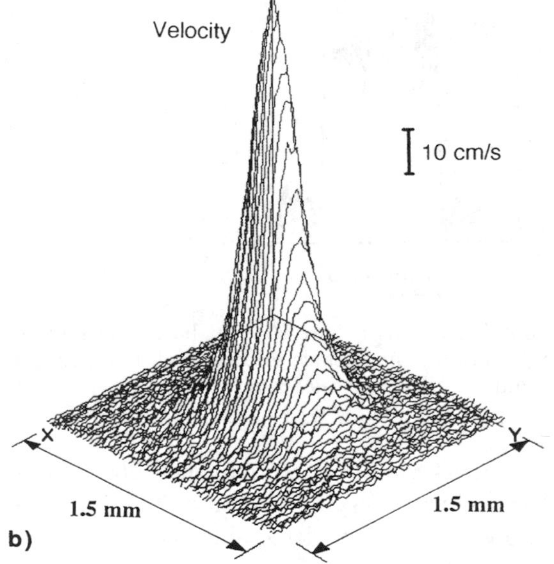

Fig. 3.30. Optical imaging device for acoustic fields in water. (**a**) Experimental setup. (**b**) Particle velocity map recorded by a Mylar membrane of thickness 3 μm lying in the focal plane ($F = 18$ mm) of a concave piezoelectric transducer with centre frequency 25 MHz

in terms of their spatial resolution (0.4 mm). Figure 3.30a represents an optical characterisation system for acoustic fields emitted in water by piezoelectric transducers [3.45]. The device comprises a metallised Mylar membrane of thickness several μm, which is immersed and from which the probe beam of the heterodyne interferometer is reflected. The latter measures absolute displacement induced as the pressure wave radiated by

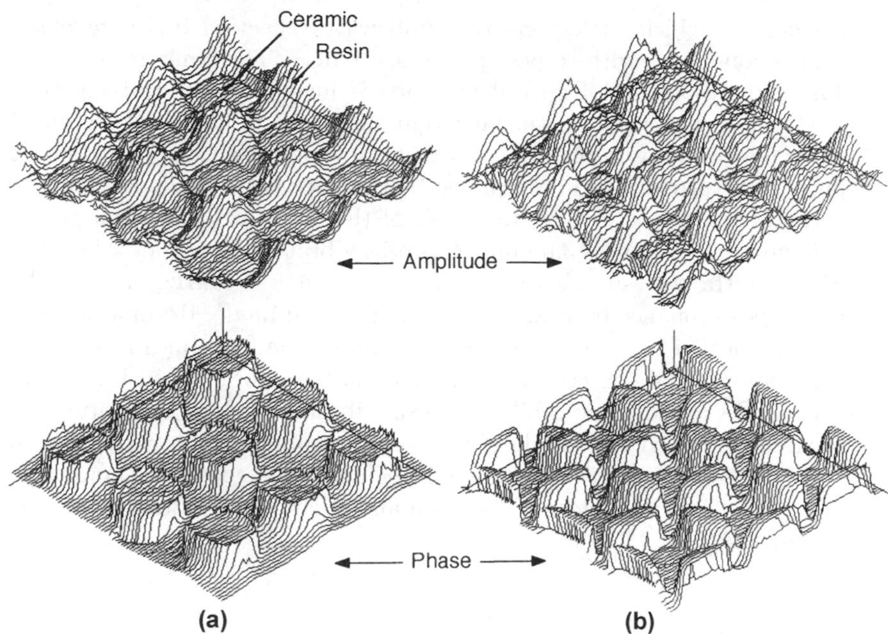

Fig. 3.31. Piezocomposite resonator. Recording made by compact heterodyne probe of the amplitude and phase of surface mechanical displacements at first and second lateral mode frequencies. (**a**) First mode $f = 0.875$ MHz. (**b**) Second mode $f = 1.61$ MHz. From [3.46]

the transducer passes through. The acoustic field is explored by moving the transducer parallel to the membrane.

Figure 3.30b is a 2-dimensional survey, in the pulse regime, of the acoustic field at the focus of a transducer (focal length 18 mm, diameter 6.35 mm, frequency 25 MHz). Spatial resolution is 20 μm and detection bandwidth is 40 MHz. Since heterodyne optics and phase coherent detection make these measurements virtually insensitive to perturbations such as low-frequency motions of the water in the tank, they are as sensitive as they would be in air (1 Å for a 50 MHz bandwidth).

(b) In order to increase the bandwidth and efficiency of piezoelectric transducers used in medical imaging and non-destructive testing, composite materials have been designed with an active piezoelectric phase and an inactive polymer phase (Sect. 1.6). A compound (of connectivity 1-3), such as the one shown in Fig. 1.25, exhibits more complex mechanical behaviour than a homogeneous PZT piezoceramic plate. In addition to the thickness vibration mode, which generates longitudinal waves in the object under investigation, lateral vibration modes also occur. These modes, related to the periodicity of the medium, give rise to spurious resonances, which correspond to Lamb wave reflection in Bragg phase tuning conditions with respect to the array of PZT ceramic rods.

The high resolution of optical methods is invaluable in studying such modes. Figure 3.31 shows the vibration amplitude in harmonic regime of a free piezocomposite plate. Measurements were made using the compact heterodyne probe. The area scanned by the laser beam (3.75×3.75 mm^2) corresponds to 3×3 elementary cells of the structure, of spatial period 1.25 mm and thickness 3.6 mm (thickness vibration frequency 430 kHz). Figure 3.31a corresponds to excitation frequency 875 kHz, close to the predicted resonance frequency of the first lateral mode. Resin located at the junction of four rods vibrates with amplitude five times as great as the amplitude of the ceramic rods. The latter, vibrating in phase, are enclosed by a nodal line, whilst the resin vibrates in phase opposition. At the frequency 1.61 MHz of the second lateral mode, resin located between two PZT rods is enclosed by a nodal line and undergoes large amplitude vibrations. The remainder of the resin and ceramic rods vibrate in phase opposition (Fig. 3.31b).

3.3 Photothermal Generation

Ever since elastic waves were first generated without mechanical contact in solids (e.g., by means of electron beams or electromagnetic waves), experimental studies have progressed through the use of powerful optical sources, like lasers, and more sensitive detectors. Theoretical models were developed to explain generation mechanisms. Compared with traditional methods (piezoelectric transducers), photothermal generation has several advantages. Apart from the fact that it requires no mechanical contact, both position and shape of the source can be modified. Elastic waves can be generated in hot materials. This technique is at present directed towards non-destructive testing, measurement of elastic constants, spectroscopy and microscopy.

In most experiments, the solid is irradiated by means of light pulses. Bulk and surface waves are thereby generated. These waves are then detected either by electromagneto-acoustic, capacitive, piezoelectric effect or by optical methods. The advantage of optical measurements, discussed in Sect. 3.2, is that they can be made at a distance and with high bandwidth, without perturbing the acoustic field. Combined optical generation and detection is potentially important for non-destructive testing.

Laser pulses focussed in a point or line on the surface of a semi-infinite solid act as a point or line source, simultaneously generating various waves. Figure 3.32 represents wave fronts at time t after the impact of a laser pulse.

In the core of the material, mechanical perturbations are localised on two circular arcs of radii $V_{\mathrm{T}}t$ and $V_{\mathrm{L}}t$. On the free surface $x_3 = 0$, the perturbation

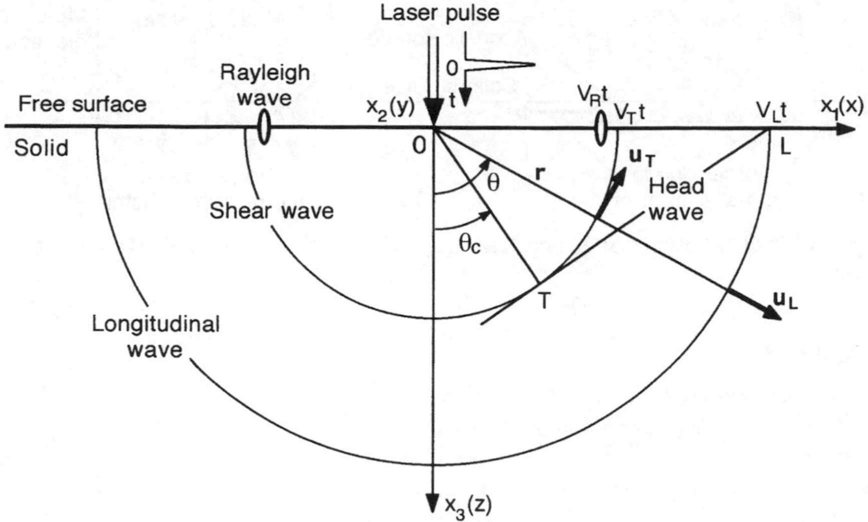

Fig. 3.32. Wave fronts generated by a point or line source in a semi-infinite isotropic solid

due to the Rayleigh wave appears at abscissa points $x_1 = \pm V_{\mathrm{R}}t$. A further perturbation occurs in the region, $|\theta| > \theta_{\mathrm{c}}$, where θ_{c} is the critical angle defined relative to the free surface normal by

$$\sin\theta_{\mathrm{c}} = \frac{V_{\mathrm{T}}}{V_{\mathrm{L}}} . \tag{3.90}$$

This is known as the *head wave*, or alternatively, the *conical wave* or *lateral wave*. Its wave fronts are parallel to the line LT. It arises because the longitudinal wave does not satisfy boundary conditions at the free surface. On this surface, the head wave propagates at the same speed as the longitudinal wave. However, it radiates energy towards the interior of the material and its amplitude therefore decreases much more rapidly than the Rayleigh wave with distance from the source.

We list the main parameters involved here, some of the notation being new.

- Laser pulse

Incident and absorbed energies	E, Q [mJ]
Duration	Δ [ns]
Area of irradiated zone	A [cm^2]
Absorbed power density	I [MW/cm^2]
Wavelength	λ_0 [μm]

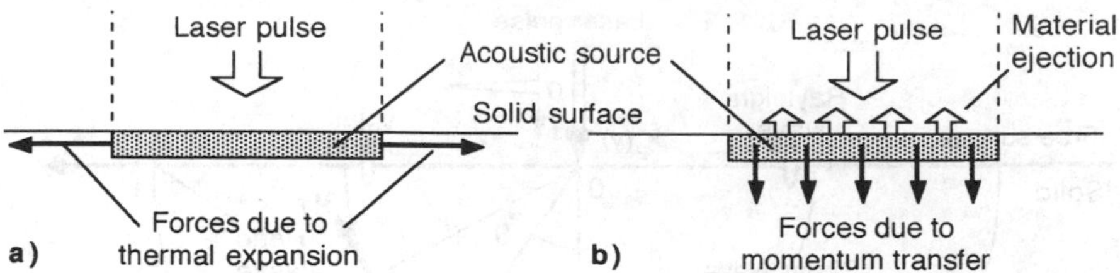

Fig. 3.33. Photothermal generation. (**a**) Thermoelastic regime. (**b**) Ablation regime

- Material

Absorption coefficient	η [%]
Linear expansion coefficient	α [K^{-1}]
Density	ρ [kg/m^3]
Heat capacity per unit mass	C [J/kg K]
Thermal conductivity	K [W/m K]
Elastic constants (Lamé)	λ, μ [N/m^2]

The most commonly used sources are the Nd:YAG laser, made by doping a $Y_3Al_5O_{12}$ (yttrium aluminium garnet) crystal with neodymium, and the CO_2 laser. Emitted wavelengths are 1.064 μm and 10.6 μm, respectively, for pulse widths lying in the range 10–100 ns and pulse rate ranging from 1 to 100 pulses per second.

Depending on the power density, the impact of a light pulse on the free surface of an opaque solid generates elastic waves via different mechanisms. These fall into two categories. In the first, the state of the surface is modified (*ablation*). In the second, it is not (radiation pressure, thermoelastic effect). When the power density of the light causes no vaporisation, local expansion through heating (*thermoelastic regime*) dominates over radiation pressure. Thermal expansion leads to forces more or less parallel to the free surface (Fig. 3.33a). If enough power density is absorbed ($I > 15$ MW/cm^2 for duralumin), the incident pulse will vaporise matter. This so-called ablation regime, in which momentum is transferred, produces forces more or less normal to the surface (Fig. 3.33b). If there is a film on the surface, normal forces are also increased.

3.3.1 Thermoelastic Regime

When low intensity radiation falls on a metallic surface, the electromagnetic field induces a conduction current near the surface. Part of the incident energy is absorbed and converted into heat by the Joule effect, whilst the rest is reflected. Because of screening by conduction electrons, these phenomena are restricted to within the *skin thickness* of the metal surface. A low intensity pulsed laser plays the role of a thermal energy source and generates a mechanical deformation. Such a thermoelastic source only produces spherical longitudinal waves if buried within the solid. The surface causes a conversion to transverse waves. Combination of longitudinal and transverse polarisations generates a Rayleigh wave which propagates at the surface of the solid.

3.3.1.1 Temperature Distribution. Electromagnetic waves penetrate to a depth equal to the so-called skin thickness γ which depends on the wavelength λ_0, electrical conductivity σ and permeability μ of the metal:

$$\gamma = \left(\frac{\lambda_0}{\pi\sigma c\mu}\right)^{1/2} .$$

For aluminium ($\sigma = 4 \times 10^7\ \Omega^{-1}\mathrm{m}^{-1}$) and radiation by a YAG laser ($\lambda_0 = 1.06\ \mu\mathrm{m}$), the penetration depth is of order 5 nm. The absorption coefficient η is given by

$$\eta = \frac{4\pi\gamma}{\lambda_0} . \tag{3.91}$$

In the same conditions, the absorbed fraction of incident energy is of order 6–7%. The depth of the thermoelastic source is determined by thermal diffusion processes. The temperature increase Θ obeys the heat equation [3.47]

$$\nabla^2\Theta - \frac{1}{\kappa}\frac{\partial\Theta}{\partial t} = -\frac{P_{\mathrm{a}}}{K} , \tag{3.92}$$

where K ($\kappa = K/\rho c$) is the thermal conductivity (diffusivity), and $P_{\mathrm{a}} = \eta P$ the power absorbed per unit volume.

Denoting the temporal variation of light power by $q(t)$, normalised so that $\int_0^\infty q(t)\,\mathrm{d}t = 1$, and assuming a Gaussian spatial distribution for the laser beam intensity (of radius a), the power P_{a} absorbed per unit volume is

$$P_{\mathrm{a}}(r,\, z,\, t) = \frac{Q}{\gamma}\mathrm{e}^{-z/\gamma}\,\frac{1}{\pi a^2}\mathrm{e}^{-r^2/a^2}\,q(t) .$$

Neglecting thermal diffusion in the air, Ready has established the temperature variation in the metal [3.48]:

$$\Delta\Theta(r,\, z,\, t) = \frac{Q}{\pi K}\sqrt{\frac{\kappa}{\pi}}\int_0^t \frac{q(t-t')}{\sqrt{t'}(4\kappa t' + a^2)}\exp\left(-\frac{z^2}{4\kappa t'} - \frac{r^2}{4\kappa t' + a^2}\right)\mathrm{d}t' . \tag{3.93}$$

If the diffusivity κ and width Δ of the laser pulse are small enough to ensure $\kappa\Delta \ll a^2$, then transverse diffusion is negligible. The radial temperature distribution follows the distribution of absorbed light intensity.

The temperature distribution is shown in Fig. 3.34 for aluminium, with pulse shape $q(t) = (t/\tau^2)\exp(-t/\tau)$ and width $\Delta = 2.4\tau = 24$ ns (absorbed energy $Q = 1$ mJ), and beam diameter $2a = 1$ mm.

The thermal wave does not penetrate further than 10 μm, even 100 ns after the beginning of the laser pulse. During the pulse, the diffusion depth

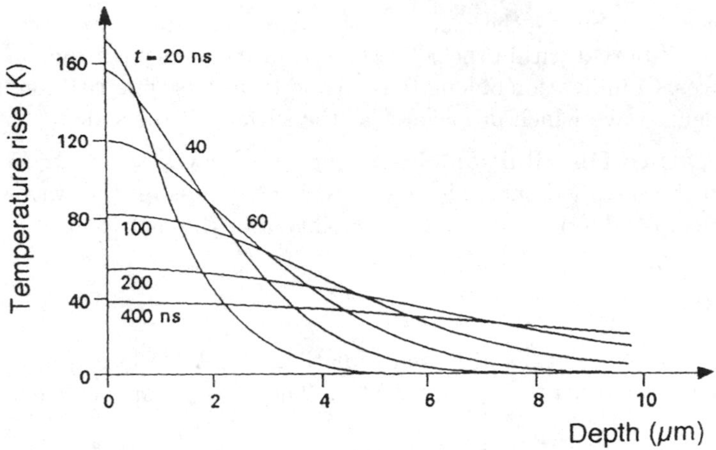

Fig. 3.34. Temperature distribution in duralumin as a function of depth and at different times. Width of incident laser pulse $\Delta = 25$ ns. Absorbed energy $Q = 1$ mJ

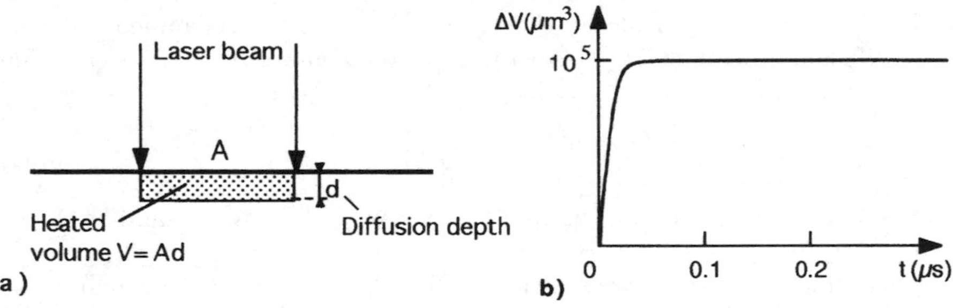

Fig. 3.35. (**a**) Thermoelastic regime. The source is localised within a thin disk of thickness d equal to the thermal diffusion depth. (**b**) Time variation of thermoelastic expansion at the surface of a duralumin plate, produced by laser pulse (absorbed energy $Q = 4$ mJ)

reaches about 2 μm, which is very small compared with the laser beam width and acoustic wavelength. The thermoelastic source is localised under the surface in a very thin disk, of thickness $d \sim 1$ μm and area $A = \pi a^2$ (Fig. 3.35a). Local heating disappears after about 400 ns. Effects of pulses arriving at a rate less than 1 MHz are therefore independent.

3.3.1.2 One-Dimensional Model. In this simple model, thermal diffusion is neglected during the laser pulse, which is assumed very short (10–100 ns). For a pulse of energy E reaching the surface $x_3 = 0$ at time $t = 0$, the incident power density per unit area is

$$I(t) = \frac{E}{A}\delta(t) , \tag{3.94}$$

where $\delta(t)$ is a Dirac pulse and A the cross-sectional area of the laser beam. The temperature increase $\Delta\Theta$ of the volume $V = Ad$ heated by absorbing fraction η of the incident power is then a step function $H(t)$:

$$\Delta\Theta(t) = \int_0^t \frac{\eta I(t')}{\rho C d}\,\mathrm{d}t' = \frac{Q}{\rho C V} H(t) \,. \tag{3.95}$$

This causes expansion ΔV of the volume, given by

$$\Delta V = 3\alpha V \Delta\Theta = \frac{3\alpha}{\rho C} Q H(t) \,, \tag{3.96}$$

which is proportional to absorbed energy Q and thermoelastic parameter $\alpha/\rho C$. For duralumin,

$$\frac{\alpha}{\rho C} = 0.84 \times 10^{-11}\,\mathrm{m}^3 J^{-1} \,, \tag{3.97}$$

and the volume increase is

$$\Delta V\ [\mu\mathrm{m}^3] = 2.5 \times 10^4 Q\ [\mathrm{mJ}] \,.$$

Figure 3.35b shows that the temporal shape of the expansion is only slightly modified when thermal diffusion is taken into account, at least, up to the rise time, remaining close to the step function.

The irradiated region is not free to expand radially because the material around it is rigid. The resulting mechanical stresses T_{ij} can be expressed in terms of strains S_{kl} and temperature increase $\Delta\Theta$ through the relation

$$T_{ij} = c_{ijkl}(S_{kl} - \alpha_{kl}\Delta T) = T_{ij}^{(0)} - \Delta T_{ij} \,, \tag{3.98}$$

where $\Delta T_{ij} = c_{ijkl}\alpha_{kl}\Delta\Theta$ represents stresses exerted by the thermoelastic source on the matter surrounding it. In the case of an isotropic solid,

$$T_{ij} = (\lambda S_{kk}\delta_{ij} + 2\mu S_{ij}) - (3\lambda + 2\mu)\alpha\Delta\Theta\delta_{ij} \,, \tag{3.99}$$

where we have used expressions for the rigidity tensor c_{ijkl} in terms of Lamé constants λ, μ (Sect. 3.2.3.1, Vol. I) and the expansion tensor $\alpha_{kl} = \alpha\delta_{kl}$. If lateral dimensions of the source are large compared with the acoustic wavelength, the strain S_{33} along the normal to the surface is dominant. Since the stress

$$T_{33} = (\lambda + 2\mu)S_{33} - (3\lambda + 2\mu)\alpha\Delta\Theta \tag{3.100}$$

normal to the irradiated surface is zero on the free surface $x_3 = 0$, the strain is given by

$$S_{33} = \frac{3\lambda + 2\mu}{\lambda + 2\mu}\alpha\Delta\Theta = \frac{\partial u_3}{\partial x_3} \,.$$

In this one-dimensional model, the piston-type source generates only longitudinal waves perpendicular to the surface, of amplitude $u_3 = S_{33}d$. By (3.95), we find

$$u_3 = \frac{3\lambda + 2\mu}{\lambda + 2\mu} \frac{\alpha}{\rho C} \frac{Q}{A} ; \tag{3.101}$$

Note that this displacement is independent of the diffusion depth d. In fact, it depends on:

- a dimensionless coefficient, given in terms of the rigidity of the material and equal to 2 in the case of a material like duralumin, for which Poisson's ratio is $\nu = 1/3$ and hence $\lambda = 2\mu$;
- the grouping $\alpha/\rho C$ of thermoelastic constants of the material;
- the absorbed energy density per unit area Q/A.

Order of Magnitude. For duralumin, using values given in (3.97) and for an absorbed energy density $Q/A = 10$ mJ/cm^2, the displacement is $u = 1.68$ nm.

The model is valid as long as lateral dimensions of the source are large compared with the acoustic wavelength. In practice, given the laser pulse width (30–50 ns), the latter is of order [mm]. This condition implies beam dimensions ($A \approx 1$ cm^2) such that the power density becomes too weak to generate an amplitude of any significance.

3.3.1.3 Point Source Model. Radiation by a thermoelastic source of arbitrary dimensions can be calculated from that of a point source. From (3.98), thermoelastic expansion produces a sudden stress increase ΔT_{ij} in the heated region. Aki and Richards [3.49] have shown that the displacement created by this explosive type of source, encountered in seismology, is given by

$$u_n(x_i, t) = \int_V \Delta T_{ij}(\xi_i, t) \otimes \frac{\partial G_{ni}}{\partial \xi_j}(x_i, \xi_i, t) \, \mathrm{d}V(\xi_i) , \tag{3.102}$$

where $G_{ni}(x_i, \xi_i, t)$ is the Green function (spatio-temporal impulse response) giving the nth displacement component, at observation point x_i and time t, due to an impulsive force parallel to axis x_i and applied at source point ξ_i and time t (see Fig. 3.36a). The symbol $\otimes$ denotes convolution with respect to time.

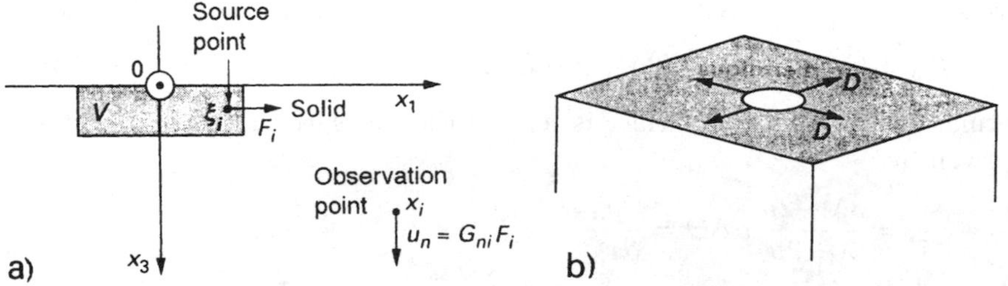

Fig. 3.36. (**a**) Coordinate system. ξ_i denotes source point coordinates, contained within volume V, and x_i denotes observation point coordinates. (**b**) Force dipoles equivalent to a surface thermoelastic source

This expression simplifies if the source is assumed to be pointlike and localised at the coordinate origin ($\xi_i = 0$):

$$u_n(x_i, t) = M_{ij}(t) \otimes \frac{\partial G^{\delta}_{ni}}{\partial \xi_j}(x_i, 0, t) , \tag{3.103}$$

where

$$M_{ij}(t) = \int_V \Delta T_{ij}(\xi_i, t)\,\mathrm{d}V(\xi_i) = V\Delta T_{ij}(0, t) \tag{3.104}$$

is the seismic moment representing the point source intensity.

For an isotropic solid, equations (3.98) and (3.99) imply that the tensor $\Delta T_{ij} = (3\lambda + 2\mu)\alpha\Delta T\delta_{ij}$ reduces to a scalar characterising the source, with a step-shaped time dependence which imitates that of the irradiated volume expansion ΔV,

$$M_{ij}(t) = (3\lambda + 2\mu)\frac{\alpha}{\rho C}QH(t)\delta_{ij} . \tag{3.105}$$

Denoting the divergence of the Green function by $g_n = \partial G_{ni}/\partial \xi_i$, and using (3.103), mechanical displacement components are given by

$$u_n(x_i, t) = D\, g^H_n(x_i, 0, t) , \qquad D = (3\lambda + 2\mu)\frac{\alpha}{\rho C}Q , \tag{3.106}$$

where $g^H_n(x_i, 0, t)$ is the divergence of the Green function corresponding to a step-shaped (rather than delta function) time variation. The expression for the factor D shows that the mechanical displacement amplitude is proportional to absorbed energy, i.e., in the thermoelastic regime, to incident light energy E. When the laser power has arbitrary time dependence $q(t)$, displacements are found by convolution:

$$u_n(x_i, t) = D\, q(t) \otimes g^H_n(x_i, 0, t) .$$

This expression can also be found by modelling stresses due to a sudden volume change ΔV in terms of three orthogonal force dipoles of intensity $D = B\Delta V$, where $B = \lambda + 2\mu/3$ is the bulk modulus [3.50]. A thermoelastic source located on the free surface of a solid occupying the half-space $x_3 > 0$ produces only tangential forces: $G_{n3} = 0$. A point source is therefore modelled by two orthogonal and horizontal force dipoles expressing expansion of surface matter away from the centre, due to laser heating (Fig. 3.36b).

In order to determine mechanical displacements generated in the solid, we must calculate the Green function g^H_n. We begin with bulk wave generation and then look at surface waves.

(a) Bulk Waves. Mechanical displacements generated in the core of the material, at distance r from impact, fall into two categories. There are those observed around the arrival times $t_\mathrm{L} = r/V_\mathrm{L}$ and $t_\mathrm{T} = r/V_\mathrm{T}$ of longitudinal and transverse elastic wave fronts (of speeds V_L, V_T); and there are those observed between these times and after the transverse wave front. This distinction is just the distinction between far and near fields. In the far field,

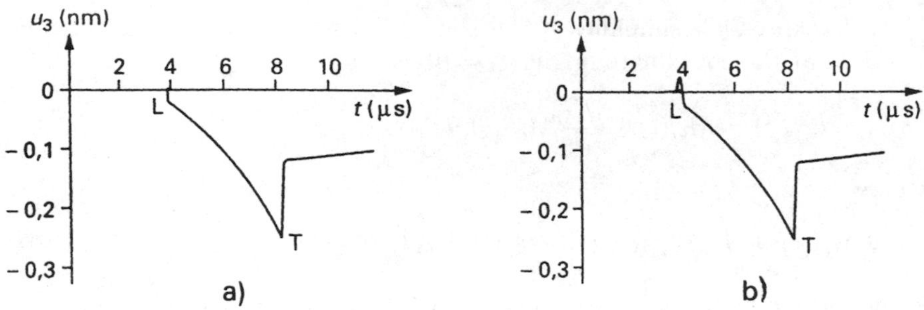

Fig. 3.37. Generation by a 40 mJ laser pulse in a 25 mm thick duralumin plate. Mechanical displacement calculated at the epicentre: (**a**) without taking heat diffusion into account; (**b**) including heat diffusion

longitudinal and transverse waves decouple, and the displacement between t_L and t_T is zero. In the near field, longitudinal and transverse waves are not decoupled: arrival is observed at times t_L and t_T, with continuous variation between these two times.

Rose [3.51] has calculated the time dependence of near field displacements on the x_3-axis, i.e., at the epicentre with respect to the source. By symmetry, the displacement is normal to the free surface, so that $u_3 = Gg_3^H$. Figure 3.37a refers to a material with Poisson ratio $\nu = 1/3$ ($V_L/V_T = 2$), such as duralumin. There is an initial negative step variation, propagating at speed V_L, which corresponds to withdrawal of matter at the epicentre in the direction normal to the surface. This is followed by a slow depression and a positive step, propagating at transverse wave speed $V_T = V_L/2$, which corresponds to a shear deformation. After this wave front, the sample tends to return to a different equilibrium state from the rest state, because the (step) source is still expanding. The slow variation between t_L and t_T is characteristic of L and T wave coupling in the near field.

The thermoelastic source penetrates several tens of [μm] into the source because of heat diffusion. Figure 3.37b represents displacement at the epicentre calculated when this heat diffusion effect is included [3.52]. The positive longitudinal displacement impulse arises from penetration of the source inside the material, which produces a small force dipole normal to the surface.

Wave Front Approximation. Directivity Patterns. Rose [3.51] has established formulas for the Green function at observation times close to wave front arrival times. This approximation applies to the far field, that is, for observation distances r which are much greater than the acoustic wavelength. In the case of a point source:

- The radial (longitudinal) displacement amplitude decreases as $1/r$ and varies with observation direction θ according to

$$g_r^H(r,\theta,t) = \frac{\Lambda}{rV_L}A(\theta)\delta\left(t-\frac{r}{V_L}\right), \quad \Lambda = \frac{1}{\pi\rho V_L^2}. \tag{3.107}$$

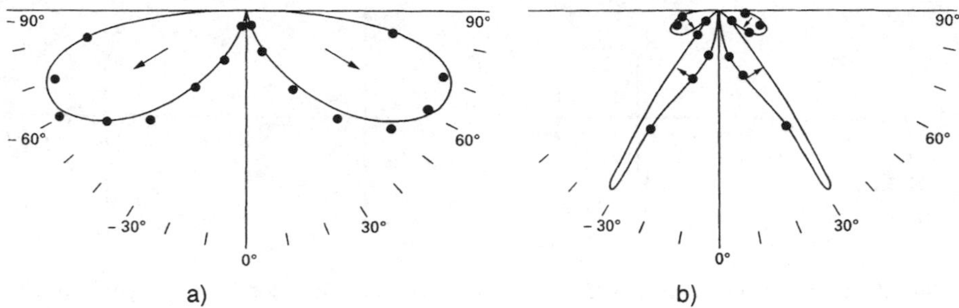

Fig. 3.38. Directivity patterns for a point thermoelastic source. (**a**) Longitudinal wave. (**b**) Transverse wave. • *experimental points* [3.61]

The directivity function $A(\theta)$ depends on the ratio of speeds $v = V_{\mathrm{T}}/V_{\mathrm{L}}$:

$$A(\theta) = \frac{\sin\theta \sin 2\theta (v^{-2} - \sin^2\theta)^{1/2}}{(v^{-2} - 2\sin^2\theta)^2 + 2\sin\theta\sin 2\theta(v^{-2} - \sin^2\theta)^{1/2}} \,. \tag{3.108}$$

This is real whatever the value of angle θ with respect to the normal to the surface, because $v = V_{\mathrm{T}}/V_{\mathrm{L}}$ is less than $\sqrt{2}/2$ for any solid. The time dependence of the longitudinal wave is the same as the time dependence $q(t)$ of the laser pulse.

- The transverse displacement also varies as $1/r$:

$$g_\theta^H(r,\theta,t) = \frac{A}{2rV_{\mathrm{T}}}\left[B_1(\theta)\delta\left(t - \frac{r}{V_{\mathrm{T}}}\right) - \frac{B_2(\theta)}{\pi(t - r/V_{\mathrm{T}})}\right] \,. \tag{3.109}$$

The directivity function

$$B(\theta) = B_1(\theta) + \mathrm{i}B_2(\theta) = \frac{\sin 2\theta\cos 2\theta}{\cos^2 2\theta + 2\sin\theta\sin 2\theta(v^2 - \sin^2\theta)^{1/2}} \tag{3.110}$$

is real for $\theta < \theta_{\mathrm{c}} = \arcsin(V_{\mathrm{T}}/V_{\mathrm{L}})$ and complex for $\theta > \theta_{\mathrm{c}}$. The time dependence of the transverse wave is not the same as that of the laser pulse because of the term $[\pi(t - r/V_{\mathrm{T}})]^{-1}$.

Figure 3.38a shows that there is no radiation of longitudinal waves along the normal to the surface. Emission is maximal in a direction depending on $V_{\mathrm{T}}/V_{\mathrm{L}}$ which is close to 65° for duralumin. For transverse waves (Fig. 3.38b), maximum energy is radiated at an angle close to 30°. Emission, which is always zero at the epicentre ($\theta = 0$), is also zero at $\theta = 45°$, because the term $\cos 2\theta$ in (3.110) then goes to zero. Polarities of principal and secondary lobes are opposite.

Comparing displacement amplitudes in the thermoelastic regime, we observe that transverse waves are more efficiently generated than longitudinal waves. Since $\rho V_{\mathrm{L}}^2 = \lambda + 2\mu$, and using (3.106), displacement maxima for a laser pulse of width Δ are given for the longitudinal wave by

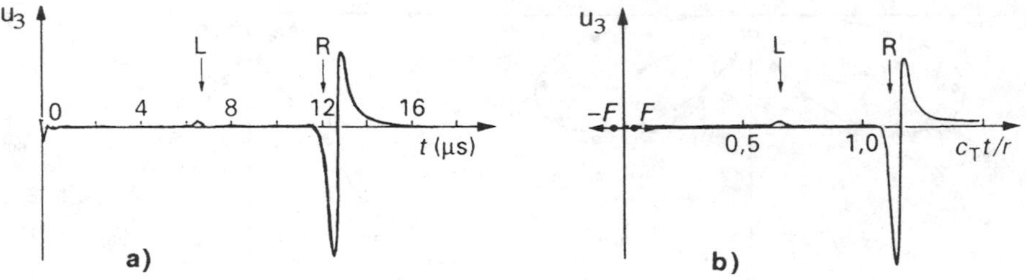

Fig. 3.39. Rayleigh wave R generated in thermoelastic regime at the surface of a material with Poisson ratio $\nu = 0.25$. (**a**) Experiment. (**b**) Theory

$$u_{\max}^{\mathrm{L}} = \frac{3\lambda + 2\mu}{\lambda + 2\mu} \frac{\alpha}{\rho C} \frac{Q}{r V_{\mathrm{L}} \Delta} \frac{A_{\max}}{\pi} , \tag{3.111}$$

and for the transverse wave by

$$u_{\max}^{\mathrm{T}} = \frac{3\lambda + 2\mu}{\lambda + 2\mu} \frac{\alpha}{\rho C} \frac{Q}{2r V_{\mathrm{T}} \Delta} \frac{B_{\max}}{\pi} . \tag{3.112}$$

For duralumin and $\Delta = 25$ ns,

$$u_{\max}^{\mathrm{L}} \,[\mathrm{nm}] = 5.2 \frac{Q\,[\mathrm{mJ}]}{r\,[\mathrm{mm}]} , \quad u_{\max}^{\mathrm{T}} \,[\mathrm{nm}] = 56 \frac{Q\,[\mathrm{mJ}]}{r\,[\mathrm{mm}]} .$$

The transverse displacement emitted by a point source is more than an order of magnitude greater than the longitudinal displacement.

(b) Rayleigh Waves. Because the thermoelastic source is located inside the material very near the surface, Rayleigh waves are generated. When the laser pulse is focussed along a line of width a parallel to x_2, Rayleigh waves are preferentially emitted along the x_1-axis perpendicular to this direction. The normal displacement u_3 is the sum of force contributions $+F$ and $-F$ applied at $\xi_1 = +a/2$ and $\xi_1 = -a/2$, respectively. An expression for the Green function G_{31}^{H} has been established theoretically by Pekeris [3.53], giving the normal displacement of an elastic half-space in response to a suddenly applied tangential force.

Figure 3.39a shows the signal detected by a wideband capacitive transducer. The impulse is bipolar with width proportional to the transit time of the Rayleigh wave across the source, i.e., to the light beam diameter. Figure 3.39b represents the normal component of displacement, calculated by means of a model with force dipoles parallel to the surface.

3.3.2 Ablation Regime

When the absorbed luminous power density I [W/m^2] is sufficiently great, the laser impact causes melting and then vaporisation of a small quantity of matter. Momentum transfer due to particle ejection creates a force normal to the surface in the irradiated zone, tending to generate longitudinal elastic waves.

The ablation regime begins at the threshold [3.48]

$$I > \left(\frac{\pi K \rho C}{4\Delta}\right)^{1/2} (\Theta_v - \Theta_i) , \quad (3.113)$$

which depends on vaporisation temperature Θ_v and initial temperature Θ_i of the material. For aluminium ($\Theta_v = 2\,600$ K, $\Theta_i = 400$ K) and a laser pulse of width $\Delta = 20$ ns, ablation occurs when the absorbed power density I is above 15 MW/cm^2. In contrast to the thermoelastic regime, the absorption coefficient η varies with incident power density. This is because there is a liquid or gaseous phase at the material surface. It can reach 90%. The normal force per unit area due to vaporisation is given by [3.48]

$$\frac{F_3}{A} = \frac{I^2}{\rho L_v [L_v + C(\Theta_v - \Theta_i)]} , \quad (3.114)$$

where L_v is the latent heat of vaporisation of the material. For absorbed power density 80 MW/cm^2 on a duralumin plate ($L_v = 284 \times 10^6$ J/kg), this stress is 2.8×10^3 N/m^2. For a light beam of radius 0.2 mm, the normal force is $F_3 \approx 3.5 \times 10^{-4}$ N. In fact, most ejected material is in liquid rather than gaseous phase. This liquid phase considerably increases the normal stress. When the impulse ceases, this phenomenon continues until the temperature of the material falls back below vaporisation temperature.

(a) Bulk Waves. Experimentally, power density is increased by keeping the energy constant and reducing the size of the focal spot. Figure 3.40 shows the signals detected at the epicentre by a capacitive transducer, over an increasing range of power densities [3.54]. The first recording corresponds to a level just beyond the ablation threshold, where thermoelastic contributions still play a significant role. Further on, the longitudinal displacement amplitude increases, whilst the jump on arrival of the transverse waves gradually fades.

The longitudinal displacement amplitude goes through a maximum and then decreases. This phenomenon is caused by a screening effect due to the plasma growing more and more opaque to the laser radiation. The sudden jump just above ablation threshold arises because of the very rapid increase in absorption coefficient due to local melting.

Directivity functions $C(\theta)$ and $D(\theta)$ for an ablative source differ from those of a thermoelastic source. They can be found by a method due to Miller and Pursey [3.55]. Assuming the force to be normal and pointlike, the result for longitudinal waves (Fig. 3.41a) is

$$C(\theta) = \frac{\cos\theta(v^{-2} - 2\sin^2\theta)}{(v^{-2} - 2\sin^2\theta)^2 + 2\sin\theta\sin 2\theta(v^{-2} - \sin^2\theta)^{1/2}} . \quad (3.115)$$

The radiation of the source is maximal at the centre. It is omnidirectional but not isotropic. For transverse waves (Fig. 3.41b),

$$D(\theta) = \frac{\sin 2\theta(v^2 - \sin^2\theta)^{1/2}}{\cos^2 2\theta + 2\sin\theta\sin 2\theta(v^2 - \sin^2\theta)^{1/2}} . \quad (3.116)$$

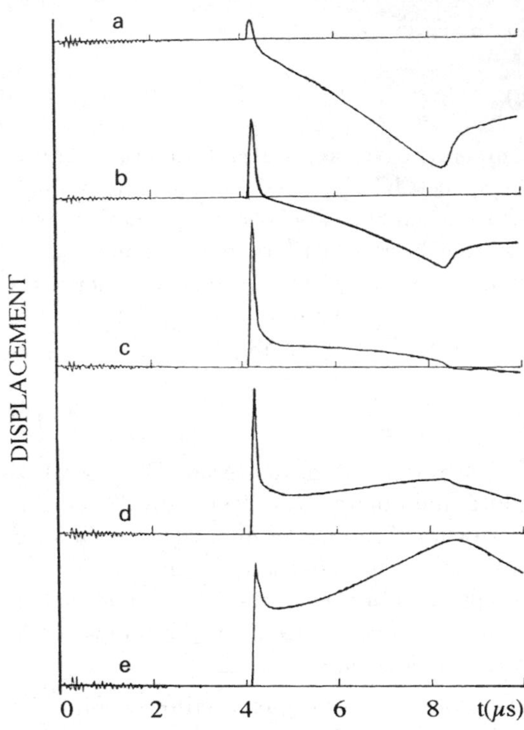

Fig. 3.40. Signals detected at the epicentre by a capacitive transducer for increasing power densities (from a to e). The first recording corresponds to a level just beyond the ablation threshold [3.54]

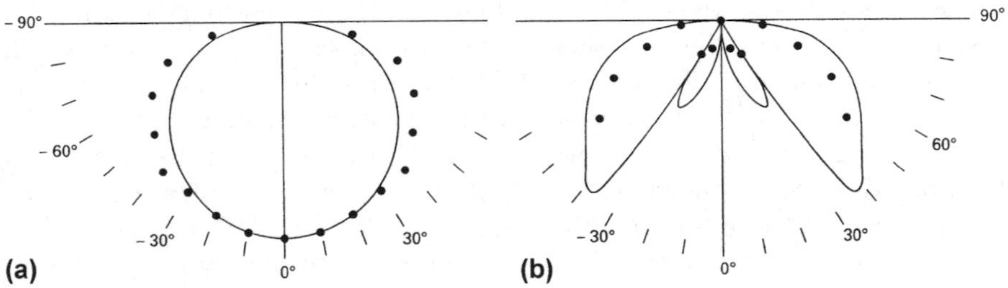

Fig. 3.41. Ablation regime. Radiation patterns for a (normal force) point source. (**a**) Longitudinal wave. (**b**) Transverse wave. • *experimental points* [3.61]

The source is more directional. Maximal energy emission occurs at an angle close to 35° in the case of duralumin ($v = V_T/V_L = 0.5$).

(b) Rayleigh Waves. Amplitudes of Rayleigh waves generated in ablation conditions are also increased. The pulse shape is the image of the one in Fig. 3.39, in a mirror placed perpendicular to the axis and passing through the centre of the pulse. This difference can be explained as follows: in the thermoelastic regime, the surface first moves outwards, whereas in ablation conditions, it begins by moving inwards.

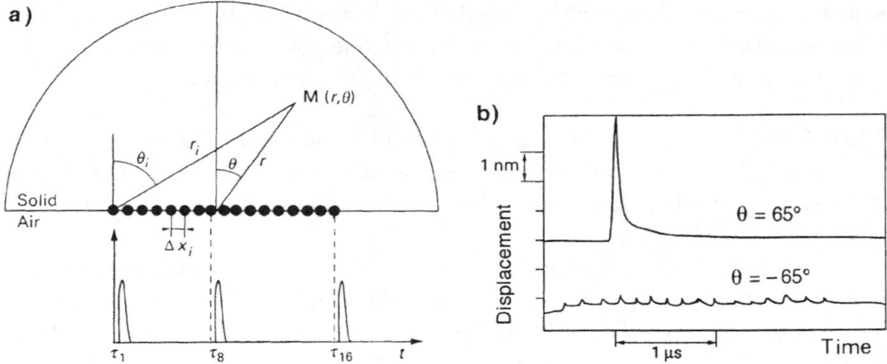

Fig. 3.42. (**a**) Focussing of elastic waves generated by a thermoelastic source array. (**b**) Mechanical displacements detected optically at a distance of 24 mm, at the focal point $\theta = 65°$, and in the symmetrical direction $\theta = -65°$ where there is no focussing

3.3.3 Increasing Efficiency

In non-destructive testing, the sample surface must not be damaged and hence the thermoelastic regime is a necessity. Optical detection is difficult owing to the small amplitudes of mechanical displacements generated. Generation efficiency can be increased if an array of sources or a mobile source is used to distribute incident laser power over a larger area [3.56]. An absorbing film or transparent layer on the surface of the material also tends to modify the nature of the thermoelastic source in a favourable way.

Phased Source Array. The idea of transducer arrays used in medical echography can be taken across to laser impact generation (Fig. 3.42a). The time of emission of each light pulse is delayed by τ_i in such a way that all acoustic impulses emitted by the sources are summed at the same time t_0 and point M inside the material:

$$\tau_i = t_0 - \frac{r_i}{V_{\mathrm{L}}} ,$$

where r_i is the distance from M to the ith source.

Figure 3.42b shows results of an experiment carried out with a YAG laser supplying 16 independent beams [3.57]. Each beam is focussed along a line of height 8 mm at a distance of 1 m. The longitudinal displacement generated by an array of aperture 24 mm in a block of duralumin is detected at a distance of 24 mm by a heterodyne optical probe (Sect. 3.2.2). In direction $\theta = 65°$, where acoustic impulses add, the peak value of the signal reaches 4 nm. In direction $\theta = -65°$, acoustic impulses arrive with delays which separate them in time and therefore have much smaller amplitudes. In order to detect defects without mechanical contact, there are two advantages in using this technique:

- a significant improvement in signal-to-noise ratio (> 20 dB);
- the direction of the acoustic beam is electronically controllable by altering delays τ_i, so as to carry out rapid scanning of the sample.

Thin Film. A thin film (of oil, for example) increases absorption of light energy. The film temperature increases and it vaporises, causing momentum transfer and a normal force, as in the ablation regime. Such a reaction tends to produce longitudinal waves.

In the case of a transparent film, such as those covering composite materials, light energy is absorbed at the material/film interface. The amplitude of the longitudinal wave generated along the normal is increased by the stress exerted by the film at the surface [3.58, 3.59]. This buried source radiates in a piston mode similar to that of piezoelectric transducers.

3.3.4 Experiments

We shall describe two experiments which illustrate the method of photothermal generation. The first involves excitation of waves in the wall of a tube, detection by heterodyne probe (Sect. 3.2.2.2) and consequent deduction of the tube thickness. The second is a simulation of tests used in seismic inspection.

Figure 3.43 illustrates excitation and detection of Lamb waves propagating in a tube of external diameter 20 mm and thickness 0.5 mm. These

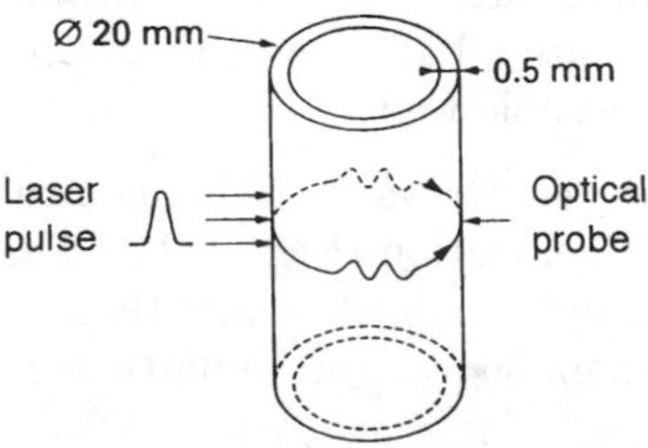

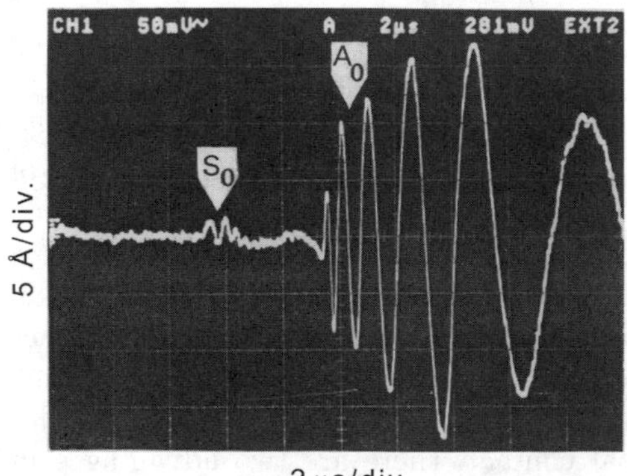

Fig. 3.43. Generation and detection of Lamb waves in a tube. A line-focussed laser pulse hits the tube along a generatrix. The probe beam is diametrically opposite the source. Given the probe bandwidth (20 MHz), dispersion in the A_0 mode is clearly observable

waves propagate in two modes, as in a plate (see Sect. 5.5, Vol. I). One (S_0) is symmetric with mainly longitudinal motion, the other (A_0) is antisymmetric and is predominantly flexural. The first mode has greater phase speed than the second. The YAG laser pulse, of energy 10 mJ and width 80 ns, is focussed along a line of length 15 mm parallel to the tube axis. The probe beam, located diametrically opposite, detects normal surface displacements. The first signal corresponds to the symmetric mode, with small normal displacement. The second signal corresponds to the antisymmetric mode, with much larger normal displacement. The dispersive effect can be clearly observed. High-frequency components propagate faster than low-frequency components. Such a signal, with spectral components extending over 0.2–2.5 MHz, can be recorded by this probe with its high bandwidth, much wider than that of a piezoelectric transducer. The tube thickness can be determined without mechanical contact by measuring the delay of this signal $e(t)\cos\phi(t)$ at maxima, zeros and minima [3.60]. Theoretical predictions based on the Rayleigh–Lamb dispersion relation are in good agreement with experimental results [3.62].

Seismic prospection experiments from one inspection pit to another can be simulated either numerically, or physically by means of scale models. In the latter case, classical piezoelectric transducers are unsuitable for representing sources and detectors used in seismic prospection, whose dimensions are much smaller than the relevant acoustic wavelengths. Laser sources and optical detectors have the advantage of being more or less pointlike, as well as requiring no mechanical contact. Figure 3.44a shows the experimental setup used at the Institut Français du Pétrole. The model (reduced by a factor of 10^5) is composed of one plexiglass and one duralumin plate [3.61]. The thermoelastic source is located on one face of the plexiglass parallelepiped and the observation point on the opposite face. Contributions of longitudinal and transverse bulk waves (P and S), together with those from Rayleigh waves (R) and head waves, are observed by recording wave fronts transmitted or reflected from the dioptric plane (Fig. 3.44b).

Problems

3.1 From Maxwell's equations, establish the equation giving the polarisation vector $\boldsymbol{D}$ and refractive index of a plane light wave propagating in a crystal characterised by tensor B_{ij}, inverse to ε_{ij}. Calculate the Poynting vector and write down the condition normalising the power carried by the wave, in terms of the transverse electric field component.

Solution. For a plane electromagnetic wave propagating in the direction of unit vector $\boldsymbol{s}$, for which all quantities vary like the electric field,

$$\boldsymbol{E} = \boldsymbol{E}_0 \exp \mathrm{i}(\omega t - \boldsymbol{k}\cdot\boldsymbol{x}) , \quad \boldsymbol{k} = \frac{\omega}{c}\boldsymbol{s} = \frac{\omega}{c_0} n\boldsymbol{s} ,$$

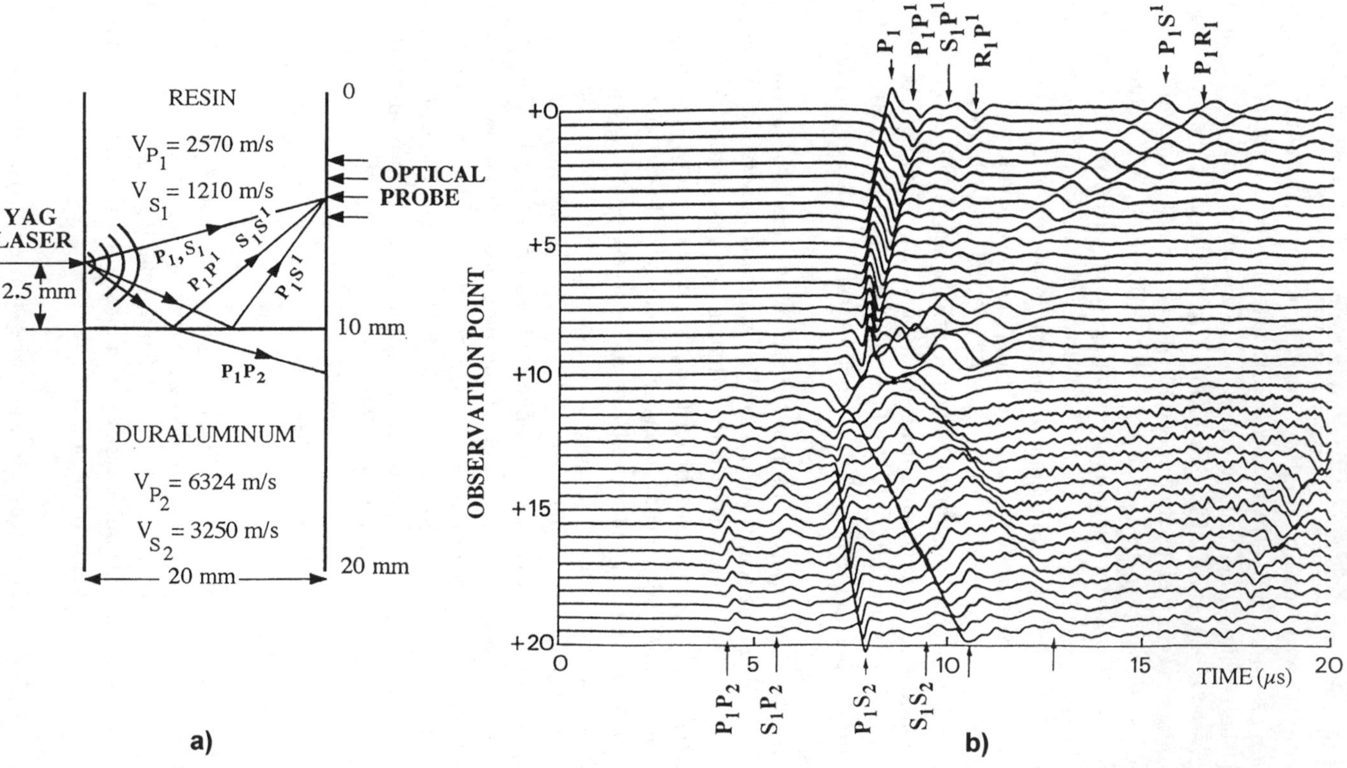

Fig. 3.44. Physical model of a seismic prospection experiment. (**a**) The emitter (a YAG laser beam) is fixed 2.5 mm above the Lucite-duralumin interface. (**b**) Each reading corresponds to a position of the probe beam of a heterodyne interferometer. After Pouet, B. [3.61]

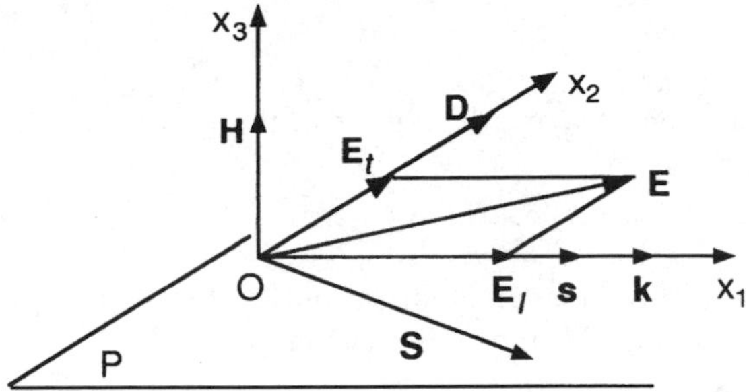

Fig. 3.45. Vectors for a plane electromagnetic wave

and in a non-magnetic medium, Maxwell's equations

$$\nabla \times \boldsymbol{E} = -\frac{\partial \boldsymbol{B}}{\partial t}\,, \quad \nabla \times \boldsymbol{H} = \frac{\partial \boldsymbol{D}}{\partial t}\,, \quad \boldsymbol{B} = \mu_0 \boldsymbol{H}\,,$$

become

$$\boldsymbol{k} \times \boldsymbol{E} = \omega\mu_0 \boldsymbol{H}\,, \quad \boldsymbol{k} \times \boldsymbol{H} = -\omega \boldsymbol{D}\,. \tag{3.117}$$

Eliminating the field $\boldsymbol{H}$,

$$\boldsymbol{s} \times (\boldsymbol{s} \times \boldsymbol{E}) + \mu_0 c^2 \boldsymbol{D} = 0 \quad \Rightarrow \quad \boldsymbol{s} \cdot \boldsymbol{D} = 0\,.$$

By the identity $\boldsymbol{a} \times (\boldsymbol{b} \times \boldsymbol{c}) = \boldsymbol{b}(\boldsymbol{a} \cdot \boldsymbol{c}) - \boldsymbol{c}(\boldsymbol{a} \cdot \boldsymbol{b})$ and the relation $\mu_0\varepsilon_0 c_0^2 = 1$, the above relation becomes in vector and tensor form,

$$\varepsilon_0 \left[\boldsymbol{E} - \boldsymbol{s}(\boldsymbol{s} \cdot \boldsymbol{E})\right] = \frac{\boldsymbol{D}}{n^2}\,, \quad (B_{ij} - B_{kj} s_i s_k) D_j = \frac{D_i}{n^2}\,, \tag{3.118}$$

where B_{ij} is the impermittivity tensor satisfying $\varepsilon_0 E_i = B_{ij} D_j$. We choose coordinates ξ_i such that propagation is parallel to the ξ_1-axis. Since $s_i = \delta_{i1}$ and $D_1 = 0$, equation (3.118) is of order 2:

$$\begin{pmatrix} B_{22} - 1/n^2 & B_{23} \\ B_{23} & B_{33} - 1/n^2 \end{pmatrix} \begin{pmatrix} D_2 \\ D_3 \end{pmatrix} = 0\,. \tag{3.119}$$

The matrix has two orthogonal eigenvectors (polarisations) with eigenvalues $1/n^2$.

Vectors $\boldsymbol{s}$, $\boldsymbol{E}$, $\boldsymbol{D}$ and the Poynting vector $\boldsymbol{S} = \boldsymbol{E} \times \boldsymbol{H}$ lie in a plane perpendicular to $\boldsymbol{H}$ (Fig. 3.45). The electric field has longitudinal and transverse components E_l and E_t, with

$$\boldsymbol{E} = \boldsymbol{s}(\boldsymbol{s} \cdot \boldsymbol{E}) + \frac{\boldsymbol{D}}{n^2 \varepsilon_0} = \boldsymbol{E}_\mathrm{l} + \boldsymbol{E}_\mathrm{t}\,.$$

Only the contribution to the Poynting vector arising from E_t is parallel to $\boldsymbol{s}$. In harmonic conditions, the mean transported power along x_1 is

$$\langle P \rangle = \frac{1}{2} \int \boldsymbol{s} \cdot \boldsymbol{S}\, \mathrm{d}x_2\, \mathrm{d}x_3 = \frac{1}{2} \int E_\mathrm{t} H\, \mathrm{d}x_2\, \mathrm{d}x_3\,.$$

By the first relation of (3.117), $H = kE_\mathrm{t}/\mu_0\omega$, it follows that

$$\langle P \rangle = \frac{k}{2\mu_0\omega} \int E_t^2 \mathrm{d}x_2 \, \mathrm{d}x_3 \; .$$

The transverse electric field of a light wave carrying mean power 1 W is

$$\boldsymbol{E}_t = \boldsymbol{e} \left(\frac{2\mu_0\omega}{k} \right)^{1/2} \exp \mathrm{i}(\omega t - \boldsymbol{k} \cdot \boldsymbol{x}) \; , \quad |\boldsymbol{e}| = 1 \; ,$$

where $\boldsymbol{e}$ is a unit vector in the direction of polarisation of the wave, i.e., parallel to $\boldsymbol{D}$. Orthogonality of modes M and N is written

$$\int \boldsymbol{E}_N^* \cdot \boldsymbol{E}_M \mathrm{d}x_2 \, \mathrm{d}x_3 = 2\mu_0 \frac{\omega_N}{k_N} \delta_{MN} \; . \tag{3.120}$$

3.2 In a uniaxial crystal, how do refractive indices vary with direction of propagation θ (polar angle, shown in Fig. 3.46)?

Solution. For vibration $\boldsymbol{D}^{(1)}$, $n_1 = n_o$, $\forall \theta$. For vibration $\boldsymbol{D}^{(2)}$,

$$\frac{x_1^2 + x_2^2}{n_o^2} + \frac{x_3^2}{n_e^2} = 1 \; ,$$

where $x_1 = 0$, $x_2 = -n \cos\theta$, $x_3 = n \sin\theta$. Hence,

$$\frac{1}{n_2(\theta)^2} = \frac{\cos^2\theta}{n_o^2} + \frac{\sin^2\theta}{n_e^2} \; .$$

3.3 Show that in the Raman–Nath regime, bending of light rays caused by variations in refractive index (the mirage effect) can be ignored, provided that $Q|\Delta\Phi| < \pi$.

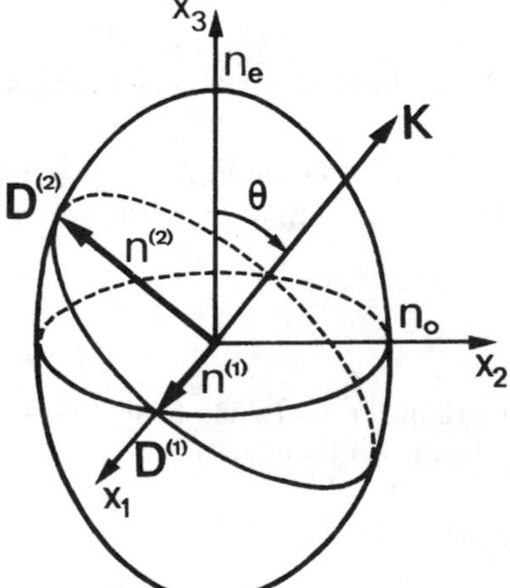

Fig. 3.46. Uniaxial crystal

Solution. The last equation in Problem 1.8, Vol. I, gives the radius of curvature

$$\frac{1}{R} = \frac{1}{n_0}\frac{\mathrm{d}n}{\mathrm{d}x_2} , \qquad \left.\frac{\mathrm{d}n}{\mathrm{d}x_2}\right|_{\max} = \frac{2\pi}{\Lambda}|\Delta n| ,$$

for sinusoidal index variation according to (3.23). The condition on the height of the light ray after travelling distance L,

$$\Delta x_2 \approx \frac{L^2}{2R} < \frac{\Lambda}{4} \quad \Rightarrow \quad \frac{\pi L^2}{n_0 \Lambda}|\Delta n| < \frac{\Lambda}{4} ,$$

can be written

$$2\pi\frac{\lambda L}{\Lambda^2}\, 2\pi\frac{L}{n_0\lambda}|\Delta n| < \pi \quad \Rightarrow \quad Q|\Delta\Phi| < \pi ,$$

where we have used (3.32) and (3.25) with $\lambda_0 = n_0\lambda$.

3.4 A monochromatic light beam enters a crystal along the x_1-axis. Plane elastic waves, all of frequency F, cross the (x_2, x_3)-plane in all directions. What pattern is observed on a screen parallel to the (x_2, x_3)-plane, at distance D? Consider the case of an isotropic solid.

Solution. The first order beam, deflected through angle $\theta \approx \lambda/\Lambda = cF/Vf$ [equation (3.30)], produces a spot at distance $d = DcF/Vf$ in the direction of propagation $\boldsymbol{n}$ of the active elastic wave. This is inversely proportional to the phase speed V of elastic waves in direction $\boldsymbol{n}$. Since this direction is arbitrary in the (x_2, x_3)-plane, the three cross-sections of the slowness surface of the crystal by the (x_2, x_3)-plane are observed on the screen (see Sect. 4.2.5, Vol. I). In the case of an isotropic solid, this yields two circles. This method, due to Schaefer and Bergmann, provides a measurement of elastic wave speeds.

3.5 Applying energy conservation, establish (3.50) which gives the intensity diffracted at the Bragg angle when phase synchronism is satisfied.

Solution. The spectral decomposition (3.27) implies that the increase $\mathrm{d}|B|$ in amplitude $|B|$ of the first order diffracted wave produced by a slab of thickness $\mathrm{d}x$ is

$$\mathrm{d}|B| = \frac{|A|}{2}\,\mathrm{d}\Phi = \frac{|A|}{2}\frac{\Delta\Phi}{L}\,\mathrm{d}x . \tag{3.121}$$

The carrier wave amplitude decreases accordingly. By energy conservation,

$$|A|^2 + |B|^2 = I \quad \Rightarrow \quad |A|\,\mathrm{d}|A| + |B|\,\mathrm{d}|B| = 0 .$$

Relation (3.121) implies

$$\frac{\mathrm{d}|B|}{\mathrm{d}x} = |A|\frac{\Delta\Phi}{2L}\,, \qquad \frac{\mathrm{d}|A|}{\mathrm{d}x} = -|B|\frac{\Delta\Phi}{2L}\,.$$

Amplitudes $|A|$, $|B|$ satisfy the same differential equation:

$$\frac{\mathrm{d}^2 f}{\mathrm{d}x^2} + \left(\frac{\Delta\Phi}{2L}\right)^2 f = 0\,.$$

Solutions satisfying $B = 0$ and $A = A_0$ at $x = 0$ are

$$|B| = |A_0| \sin\frac{|\Delta\Phi|}{2L}x\,, \qquad |A| = |A_0| \cos\frac{|\Delta\Phi|}{2L}x\,.$$

At the $x = L$ end of the elastic beam, the relative intensity of the diffracted beam is

$$\frac{I_\mathrm{D}}{I_0} = \left|\frac{B}{A_0}\right|^2 = \sin^2\frac{|\Delta\Phi|}{2}\,.$$

In the expression $|\Delta\Phi| = 2\pi L|\Delta n|/\lambda_0$, the change Δn in refractive index is related to the power density P and the figure of merit $M = p^2 n^6/\rho V^3$:

$$|\Delta n| = \frac{n^3}{2}pS = \sqrt{\frac{MP}{2}} \quad\Rightarrow\quad |\Delta\Phi| = \frac{2\pi L}{\lambda_0}\sqrt{\frac{MP}{2}}\,. \tag{3.122}$$

The relative intensity of the diffracted beam is therefore given by (3.50).

3.6 Check that values for the figure of merit of lead molybdate are very close for the two conditions in Table 3.3 (light polarised parallel or perpendicular to the wave vector plane).

Solution. The strain S_{33} of the longitudinal wave produces variations

$$\Delta\varepsilon_{22} = \Delta\varepsilon_{11} = -\varepsilon_{11}^2 p_{13} S_{33}\,, \qquad \Delta\varepsilon_{33} = -\varepsilon_{33}^2 p_{33} S_{33}\,.$$

- Parallel polarisation $(0,\,0,\,E_3) \Rightarrow$ index n_e : $\Delta D_3 = \Delta\varepsilon_{33}E_3 \Rightarrow$ elasto-optic constant $p_{33} \Rightarrow M = n_\mathrm{e}^6 p_{33}^2/\rho V^3 = 23.9 M_0$.
- Perpendicular polarisation $(E_1,\,0,\,0) \Rightarrow$ index n_o: $\Delta D_1 = \Delta\varepsilon_{11}E_1 \Rightarrow$ elasto-optic constant $p_{13} \Rightarrow M = n_\mathrm{o}^6 p_{13}^2/\rho V^3 = 23.5 M_0$.

3.7 For the three values $R_\mathrm{max} = 0,\,0.5,\,1$ of maximal diffraction efficiency, calculate the maximal phase asynchronism $\Delta\phi$ corresponding to a 3 dB reduction in diffraction efficiency, i.e., a ratio $R = R_\mathrm{max}/2$.

Solution. Interaction efficiency is maximal when $\Delta\phi = 0$:

$$R_\mathrm{max} = \sin^2\frac{\pi x}{2}\,, \qquad x = \sqrt{\frac{P}{P_1}}\,.$$

By (3.55), efficiency R is halved when

$$\left(\frac{\pi x}{2}\mathrm{sinc}\frac{\pi z}{2}\right)^2 = \frac{1}{2}\sin^2\frac{\pi x}{2}\,, \quad z = \sqrt{x^2+y^2}\,, \quad y = \frac{\Delta\phi}{\pi}\,.$$

For given x, maximal phase asynchronism follows from

$$\mathrm{sinc}\frac{\pi z}{2} = \frac{1}{\sqrt{2}}\mathrm{sinc}\frac{\pi x}{2}\,, \quad \text{or} \quad \sin\frac{\pi z}{2} = \frac{z}{x\sqrt{2}}\sin\frac{\pi x}{2}\,.$$

Results are:

$$R_{\max} \approx 0 \quad \Rightarrow x \ll 1\,, \quad \mathrm{sinc}\frac{\pi z}{2} = \frac{1}{\sqrt{2}} \Rightarrow z_0 = \pm 0.886 \text{ and } \left.\frac{\Delta\phi}{\pi}\right|_{\max} \approx 0.9\,,$$

$$R_{\max} \approx 0.5 \Rightarrow x = 0.5\,, \sin\frac{\pi z}{2} = z \quad \Rightarrow z_{0.5} = \pm 1 \text{ and } \left.\frac{\Delta\phi}{\pi}\right|_{\max} \approx 0.87\,,$$

$$R_{\max} \approx 1 \quad \Rightarrow x = 1\,, \quad \sin\frac{\pi z}{2} = \frac{z}{\sqrt{2}} \Rightarrow z_1 = \pm 1.28 \text{ and } \left.\frac{\Delta\phi}{\pi}\right|_{\max} \approx 0.80\,.$$

3.8 Prove (3.59) giving angles α_{I} and α_{D} of incidence and diffraction for the anisotropic acousto-optic interaction in a uniaxial crystal (Fig. 3.15). Show that α_{I} goes through an extremum for a value of λ_0/Λ such that $\alpha_{\mathrm{D}} = 0$.

Solution. Projecting the wave vector diagram onto x_1 and x_2 axes,

$$k_{\mathrm{I}}\sin\alpha_{\mathrm{I}} = k_{\mathrm{D}}\sin\alpha_{\mathrm{D}} - K\,,$$
$$k_{\mathrm{I}}\cos\alpha_{\mathrm{I}} = k_{\mathrm{D}}\cos\alpha_{\mathrm{D}}\,,$$

then squaring and adding, yields

$$k_{\mathrm{I}}^2 = k_{\mathrm{D}}^2 + K^2 - 2k_{\mathrm{D}}K\sin\alpha_{\mathrm{D}}\,, \quad k_{\mathrm{D}}^2 = k_{\mathrm{I}}^2 + K^2 + 2k_{\mathrm{I}}K\sin\alpha_{\mathrm{I}}\,.$$

Algebraic values of angles α_{I} and α_{D} are given by

$$\sin\alpha_{\mathrm{I}} = \frac{1}{2k_{\mathrm{I}}}\left(\frac{k_{\mathrm{D}}^2 - k_{\mathrm{I}}^2}{K} - K\right)\,, \quad k_{\mathrm{I}} = \frac{2\pi}{\lambda_0}n_{\mathrm{e}}\,,$$

$$\sin\alpha_{\mathrm{D}} = \frac{1}{2k_{\mathrm{D}}}\left(\frac{k_{\mathrm{D}}^2 - k_{\mathrm{I}}^2}{K} + K\right)\,, \quad k_{\mathrm{D}} = \frac{2\pi}{\lambda_0}n_{\mathrm{o}}\,, \quad K = \frac{2\pi}{\lambda}\,,$$

and these lead to (3.59). When $n_{\mathrm{e}} > n_{\mathrm{o}}$, the sine of the angle of incidence

$$\sin\alpha_{\mathrm{I}} = -\frac{1}{n_{\mathrm{e}}}\left[(n_{\mathrm{e}}^2 - n_{\mathrm{o}}^2)\frac{\Lambda}{\lambda_0} + \frac{\lambda_0}{\Lambda}\right]\,,$$

has form $r + a/r$, where $r = \lambda_0/\Lambda$. It has a minimum when $r = \sqrt{a}$, i.e.,

$$\frac{\lambda_0}{\Lambda} = (n_{\mathrm{e}}^2 - n_{\mathrm{o}}^2)^{1/2} \quad \Rightarrow \quad \sin\alpha_{\mathrm{I}} = -\frac{(n_{\mathrm{e}}^2 - n_{\mathrm{o}}^2)^{1/2}}{n_{\mathrm{e}}}\,, \quad \alpha_{\mathrm{D}} = 0\,.$$

3.9 For the interaction conditions of Table 3.3, find elasto-optic coefficients and refractive indices entering the figure of merit of paratellurite TeO_2.

Solution.

1. The elastic wave polarised along $[1\bar{1}0]$ and propagating along $[110]$ is accompanied by strains

$$S_{11} = \frac{\partial u_1}{\partial x_1}\,, \quad S_{22} = \frac{\partial u_2}{\partial x_2} = -S_{11}\,, \quad S_{12} = \frac{1}{2}\left(\frac{\partial u_1}{\partial x_2} + \frac{\partial u_2}{\partial x_1}\right) = 0\,,$$

and $S_{i3} = 0$ for $i = 1, 2, 3$, since $\partial/\partial x_1 = \partial/\partial x_2$ and ${}^\circ u_2 = -{}^\circ u_1$. Given the speed $V = [(c_{11} - c_{12})/2\rho]^{1/2}$, the elastic power density is $P = \langle \mathcal{E} \rangle\, V$, where

$$\langle \mathcal{E} \rangle = \frac{1}{2} c_{ijkl} S_{ij} S_{kl} = \frac{1}{2}[c_{11} + c_{22} - 2c_{12}]S_{11}^2 = (c_{11} - c_{12})S_{11}^2\,.$$

P can then be put into the form (3.47) by setting $S = 2S_{11}$:

$$P = 2\rho V^3 S_{11}^2 = \frac{1}{2}\rho V^3 S^2\,.$$

By (3.10), $\Delta\varepsilon_{il} = -\varepsilon_{ii}\varepsilon_{ll}(p_{il11} - p_{il22})S_{11}$, so that $\Delta\varepsilon_{33} = 0$, because $p_{31} = p_{32}$. Hence,

$$\Delta\varepsilon_{ij} = \begin{pmatrix} \Delta\varepsilon_{11} & 0 & 0 \\ 0 & -\Delta\varepsilon_{11} & 0 \\ 0 & 0 & 0 \end{pmatrix}\,, \qquad \Delta\varepsilon_{11} = -\varepsilon_{11}^2(p_{11} - p_{12})S_{11}\,.$$

Optical axes of the crystal are unchanged. A light wave incident in the Z direction (index n_o), of polarisation $\boldsymbol{e}_\mathrm{I} = (\cos\alpha,\, \sin\alpha,\, 0)$, is diffracted with polarisation $\boldsymbol{e}_\mathrm{D} = (\cos\alpha,\, -\sin\alpha,\, 0)$. The coupling coefficient (3.44),

$$\kappa = \frac{\pi}{2\lambda_0}\frac{\Delta\varepsilon}{n_\mathrm{o}}\,, \qquad \Delta\varepsilon = e_i^\mathrm{I}\Delta\varepsilon_{il}e_l^\mathrm{D} = \Delta\varepsilon_{11}\,,$$

is related to $S = 2S_{11}$ via elasto-optic coefficient $p = (p_{11} - p_{12})/2$:

$$\kappa = -\frac{\pi}{2\lambda_0}\frac{\varepsilon_{11}^2}{n_\mathrm{o}}(p_{11} - p_{12})S_{11} = -\frac{\pi}{2\lambda_0}n_\mathrm{o}^3 pS\,.$$

2. For the longitudinal elastic wave along $[001]$, $S = S_{33}$ and

$$\Delta\varepsilon_{ij} = -\varepsilon_{ii}\varepsilon_{ll}p_{il33}S_{33}\,,$$

so that

$$\Delta\varepsilon_{11} = \Delta\varepsilon_{22} = -\varepsilon_{11}^2 p_{13}S\,, \quad \Delta\varepsilon_{33} = -\varepsilon_{33}^2 p_{33}S\,, \quad \Delta\varepsilon_{ij} = 0 \text{ for } i \neq j\,.$$

The light wave with polarisation $(E_1,\, 0,\, 0)$, index n_o, is diffracted into a wave of polarisation $(\Delta D_1 = \Delta\varepsilon_{11}E_1,\, 0,\, 0)$, with modulus $p = p_{13}$.

3.10 Find the optical bandwidth $\Delta\lambda$ of a collinear acousto-optic filter, taking the values $\Delta n = 0.01$, interaction length $L = 35$ mm and $\lambda_0 = 0.5$ µm.

Solution. Using the notation in Appendix B, the phase asynchronism

$$\Delta\phi = L\Delta k = L\left(\frac{2\pi}{\lambda}|\Delta n| - \frac{2\pi}{\Lambda}\right)$$

is zero when $\lambda = \lambda_0 = \Lambda|\Delta n|$. The optical bandwidth is given by the maximum possible value of the factor $\Delta\phi/\pi$:

$$2\pi L\frac{\Delta\lambda}{\lambda_0^2}|\Delta n| = 2\Delta\phi_{\max} \Rightarrow \Delta\lambda = \frac{\lambda_0^2}{L|\Delta n|}\left.\frac{\Delta\phi}{\pi}\right|_{\max}.$$

The -3 dB cutoff frequency is reached when $\Delta\phi/\pi \approx \pm 0.85$, so that, for the above values, $\Delta\lambda = 8$ Å.

4. Signal Processing Components

There are many ways to process a time-varying physical quantity like a signal: it can be amplified, retarded relative to some other signal chosen as a reference, compressed, stored, convoluted with another signal, sampled, temporally or spatially filtered, i.e., its spectrum can be modified, and so on.

Acousto-electronic components described in this chapter use elastic waves to process electrical signals whose centre frequencies lie in the range 1 MHz to 10 GHz. Before beginning this task, let us recall that any device, be it electrical, mechanical or optical, obeys certain general relations whatever treatment it may carry out, provided that it operates linearly and time-invariantly. Linearity ensures that if $y_1(t)$ and $y_2(t)$ are the device response to any two signals $x_1(t)$ and $x_2(t)$, its response to $\lambda x_1(t) + \mu x_2(t)$ will be $\lambda y_1(t) + \mu y_2(t)$, for any constants λ and μ. Time invariance implies that its response to signal $x(t+\tau)$ is $y(t+\tau)$ for any time delay τ.

The component is represented by one of two responses. The first is its *impulse response* $h(t)$. This is its response to a Dirac pulse, which means physically a pulse whose duration is much smaller than any characteristic time of the device. The second is its *frequency response* $H(f)$. This is its response to a series of harmonic (sinusoidal) signals, whose frequencies vary in principle from $-\infty$ to ∞, but in practice, over a range including the passband of the system. Time response $h(t)$ and frequency response $H(f)$ are related by Fourier transform:

$$H(f) = \int_{-\infty}^{\infty} \mathrm{e}^{-2\pi \mathrm{i} f t} h(t)\,\mathrm{d}t\,, \quad h(t) = \int_{-\infty}^{\infty} \mathrm{e}^{2\pi \mathrm{i} f t} H(f)\,\mathrm{d}f\,. \tag{4.1}$$

In particular, the first relation here indicates that each spectral component of the Dirac pulse $\delta(t)$ is equal to 1. This pulse is obtained by superposing infinitely many unit amplitude harmonic signals.

The action of the device on a general input signal $x(t)$ is therefore expressed either in the time or the frequency domain. In the time domain, the output signal $y(t)$ is the convolution product of input signal $x(t)$ and impulse response $h(t)$:

$$y(t) = x(t) \otimes h(t) = \int_{0}^{t} x(\tau) h(t-\tau)\,\mathrm{d}\tau\,. \tag{4.2}$$

In the frequency domain, the spectrum $Y(f)$ of the output signal [Fourier transform of $y(t)$] is the ordinary product of the spectrum $X(f)$ of input signal $x(t)$ with frequency response $H(f)$:

$$Y(f) = H(f)X(f) \,. \tag{4.3}$$

These ideas have already been applied in our discussion of transducers (e.g., in Sect. 2.2, the impulse response of a simple interdigital transducer).

Components will be presented according to the functions they are designed to carry out. The first are concerned with delay. They are generally characterised by their frequency response. Then come bandpass filters, which only allow through signals whose frequencies lie within a certain range. There are several kinds. Some work with surface travelling waves, others with stationary bulk or surface waves. Depending on their structure, they can be described either in terms of their impulse or frequency response. Such filters, particularly narrowband filters, have witnessed considerable development in connection with communication technologies, like mobile telephones. The third function we examine is time compression by filters matched to given signals. We recall the aims and operation of such filters, whose output signal is shorter but whose amplitude is greater than those of the input signal, for the same noise level. Historically, these Rayleigh wave components were invented before the bandpass filter. They elegantly resolved the problem of extracting noise from radar signals. Two other types of component will be presented. The first carry out spectral analysis of their input signal and the second convolute two signals. The remarks made at the beginning of this introduction do not apply to the latter components, because they involve non-linear effects. Let us now discuss elements and structures of all these components.

4.1 Elements and Structures

Elements involved in elastic wave components are: the transducer, the substrate, the multistrip coupler and the reflector. Figure 4.1 shows some examples of how these elements may be associated.

- **(a)**, **($\mathbf{a}_1$)** Simple travelling wave delay line (or filter). The structure contains only an emitter transducer $\mathrm{T_e}$, a substrate S and a receiver transducer $\mathrm{T_r}$. The substrate in ($\mathbf{a}_1$) serves as both generating and propagating medium.
- **(b)** Travelling Rayleigh wave filter equipped with multistrip coupler C.
- **(c)** Transmission line involving an acousto-optic interaction.
- **(d)**, **($\mathbf{d}_1$)** Stationary wave filter (resonator) made with two straight reflectors R and one transducer (dipole) or two transducers (quadripole).
- **($\mathbf{d}_2$)**, **($\mathbf{d}_3$)** Longitudinally and transversely coupled dipole resonators.

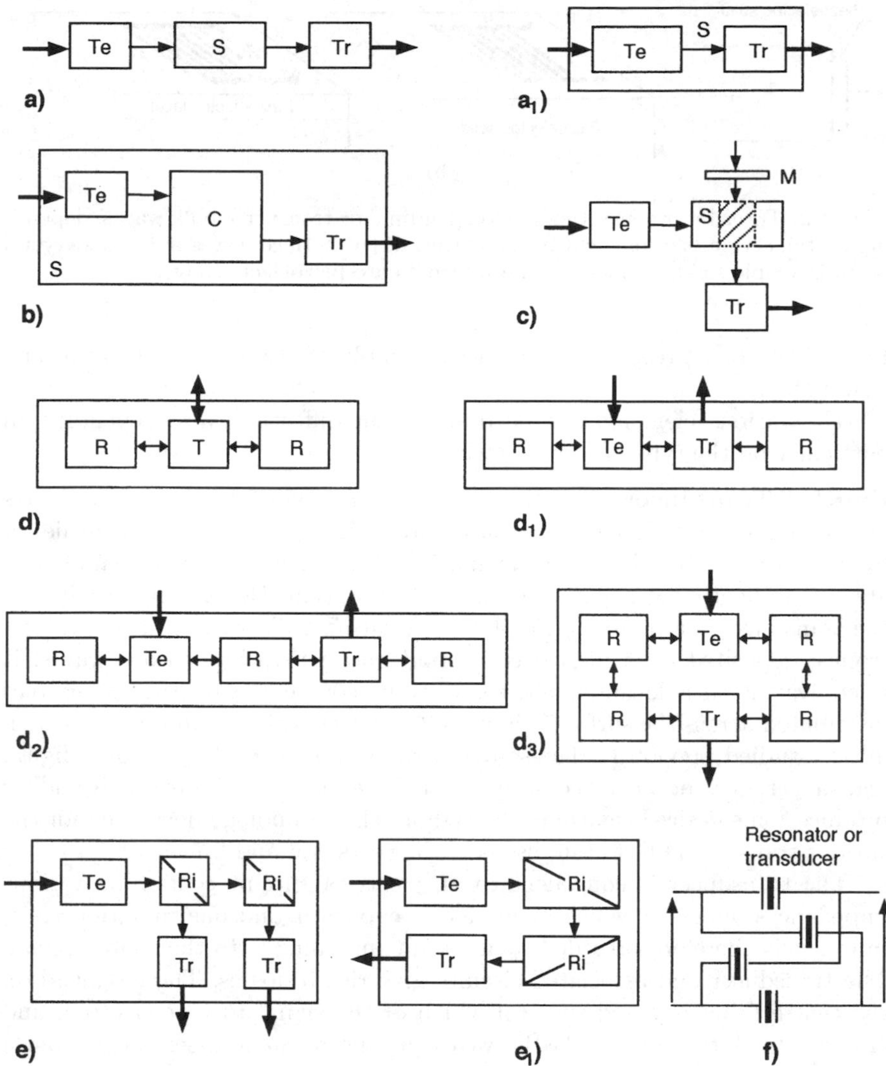

Fig. 4.1. Structures of elastic wave components. T_e emitter transducer, T_r receiver transducer, S substrate, C coupler, R straight reflector, R_i inclined reflector. (**a**), (**a**$_1$) Simple travelling wave delay lines or filters. (**b**) Filter equipped with multistrip coupler. (**c**) Transmission line with acousto-optic interaction. (**d**), (**d**$_1$) Stationary wave dipole and quadripole resonators. (**d**$_2$), (**d**$_3$) Longitudinally and transversely coupled dipole resonators. (**e**), (**e**$_1$) Filters including two selective reflectors. (**f**) Lattice connection of resonators or transducers

(**e**), (**e**$_1$) Filters containing two inclined selective reflecting gratings R_i.

Variants of these basic structures come in as soon as some coupling or assembly is required: longitudinal or transverse coupling of two resonators,

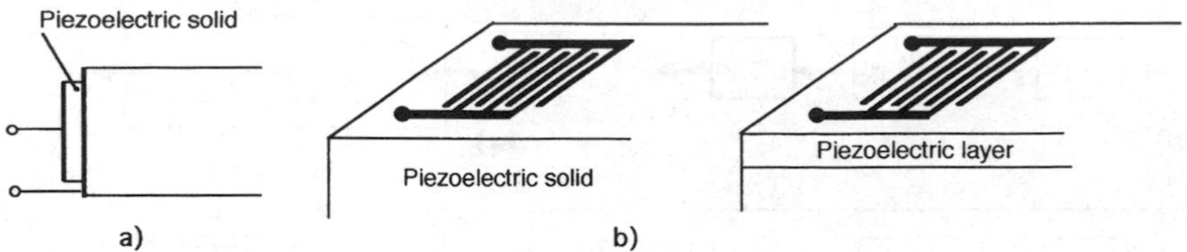

Fig. 4.2. Transducer generating (**a**) longitudinal or transverse bulk waves, depending on the crystallographic cut of the material, (**b**) surface waves in a monocrystal or an isotropic solid, by means of an intermediate piezoelectric layer

Fig. 4.1d$_2$ and d$_3$, respectively; lattice assembly of resonators or transducers, Fig. 4.1f.

The various elements can be designed in different ways, depending in particular on the type of waves involved.

Emitter Transducer. This was discussed in Chaps. 1 and 2. It converts an electrical input signal into elastic waves. Energy conversion is achieved by using the electric field of the signal to set a piezoelectric material of suitable crystallographic cut into mechanical vibration. Depending on whether the transducer is a piezoelectric plate generating bulk waves or a set of electrodes, deposited on a piezoelectric solid and generating surface waves, it is equivalent to a localised ultrasonic wave source or an array of sources distributed across the surface. Figure 4.2 represents these two types of transducer, studied previously. The source distribution created by an interdigital transducer may be arranged so as to make emission unidirectional and/or produce some desired change in the signal. The technology developed for the various transducers has been discussed in Sects. 1.6 and 2.7.

The transducer is equivalent to an electrical circuit containing various impedances, amongst which at least one capacitor and one radiation resistance. It is therefore preceded by a circuit matching it to the source signal. The transducer and associated circuits give rise to losses. These depend on the centre frequency and spectral width of the signal to be converted, and also on the nature of the elastic waves produced. Such losses range from a few dB for narrow bandwidths, to 10 dB and more, depending on the width of the passband. Elastic wave frequencies in these processors fall between a few MHz and several GHz.

Substrate. The elastic wave propagates either inside the substrate or close to its surface, but always at a speed around 10^5 times smaller than the electromagnetic wave. This property of elastic waves, which propagate only a few millimetres in 1 μs, is used to delay the signal (relative to any other, not converted into an elastic wave) or to make it interact with another external quantity.

We shall assume here that the substrate is a solid propagating medium. It may be homogeneous and isotropic, like a block of silica, or anisotropic, like an aluminium oxide monocrystal; or it may be a layered structure, for example, a film on a substrate. In the case of a crystal, the crystallographic cut is chosen according to the type of wave. The phase speed is constant, whatever the frequency. This is so because wavelengths are large compared with interatomic distances (Sect. 4.1, Vol. I). For a layered medium, frequency dependence of wave speeds is not the same for film and substrate. Signal delays are frequency dependent. This dispersion effect also exists in guide-shaped homogeneous materials, e.g., for bulk waves propagating inside a plate (Sect. 5.5, Vol. I).

There are many variants on this theme:

- The propagating medium may be a rod, a polygonal block (the wave path, reflecting from the polygon faces, is made up of broken lines and the delay is large, e.g., 1 ms), a long metallic ribbon (dispersive system). Wave polarisations can be transformed by reflection (Sect. 4.4.2.2, Vol. I).
- The propagating medium may also be the conversion medium. It is then piezoelectric. This is often the case for Rayleigh waves.

The attenuation coefficient for elastic waves (Fig. 4.31, Vol. I) depends on type of material and wave polarisation. It increases with frequency. Defects, such as holes or impurities, become significant when their dimensions are of the same order as the wavelength. All being equal, losses are greater in polycrystalline solids than in monocrystals.

Interaction. During propagation, elastic waves can interact with an external wave such as a light wave, a space-charge wave associated with electrons in motion, or another elastic wave.

In the case of an interaction with *light*, collinear or otherwise, the result is a diffracted light wave carrying the characteristics (i.e., frequency, power) of the elastic wave, as well as those of the incident light wave (i.e., spatial and temporal modulation). Diffraction is caused by the wave of refractive index variations which moves along with any elastic wave (Sect. 3.1.4). In Fig. 4.1c, M symbolises some modulation related, in general, to the shape of the signal to be processed. For example, it may be an amplitude mask, a spatial replica of the signal code.

In the case of an interaction with *electrons* moving in the same direction as the elastic wave, coupling has the effect of attenuating or amplifying the elastic wave, depending on whether these carriers are moving more slowly or more quickly than the sound. This coupling effect is similar to the one produced in a vacuum travelling wave tube when an electromagnetic wave, slowed down by the helix, interacts with the electron beam. Coupling between an elastic wave and charge carriers requires a piezoelectric material in order to associate an electric field with the elastic wave, and a semiconducting material in which carriers can move [4.1]. A single crystal can have both properties. Examples are cadmium sulfide, zinc oxide, gallium arsenide and selenium. Coupling then occurs inside the material. However, more possibilities are available when each property is provided by a distinct material. Coupling is

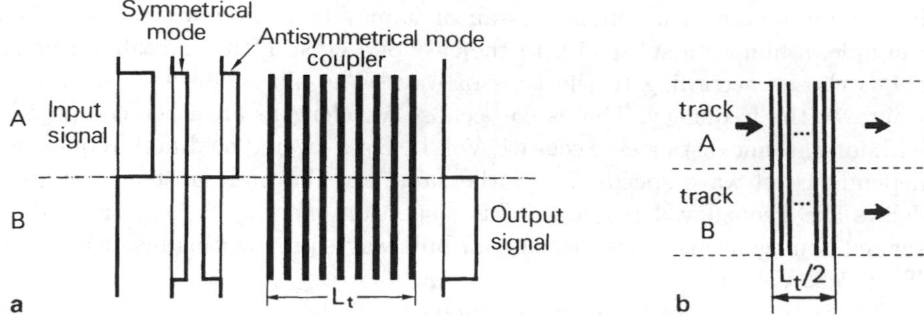

Fig. 4.3. Rayleigh wave multistrip coupler. The two modes propagate under the coupler at different speeds. Its length is chosen so that the phase difference between the two modes at the output is: (**a**) π, with total beam transfer from channel A to channel B; (**b**) $\pi/2$, with half the beam output from each channel

ensured by placing the two materials close together. The second solution is particularly suitable for surface waves, especially when the semiconductor is deposited in the form of a film on a piezoelectric substrate. Elastic waves can then be amplified by as much as several tens of [dB]. However, we shall not discuss this amplification effect any further since it has never been exploited outside the laboratory.

If the amplitude of two elastic waves is large enough to produce non-linear behaviour in the solid, contrary to what we have been assuming so far, coupling between them gives rise to a product of their associated signals.

The main characteristics of materials used as substrates are given in Fig. 5.27, Vol. I, and Tables 1.1, 2.1.

Multistrip Coupler. This element, sometimes introduced between two transducers, is made from a series of metallic strips produced on a piezoelectric solid [4.2, 4.3]. In its simplest form (Fig. 4.3a), the strips are parallel and of the same width. It works as follows: a Rayleigh wave, launched along channel A, induces a voltage between the strips. As this voltage also appears across channel B, it generates a Rayleigh wave there. Transfer from one channel to another is total for a number N_t of strips, which can be found approximately by decomposing the input signal on channel A into symmetric and antisymmetric modes propagating under the coupler at different speeds. The symmetric mode is not affected by the metallic bars since no current circulates in the direction of propagation. It travels at the speed V_{R} of waves moving on an unmetallised free surface [strip thickness and type of metal (aluminium) are such that mechanical loading of the substrate is negligible]. The antisymmetric mode propagates at speed V_∞ since the electric potential associated with it is cancelled by electric charges moving along the metallic strips.

The phase difference between these two modes is equal to π for a length L_{t} of the coupler such that

$$2\pi f\left(\frac{1}{V_\infty}-\frac{1}{V_\mathrm{R}}\right)L_\mathrm{t}=\pi\,,\quad \text{or}\quad L_\mathrm{t}=\frac{1}{2f}\frac{V_\mathrm{R}V_\infty}{V_\mathrm{R}-V_\infty}\,.$$

In terms of the coupling coefficient and wavelength,

$$L_\mathrm{t}\approx\frac{\lambda}{K_\mathrm{R}^2}\,. \tag{4.4}$$

If p is the grating pitch, the number of strips N_t is

$$N_\mathrm{t}\approx\frac{\lambda}{pK_\mathrm{R}^2}\,. \tag{4.5}$$

Thus the two modes cancel in channel A and add together in channel B, so that the wave which enters via channel A, under a coupler of length L_t, exits via channel B. In fact the simple formula (4.4) implicitly assumes the coupler to be a uniform film which is perfectly insulating in the propagation direction and perfectly conducting in the perpendicular direction. In practice, such anisotropy is obtained by means of metallic strips. Periodicity of a real coupler therefore introduces frequency cutoffs and reduces coupling efficiency. The first cutoff occurs at frequency $f_0 = V_\mathrm{R}/2p$.

We may view the presence of these strips as equivalent to sampling of a sinusoidal signal in channel A and its restitution by a periodic structure in channel B. The reduction factor is equal to [4.4]

$$\left(\frac{\sin\theta/2}{\theta/2}\right)^2\,,\quad \text{where}\quad \theta=\frac{2\pi\alpha p}{\lambda}\,,$$

and αp represents the active width of a metallic strip. Then (4.5) becomes

$$N_\mathrm{t}=\frac{\lambda}{K^2\alpha p\left(\dfrac{\sin\theta/2}{\theta/2}\right)^2}=\frac{\pi}{K^2}\frac{\theta}{1-\cos\theta}\,. \tag{4.6}$$

However, in order to find agreement with experimental results, parameters α and K^2 must be adjusted. α is not just the fraction η of metallised surface ($\alpha = r\eta$), and the value of K^2 must be corrected by a factor F:

$$N_\mathrm{t}=\frac{\pi}{FK^2}\frac{\pi r\eta f/f_0}{1-\cos(\pi r\eta f/f_0)}\,,\quad f_0=\frac{V}{2p}\,. \tag{4.7}$$

Maximal coupler efficiency (maximum FK^2) is obtained when $\eta = 1/2$ (as for the interdigital transducer). For Y-cut, Z-propagating lithium niobate, with $\eta = 1/2$, $r = 1.7$ and $FK^2 = 0.043$, this yields [4.5]

$$N_\mathrm{t}=\frac{195f/f_0}{1-\cos(153f/f_0)}\,, \tag{4.8}$$

where the cosine argument is given in degrees.

Figure 4.4 represents the strip number N_t required for 100% transfer between channels, as a function of normalised frequency f/f_0. Variation is slow near the minimum ($f/f_0 = 0.873$, $N_\mathrm{t} = 101$). At given frequency f, the

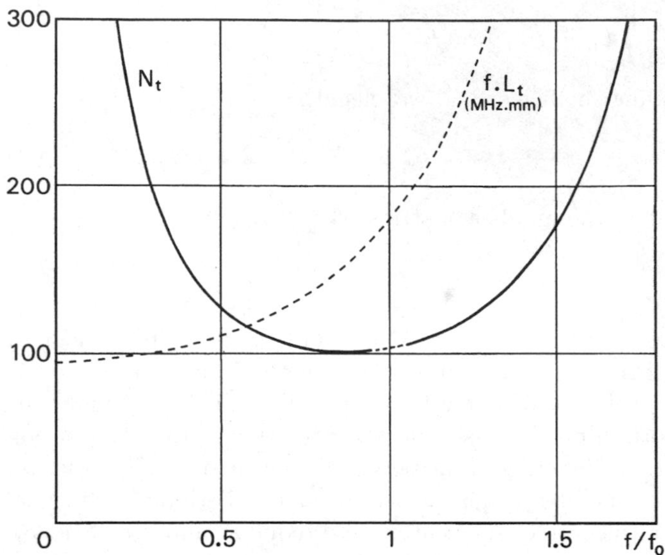

Fig. 4.4. Multistrip coupler. Strip number N_t required for 100% transfer from one channel to the other (*continuous curve*) and product of coupler length with frequency in [MHz mm] (*dashed curve*) as functions of frequency, normalised to the first cutoff frequency f_0

coupler length is smaller, provided that the spatial period is reduced, i.e., the frequency $f_0 = V/2p$ is increased. This is shown by the curve for the product of coupler length and frequency. Relative to the value corresponding to minimum N_t, the coupler length is reduced by 30% when $f_0 = 2f$. This reduction can be significant if the dimensions of the device are an important factor. Characteristics are plotted with a dotted curve near the minimum because the simplified theory above does not allow for the fact that cumulative reflections produce a cutoff in this zone.

The photographs in Fig. 4.5 show how well this simple coupler works. They refer to a coupler comprising 140 strips on lithium niobate. Transfer losses between channels are 0.5 dB at 45 MHz, and rejection between the two output channels is greater than 30 dB.

The total transfer multistrip coupler has the advantage of laterally displacing only the Rayleigh waves. Bulk waves from the emitter are not sensed by the receiver. Diffraction of small sources is reduced. The receiver is 'illuminated' by a wide and uniform beam. One disadvantage is that it requires a fairly long, strongly piezoelectric substrate such as lithium niobate.

Transfer is partial if the coupler length is less than L_t. For length $L_t/2$, half of a beam entering channel A leaves by the same channel, and the other half by channel B (Fig. 4.3b). The part leaving channel B is $\pi/2$ out of phase with the part leaving channel A (Problem 4.1). This coupler with

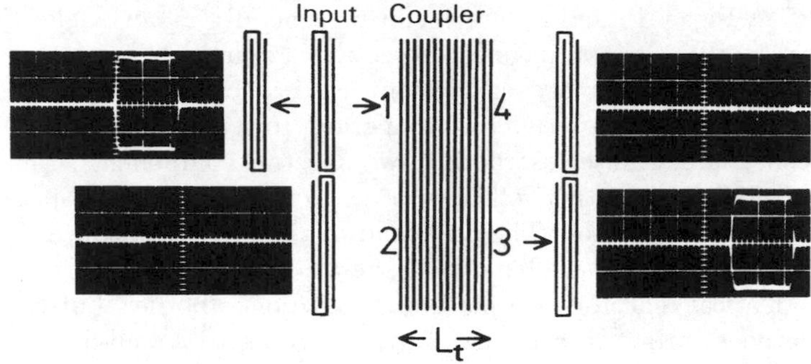

Fig. 4.5. Response of a multistrip coupler of length L_{t} (140 strips on lithium niobate). The incoming signal at 1 emerges at 3. Fig. 6 in [4.3], kindly communicated by E.G.S. Paige

3 dB transfer is useful when echoes produced by reflection from the output transducer are unacceptable (see Sect. 4.3.2).

There are other versions of these couplers in which strips are not necessarily identical and parallel, and which can not only translate but also compress a beam (Sect. 4.7). It can also act as a filter if need be [4.6].

Reflector. The simplest reflector is just a discontinuity in the impedance of the medium carrying the wave. However, reflection can only be partial. The incident wave transforms to another type on the obstacle (Sect. 4.4, Vol. I). The word 'reflector' refers to two situations here. The first is the almost total reflection a bulk wave undergoes on encountering a solid/vacuum (or solid/low pressure gas) interface. The main, parallel and polished faces of a plate behave with respect to longitudinal and transverse bulk waves like two excellent mirrors. A thin plate is thus a good 1-dimensional resonator. Such a resonator is easily excited if made of some piezoelectric material (discussed in Sect. 1.4.3.1). It is in fact an elementary filter. Filters with various characteristics are made by putting together several resonators, either in π or lattice arrangements (Fig. 4.1f), or else on the same substrate with elastic-type coupling (monolithic filter, Sect. 4.4.2).

The second situation, shown in Fig. 4.1d, refers to reflection of surface waves from a strip or groove grating. If the strips are periodic, with period p, the grating reflects any normally incident wave in the opposite direction, provided it has frequency close to $f_0 = V/2p$. Reflection is almost total if there are enough reflecting strips (Fig. 2.34). Two such gratings (straight reflectors) act as mirrors for Rayleigh waves and constitute a cavity. The cavity resonates when excited by a transducer placed between the reflectors. This resonator can be used like the bulk wave resonator to construct various filters, by associating independent elements (Fig. 4.1f) or coupled elements (Fig. 4.1d_2, d_3).

Reflectors R_i in Fig. 4.1e and e_1 are inclined, generally at an angle of 45°, with respect to the wave propagation direction. Figure 4.1e comprises two reflection gratings of different period, connected in cascade. Each reflects waves whose frequencies are tuned to its period, towards its associated transducer. It constitutes a filter extracting two spectral components from the signal. With the help of further reflector–transducer couples, more spectral components can be extracted. The passband of the input transducer is clearly broader than the sum of the bandwidths of all output transducers. In Fig. 4.1e_1, two identical reflectors are placed opposite one another, but the period of their grooves grows in a regular way, according to a well-defined rule, as their separation from the transducers increases. Consequently, the various spectral components of the signal are reflected by different zones. Low-frequency components travel a longer path than high-frequency components. The latter can be brought back together with the former if the input transducer emits them at a later time, which obeys an inverse law to the one governing growth of the groove period. Such a dispersive reflector component compresses appropriately matched signals. In this case, the two transducers function only as converters. A transmission line with no reflector (Fig. 4.1a_1) also constitutes a signal-matched filter if one (or both) transducer(s) produce dispersion corresponding to the spectrum of that signal. Although such simple matched filters perform less well than reflector filters, they are cheaper.

To end this section, let us just note that a grating can act as a guide for waves such as transverse waves, which are not intrinsically surface waves, by modifying the elasticity of the surface. This effect only occurs if the wave frequency lies outside the stop band of the grating. Note also that for transverse waves, like Bleustein–Gulyaev waves, naturally held close to the surface by piezoelectricity of the substrate, a sufficiently wide and deep groove is in theory a good reflector, so that two such grooves parallel to one another form a cavity [4.7].

Receiver Transducer. The signal output from the component must be electrical. Depending on the type of wave, the receiver is a transducer, either localised or distributed over the surface, and analogous to the one at the input of the transmission line. It may be involved in processing the signal. Its passband may complement that of the emitter transducer. When the transmission line (travelling or stationary wave) includes an interaction between the elastic wave and an external wave, it is the result of interaction which must be converted into an electrical signal by a suitable transducer, e.g., a photodiode, if a light wave carries the information, or a pair of long electrodes, if conversion is made by integration (Sect. 4.7). The characteristics of this receiver and its associated matching circuit are chosen in such a way that the transformation the signal has undergone remains unaffected.

4.2 Delay Lines

In this section, we shall describe three types of delay line, based on simple bulk waves, bulk waves with elasto-optic interactions, and Rayleigh waves, respectively. The second type was developed for a specific application and is less often used than the others, but the principle is an interesting illustration of the acousto-optic interaction. Bulk wave transmission lines, made of ceramic transducers cemented onto the end of a rod of isotropic material such as glass or silica, and operating at relatively low frequencies ($f < 50$ MHz), will not be considered.

4.2.1 Bulk Wave Delay Lines

A signal propagating in the form of electromagnetic waves in a cable can be delayed relative to one propagating in vacuum or in air. However, to delay it by 2 µs, for example, about 400 m of cable would be required. Apart from the considerable volume of cable required, prohibitive losses would be involved (greater than 100 dB at 1 GHz). Elastic wave transmission lines of volume a few [cm^3] will delay a signal by more than 10 µs, with reasonable losses, at frequencies above 5 GHz.

Structure. The structure here (Fig. 4.6a) comprises an input transducer, such as a zinc oxide film (ZnO), which launches longitudinal waves, a sapphire monocrystal (Al_2O_3, $V_L = 11\,200$ m/s), whose length determines the delay, and an output transducer (e.g., ZnO). ZnO films are generally chosen with different thicknesses (a fraction of a wavelength), and hence different resonance frequencies, with a view to increasing the bandwidth. The nature and thicknesses of metallic films used as electrodes (e.g., chromium, platinum, aluminium, Fig. 4.6b) also help to determine the bandwidth. The internal electrode may play the role of impedance matcher or transformer (see Sect. 1.5.2). Sapphire constitutes a good propagating medium. It is available at reasonable cost in the form of very long monocrystals (> 20 cm) and its attenuation coefficient (1 dB/µs at 2 GHz) is relatively low, as can be seen from Fig. 4.31, Vol. I, which shows propagation losses for various materials. These losses increase as the square of the frequency.

The crystallographic axis of symmetry A_6 of a zinc oxide film is normal to the surface on which it is deposited. Hence, only longitudinal waves are excited. When transverse waves are needed, one technique involves indium soldering a suitably cut monocrystalline plate ($LiNbO_3$) and then thinning it by ion bombardment (down to 1 µm). This technique has been applied to construct delay lines based on acousto-optic interaction. It allows a much wider choice of propagating medium.

A strip transmission line is connected between the cable carrying the signal and each transducer. This strip, made on a ceramic support, matches the signal to the transducer generator (Fig. 4.7). The connection between

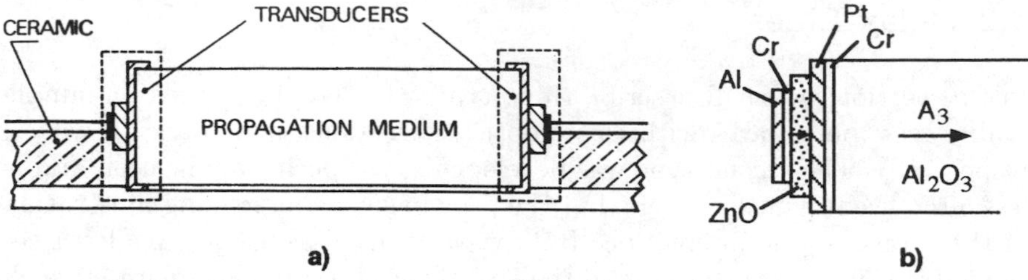

Fig. 4.6a,b. Structure of a bulk wave delay line ($f > 1$ GHz)

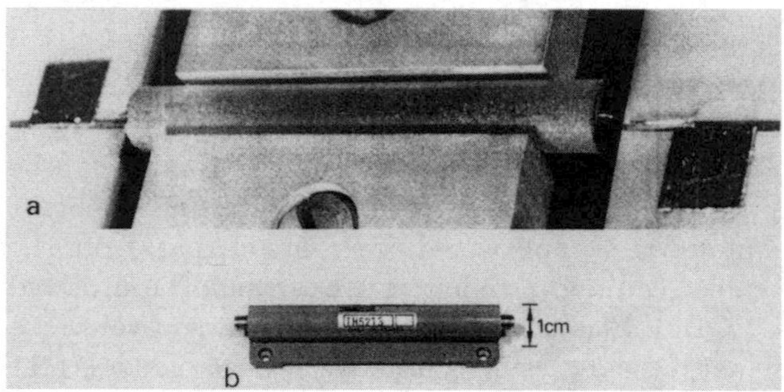

Fig. 4.7. Bulk wave delay line. (**a**) Aluminium oxide monocrystal, diameter 3 mm, carrying a zinc oxide thin film transducer at each end. (**b**) Coaxial structure

transducer and strip is made with a gold wire (diameter 50 μm), welded onto the external electrode (often aluminium) by thermocompression. The electrode diameter, less than 1 mm, fixes the capacitance of the transducer and its piezoelectrically active part.

Characteristics. The frequency response of a delay line can be predicted from one of the 1-dimensional models for the transducer described in Chap. 1. Its impedance at the centre frequency is essentially made up of its radiation resistance and static capacitance. The following table indicates the domain of operation of this kind of line.

Centre frequency f_0	0.5–10 GHz
Relative bandwidth	$\leq 50\%$
Time delay τ	0.5–10 μs
Temperature dependence of delay	
(−50 to 120°C) $(1/\tau)\mathrm{d}\tau/\mathrm{d}\theta$ (Al_2O_3)	25×10^{-6} K^{-1}
Standing wave ratio	≤ 2
Radiated signal level	−30 dB
First echo level	−15 to −30 dB
Insertion loss	20–80 dB

The last few figures include:

- transduction loss, which increases with relative bandwidth;

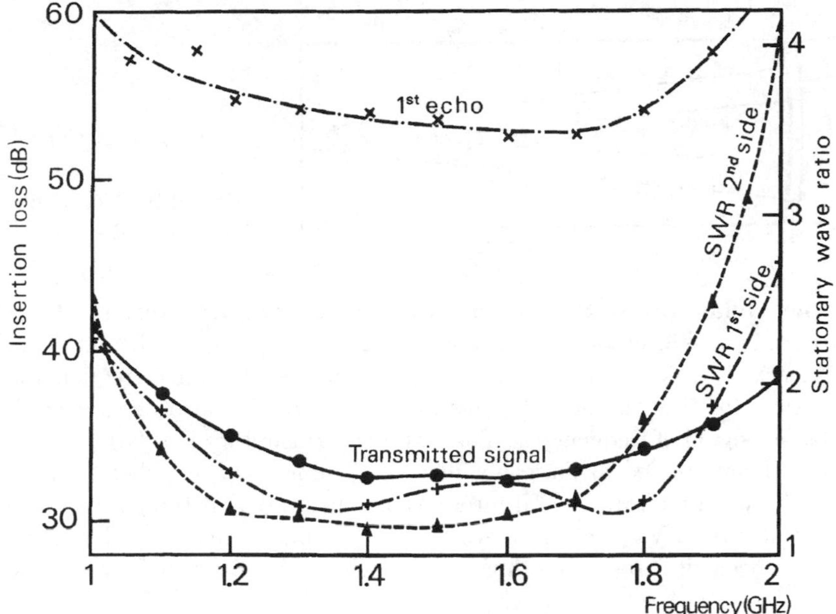

Fig. 4.8. Frequency response of a 2 μs delay line with centre frequency 1.5 GHz

- propagation loss proportional to the delay and the square of the frequency;
- diffraction loss (beam spreading), a function of transducer dimensions, frequency and delay (Sect. 1.3.2, Vol. I).

Figure 4.8 gives an example of the frequency response of a 2 μs delay line with centre frequency 1.5 GHz. Losses amount to 33 dB for a bandwidth of 700 MHz. The standing wave ratio (Sect. 1.1.2.1, Vol. I), at the input and output, is less than 2. The first echo spurious signal has level −20 dB relative to the transmitted signal. The unretarded signal, reaching the output by direct radiation, is at −30 dB of the useful signal.

Figure 4.9 shows a delay line based on reflection. Two transducers are fixed onto the concave end of the line. One emits the elastic signal, which reflects off the polished, plane end, and possibly also off the concave end, before being detected by the receiver transducer.

One variant delay line is the *echo delay line*, which contains a single transducer and has one polished, free end. In addition, shaping the ends of the propagating solid so that they focus the beam at each reflection and reduce echo, extremely long delays are accessible, but at the expense of bandwidth. For example, a 100 μs delay can be obtained for a bandwidth of order 5% ($f_0 = 1.5$ GHz, losses 55 dB).

These bulk wave delay lines have applications in radar (calibration, jamming by false echo emission). However, surface wave delay lines now compete for centre frequencies up to a few GHz.

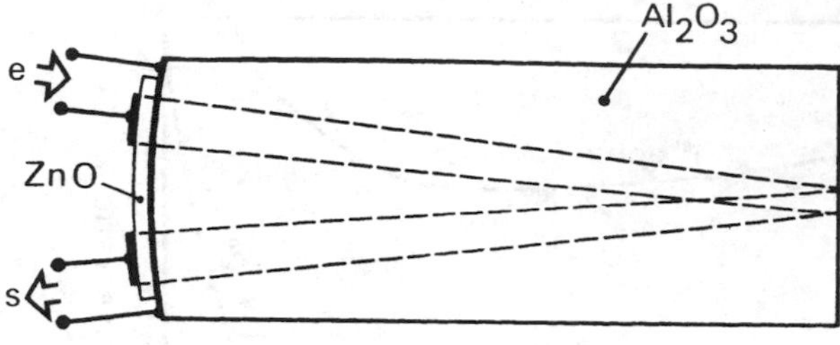

Fig. 4.9. Delay line with reflecting end

Variable Delay Lines Based on Elasto-optic Interaction. In Bragg conditions, light diffraction by elastic waves produces a single deflected beam of frequency $F - f$, where F and f are light and elastic wave frequencies, respectively. Beating the deflected beam with a light beam of frequency F gives back a signal of frequency f. The latter signal is retarded relative to the electrical signal applied to the transducer, by a time $\tau = l/V$, where l is the distance between transducer and interaction zone, and V is the elastic wave propagation speed. Varying l changes the delay. Continuously variable delay lines have been designed according to the schematic diagram in Fig. 4.10 [4.8].

The laser beam is split and the diffracted part of one beam superposed upon the transmitted part of the other. Only one of the two resulting beams is received at a photodiode. The current intensity is proportional to the square of the amplitude and contains the term of the same frequency as the signal applied to the transducer (Sect. 3.2.2.2). The laser beam is fixed, so the delay varies continuously as the crystal is moved.

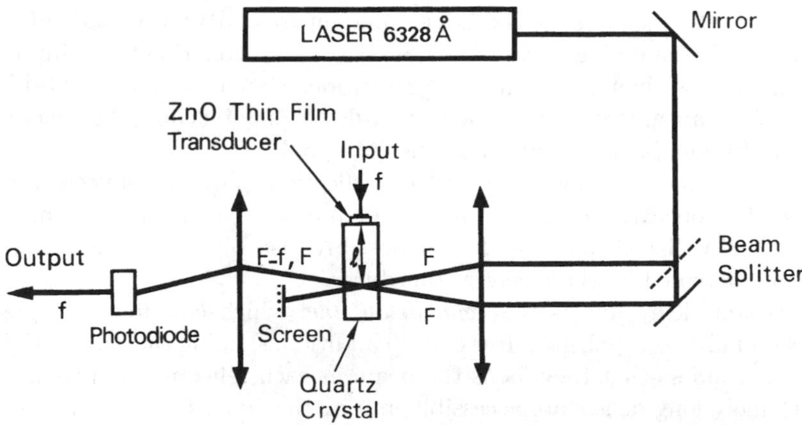

Fig. 4.10. Continuously variable delay line. The delay of the output signal relative to the electrical signal at the transducer is $\tau = l/V$, which can be continuously varied by moving the crystal

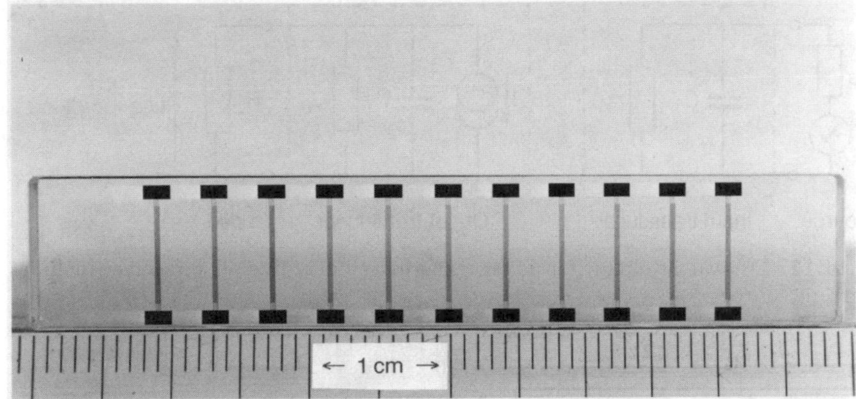

Fig. 4.11. 10-output Rayleigh wave delay line. Each transducer has 8 fingers of width 26 μm

The delay can vary from several tens of ns to several μs. Losses are rather high, of course, being several tens of dB for centre frequencies of several hundred MHz, and hence, bandwidths of 200–300 MHz.

4.2.2 Rayleigh Wave Delay Lines

The fact that Rayleigh waves propagate at solid surfaces, and that many geometries are available in the design of interdigital transducers used to generate and detect them, has led to a great variety of delay lines. Because the waves are accessible at the surface, multiple output lines can be built. The desired frequency response is obtained via the transducer geometry.

The substrate is a piezoelectric monocrystal, such as quartz (Y- or ST-cut and X-propagating), or lithium niobate (Y-cut, Z-propagating), if temperature dependence of the delay presents no particular problem. The delay is 1 μs for a 3.15 mm path on quartz and a 3.49 mm path on lithium niobate. Figure 4.11 shows one of the first multiple output quartz delay lines (10 outputs, $f_0 = 31$ MHz, bandwidth > 6 MHz). It was built in order to demonstrate the possibility of surface sampling and hence show that this type of delay line could generate a coded signal (Sect. 4.5.4). Each transducer has four pairs of fingers of width 26 μm. Losses are around 35 dB and vary by only a few dB between the first and tenth receiver (because the substrate is only weakly piezoelectric and lightly loaded mechanically).

When certain characteristics are required of the delay line, we must take into account those effects investigated when discussing the various transducer models (e.g., piezoelectric regeneration, mechanical reflection, diffraction). The relative importance of these effects will depend on the type and length of substrate, and also on the shape and duration of the signal to be delayed. We must also allow for effects caused by introducing circuits at the input

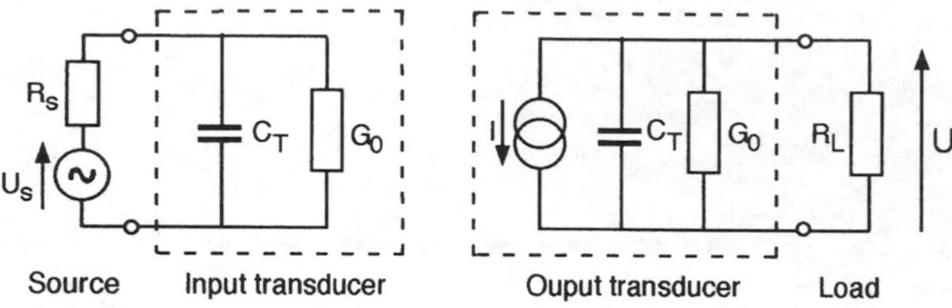

Fig. 4.12. Equivalent circuit for a two-transducer delay line. The receiver includes a current source

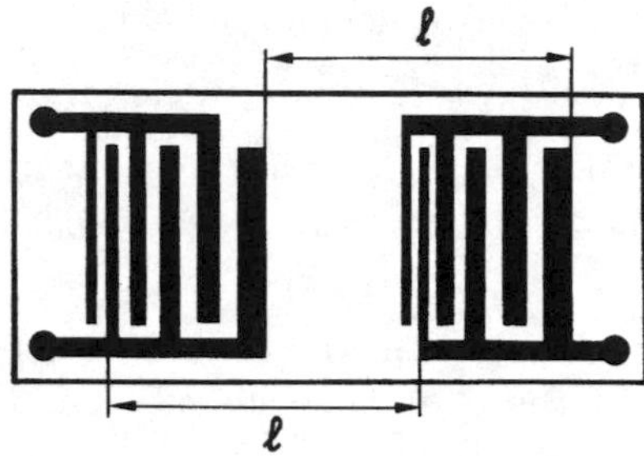

Fig. 4.13. Broadband delay line. Translation through distance l superposes the two transducers. The delay is l/V_{R}

and output of the delay line. Let us consider a delay line with two simple transducers, whose equivalent circuit is shown in Fig. 4.12.

Close to the centre frequency, the impedance of each transducer is made up of a capacitance C_{T} and a conductance $G_0 = \pi^2 N K^2 C_{\mathrm{T}} f_0$ (Fig. 2.15). Effects of the input transducer capacitance can be compensated by adjoining an inductance L. The resonant circuit thereby constructed has quality factor $Q_{\mathrm{e}} = C_{\mathrm{T}}\omega_0/G_0 = 2/\pi N K^2$. Using the results at the end of Sect. 2.2.2, the maximal bandwidth of the delay line is proportional to the electromechanical coupling coefficient K ($\approx 1.1K$), and the optimal number of finger pairs is inversely proportional to K. From Fig. 5.27, Vol. I, the bandwidth for a quartz substrate ($K = 4\%$) is narrower than for a lithium niobate substrate ($K = 21.9\%$), but the required number of finger pairs is greater ($N_{\mathrm{opt}} = 19$ as compared with $N_{\mathrm{opt}} = 3.5$).

Far from the synchronism frequency, input impedance is that of a first order system (RC), characterised by a frequency response of slope 20 dB per decade [4.9].

Broadband delay lines (50%) have been built with the geometry shown in Fig. 4.13. The interdigital distance varies but the output transducer is deduced from the input transducer by a translation defining the delay, which is identical for all frequencies.

Rayleigh waves propagate with reasonable losses on curved surfaces whose radius of curvature is much greater than the wavelength. This property has been exploited in wrap-around delay lines (Fig. 4.14) and helical delay lines.

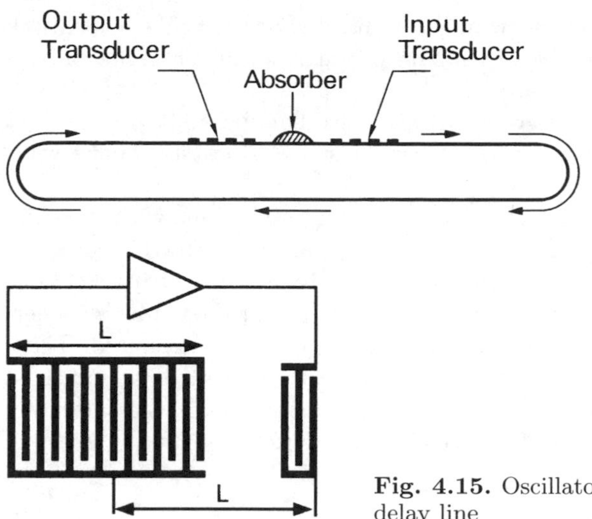

Fig. 4.14. Wrap-around delay line. The radius of curvature is much greater than the wavelength

Fig. 4.15. Oscillator based on a Rayleigh wave delay line

In principle, large delays (up to 1 ms) are accessible using materials with low propagation speeds. A 10 cm loop on a (111)-cut $Bi_{12}GeO_{20}$ crystal, with propagation along $[1\bar{1}0]$ at speed 1 708 m/s, produces a delay of 58.5 μs. The wave can be guided by a strip deposited on the surface, in order to reduce beam spreading and the resulting losses (Problem 5.6, Vol. I).

Rayleigh wave delay lines are often used as the main element in oscillators ($f \geq 1$ GHz). This is illustrated in Fig. 4.15. The delay line–amplifier system oscillates if the amplifier gain compensates line losses and, in addition, every signal from any point on the loop returns there in phase. Distance L is chosen equal to the transducer length so that the phase condition is only satisfied at the transducer synchronism frequency [4.10]. Compared with classical oscillators based on bulk wave resonators [4.11], these oscillators have the advantage of operating at higher frequencies (> 1 GHz). There are also oscillators based on surface wave resonators (Sect. 4.3.4).

4.3 Surface Wave Filters

The bandpass filters studied here are those which only let through signals whose frequency lies within a certain range (f_{min}, f_{max}). From a technological point of view, there is a wide variety of such filters [4.12]: passive filters with LC reactive quadripoles, active filters with operational amplifiers and RC circuits (the amplifier serves to eliminate inductances which are bulky and expensive at low frequencies), filters with mechanical, electromechanical or electromagnetic resonant cavities (L and C elements are distributed and not localised, as in an LC unit), transversal filters involving signal sampling and weighted sample summing (non-recursive digital filters, charge transfer filters), recursive digital filters (which contain a loop ensuring a feedback effect: the output signal is a weighted sum of previous elements, not only of the input signal, but also the output signal).

The following sections deal with surface wave bandpass filters. The first three discuss travelling wave filters and the last describes stationary wave filters.

A simple travelling wave filter contains a transmission line with interdigital emitter transducer and interdigital receiver transducer. In this category, we must distinguish those filters, generally broadband, in which reflections are eliminated (or pushed out of the useful band), and those filters, generally narrowband, in which such reflections are deliberately exploited. These filters behave differently. The first behaves like a *transversal filter*, at least according to the discrete source model. Indeed, it operates in the same way as the structure shown in Fig. 2.10, which has impulse response:

$$h(t) = \sum_{n=1}^{N} a_n \delta(t - t_n) . \tag{4.9}$$

Its frequency response $H(f)$ is the Fourier transform:

$$H(f) = \sum_{n=1}^{N} a_n \mathrm{e}^{-2\pi \mathrm{i} f t_n} . \tag{4.10}$$

Physically, each finger pair of a transducer constitutes a source of waves. Their amplitude is determined by finger overlap length and their phase by position relative to the receiver (reduced here, by hypothesis, to two fingers). Response $h(t)$, like $H(f)$, is therefore the sum of contributions from the various sources. It results from interference between travelling waves emitted by these sources. Depending on whether interference is destructive or constructive, the output signal is zero or maximal. The frequency response has no poles since there are no resonances (arising from energy exchange between two reactive elements), but it does have zeros.

The above observation does not apply to the travelling wave filter if reflections from fingers are taken into account. Indeed, each transducer behaves partially as one or several cavities.

The filtering mechanism in stationary wave filters (that is, resonators: LC units or other cavities) can generally be explained by frequency domain arguments. Large impedance variations between source and load, which serve to favour the useful signal and reduce any others, are produced by resonances, at frequencies chosen either in a single cavity or in several, connected in series or parallel. Consequently, transfer functions for these ideal filters include both zeros and poles and take the form of a ratio of polynomials:

$$H(s) = K \frac{\prod_{i=1}^{r}(s - z_i)}{\prod_{j=1}^{n}(s - p_j)} , \quad \text{with} \quad r < n \text{ , i.e., } H(s \to \infty) = 0 .$$

The impulse response $h(t)$ is found by taking the Laplace transform of $H(s)$. Depending on their orders, it contains terms $c_j e^{p_j t}$, $c_j t e^{p_j t}$, ... from the poles, and their derivatives from the zeros [4.13].

These comparisons are made in order to find out which methods previously developed for classical filters, with localised or distributed elements, can be applied to elastic wave filters.

However, as the range of applications is extended, it has been found necessary to develop techniques specific to surface wave devices (see Sect. 2.5). Finer analysis has also been required, to take second order effects into account (e.g., reflection, finger mass loading).

4.3.1 Filters with Bidirectional Transducers

The desired frequency response is defined, on the whole, by the emitter or the receiver structure, or both. In the simple case where the receiver, for example, has few fingers and hence wide bandwidth, and those fingers have constant length greater than the width of the elastic beam from the emitter, the impulse response and therefore also the frequency response are determined by the emitter structure (Fig. 4.16).

In principle, a $\sin x/x$ variation in the finger overlap length of the two comb electrodes constituting the emitter produces a rectangular frequency response. Indeed, for a box-shaped frequency response of width B, centred on frequency $\pm f_0$,

$$H(f) = \frac{1}{2}\Pi\left(\frac{f-f_0}{B}\right), \quad f > 0, \tag{4.11}$$

the impulse response is

$$h(t) = B\frac{\sin \pi Bt}{\pi Bt}\cos 2\pi f_0 t\,. \tag{4.12}$$

Using the method of Sect. 2.1.2, discrete sources are distributed at times t_n such that

$$\phi(t_n) = 2\pi f_0 t_n = n\pi \quad \Rightarrow \quad t_n = \frac{n}{2f_0}\,,$$

hence at equidistant abscissa points $x_n = n(V_\mathrm{R}/2f_0)$, and their amplitude is proportional to

Fig. 4.16. Schematic diagram of a filter with two in-line transducers. The finger overlap length for one transducer varies so as to produce the desired frequency response

$$A_n = \frac{\sin(n\pi B/2f_0)}{\pi B/2f_0} \,. \tag{4.13}$$

It is also possible to vary finger lengths, according to the $|\sin x/x|$ rule, in just one of the two comb electrodes constituting the emitter. Phase inversions at the zeros of $\sin x/x$ are effected by separating two adjacent fingers on the same electrode by half a wavelength.

This type of elementary filter is quite satisfactory if internal reflections inside the transducer are reduced, e.g., by splitting fingers, if also the phase speed is made uniform by means of dummy fingers (Sect. 2.1.1), and if the receiver load does not give rise to significant triple transit echo. The envelope of the impulse response correctly reproduces the transducer pattern. However, the frequency response naturally deviates from the box function and contains oscillations (ripple) because the $\sin x/x$ function is only determined by a limited number of lobes (Fig. 4.17a). Since every transducer has limited length, and its impulse response is necessarily finite, a truly rectangular frequency response cannot be achieved. Increased transducer length leads to a passband with steeper flanks (the product of their width, the transition zone, and the duration of the impulse response remaining constant). Ripple is not suppressed, however, but approaches the sides of these flanks.

One way of reducing the amplitude of these oscillations, although to the detriment of flank steepness, is to use a window function with less steep slopes (Fig. 4.17b). This is done by varying finger overlap length according to a classical weighting function (e.g., Hamming, Dolph–Chebyshev or Kaiser functions).

The bandpass filter with symmetric frequency response is made with constant interdigital distance because the phase of its impulse response increases linearly. Techniques used for digital filters can therefore be exploited insofar as interactions between sources are negligible. Bandpass filters with non-symmetric frequency response require non-constant intervals, and digital filtering techniques cannot be applied. In this case, we must have recourse to one of the methods mentioned previously, depending which characteristics we seek with regard to losses, bandwidth, rejection and so on. These methods are: impulse response, simple or complex equivalent circuits, coupled modes, and the P-matrix. Whichever is chosen, filter design consists in determining finger positions and overlap lengths for neighbouring fingers in such a way as to produce the sought impulse response.

Figure 4.18 shows an example of frequency responses obtained using filters with bidirectional transducers.

Apart from transducer arrangement and weighting technique (Sect. 2.2), other aspects must be considered. One is the choice of material (quartz, lithium niobate or tantalate), fixed by temperature stability and bandwidth requirements. Another is input and output impedances, and the way they are matched to those of the signal generator and load. Substrate dimensions and time stability in specific operating conditions may also be relevant. Many

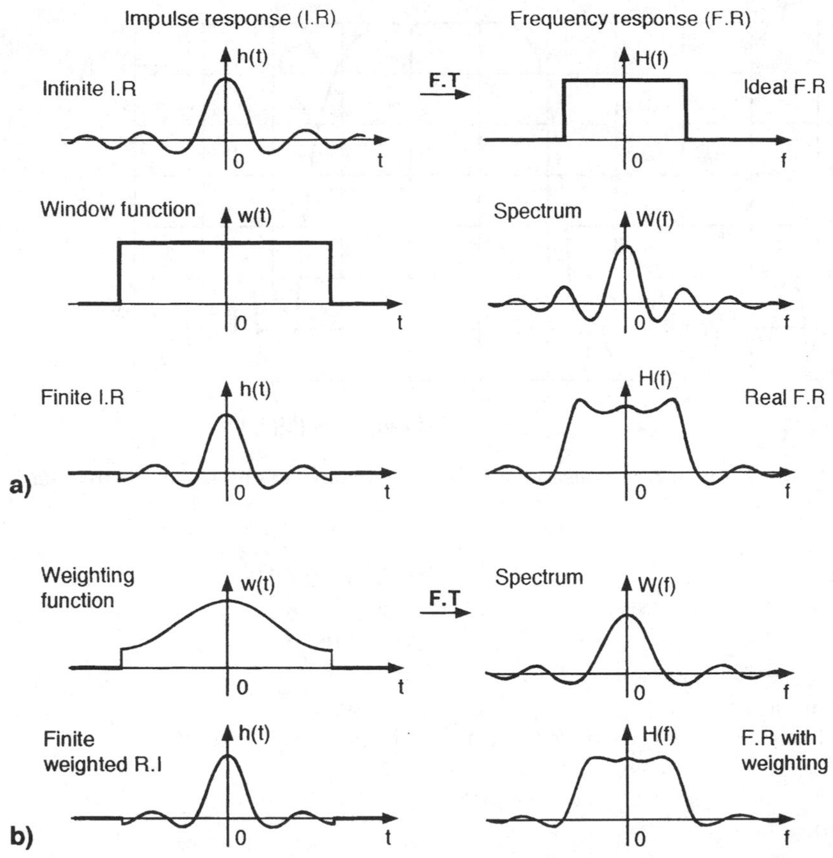

Fig. 4.17. Impulse response and frequency response. (**a**) $\sin x/x$ curve, defined by a limited number of lobes, arises from the simple product of an infinite $\sin x/x$ curve with a box function (window function). The Fourier transform of this simple product is the convolution product of the Fourier transforms of each term. Hence, corresponding to the finite impulse response $h(t)$ of a real transducer is a frequency response whose shape grows closer to a box function as the number of lobes of $h(t)$ increases. However, oscillations persist both inside and outside the passband. (**b**) One way of reducing these oscillations (ripple) is to use a window function with less steep slopes

models and computational programs have been established and improved to satisfy the main requirements [4.6, 4.14], since the first filters were made in 1969 [4.16]. They are often based on the Remez algorithm [4.15].

The operating domain of this type of filter (including variants described below) is as follows:

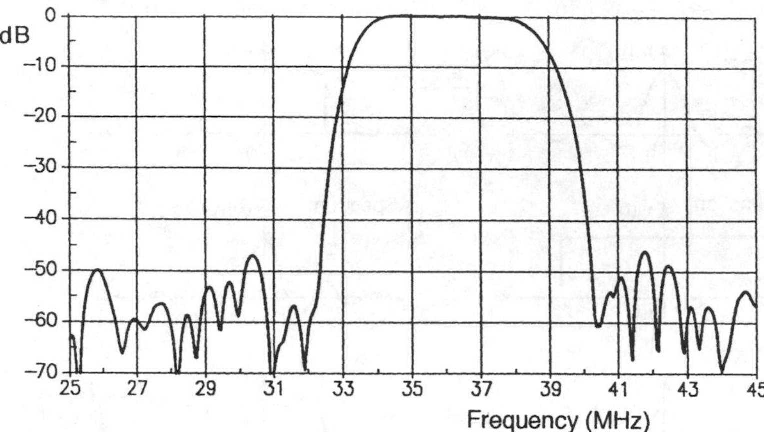

Fig. 4.18. Frequency response of a Siemens–Matsushita intermediate frequency television filter

Centre frequency	20 MHz–2 GHz
Relative bandwidth	$< 40\%$
Insertion losses	20–40 dB
Accessible rejection level	50 dB
Ripple (in-band)	± 0.2 dB
Group delay variations (in-band)	± 10 ns

Variants

There are of course other filtering configurations where transducers are arranged in this in-line way. Before giving examples, let us note that the weighting technique which consists in varying overlap lengths between neighbouring fingers is sometimes replaced, at least locally, and if the envelope of the required impulse response varies slowly, by another weighting method, viz., variation of the spatial density of sources [4.17]. Figure 4.19 refers to a filter (on ST-cut quartz) in which both weighting methods are used: selective withdrawal of the sources for the emitter (by earthing adjacent fingers), and variation of overlap length for the receiver.

Double Receiver. The emitter transducer is placed between two receiver transmitters mounted in parallel. Conversion losses are reduced because both beams, emitted in opposite directions, are detected.

Double Reflector. One reflector is placed at each end of the filter. The one to the left of the emitter returns the beam which is not usually used to the receiver on its right. The reflector to the right of the receiver returns those waves which have passed through it. The presence of these reflectors reduces the filter bandwidth to varying degrees, depending on their efficiency.

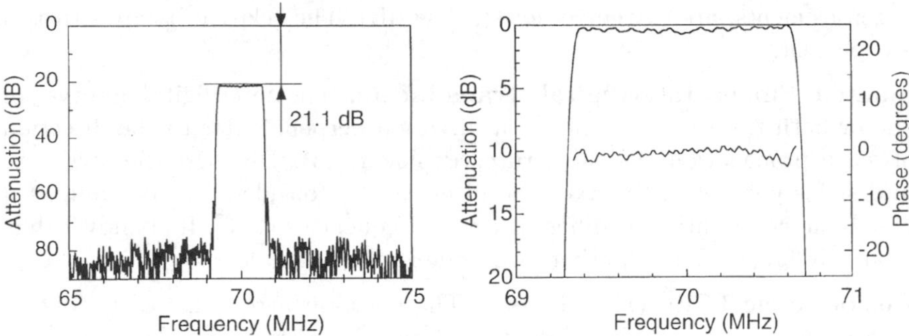

Fig. 4.19. Frequency response of an intermediate frequency filter for a mobile telephone base station. Kindly communicated by J.M. Hodé, Thomson Microsonics

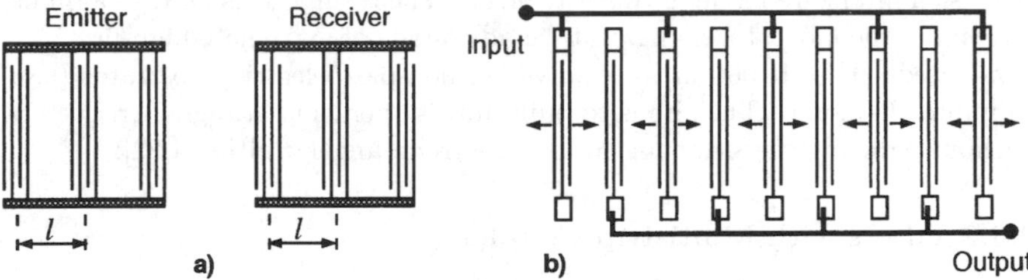

Fig. 4.20. (**a**) Filter composed of two identical ladder transducers. (**b**) Filter with interdigitated interdigital transducers

Transducers with Ladder Structure. Each transducer is made up of elements like those in the delay line of Fig. 4.11, spatially separated but connected together [4.18]. In the simplest case, each transducer looks like a ladder whose rungs are the equidistant elements (Fig. 4.20a). Let us assume that the elementary transducers (the rungs) are the same and that the emitter consists of a single rung. Suppose also that the constant distance between receiver elements is $n\lambda_0$, where n is a whole number and $\lambda_0 = V/f_0$ the synchronism wavelength. Voltages (U_1, U_2, U_3) induced by any sinusoidal signal applied to the emitter, across the terminals of the three receiver elements, add vectorially. Depending on whether they are in phase (at frequency f_0 and its multiples), $\pm\pi/3$ or $\pm\pi$ out of phase, the resulting voltage is $3U_1$, 0 or U_1, respectively. The shape of the frequency response is a large arc with a smaller arc on each side. The ratio of their heights increases with the number of receiver elements, and of course, the number of emitter elements.

Interdigitated Interdigital Transducers. This arrangement is obtained from the last by placing one emitter element between two receiver elements, as shown in Fig. 4.20b [4.19]. A reflector may also be placed at each end of the filter. The two beams from each emitter (except the end ones) are therefore sensed by receivers, and these receive one beam at each of their two ports. Losses in this type of filter, which generally contain more than a dozen elements, are extremely low (a few dB). The relative bandwidth is a few percent.

Slanted Finger Interdigital Transducers. The interdigital interval of one or both transducers varies in a direction perpendicular to the direction of wave propagation (fan structure, Problem 4.3). Dividing the structure into strips parallel to the axis, the filter can be considered to be composed of sub-filters operating at different centre frequencies [4.20]. Its bandwidth is relatively broad; the weighting technique is more flexible.

Piezoelectric Film Transducers. The question here is no longer one of transducer geometry, but rather of the type of substrate. The latter is some cheap and ordinary material such as a glass, coated with a crystallographically orientated piezoelectric thin film, usually zinc oxide. Medium frequency television filters are made in this way [4.21]. The technique is used to a different end in the laboratory, viz., to take advantage of extremely high propagation speeds ($V > 10\,000$ m/s) of waves in non-piezoelectric substrates (e.g., sapphire, diamond). The aim is to build filters operating at higher frequency, without reducing the wavelength, i.e., electrode finger widths [4.22].

4.3.2 Filters with Multistrip Coupler

When filter specifications demand a very low level of rejection, but allow temperature variations in its characteristics, compatible with the coefficient of a strongly piezoelectric material such as lithium niobate, a multistrip coupler can be inserted between the two transducers. Of course, this does tend to make the filter less compact.

This coupler, comprising two channels and of length L_t, was described in Sect. 4.1. It serves the following purposes: to suppress any bulk waves which may be generated in the input transducer; and to deliver a uniform beam at the receiver, whatever the source dimensions (finger overlap length in the emitter transducer). In consequence, weighting is possible at the receiver. The frequency response of the filter is then the product of frequency responses of the two transducers. If they are identical, the impulse response $h(t)$ of each transducer is the Fourier transform of $\sqrt{H(f)}$. Of course, the coupler operates well in the filter only outside cutoff. Such an element is used when rejection requirements are fairly severe (> 55 dB).

The coupler with 3 dB partial transfer, of length $L_\mathrm{t}/2$, is particularly useful in suppressing (triple transit) echoes arising from reflection at the receiver. Indeed, if a transducer more or less identical to the receiver is placed at the output of channel A, as shown in Fig. 4.21, it reflects waves in the same way as the receiver. The two reflected beams, of the same amplitude but $\pi/2$ out of phase (Problem 4.1), are recombined in channel B of the coupler, into a single beam which is absorbed there.

The frequency response of such a filter is shown in Fig. 4.22. The filter is made on lithium tantalate, with a 240-strip coupler and reflector weighted

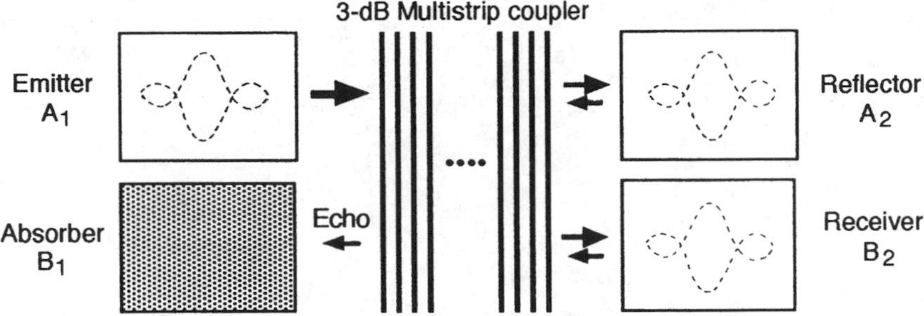

Fig. 4.21. Reflections at the output transducer (triple transit signals) are cancelled by introducing a 3 dB coupler between emitter and receiver transducers and inserting a reflector transducer. From [4.23]

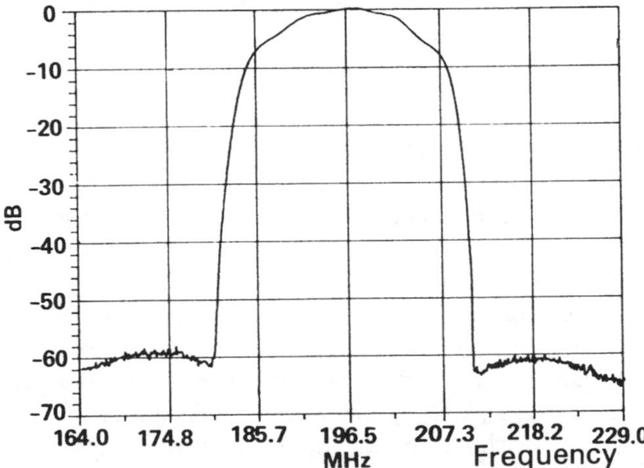

Fig. 4.22. Response of a filter containing a 3 dB coupler equipped with a reflector chosen to suppress the triple transit signal. Taken from [4.24]. ©1982 IEEE

by selective finger withdrawal [4.24]. The filter is designed according to the following specifications: centre frequency 196.7 MHz, insertion losses 20 dB, Gaussian passband shape, bandwidth 10.9 MHz at 2 dB, 21.8 MHz at 8 dB, and 32.7 MHz at 50 dB, triple transit −50 dB, rejection 60 dB.

The multistrip coupler can be used in other arrangements [4.25], to transform an ordinary transducer into a unidirectional transducer [4.26], and to narrow or broaden a wave beam [4.27].

4.3.3 Filters with Unidirectional Transducers

The advent of unidirectional transducers (Sect. 2.5.3) led to filter designs involving much lower losses, compared with the previous type of filter, and also with much weaker triple transit signal. As already mentioned, losses from

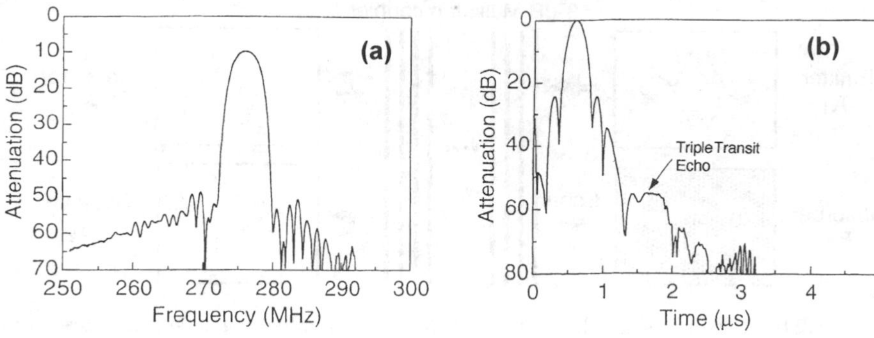

Fig. 4.23. (**a**) Frequency response and (**b**) impulse response of an ST-cut quartz filter (Thomson Microsonics) with SPUDT-type transducers (Sect. 2.5.3). Losses are low (10 dB) and yet the triple transit level is also very low (< -55 dB)

any filter made with ordinary transducers, emitting in both directions, are of course greater than 6 dB. In practice, they are always at least 15–20 dB. The variants described in Sect. 4.3.1 reduce these losses, but generally increase filter dimensions, and their design is made more difficult when secondary effects are taken into account. The transducers we discuss now are intrinsically unidirectional.

Figure 4.23 shows the frequency response ($f_0 = 276$ MHz) and impulse response of a close ST-cut quartz filter, with two transducers. These are made unidirectional by varying finger widths and weighted by selective removal of sources. Losses are 10 dB and the spurious triple transit level (-55 dB) registered in the time response is extremely low (a level which is not accessible using simple transducers, without incurring losses above 20 dB – see Fig. 2.40).

Figure 4.24 shows the frequency response of a filter on a lithium niobate crystal, where the two SPUDT transducers possess fan geometry (i.e., slanted fingers). The substrate and this geometry produce relatively high bandwidth with low losses.

Figure 4.25 is the frequency response of an ST-cut quartz filter in which transducers have partially resonant structure (RSPUDT). Resonance effects are revealed through the inverted curvature of the flanks in this response. This structure, rather than the substrate, explain low losses here. There is a further advantage: filters designed in this way are two or three times less bulky than those based upon a simple SPUDT technique. This last point is becoming more and more relevant in design of portable devices such as telephones.

Globally speaking, these filters, still an active area of research, have the following characteristics:

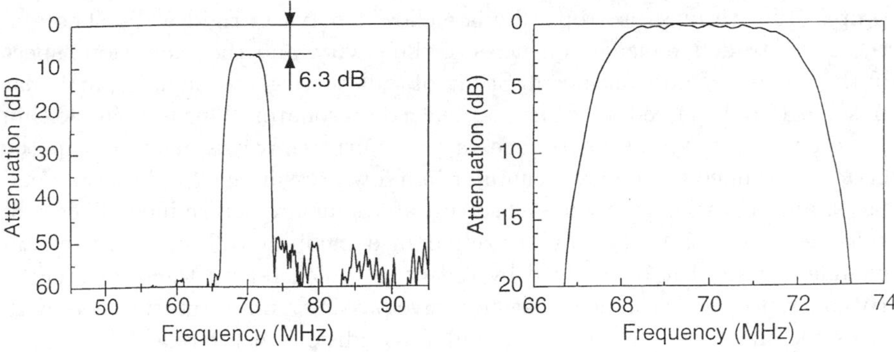

Fig. 4.24. Frequency response of a filter built from SPUDT-type unidirectional transducers (Sect. 2.5.3) with slanted fingers

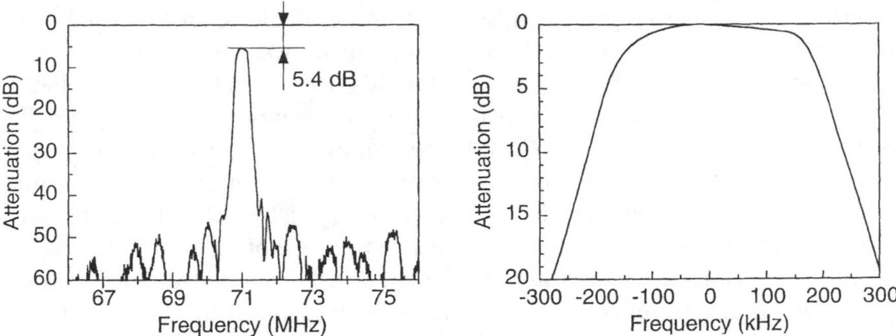

Fig. 4.25. Frequency response of an ST-cut quartz filter in which transducers have partially resonant structure (R-SPUDT). Losses and volume are reduced (by a factor of 2 to 3) by exploiting internal resonances

Centre frequency	40–1 000 MHz
Relative bandwidth	$< 10\%$
Insertion losses	< 10 dB
Accessible rejection level	50 dB
Ripple (in-band)	± 0.5 dB
Group delay variations (in-band)	± 50 to ± 500 ns

Other structures exist for travelling wave filters: filters with slanted groove gratings (Sect. 4.5.3.2); filters built from delay lines mounted in parallel and operating at different centre frequencies (Problem 4.2). Two configurations are given in Problems 4.4–5.

4.3.4 Stationary Wave Filters

A stationary wave filter is made from one or more resonators (resonant cavities). Resonators were discussed in Sect. 1.1.2, Vol. I, when investigating wave propagation through isotropic fluids enclosed between rigid walls. The aim there was to define eigenfrequencies of the cavity, and the excitation aspect of the waves was not considered. Study of bulk wave generation, in particular in solids (Chap. 1), led us to the piezoelectric resonator. This is a simple thin plate or rod of crystal whose faces act as mirrors. Waves generated in the crystal (assumed to be in a vacuum or in a low pressure gas) reflect from end faces, and a stationary regime is set up at frequency determined by crystal thickness (Sect. 1.4.3.1). Waves excited in a medium will therefore remain confined there, if it is bounded by reflectors which return them coherently. By coherently, we mean here that any wave passing a point returns there with the same phase, having satisfied boundary conditions at the reflectors. This is a quite general observation. It is thus possible in principle to make resonators for surface waves as well as for bulk waves. However, bulk wave resonators appeared in 1920 (see the chronology at the beginning of Vol. I), and have undergone much development since then, as we shall see in Sect. 4.4. The first studies of surface (Rayleigh) wave resonators were only undertaken in 1970 [4.28]. There are two explanations for this delay. Firstly, the interdigital transducers required to excite them were only invented in 1965. Secondly, since these waves contain phase shifted components (in the simplest case, two components $\pi/2$ out of phase), their reflectors are complex. They cannot reflect from a single discontinuity without being distorted (although transverse surface waves of Bleustein–Gulyaev type are not in principle subject to this restriction). Much research in design and fabrication of surface wave resonators, and ways of coupling them, has been inspired by attempts to account for reflections inside interdigital transducers (Sect. 2.1.1), to modify the direction of waves using arrays, and to produce very low loss filters. By coupling resonators in various ways, broader passband filters can be made. If a filter is constructed with two LC units, the bandwidth can be modified by varying the distance (coupling) between the two inductances. Reflector characteristics have already been examined in Sect. 2.5.1. Compared with bulk wave resonators, the surface wave resonator has the advantage of being less sensitive to the way it is fixed, as well as operating at higher frequencies (several [GHz]) and coupling in various ways.

4.3.4.1 One or Two-Port Resonator. Figure 4.26 is a schematic representation of a 1-port Rayleigh wave resonator. Waves launched by transducer T are reflected at mirrors R. The latter are gratings made of metallic strips or grooves (groove resonators perform better but are not produced in such large quantities, owing to their higher cost). Assuming the transducer to be made of a few weakly reflective fingers, stationary wave conditions are established at a frequency depending on both the period $\lambda/2$ and the spacing of the grating reflectors. The reflection centre of the gratings, assumed identical, is located at a position given by (2.149). Positions of transducer fingers and elements of the two gratings must correspond to the predicted distribution of vibrational antinodes and nodes, e.g., of the electrical potential.

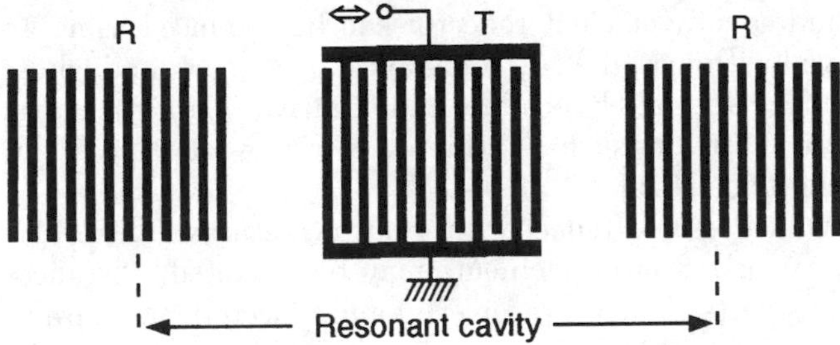

Fig. 4.26. Rayleigh wave resonator. The substrate is piezoelectric. Two mirrors R, groove or metallic strip gratings, reflect surface waves launched by transducer T. The length of the equivalent cavity is greater than the distance between the two gratings

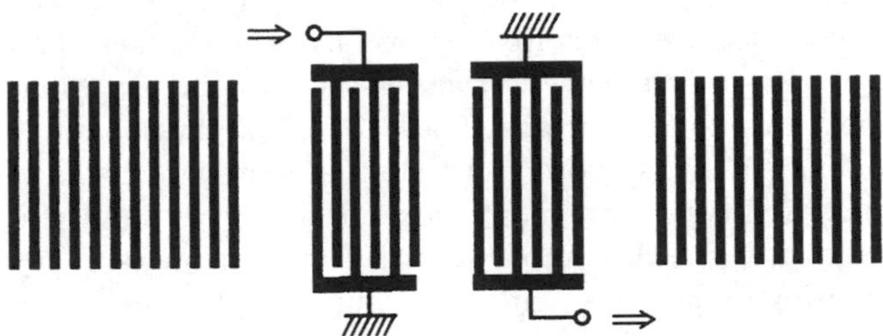

Fig. 4.27. Two-port Rayleigh wave resonator. One of the two transducers excites the cavity at a resonance frequency, whilst the other extracts energy

Transducer fingers are located at antinodes. Elements of the two gratings are placed at points determined by their state (zero potential if shorted). Given the values of reflection coefficients (for aluminium strips of width $\lambda_0/4$ and relative thickness $h/\lambda_0 = 0.01$ on ST-cut quartz, $r \approx -0.005$), there are a few hundred strips in each grating. They have widths of order $50\lambda_0$, only slightly different from the transducer width. The behaviour of this cavity is described in terms of the frequency dependence of its impedance, which can be calculated near the resonance frequency by means of a three-element circuit diagram.

Figure 4.27 is a schematic diagram of a 2-port (i.e., two-transducer) Rayleigh wave resonator. This resonator is studied with the help of circuit diagrams or more physical analysis (Sect. 2.6.2.3). Intuitively, we expect the frequency response to include a narrow peak corresponding to the cavity resonance, which adds to the response of the transducers [$(\sin x/x)^2$ for identical transducers].

The above description is incomplete. The following points need development.

- We need to understand the main cause, either electrical or mechanical, of reflection and the coefficient characterising it (both amplitude and phase).

Indeed, a shorter and equally efficient reflector can be produced using an array of elements, still with period $\lambda_0/2$, a distance $\lambda_0/4$ apart, and whose reflection coefficients are alternately positive and negative. The sign change is obtained on certain substrates such as lithium niobate by shorting or not shorting the metallic strips [4.29].

- Periods of the transducer(s) and reflectors are not necessarily tuned. The distribution of sources or reflecting elements may be weighted. Distances separating the various parts, emitter, receiver(s) and reflectors, are parameters which modify the frequency response. The resonant cavity can be localised within the transducer.
- A resonator can vibrate in modes with different frequencies, as Fig. 4.28 shows. Operation over two modes broadens the passband.

One further observation of a rather different kind will be important. The cavity structure also resonates when excited by surface waves other than Rayleigh waves, e.g., pseudo-surface waves, transverse waves and even longitudinal waves confined close to the surface [4.30, 4.31]. The crystal cut of the substrate is then different. It is chosen so that the emitter transducer radiates primarily the desired type of wave. Pseudo-surface waves are useful because of their much greater speed, compared with Rayleigh waves. They allow much higher operating frequencies, without requiring higher resolution technology.

4.3.4.2 Associations of Several Resonators. A filter is generally composed of several resonators which are mechanically or electrically coupled. There are two types of acoustic coupling: longitudinal coupling (in-line coupling) and transverse coupling (guide coupling) [4.32].

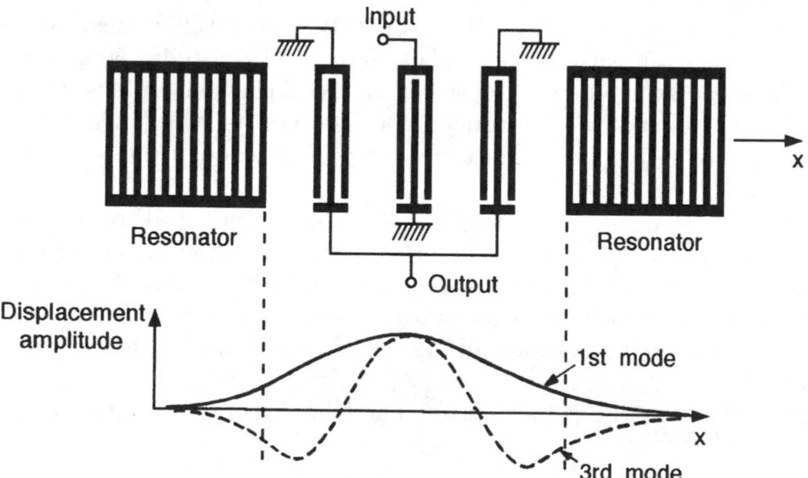

Fig. 4.28. Arrangement for exciting the first two symmetric longitudinal vibration modes of a resonant cavity. The second, antisymmetric, mode is not coupled

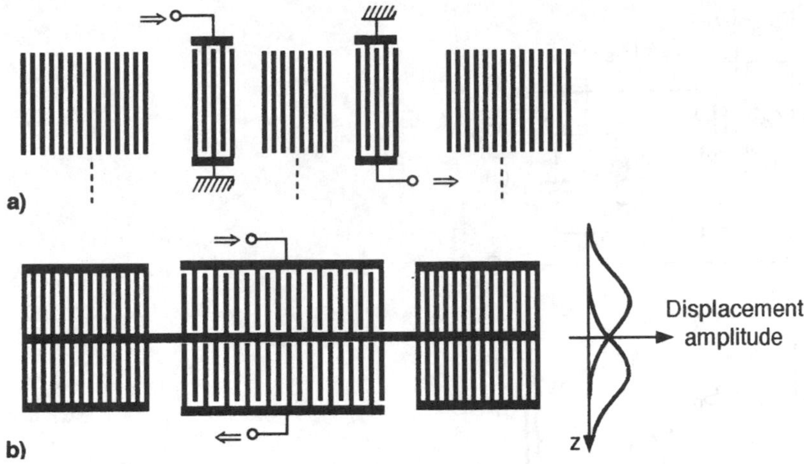

Fig. 4.29. Double-pole filter composed of two coupled resonators. (**a**) Longitudinal coupling. (**b**) Transverse coupling

Acoustic Coupling. Figure 4.29 shows schematically a double-pole filter comprising either two longitudinally coupled resonators, or two transversely coupled resonators. Transverse coupling, less visible than longitudinal coupling, is effected by evanescent waves from each resonator, considered as a guide. These evanescent waves propagate along the central metallic strip (see Problem 1.5, Vol. I). The arrangement should be compared with the monolithic filter in which two bulk wave resonators are coupled by the substrate (Sect. 4.4.2). Indeed, it is described by an almost identical equivalent circuit. The fact that the transducers are not in-line coupled, and hence that coupling is zero outside the stop band, means that these filters have very narrow relative passband (0.03–0.15%), very high rejection and very steep flanks. Filters with longitudinally coupled resonators have broader pass bands (a few percent) and low losses (< 3 dB), if made on a substrate with high coupling coefficient, such as $Y+36°$-cut X-propagating lithium tantalate, where $K^2 = 4.7\%$. 4-pole filters can be constructed by associating two of these filters.

Electrical Coupling. Apart from these filters, based on acoustic coupling of neighbouring resonators, there are also filters fabricated by electrically connecting independent resonators in ladder, lattice or bridge structures (Fig. 4.30). Such assemblages have been in use for a long time now with bulk wave resonators. In this type of filter, we exploit the frequency dependence of the electrical impedance of the surface wave resonator. It is similar to the frequency dependence of a bulk wave resonator (Fig. 1.15). The impedance, which is purely imaginary for a lossless resonator, is very low (a few Ω) at resonance and high (several kΩ) at antiresonance. Figure 4.30b represents the frequency dependence of the series resonator reactance $X_s(\omega)$ and par-

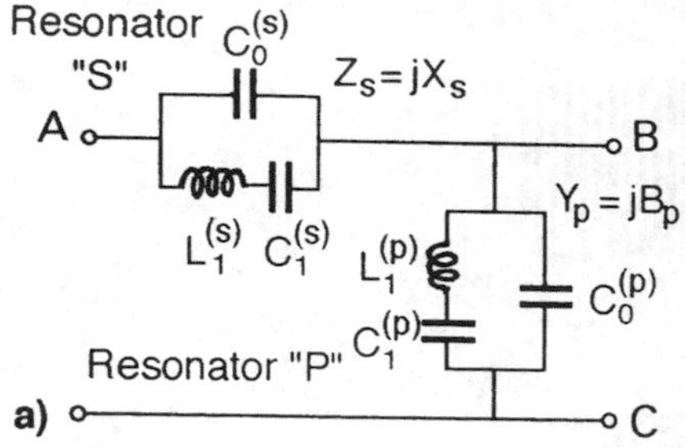

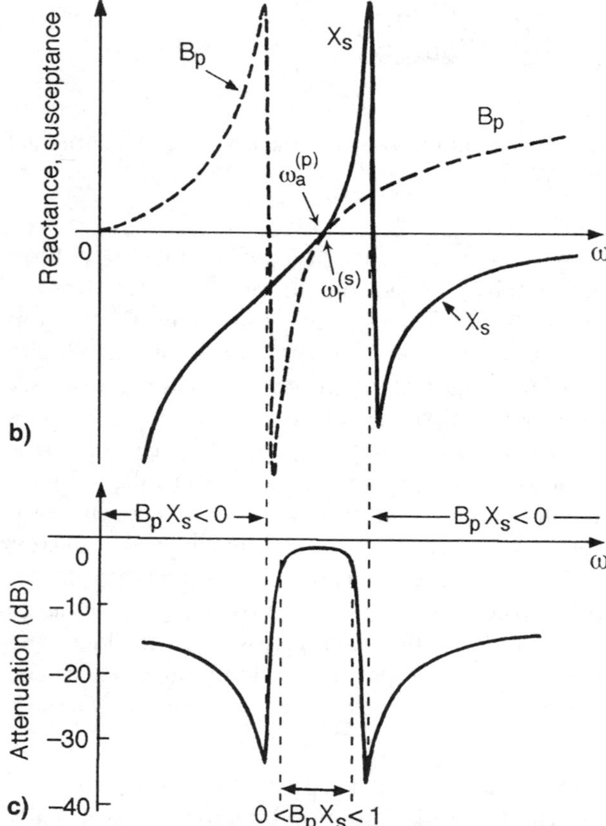

Fig. 4.30. (**a**) One unit of a ladder filter. (**b**) Resonance frequency of series resonator S is close to antiresonance frequency of parallel resonator P. (**c**) Frequency response of the unit

allel resonator susceptance $B_{\mathrm{p}}(\omega)$, for the ladder structure in Fig. 4.30a. It explains the shape of the frequency response (Fig. 4.30c) of one unit when the antiresonance frequency $\omega_{\mathrm{a}}^{(\mathrm{p})}$ of resonator P, connected in parallel between B and C, and the resonance frequency $\omega_{\mathrm{r}}^{(\mathrm{s})}$ of resonator S, connected in series, are almost equal. The passband is defined by the condition

$$0 < B_{\mathrm{p}}(\omega)\, X_{\mathrm{s}}(\omega) < 1\,.$$

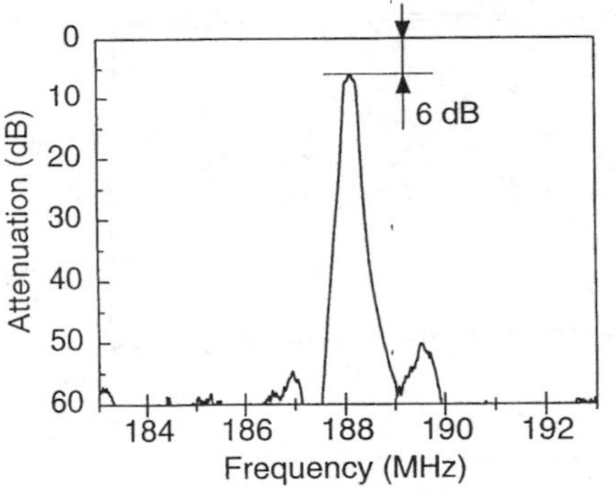

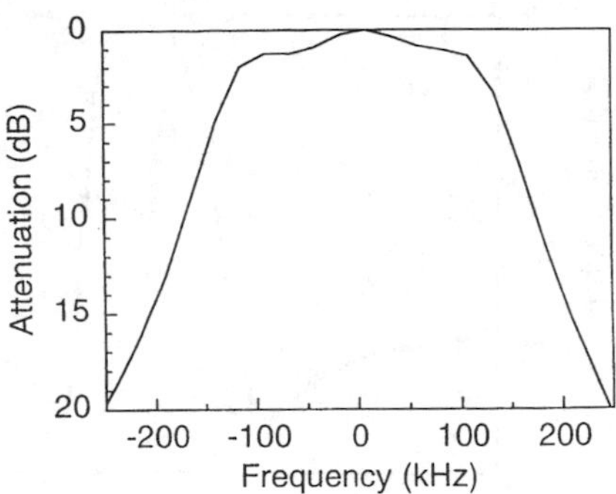

Fig. 4.31. 4-pole filter containing two double-pole elements [intermediate frequency Thomson Microsonics filter for the GSM (Groupe Spécial Mobile) system]. Each element comprises two transversely coupled resonators. Coupling between elements is controlled by means of an external inductance

On either side (stop band) $B_p(\omega)\,X_s(\omega) < 0$, and in the intermediate zone $B_p(\omega)\,X_s(\omega) > 1$.

Insertion losses for one ladder unit are small (1–2 dB), for there are neither propagation nor conversion losses. Outside the passband, transmittance depends on the ratio of static capacitances $C_0^{(p)}/C_0^{(s)}$. Rejection can be improved in this type of filter by cascading several units.

Impedance variations in a long transducer can also be exploited in designing this kind of filter [4.33].

4.3.4.3 Examples of Frequency Responses. Amplitude and phase of the impulse response of a resonator filter cannot be determined separately. This drawback relative to the transversal filter is balanced by an advantage, viz., lower losses. It can therefore operate at higher frequencies. Figure 4.31 shows the frequency response of a 4-pole filter ($f_0 = 188$ MHz) on ST-cut quartz, comprising two double-pole elements with transverse internal coupling (an inductance is introduced between the earth and the connection between elements to control external coupling). Such technology affords compact filters, but they have more restricted relative passband than filters using R-SPUDT technology.

Figure 4.32 is the frequency response of a (Motorola) filter with centre frequency 813.5 MHz. The usual characteristics of this type of filter, with ladder-mounted resonators, are as follows: between 806 and 821 MHz, losses 3 dB, ripple 1.5 dB, group delay variation 10 ns, rejection (outside passband) 50 dB.

Figure 4.33 shows the frequency response of a Fujitsu filter (also ladder structured) with centre frequency 2.45 GHz [4.34].

Considering their performances, cost and dimensions (the Fujitsu filter measures $3.8 \times 3.8 \times 1.5$ mm^3), surface wave filters constitute an essential step in the development of mobile telecommunications [4.35, 4.36]. Radio

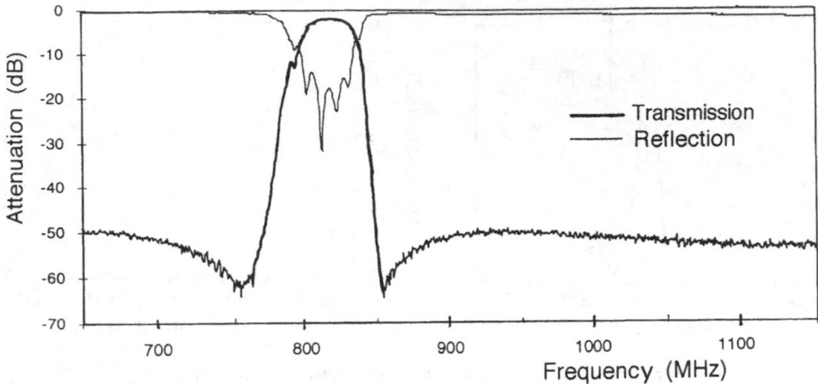

Fig. 4.32. Frequency response of a Motorola filter (f_0 = 813.5 MHz). Kindly communicated by F. Hickernell

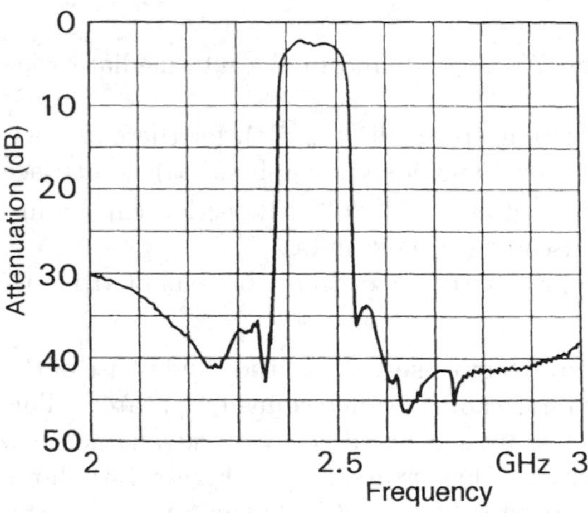

Fig. 4.33. Frequency response of a Fujitsu filter (f_0 = 2.45 GHz, bandwidth 97 MHz)

frequencies extend from 800 to over 2 000 MHz (e.g., GSM system: emission band 890–915 MHz, reception 935–960 MHz). Intermediate frequencies vary from a few tens to a few hundred MHz. Surface wave filters are involved in the various stages. Filter characteristics are, of course, a prerequisite for each system, and manufacturers (about 20 around the world) sell their products by catalogue.

Whichever structure is adopted, the choice of piezoelectric crystal and its cut are all-important. These determine temperature stability, losses and passband of the filter. The three fundamental parameters (electromechanical coupling coefficient, phase speed and temperature coefficient) corresponding to the various cuts are given in Fig. 5.27, Vol. I, for Rayleigh waves, and Table 2.1 for pseudo-surface waves. Because of its weak electromechanical coupling, quartz is used for relatively narrow passband filters ($< 5\%$). It has high temperature stability, and for this reason is favoured in intermediate frequency filters. Lithium niobate is reserved for broadband applications, in which its strong electromechanical coupling is an advantage and its large temperature coefficient creates no particular difficulties. Lithium tantalate, with intermediate properties, is suitable for medium bandwidth filters.

Filters must also be matched to both source and load. In mobile phone applications, where weight and size must be reduced as far as possible, transducer geometry is determined, allowing for the substrate, so that the passband impedance of the filter is 50 Ω.

4.4 Bulk Wave Filters

There are no filters based on travelling bulk waves. The stationary wave filters described in this section are therefore composed of resonators. The bulk wave resonator is not a recent invention. In the 1920s, it was already the essential element used in stabilising frequencies of radio broadcast signals. And yet it is discussed on several occasions in the present work. This is because it has witnessed continual development as a transducer (Chap. 1), a sensor (Chap. 5), and a filter, discussed in the present section.

4.4.1 Quartz Resonators

The equivalent circuit diagram for a free piezoelectric resonator was established in Sect. 1.4.3. The quality factor of a real resonator is given by a sum:

$$\frac{1}{Q} = \frac{1}{Q_\mathrm{i}} + \frac{1}{Q_1} + \frac{1}{Q_2} + \ldots + \frac{1}{Q_n} ,$$

where Q_i is the intrinsic quality factor of the material, defined in Sect. 4.2.6, Vol. I. For a given cut, it takes the form $Q_\mathrm{i} = c_{ij}/\omega\eta$, where η is a viscosity coefficient. Terms Q_1, Q_2, ..., Q_n express attenuation caused by electrodes, mountings, the surrounding air, and so on.

Classic quartz resonators, manufactured in almost countless quantities (hundreds of millions per year), are used in the frequency range from 1 kHz to 300 MHz (beyond 30 MHz, thin quartz plates often vibrate in partial mode, i.e., at a frequency slightly below some harmonic, see Sect. 1.4.3, because they are rarely much less than 50 μm thick). Such widespread use can be explained by the properties of quartz, a hard crystal, barely affected by environmental fluctuations. It combines a high quality factor ($> 10^5$, when $f > 1$ MHz), excellent frequency stability under temperature change (drift $< 10^{-6}$/K for an AT-cut), as well as in time. The passband of a resonator filter is limited by the difference between resonance and antiresonance frequencies, i.e., by the electromechanical coupling coefficient, $\Delta f/f \leq K^2/2$ [see (1.59)]. Quartz filters are thus highly selective. Piezoelectric ceramics have larger values of K ($K^2 \approx 0.1 \Rightarrow \Delta f/f \leq 5\%$), but their poor temperature behaviour and losses

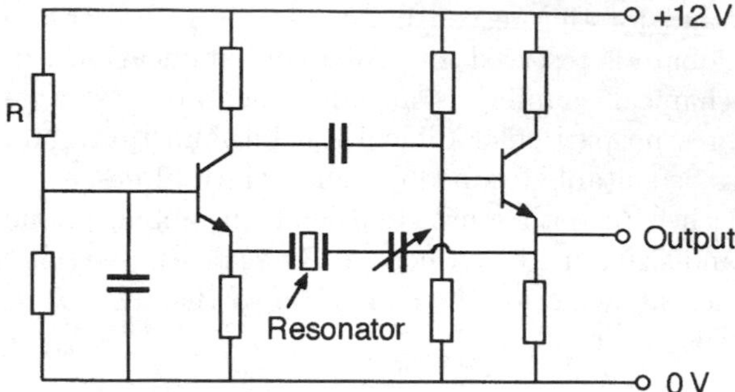

Fig. 4.34. Elementary circuit diagram for an oscillator whose frequency is determined by a quartz resonator. The crystal is placed between amplifier input and output to ensure selective feedback. The amplifier supplies the necessary energy (via DC supply) to maintain vibrations in the crystal. The power required is less than 1 mW for high quality quartz. However, if its frequency variations are to remain fairly constant, for example, less than $10^{-8}f$, over a wide temperature range (−20 to 70°C), the oscillator is placed in a thermostatically controlled housing whose temperature is locked at the inversion point of the $f(\theta)$ curve of the crystal by a feedback loop. The latter consumes several Watts

limit applications. Lithium tantalate $LiTaO_3$ replaces quartz in broader band applications ($\Delta f/f \leq K^2/2 = 2.5 \times 10^{-2}$) and when stability requirements are less stringent (drift $\approx 10^{-5}$/K).

A quartz resonator kept vibrating by an amplifier is a clock (Fig. 4.34). Demand for this kind of compact, accurate and stable clock has greatly increased with the development of positioning systems such as radio positioning and radio navigation.

Certain quartz oscillators now perform almost as well as the rubidium clock [4.37]. In all cases, these atomic clocks contain a quartz crystal whose frequency is controlled by feedback. However, quartz oscillators are merely secondary standards since their frequency is imposed by another technically achieved dimension (the primary standard is an oscillator whose frequency is linked by feedback circuit to the frequency $f = 9\,192\,631\,770$ Hz of a transition between two quantum levels of caesium 133). Specialists in the measurement of frequencies, and hence time, distinguish the short term (< 100 s), medium term and long term (> 1 month) when considering stability.

Let us therefore review characteristics of these resonators in the light of recent progress.

4.4.1.1 Cut and Vibration Mode. In the first resonators, derived from the Curie plate, the X cut was used (i.e., small plates in which the normal to the large faces is parallel to the crystallographic X-axis), and also nearby cuts ($X + 5°$), vibrating lengthwise or in thickness compression. The (series) resonance frequency varies with temperature. For temperature θ close to θ_0,

$$\frac{f - f_0}{f_0} = A(\theta - \theta_0) + B(\theta - \theta_0)^2 + C(\theta - \theta_0)^3 ,$$

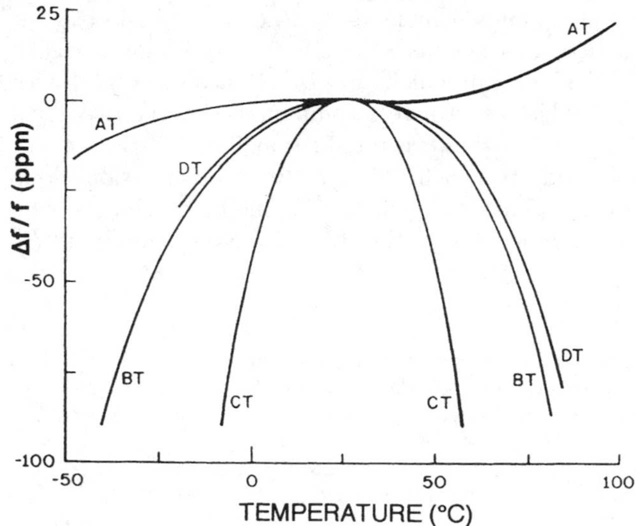

Fig. 4.35. Relative variation of resonance frequency as a function of temperature for AT, BT, CT and DT cut quartz. Taken from [4.38]

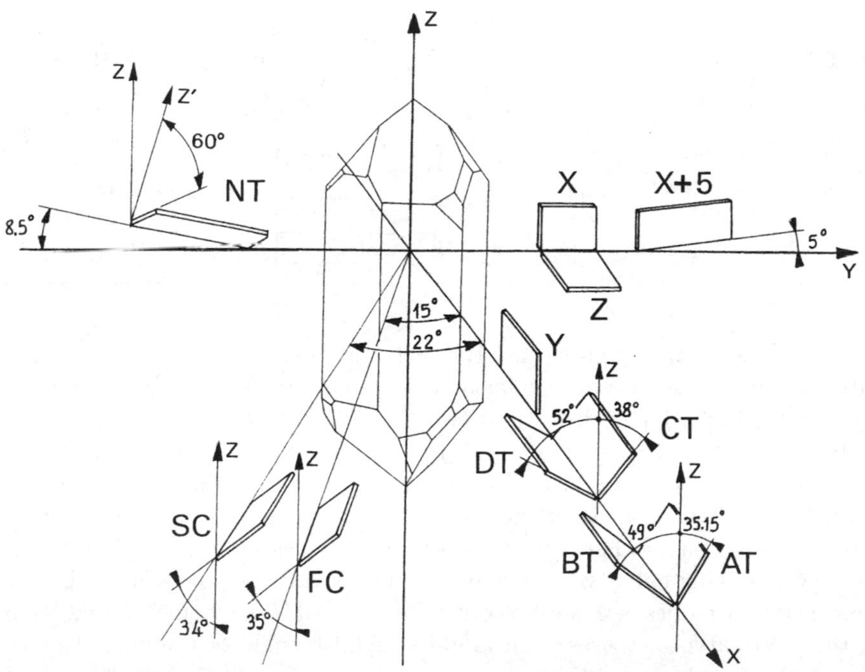

Fig. 4.36. Quartz crystal. Classic cuts for resonators. FC and SC denote doubly-rotated cuts. They arise from a Y cut rotated about the Z-axis (15° for FC and 22° for SC), and then rotated through 34° about the new X-axis (X')

where f_0 is the frequency at θ_0, and coefficients A, B, C depend on both cut and vibration mode. Therefore, rather insensitive cuts were sought for each given vibration mode. The AT cut, in which $A = B = 0$, has been used with a thickness shear vibration. This cut, still very common, is characterised by a cubic $f(\theta)$ curve (Fig. 4.35). It is a Y cut, rotated through $35°15'$ (Fig. 4.36). Only the slow transverse mode (speed 3 320 m/s), with polarisation along X, can be excited electrically. (This is called the C mode by specialists, who refer to the longitudinal mode as A and the other transverse mode as B: $V_A > V_B > V_C$.)

Table 4.1. Quartz resonators. Classic cuts and vibration modes. Nodal lines pass through *points* • and nodal planes through *dashed lines*. To excite a mode, electrodes must be disposed so that the electric field generates the corresponding deformations

Cut	Vibration mode	Deformation	Frequencies
XY, NT	Flexural	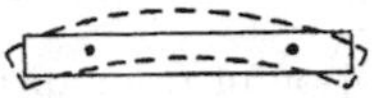	1–100 kHz
X, $X + 5°$	Extension		50–200 kHz
CT, DT	Surface shear	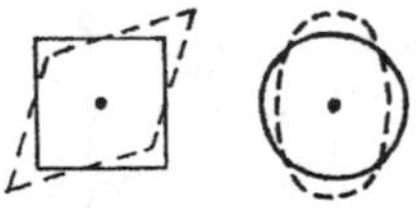	150–800 kHz
AT, BT,	Thickness shear:		
FC, SC	fundamental		1–300 MHz
	partial 3	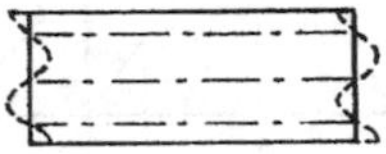	→ 2 GHz

Other singly-rotated cuts, viz., BT, CT and DT are also rotated Y cuts. Table 4.1 indicates the vibrational modes they carry. For example, the BT cut vibrates in thickness shear mode B (speed 5 070 m/s).

In the 1980s, new cuts were found: FC, IT and SC [4.39]. These are called *doubly-rotated cuts* because they are considered to be obtained from a Y cut rotated through angle Φ about the Z-axis and then angle Θ about the new X-axis (i.e., the X'-axis). For the slow shear mode C, as for the simple AT cut, these cuts exhibit cubic variation of frequency with temperature, whereas the other two modes are piezoelectrically coupled. The point of inflection on the $f(\theta)$ curve increases with angle Φ, whilst angle Θ remains close to 34° for these three cuts (Table 4.2). In a given temperature range, relative frequency variations are less than for the AT cut. These cuts also have other useful properties. For example, the SC (stress compensated) cut is extremely insensitive to stresses in the crystal plane.

Table 4.2. Quartz resonators. Characteristics of doubly-rotated cuts [4.40]. C_0 and C_1 are static and motional capacitances of the resonator (see Sect. 1.4.3.1)

Cut	Angles Φ, Θ [deg]		Coupling coefficient [%]	Speed of C$\parallel$ X mode [m/s]	C_0/C_1	$T_{\text{inflection}}$ [°C]
FC	15	34.6	6.8	3 470	270	55
IT	19	34.3	5.8	3 550	360	75
SC	22	34	4.9	3 600	520	95

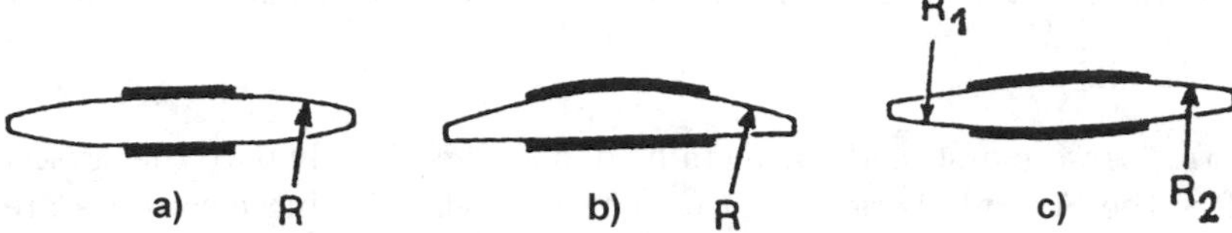

Fig. 4.37. Thickness shear mode resonators. (**a**) Bevelled plane shape. (**b**) Plane–convex shape. (**c**) Biconvex shape

4.4.1.2 Structures. The most common shapes for ordinary quartz resonators (Table 4.1) are rods with square or rectangular cross-section (extensional modes), or circular cross-section (flexural modes), and plates (plane or thickness shear modes). These are suspended by nodal points using springs. Resonator shapes involve more intricate design when intended for professional use. Thickness shear mode resonators with the best performance, operating in the range 1–10 MHz, for example, are bevelled or have one or two convex faces (Fig. 4.37). Such geometries, with thickness decreasing from the centre out to the edge of the resonator, are designed to localise (or trap) acoustic energy at the centre. The idea is to reduce vibration amplitudes towards the edges, which are used for mountings, and also to decrease amplitudes of spurious modes and the size of the crystal. The convex face may have two curvatures, with radii chosen according to lateral anisotropy of the crystal, and electrodes may have an elliptical shape [4.41].

Ultrastable Oscillators. Remarkable progress has been made with regard to vibration spectrum and stability by means of an original mounting technique [4.42]. The central part, the only part to vibrate (in SC mode), is supported by four bridges (Fig. 4.38). The three elements: resonant part, bridges and ring, form a monolithic setup. It is fabricated by ultrasonic or chemical milling of a crystalline disk. Bridges are orientated in directions with minimal sensitivity to stresses. Electrodes are no longer deposited on the resonator itself, but rather on crystals located very close (a few μm) on either side, either by vacuum evaporation or sputtering. This was a major step forward as regards stability [4.43]. (In fact, the first quartz resonators, less sophisticated than these, also had electrodes separated from the crystal, by a thin layer of

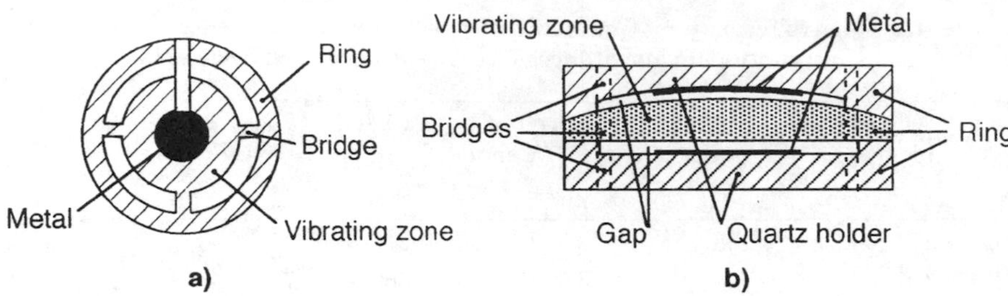

Fig. 4.38. Resonators supported by four bridges. (**a**) QAS (Quartz Auto-Suspendu) resonator. (**b**) BVA (Boitier à Vieillissement Amélioré) resonator. Electrodes are carried by the active part in the first, but not in the second. External diameter < 15 mm

air; electrodes were first made by metallic deposition directly onto the crystal only after the Second World War.) Quartz crystals carrying electrodes are cut in the same way as the resonator (SC cut) so as to minimise stresses. As there is no contact between electrodes and vibration zone, stresses due to electrodes are also suppressed. In addition, there is no metal diffusion in the active part, so the resonator is insensitive to aging due to such diffusion (always produced in the long term when crystal faces are metallised). These resonators have been called QAS (Quartz Auto-Suspendu) and BVA (Boitier à Vieillissement Amélioré) resonators, depending on whether the active centre part is metallised or not [4.44].

BVA resonators and oscillators made from them have been studied by Besson, R.J., and his team at the Ecole Nationale Supérieure de Mécanique et des Microtechniques in France. They have remarkable characteristics, as can be seen from Table 4.3.

Table 4.3. Characteristics of BVA quartz oscillators. σ_y is the standard deviation of the relative frequency variation

Frequency	5 MHz
Standard deviation σ_y $(10 < \tau < 1\,000$ s)	$4 \times 10^{-14} < \sigma_y(\tau) < 2 \times 10^{-13}$
Spectral purity	-130 dBc/Hz at 1 Hz, -160 at 100 Hz
Aging	2×10^{-12} to 5×10^{-11}/day
Thermal sensitivity	-5×10^{-13}/°C from -20 to $+60$°C
Accelerometric sensitivity	10^{-10}/g

Relative variation $\Delta f/f$ of the resonance frequency is therefore 10^{-14} in the short term (1 s) and 10^{-11} per day. Hence, for $f = 10$ MHz, we have $\Delta f = 10^{-7}$ Hz/s, or 10^{-4} Hz/day.

These oscillators are currently the best quartz clocks. In fact, although suspension and excitation techniques play a significant role, crystal quality is all important. It is not possible here to describe the main stages of fabrication (by hydrothermal synthesis) and preparation (lapping, polishing, spectrometric identification of impurities, elimination by electromigration, reduction of internal stresses by annealing, surface outgassing in vacuum). The technology involved in preparing and inspecting crystals has thus considerably improved over the past 20 years. These improvements are partly due to invention of new instruments and measurement techniques (scanning microscopy, spectrometry, interferometry and X-ray topography), and partly to a better understanding of the way resonators work. Concerning the last point, 3-dimensional models have been developed to take curvature of faces into account, as well as coupling between modes, effects due to electrodes and even certain nonlinear effects [4.45].

Ultrastable oscillators are still the subject of much study. Attempts are being made to reduce their size ($1\,000 \rightarrow < 100$ cm^3) with the help of smaller crystals operating at higher frequencies (10 MHz) [4.46].

High-Frequency Resonators (f > 100 MHz). The minimal thickness $d_{\min}$ accessible by mechanical grinding is only slightly less than 50 μm. This limits the fundamental resonance frequency to $f_{\max} = V/2d_{\min}$. For AT-cut quartz, $f_{\max} = 50$ MHz requires $d_{\min} \approx 33$ μm. The resonator can of course be excited in a partial mode, but this solution has a disadvantage in certain applications, viz., the bandwidth is divided by the square of the order of the partial (Problem 1.5).

One way of reducing thickness, whilst still being able to manipulate the resonator, is to bombard locally one face of the plate, of thickness about 50 μm, with an (argon) ion beam. Figure 4.39 depicts the basic principle of this method [4.47]. The disk to be thinned is covered with a glass mask containing an opening which determines the bombardment area. It is placed on a water-cooled turntable. The angle between the axis of rotation of the turntable and the ion beam is chosen (near 35°) to ensure homogeneous milling of the surface.

The milling rate, a fraction of 1 μm/min, depends on ion energy and current intensity. Membranes of thickness a few [μm] and diameter several [mm] can be made in this way at the centre of disks with ordinary thicknesses, which therefore remain easy to manipulate. Electrodes consist of a metallic (aluminium) film deposited by sputtering through a metallic mask, on either side of the membrane. Simple resonators vibrating in fundamental mode at 300 MHz and possessing quality factors greater than 10^4 are now mass-produced. In the laboratory, (AT-cut) resonators with fundamental resonance frequencies greater than 500 MHz and resonators vibrating in the third partial at close to 2 GHz have been produced [4.48]. The thickness of an AT-cut shear mode resonator operating at f = 240 MHz is 6.7 μm. Coupled resonators can be built on the same membrane (Sect. 4.4.2).

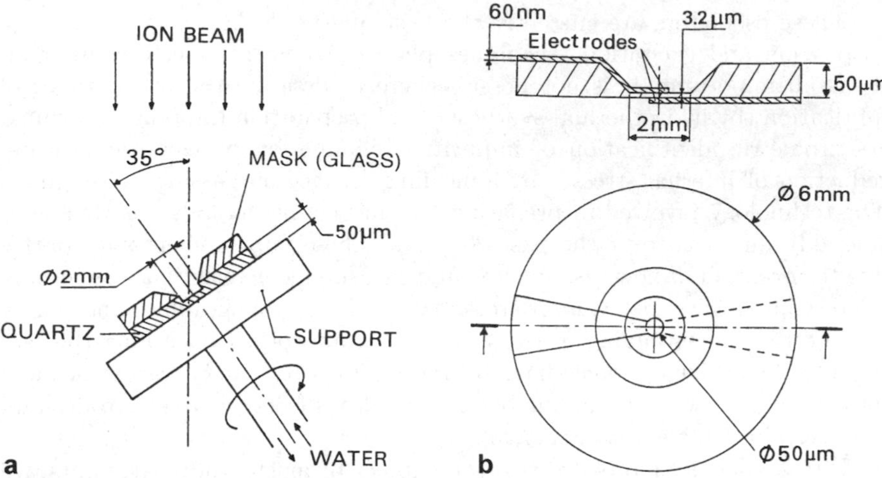

Fig. 4.39. High-frequency resonator. (**a**) Thinning by ion bombardment. The part of the crystal surface not protected by the glass mask is bombarded by an argon ion beam at an angle of 35°. The milling rate is a fraction of 1 μm/min. (**b**) Structure of a resonator ($f \approx 500$ MHz)

Another method for producing thin membranes is chemical etching [4.49]. This exploits the fact that quartz is soluble in certain solutions, e.g., hydrofluoric acid. The milling rate depends not only on the state of the substrate, concentration of the solution, temperature and nature of the material, but also on the crystalline orientation of the face. The last fact in particular means that the technique has more limited application than ion bombardment. However, progress has been made here, since chemical etching forms the basis for developments in micromechanics. Quartz resonators (SC-cut) operating at several hundred MHz have been achieved using this technique.

Miniature Resonators. Quartz resonators have applications, not only in telecommunications, instrumentation and metrology, but also with a direct relevance to the general public, for example, in computer games, domestic appliances, personal computers, and radio and television receivers. Quartz resonators ensure the stability of clocks, digital processors and ordinary watches. Naturally, the quartz in a watch is subject to quite different requirements (cost and size) to the quartz in an ultrastable oscillator. To begin with, it vibrates at much lower frequency (32 768 Hz $= 2^{15}$ Hz, as compared with 5 or 10 MHz), and they must be mass-produced. Specific technologies for crystal lapping, metallisation and testing are applied [4.50, 4.51, 4.52]. For example, watch resonators are U-shaped, like tuning forks. They are produced in large quantities by slicing a quartz cube of volume several cm^3, with chosen crystallographic orientation, into (Z-cut) wafers a fraction of a mm thick. Electrodes are then vacuum deposited in the form of a gold film onto the two faces of these wafers. After photolithography, a hundred or so tuning fork shapes are cut from each wafer by chemical etching and their lateral faces are metallised (other electrodes). The frequency of each resonator is adjusted by laser-induced metal evaporation. Such clock resonators are produced in their hundreds of millions each year. The order of magnitude of their parameters is quite surprising. As an example, taken from the Micro Crystal catalogue: cylindrical metallic housing of length 5.1 mm, diameter 1.5 mm, 35 kΩ $< R_{\text{series}} <$ 50 kΩ, motional capacitance $C_1 \approx 2$ fF, static capacitance $C_0 \approx 1$ pF, maintenance power consumption 1 μW.

These photolithographic techniques, borrowed from microelectronics, and chemical etching techniques, first developed in micromechanics, are of course used to make miniature resonators other than for clocks. Resonators of this kind, operating in different, higher-frequency modes (25 MHz, 50 MHz) and mounted in hermetically sealed evacuated ceramic housings, can be found in microprocessor clocks, telephones, synthesisers and other instruments. E.g., [4.53]: AT-cut Micro Crystal resonator vibrating in a thickness mode at a frequency in the range 30–100 MHz. The active part is a rectangular membrane, thinned by chemical etching in a wafer of thickness 1.5 mm. The chemical etching rate along any axis other than Z is 10 to 100 times slower than along the Z-axis.

Lithium Tantalate and Aluminium Phosphate Resonators. The bandwidth of a filter is determined by the electromechanical coupling coefficient of the vibrating material ($\Delta f/f \leq K^2/2$). Quartz resonators therefore have relatively narrow bandwidth ($K^2 \approx 0.64\%$). Hence, when temperature stability requirements on the frequency do not necessitate the use of quartz, a lithium tantalate crystal can be used ($LiTaO_3$, class $3m$, $K^2 \approx 20\%$). Table 4.4 gives cuts and vibration modes. Aluminium phosphate (also called berlinite) and gallium phosphate are still being investigated, but langasite (lanthanum silicogallate $La_3Ga_5SiO_{14}$) is beginning to find applications.

Table 4.4. Cuts and vibration modes of lithium tantalate, aluminium phosphate and gallium phosphate resonators

Material	Cut	Vibration mode	Coupling coefficient	Speed [m/s]	Frequency
$LiTaO_3$	$Y + 135°$	Extension	0.25	5 900	0.1–1 MHz
$LiTaO_3$	X	Thickness shear	0.45	4 200	5–150 MHz
$AlPO_4$	AT	Thickness shear	0.12	2 870	10–300 MHz
$GaPO_4$	AT	Thickness shear	0.17	2 700	10–300 MHz

Relative variation of frequency with temperature is a parabola for (X-cut) lithium tantalate, with minimum close to zero at about 30°C, and a cubic for (AT-cut) aluminium phosphate, with point of inflection near 50°C.

Let us note that good resonators can be produced with non-piezoelectric materials like sapphire [4.54] and yttrium aluminium garnet (YAG). However, they are more difficult to excite (e.g., by capacitive effect) [4.55].

4.4.2 Monolithic Filters

In the 1960s, monolithic filters came to supplement the list of quartz filters composed of discrete resonators (ladder and lattice filters). In these integrated structures, resonators vibrating in thickness shear mode are arranged close together on the same (AT-cut) monocrystal. They are coupled by evanescent mechanical vibrations [4.56]. A structure with n resonators placed close together on the same thin plate has n coupled principal modes. Electrodes on end resonators serve as electrical ports, whilst those on the other $n-2$ resonators are shorted. The symmetric 2-resonator filter (Fig. 4.40) is the most common. It has two eigenmodes of slightly different frequency: a symmetric mode in which resonators vibrate in phase and an antisymmetric mode in which they vibrate in phase opposition. The symmetric (antisymmetric) mode is separately excited if electrodes are connected in parallel (cross-connected). Frequency responses are the same as for simple resonators; eigenfrequencies and elements of equivalent circuits are easily measured. In normal operation, only one resonator is excited. The vibration mode is then a linear combination of the two above modes. It can be shown, using Barlett's theorem [4.12], that the frequency response of this double-pole monolithic filter is that of a lattice in which resonators have resonance frequencies equal to those of the two modes. Equivalent circuits are derived using those found from eigenmode measurements.

Monolithic filters are more compact and cheaper than discrete resonator filters, and are commonly used in radio communications to filter intermediate frequencies in the range 5–250 MHz [4.57]. Beyond 50 MHz, ion milling or chemical etching are required. Figure 4.41 shows the frequency response of a 4-pole filter designed for the GSM and DCS 1800 telephone systems.

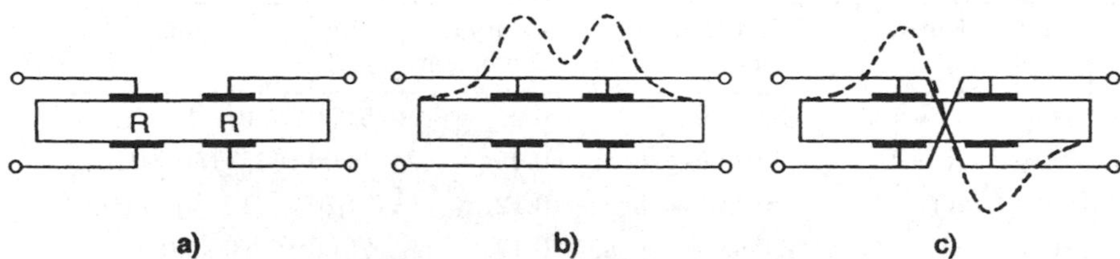

Fig. 4.40. Monolithic filter with two thickness shear mode resonators. (**a**) Resonators R, produced on the same (AT-cut) crystal, are coupled by evanescent waves. The unit is characterised by two principal modes which can be separately excited. (**b**) Excitation of symmetric mode. (**c**) Excitation of antisymmetric mode. Normal vibration, i.e., with only acoustic coupling, is a combination of these two modes

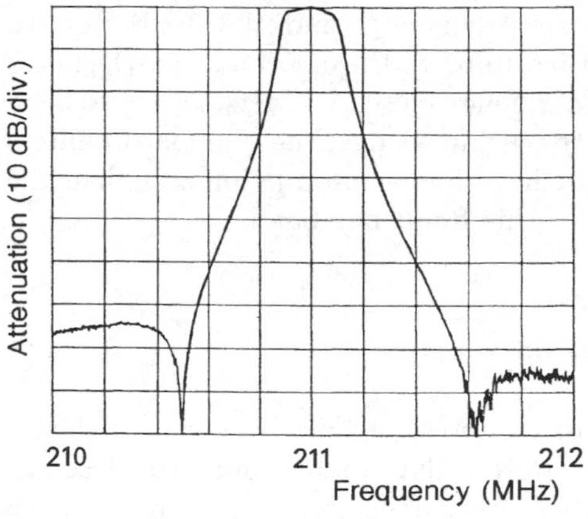

Fig. 4.41. Frequency response of a monolithic filter ($f_0 = 211$ MHz, insertion losses 3.5 dB). Kindly communicated by Michel, J.M. (Thomson-CEPE)

An interesting technique exists for cutting miniature resonators out of lithium tantalate. This is illustrated in Fig. 4.42, which shows a monolithic filter made of two rods vibrating in extension. The various elements, viz., mounting frame of length 8 mm and width 3 mm, rods, supporting bridges between frame and rods, and couplings between the rods, were all cut out by laser from a lithium tantalate wafer [4.58]. The frequency response of four cascaded monolithic units is shown in Fig. 4.42. The centre frequency

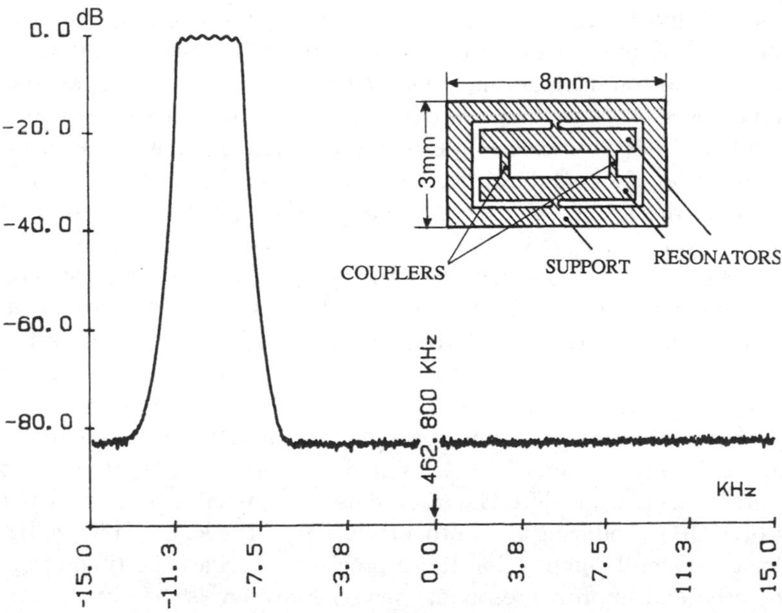

Fig. 4.42. Frequency response of a filter composed of four monolithic units. Each unit contains two resonators cut by laser from a lithium tantalate wafer

is 453.200 kHz (average radio receiver frequency) and the 3 dB relative bandwidth is 0.63%. Rods of length 0.6 mm, width 0.06 mm and thickness 0.15 mm, vibrating in extension at a frequency close to 4 MHz, have also been cut out by this technique. However, to our knowledge, these miniature filters have not been developed because they have higher production cost than the piezoceramic filters currently made for this frequency band.

4.5 Filters Matched to Signals

In radar and sonar, it is common to compress signals by means of filters which have been matched to them. Recall that radar, and indeed sonar, consists in emitting a signal – a pulse with carrier frequency – in a known direction, and then detecting any echo returning by reflection off the object. The time required for the echo to return yields the distance between emitter and object. The frequency change gives the speed of the object. Radar surveys the atmosphere and emits electromagnetic waves; sonar explores under water and emits elastic waves. Frequency and duration of signals are quite different in the two cases (kHz and ms for sonar, MHz and μs for radar). The two detection methods nevertheless resemble one another.

The aim in compressing the pulse with a filter matched to the signal is to increase the range of the radar (or sonar) at given power, without affecting resolution, or conversely, to increase resolution at fixed range. Simple radars emit pulses with fixed carrier frequency, and with width determined by the desired detection accuracy (resolution). Since the pulse level is limited by emitter power (klystron, magnetron, travelling wave tube, etc.), the range is predetermined as soon as the pulse width has been chosen. This constraint on range and accuracy is broken by pulse compression. Although the range is indeed a function of signal duration, resolving power depends only on the spectral width of this signal. And these two quantities, duration and spectral width, can be chosen separately.

Woodward has shown [4.59] that the greatest spectral width corresponds to the best resolution. For a given pulse duration, spectral width can be increased by modulating the carrier frequency, for example. But the echo, reflected by the target, is tainted with noise, in both modern and classical radar. Noise elimination by bandpass filter is easier when the signal has a narrow spectrum. Extracting a wide-spectrum signal from noise requires a more complex detector, matched to the signal and taking the type of noise into account. In the case of white Gaussian noise, the signal is best extracted by means of a filter producing the autocorrelation of the signal [4.60, 4.61]. The resulting temporal compression more precisely specifies the time of arrival of the echo and improves resolving power. Noise levels, barely affected by the filter matched to the signal, are the same at output as at input.

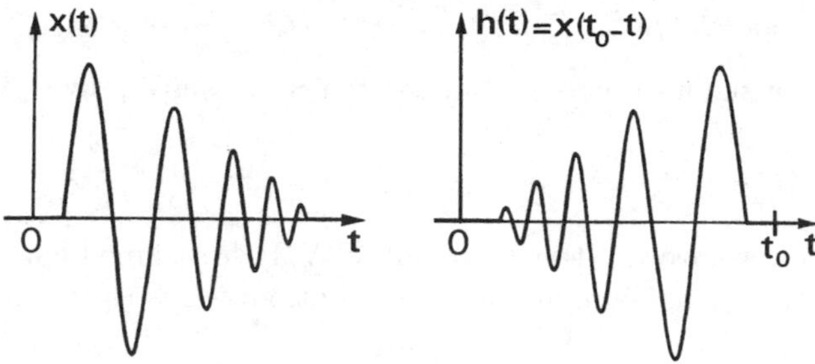

Fig. 4.43. The impulse response $h(t)$ of the filter matched to a signal $x(t)$ is the time reversal of $x(t)$ delayed by a constant t_0

These remarks on extracting information from noise, illustrated here by a radar signal, are of course relevant in the much wider context of pattern recognition.

4.5.1 Response and Signal-to-Noise Ratio

As noted in the introduction to the present chapter, the response $y(t)$ of a linear system to any real input signal $x(t)$ is the convolution product (4.2) of $x(t)$ with the impulse response $h(t)$ of the system.

The problem can therefore be stated as follows. Given $x(t)$, how should $h(t)$ be chosen to maximise the signal-to-noise ratio? The latter is defined as the ratio at the filter output of the maximal instantaneous power of $y(t)$ to the noise power (assuming white noise). Let us first formulate the answer to this question, and then give a proof. The impulse response $h(t)$ of the filter matched to the signal $x(t)$ is

$$h(t) = Cx(t_0 - t) , \tag{4.14}$$

that is, the image of the signal in a mirror placed perpendicular to the time axis (Fig. 4.43).

A real matched filter always necessitates a delay. Indeed, if t_0 were zero, the response $h(t)$ would precede the cause, that is, the Dirac impulse applied at the time origin. Consequently, the output signal is the autocorrelation of $x(t)$:

$$y(t) = x(t) \otimes h(t) = C \int_0^t x(\tau)\, x(t_0 - t + \tau)\, \mathrm{d}\tau . \tag{4.15}$$

Transformed to the frequency domain, relation (4.14) becomes

$$H(f) = CX^*(f)\mathrm{e}^{-2\pi \mathrm{i} f t_0} . \tag{4.16}$$

Hence, up to a phase difference due to delay t_0, the frequency response $H(f)$ of the filter matched to signal $x(t)$ with spectrum $H(f)$ is the complex conjugate of this spectral function. The output signal therefore has spectrum:

$$Y(f) = H(f)X(f) = C|X(f)|^2 \mathrm{e}^{-2\pi \mathrm{i} f t_0} . \tag{4.17}$$

In addition, the maximum signal-to-noise ratio, written S/N, is given by

$$\left.\frac{\mathrm{S}}{\mathrm{N}}\right|_{\max} = \frac{2E}{N_0} , \tag{4.18}$$

where N_0 is the noise power spectral density in units [W/Hz], assumed constant in the positive frequency domain, at the filter input, and E is the total energy of the signal.

Demonstration. By definition, the signal-to-noise ratio is

$$\frac{\mathrm{S}}{\mathrm{N}} = \left.\frac{\text{maximal instantaneous power}}{\text{mean noise power}}\right|_{\text{output}} = \frac{y(t_0)^2}{b^2} , \tag{4.19}$$

where t_0 is the time at which $y(t)$ is maximal. The mean noise power b^2 at the filter output, of frequency response $H(f)$, can be expressed in terms of input and output noise spectral densities $S_X(f)$ and $S_Y(f)$. Since $S_Y(f) = S_X(f)|H(f)|^2$ [4.4], it follows that

$$b^2 = \int_{-\infty}^{\infty} S_Y(f)\,\mathrm{d}f = \int_{-\infty}^{\infty} S_X(f)|H(f)|^2\,\mathrm{d}f .$$

To simplify, we assume stationary white noise, so that it is independent of frequency:

$$S_X(f) = \frac{N_0}{2} \quad \Rightarrow \quad b^2 = \frac{N_0}{2}\int_{-\infty}^{\infty} |H(f)|^2\,\mathrm{d}f . \tag{4.20}$$

$N_0/2$ is the bilateral noise spectral density (including positive and negative frequencies). In the frequency domain, $y(t_0)$ is given by

$$y(t_0) = \int_{-\infty}^{\infty} X(f)\mathrm{e}^{2\pi \mathrm{i} f t_0} H(f)\mathrm{d}f , \tag{4.21}$$

and the signal-to-noise ratio becomes

$$\frac{\mathrm{S}}{\mathrm{N}} = \frac{\left|\displaystyle\int_{-\infty}^{\infty} X(f)\mathrm{e}^{2\pi \mathrm{i} f t_0} H(f)\mathrm{d}f\right|^2}{\displaystyle\frac{N_0}{2}\int_{-\infty}^{\infty} |H(f)|^2\,\mathrm{d}f} .$$

The Schwartz inequality,

$$\left|\int_{-\infty}^{\infty} X(f)\mathrm{e}^{2\pi \mathrm{i} f t_0} H(f)\mathrm{d}f\right|^2 \le \int_{-\infty}^{\infty} |X(f)|^2\,\mathrm{d}f \int_{-\infty}^{\infty} |H(f)|^2\,\mathrm{d}f ,$$

then implies

$$\frac{\mathrm{S}}{\mathrm{N}} \le \frac{2}{N_0}\int_{-\infty}^{\infty} |X(f)|^2\,\mathrm{d}f . \tag{4.22}$$

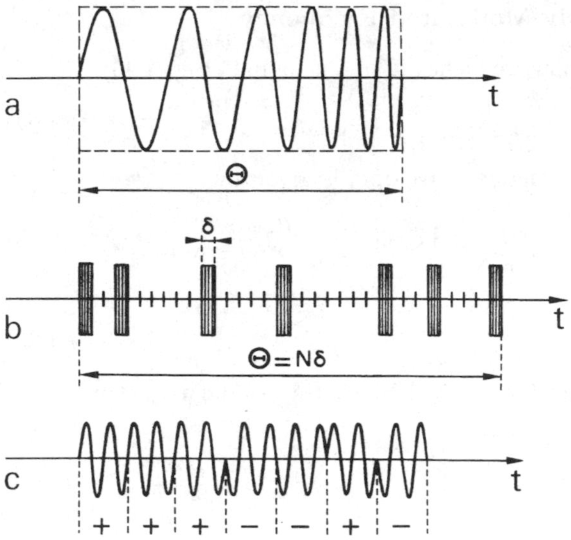

Fig. 4.44. Coded signals. (**a**) Frequency modulation. (**b**) Amplitude modulation. (**c**) Phase inversion

Equality holds when (4.16), identical to (4.14), is satisfied. If E is the energy of the input signal, and using Parseval's relation,

$$E = \int_{-\infty}^{\infty} x(t)^2 \mathrm{d}t = \int_{-\infty}^{\infty} |X(f)|^2 \, \mathrm{d}f \,, \tag{4.23}$$

maximal signal-to-noise ratio is given by (4.18). It is therefore the ratio of signal energy, regardless of shape, to noise spectral density (in energy units [W/Hz]). Of course, the duration of the signal is finite, and since the filter must be causal, we have $t_0 \geq \Theta$.

In practice, signal $x(t)$ occurs in various shapes. As already mentioned, the spectrum of a finite width pulse is broadened by frequency modulation, or any other form of coding. Time variation of one of the three characteristic wave quantities, amplitude, phase and frequency, is either continuous (analogue coding) or discrete (digital coding).

Figure 4.44 shows three examples of signals: rectangular pulse of width Θ and modulated carrier frequency; pulse train divided into N intervals (digits), occupied according to a given sequence, by elementary pulses of width $\delta = \Theta/N$; pulse of width Θ with phase inversions.

Elastic wave transmission lines processing these three signal shapes will be described in the following sections. Closer attention will be paid to dispersive transmission lines matched to signals with linearly modulated frequency. The aim of Problem 4.6 is graphical determination of the response of the filter matched to the pulse train in Fig. 4.44b.

4.5.2 Signal with Linearly Modulated Frequency

Linear variation of the angular frequency ω of the signal (Fig. 4.45) is

$$\omega = \frac{\mathrm{d}\phi}{\mathrm{d}t} = \omega_0 + \mu t \; , \quad -\frac{\Theta}{2} \le t \le \frac{\Theta}{2} \; , \tag{4.24}$$

and the signal $x(t)$, with rectangular envelope, is given by

$$x(t) = \Pi\left(\frac{t}{\Theta}\right) \cos\phi(t) = \Pi\left(\frac{t}{\Theta}\right) \cos\left(\omega_0 t + \frac{\mu}{2}t^2\right) \; , \tag{4.25}$$

since

$$\phi(t) = \omega_0 t + \frac{\mu}{2}t^2 \; . \tag{4.26}$$

The impulse response of the filter matched to this signal is, according to (4.14),

$$h(t) = C\Pi\left(\frac{t}{\Theta}\right) \cos\left(\omega_0 t - \frac{\mu}{2}t^2\right) \; , \tag{4.27}$$

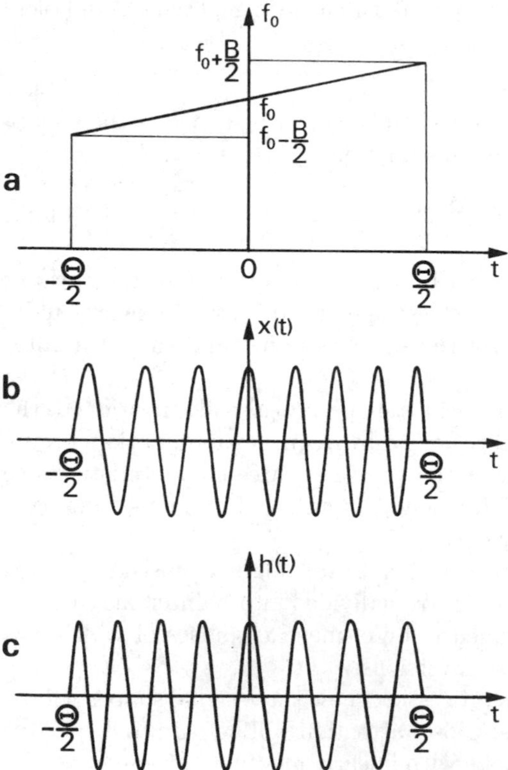

Fig. 4.45. Linear frequency modulation. (**a**) Instantaneous frequency $f = (1/2\pi)\mathrm{d}\phi/\mathrm{d}t$. (**b**) Signal $x(t) = \Pi(t/\Theta)\cos\phi(t)$. (**c**) Impulse response $h(t) = x(-t)$ of the filter matched to this signal

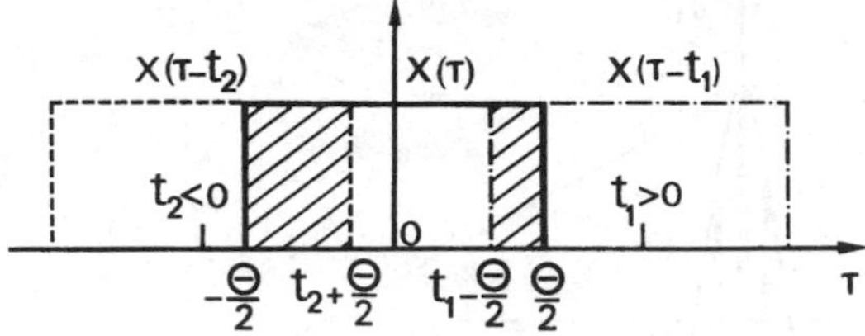

Fig. 4.46. Autocorrelation of a pulse with linearly modulated frequency. Depending on the sign of t, the integration interval is $[t-\Theta/2,\ \Theta/2]$ or $[-\Theta/2,\ t+\Theta/2]$

omitting the delay t_0 of the matched filter for the moment.

4.5.2.1 Compressed Pulse. The response $y(t)$ of the filter matched to signal $x(t)$ is the autocorrelation of $x(t)$ given in (4.15). Figure 4.46 shows that

$$\begin{aligned} &y(t) = 0\,, && \text{for} \quad |t| > \Theta\,, \\ &y(t) = C\int_{\tau_1}^{\tau_2} \cos\left(\omega_0\tau + \mu\frac{\tau^2}{2}\right)\cos\left[\omega_0(\tau - t) + \mu\frac{(\tau-t)^2}{2}\right] \mathrm{d}\tau\,, \\ &\tau_1 = t - \frac{\Theta}{2}\,, \quad \tau_2 = \frac{\Theta}{2}\,, && \text{for} \quad 0 < t < \Theta\,, \\ &\tau_1 = -\frac{\Theta}{2}\,, \quad \tau_2 = t + \frac{\Theta}{2}\,, && \text{for} \quad -\Theta < t < 0\,. \end{aligned} \tag{4.28}$$

Transforming the cosine product under the integral yields a term of frequency $2\omega_0$ which we drop since, in practice, it is suppressed. Hence,

$$\begin{aligned} y(t) &= \frac{C}{2}\int_{\tau_1}^{\tau_2} \cos[\omega_0 t + \mu t(\tau - t/2)]\,\mathrm{d}\tau \\ &= \frac{C}{2}\left[\frac{\sin[\omega_0 t + \mu t(\tau - t/2)]}{\mu t}\right]_{\tau_1}^{\tau_2}\,, \end{aligned}$$

and it follows that

$$y(t) = \frac{C}{\mu t}\sin\left[\frac{\mu t}{2}(\tau_2 - \tau_1)\right]\cos\left[\omega_0 t + \frac{\mu t}{2}(\tau_2 + \tau_1 - t)\right]\,. \tag{4.29}$$

From (4.28),

$$\tau_2 + \tau_1 = t\,, \quad \tau_2 - \tau_1 = \Theta - |t|\,,$$

and the carrier wave has fixed frequency equal to centre frequency $f_0 = \omega_0/2\pi$:

$$y(t) = C\frac{\sin[(\mu t/2)(\Theta - |t|)]}{\mu t}\cos\omega_0 t\,, \quad \text{for} \quad -\Theta < t < \Theta\,. \tag{4.30}$$

The output signal $y(t) = s(t)\cos\omega_0 t$ of the filter matched to signal $x(t)$ is called the *compressed pulse*. In the interval $[-\Theta,\ \Theta]$ where it exists, its envelope is a $\sin\alpha/\alpha$ curve multiplied by straight lines $\Theta - |t|$ which bound the autocorrelation triangle of the rectangular envelope of the input signal:

$$s(t) = C\frac{\sin\alpha}{\alpha}\frac{\Theta - |t|}{2}\,, \quad \alpha = \frac{\mu t}{2}(\Theta - |t|)\,. \tag{4.31}$$

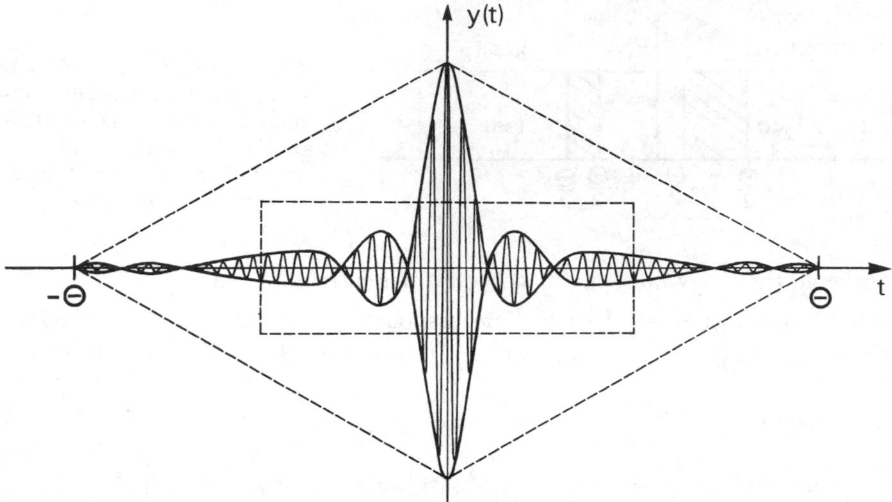

Fig. 4.47. Compressed pulse at the output of a filter with $B\Theta = 10$. The *dashed rectangle* is the envelope of the input signal whose frequency varies linearly. Noise is not shown, but assuming a lossless filter, it has the same level at output as at input

Denoting the frequency sweep of the input signal by $B = \mu\Theta/2\pi$, envelope $s(t)$ depends only on the product $B\Theta = R$. Indeed, in terms of reduced variable t/Θ,

$$s(t) = C\frac{\Theta}{2}\frac{\sin[(\pi Rt/\Theta)(1 - |t|/\Theta)]}{\pi Rt/\Theta} . \tag{4.32}$$

Figure 4.47 represents the output signal for a filter with $B\Theta = 10$.

For higher values, the compressed pulse envelope approaches a $\sin\alpha/\alpha$ curve for the first sidelobes:

$$s(t) \approx C\frac{\Theta}{2}\frac{\sin\alpha}{\alpha} , \quad \alpha \approx \pi R\frac{t}{\Theta} = \pi Bt . \tag{4.33}$$

The 3 dB ($1/\sqrt{2}$) width of the central peak is

$$\Delta t = \frac{0.885}{R}\Theta = \frac{0.885}{B} . \tag{4.34}$$

If this central peak is considered as the output signal, the filter appears to compress the input signal in time. Its duration Θ is compressed by a factor equal to $1.13R = 1.13B\Theta \approx B\Theta$. Assuming losses in this passive filter to be zero, energy conservation roughly determines the height of the output pulse at $\sqrt{B\Theta} = C\Theta/2$, so that

$$C \approx \sqrt{\frac{2\mu}{\pi}} . \tag{4.35}$$

The product $B\Theta$ (frequency sweep $\times$ duration of input signal) measures the temporal *pulse compression ratio* and also the gain due to the matched filter, in terms of signal-to-noise ratio. Indeed, the signal-to-noise ratio $(\mathrm{S/N})_\mathrm{o}$ of an ordinary radar, equipped with a filter of bandwidth B, centred on the fixed carrier frequency, is given by (Problem 4.7)

$$\left(\frac{\mathrm{S}}{\mathrm{N}}\right)_\mathrm{o} = \frac{2E}{N_0 B\Theta} . \tag{4.36}$$

Using (4.18), established for the pulse compression radar, gain due to matched filtering is

$$\frac{(\mathrm{S/N})_\mathrm{m}}{(\mathrm{S/N})_\mathrm{o}} = B\Theta . \tag{4.37}$$

The product $B\Theta = R$ is a key quantity. However, the same value of R can be obtained with large B and small Θ, e.g., $B = 20$ MHz, $\Theta = 5$ μs, $R = 100$, as with small B and large Θ, e.g., $B = 100$ kHz, $\Theta = 1$ ms, $R = 100$. The first example here corresponds to a medium frequency radar signal, the second to echo-sounding. Needless to say, the associated matched filters for these signals are made in quite different ways.

Relative to the main peak, sidelobes have levels (if $R \gg 1$) of -13.3 dB for the first and -18 dB for the second. These levels are generally considered to be too high. In radar, in order to achieve the resolution mentioned earlier ($\Delta t \approx 1/B$), with poorly reflecting targets, giving rise to peaks of height close to these levels, sidelobes associated with a target must not exceed -30 dB. Two methods are used, based on reduction of sudden changes in the amplitude and phase of the signal at its extremities (see Sect. 4.3.1). The first method consists in weighting the impulse response of the filter, by increasing amplitudes in the central part and reducing them at the edges. Several weighting functions (e.g., Hamming, Taylor, Dolph–Chebyshev) give satisfactory results. However, they somewhat alter the signal-to-noise ratio, since the filter is no longer exactly matched to the signal, and also the resolution, because the central peak is broadened. The second method does not deteriorate signal-to-noise ratio. It uses a signal whose frequency no longer varies linearly, together with a filter matched to that signal. This corresponds to phase weighting. The frequency varies more in the central part than in the outer parts.

4.5.2.2 Signal Spectrum. Frequency Response of the Filter. The difference between the variable frequency signal and the impulse response of the filter matched to it is just that, if the signal frequency increases with time, the filter response frequency decreases, and vice versa. As a consequence, the frequency response follows from the signal spectrum [equation (4.16)]. The spectrum can be found approximately by the stationary phase method (Appendix G), or exactly, by Fourier transform of a signal whose frequency is modulated symmetrically about the centre frequency ω_0 (Problem 4.8). The following result is obtained (Problem 4.9). The amplitude of the spectrum of a signal with linearly modulated frequency is a box function centred on centre frequency $f_0 = \omega_0/2\pi$, with width equal to frequency sweep B. It has phase

$$\Psi(\omega) = -\frac{\omega^2}{2\mu} + \frac{\pi}{4} . \tag{4.38}$$

Curves in Fig. 4.48 show that the amplitude approaches a rectangle as R increases.

These curves are reminiscent of the curves in optics [4.62] which describe, at finite distance, diffraction of light by a slit with straight, parallel sides (Fig. 4.49).

The light intensity received on a screen placed at distance b from a slit of width d, illuminated by a source at distance a, depends on the ratio

$$R = \frac{d^2(a+b)}{\lambda ab} .$$

This ratio plays the role of product $B\Theta$. Coordinate z corresponds to frequency and width Δ of the geometrical image of the slit corresponds to frequency sweep B.

The analogy between a filter matched to a pulse of varying carrier frequency, which it compresses in time, and a lens which focusses a beam of parallel light in space, can indeed be established. The filter could be described as a 'temporal lens'.

When all points of the slit vibrate in phase ($a = \infty$), curves in Fig. 4.48 yield the radiation pattern (with $R = d^2/\lambda b$ and $B \to d$), at finite distance b, of a rectilinear transducer of width $d \ll b$.

The variable part of the phase is found by the stationary phase method:

$$\Psi(\omega) = -\frac{(\omega - \omega_0)^2}{2\mu} . \tag{4.39}$$

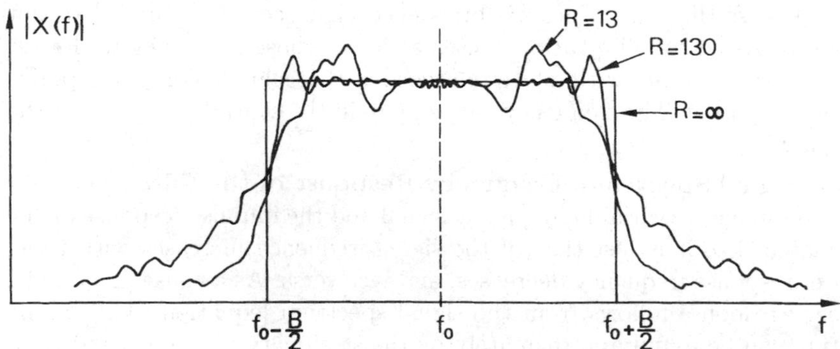

Fig. 4.48. The amplitude of the spectrum of a signal with linearly modulated frequency approximates a rectangle when the product $B\Theta = R$ increases. (Figure 6.6 in [4.61])

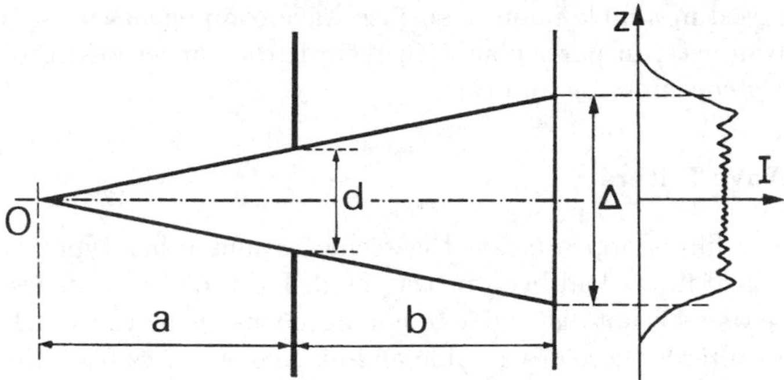

Fig. 4.49. At finite distance from a slit, the curve of light intensity distribution on a screen is similar to the square of the spectrum of a signal with linearly modulated frequency

In fact, it includes a second term which we have neglected. This term gives rise to small relative amplitude fluctuations of order $1/\pi\sqrt{R}$ in the bandwidth.

The frequency response $H(\omega)$ of the matched filter follows from the signal spectrum:

$$H(\omega) = CX^*(\omega)\mathrm{e}^{-\mathrm{i}\omega t_0} = \sqrt{\frac{\mu}{2\pi}}X^*(\omega)\mathrm{e}^{-\mathrm{i}\omega t_0} \,.$$

Its phase $\Phi(\omega)$ is therefore almost equal to $-\Psi(\omega) - \omega t_0$:

$$\Phi(\omega) = \frac{(\omega - \omega_0)^2}{2\mu} - \omega t_0 \,. \tag{4.40}$$

This implies a group delay $\tau_{\mathrm{g}} = -\mathrm{d}\Phi/\mathrm{d}\omega$ given by

$$\tau_{\mathrm{g}} = t_0 - \frac{\omega - \omega_0}{\mu} = t_0 - \Theta\frac{f - f_0}{B} \,, \tag{4.41}$$

which varies linearly with frequency.

Any filter matched to a signal with variable frequency is necessarily dispersive. The slope of its delay–frequency characteristic is opposite to the slope of the instantaneous frequency modulation of the signal. This dispersive effect explains why the pulse is compressed. In the case considered here, low-frequency components forming the beginning of the signal are retarded longer by the filter than high-frequency components arriving at the end of the signal.

The spectrum $Y(\omega)$ of the output signal $y(t)$ of the filter is given by (4.17).

The above results and comments are valid no matter how the filter is made. In the area of matched filtering, as in other fields, there is a general trend towards digital techniques at the expense of analogue techniques, related to increased computation rates in today's computers [4.63]. Indeed, the digital approach can process a wide range of signal shapes and is therefore winning out in low-frequency systems such as sonar. However, as already observed for filters used in mobile phones, surface wave components possess certain decisive advantages, in particular with regard to compactness, processing speed, energy consumption and cost.

4.5.3 Rayleigh Wave Filters

Before Rayleigh wave filters appeared on the scene, various other types of elastic wave (dispersive) filters had been investigated. For radio frequencies, propagation of these waves is not dispersive in homogeneous media (Sect. 4.1, Vol. I). The simplest filter is therefore a guide and dispersion arises from the geometry of the propagating medium. The guide is a thin plate or strip in which a TH (Transverse Horizontal) wave propagates in a dispersive mode, i.e., a mode of order greater than zero (Fig. 5.28b, Vol. I), or a Lamb wave in the first mode S_0. Compression ratios of 100 are accessible for centre frequencies of a few [MHz] (TH wave) or tens of [MHz] (Lamb wave). However, these transmission lines suffer various weak points (e.g., spurious signals due to the zero order mode, non-linear effects limiting frequency sweep).

Dispersion can also be obtained in a homogenous medium by requiring (unguided) waves to follow different paths according to their frequency. This method is easy to implement with Rayleigh waves, but problematic with bulk waves. One technique is to take a thick (silica) plate and to affix arrays of resonators on the perpendicular edge faces; emitters are placed on one edge and receivers on the other. A value of $R = 400$ has been obtained in this way for $f_0 = 40$ MHz.

Love waves can be exploited in an inhomogeneous medium comprising a semi-infinite substrate and a layer (Sect. 5.4.1.2, Vol. I). The phase speed of the waves is not the same in the substrate (speed V') and the layer (speed $V < V'$). Various pairs of materials, such as silica on silicon, have been tested [4.64]. However, it is difficult to produce a uniform homogeneous layer and a transducer on the edge face. Some such filters ($f_0 = 30$, 100 MHz, $R > 100$) have been made in the laboratory.

A layered structure is also dispersive for Rayleigh waves, but it is much easier to produce dispersion on a homogeneous substrate. This can be done either by an interdigital transducer, or by a variable interval reflector. The complexity of Rayleigh waves (Sect. 5.3, Vol. I) is compensated by the simplicity of interdigital transducer technology. These transducers are equivalent to a distribution of discrete ultrasound sources (receivers), whose relative phases and intensities are independently determined by position and active length of fingers (Sect. 2.1.3). In addition, they reflect from gratings formed of strips or grooves, provided the grating pitch is tuned to their wavelength. A wide range of impulse responses can be engineered by building the appropriate geometries into these transducers and reflection gratings.

4.5.3.1 Dispersive Transducer Filters. Figure 4.50 depicts the operating principle of a filter matched to a variable-frequency signal by means of

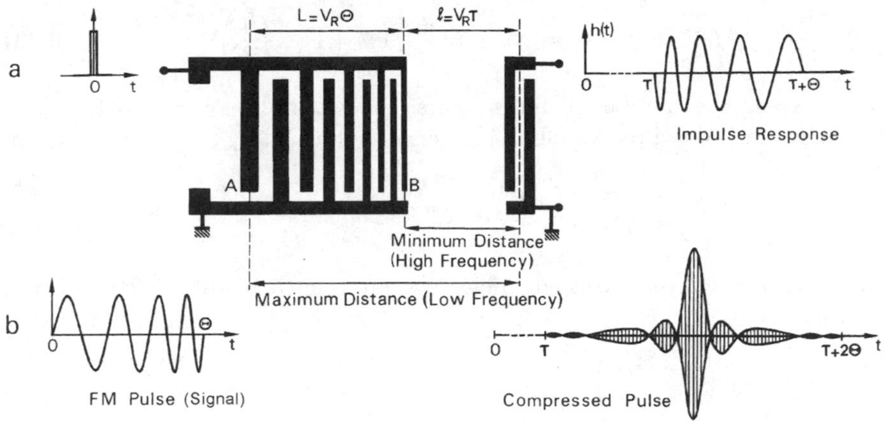

Fig. 4.50. Rayleigh wave dispersive transmission line (filter) matched to a frequency modulated signal. (**a**) Impulse response. Compare with the signal, of which it is indeed the time reversal. (**b**) Compressed pulse. High-frequency components, at the end of the signal, excite end B of the transducer at the same moment as Rayleigh waves arrive which were generated at the other end A by low-frequency components at the beginning of the signal (the beginning of the signal is located close to the time origin)

its transducer geometry. The substrate is assumed piezoelectric. Dispersion arises through varying finger spacings in the emitter transducer along the direction of propagation. Low-frequency components are preferentially emitted by sections with large interdigital spacing, and high-frequency components by sections with small interdigital spacing. In this way, the Rayleigh wave source moves for different frequencies. Since the receiver transducer, of high bandwidth, is well localised, the path travelled by elastic waves is frequency dependent. This receiver only converts the wave packet coming from the emitter. This is because, as they propagate under the emitter, waves of higher and higher frequencies are superposed upon the low-frequency waves initially emitted. Energy is concentrated by the emitter alone. Comparing the signal with the transducer pattern, or its impulse response, we find that $h(t) = x(t_0 - t)$ does indeed hold.

Designing such a filter amounts to determining abscissae x_n and overlap lengths w_n, using (2.30) and (2.31). The phase of the impulse response of the filter matched to a signal with linearly modulated frequency is given by (4.27):

$$\phi(t) = 2\pi f_0 t - \pi \frac{B}{\Theta} t^2 , \quad -\frac{\Theta}{2} < t < +\frac{\Theta}{2} . \tag{4.42}$$

Discrete sources are placed at points $x_n = V_{\mathrm{R}} t_n$, where times t_n are defined by the condition $\phi(t_n) = n\pi$:

$$\frac{B}{\Theta}t_n^2 - 2f_0t_n + n = 0 \Rightarrow t_n = \Theta\frac{f_0}{B}\left[1 - \left(1 - \frac{nR}{N^2}\right)^{1/2}\right] . \qquad (4.43)$$

$N = f_0\Theta$ is the number of finger pairs. For example, $N = 400$ for $f_0 = 100$ MHz, and $\Theta = 4$ μs. Values of integer n are limited by the signal duration Θ:

$$-\left(N + \frac{R}{4}\right) < n < N - \frac{R}{4} .$$

Since $e(t_n) = \text{Const.}$, adjacent finger overlap length, deduced from (2.31), varies:

$$w_n = w_0\left(1 - \frac{nR}{N^2}\right)^{-3/4} .$$

In this arrangement, the role of the receiver transducer is merely to restore an electrical signal. Its bandwidth Δf must be greater than the frequency sweep. It therefore contains a restricted number of finger pairs:

$$N_{\mathrm{r}} = \frac{0.885 f_0}{\Delta f} .$$

The above calculation assumes ideal conditions and neglects several effects. In particular, there is a change in speed of Rayleigh waves as they pass under the metallic fingers, where the tangential component of the electric field associated with the wave goes to zero. In addition, electrode masses modify surface impedance and cause reflection, and sources are coupled by inverse piezoelectric effect. Nevertheless, experiment confirms the above result for weakly piezoelectric crystals like quartz.

The photograph in Fig. 4.51a shows one of the first transmission lines built in this way, made with Y-cut, X-propagating quartz. The varying interdigital distance is visible (and explains the choice of picture), because the system operates at low frequency (although the aim here is merely to illustrate the principle). Operating frequency is around 14 MHz, with $B = 5.5$ MHz and $\Theta = 4$ μs, and the compression ratio is 22. Input impedance is equivalent to a capacitance of order 20 pF and a shunt resistance of 70 kΩ. The receiver transducer has only 6 fingers, of width 56 μm. The output signal has the predicted shape (Fig. 4.51b). Sidelobe levels are around −16 dB below the central peak for this compressed pulse. The photograph in Fig. 4.51c gives the frequency response for this line.

For signals with large relative frequency variation, conversion losses are more significant. This is because it is difficult to match transducer to generator over a broad band of frequencies. Radiation conductance [equation (2.43)] is lower ($R_{\mathrm{a}} < 50\ \Omega$), and the capacitive part of the transducer grows at the expense of its active part. It is then better to use a filter with two dispersive transducers, each ensuring, for example, 50% of the dispersion. Moreover, between central peak and sidelobe levels, users like radar operators require a difference greater than the 13.3 dB provided by the simple filter matched

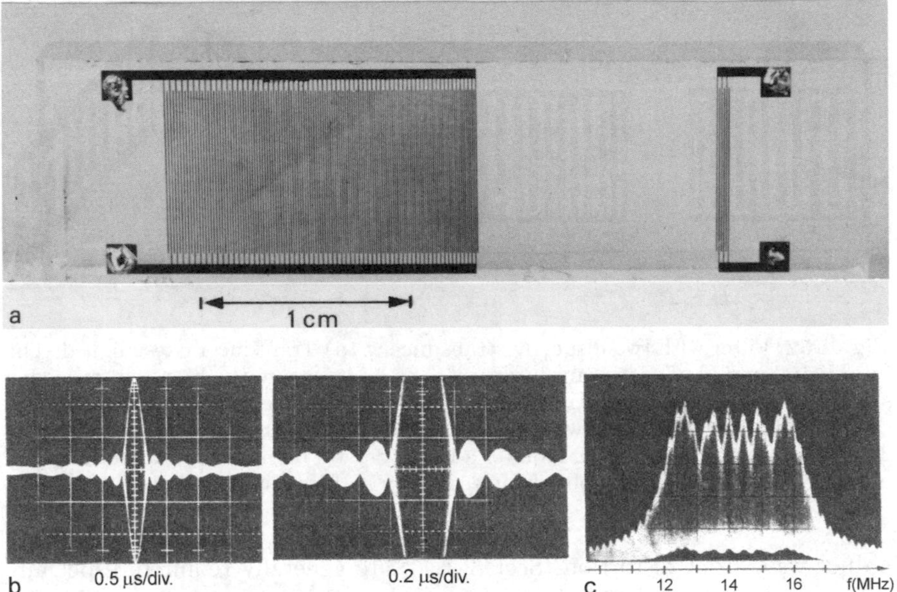

Fig. 4.51. Rayleigh wave filter with a single dispersive transducer. (**a**) Quartz substrate. $f_0 = 14$ MHz, $B = 5.5$ MHz, $\Theta = 4\mu s$, so that $B\Theta = 22$. (**b**) Output signal in response to the signal (with linearly modulated frequency) to which the filter is matched. 3 dB width 0.17 μs. Sidelobe level −16 dB. (**c**) Frequency response. Compare with curves in Fig. 4.48

to the signal with linearly modulated frequency when $B\Theta \gg 1$. A greater difference can be obtained by one of the two methods cited above.

The first consists in varying the finger overlap length in at least one of the transducers, without changing their positions, according to some weighting function (e.g., Hamming, Taylor, Dolph–Tchebyshev). The disadvantage here is a broadening of the compressed pulse, increased losses and a slight reduction in signal-to-noise ratio, since the filter is no longer perfectly matched to the signal. The second method involves using a signal whose frequency is modulated in a non-linear way. It has the advantage of not altering the signal-to-noise ratio. The time variation of the instantaneous frequency could be an S-shaped curve, for example, resulting from a straight line and a sine period. Sidelobe levels depend on the relative amplitude of the sine curve. As a result, various arrangements of dispersive transducers exist for this type of filter, particularly as transducer axes need not be aligned. Figure 4.52 shows a classic arrangement. Both transducers are dispersive but finger overlap lengths vary on only one of them. Design is simpler than in the case where both transducers are weighted. The filter response is the product of the responses of each transducer. Inactive fingers, which produce more uniform metallisation and

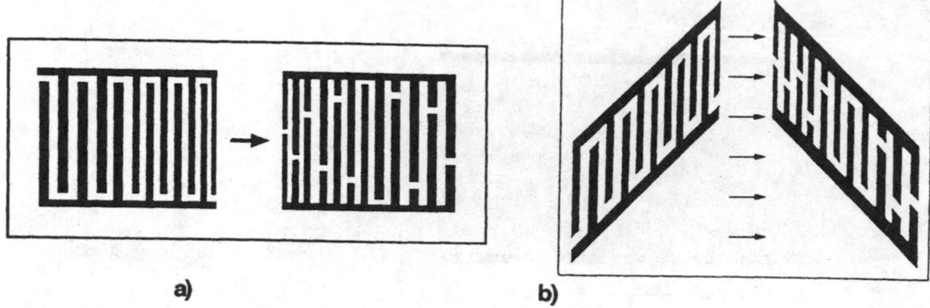

Fig. 4.52. Filter with two dispersive transducers. (**a**) Transducer axes aligned. The interdigital period of each transducer varies so as to ensure the required frequency sweep globally. Finger overlap lengths are weighted in one transducer to yield the desired difference of level between peak and sidelobes in the compressed pulse. (**b**) Transducer axes tilted. Tilting requires a larger crystal but allows local modification of phase speed and hence a phase correction by surface metallisation

reduce wave front distortion (Sect. 2.1.1), are generally required, especially if the substrate is strongly piezoelectric (e.g., $LiNbO_3$).

The configuration with tilted transducer axes (Fig. 4.52b) has two advantages. Firstly, it reduces the distance travelled by waves under the electrodes, whatever their frequency. Secondly, since waves of different frequencies propagate along different parallel paths, a selective phase correction can be made by locally metallising one of the paths. These devices are more costly, requiring larger crystals.

Figure 4.53 shows a device ($f_0 = 22$ MHz, $B = 5$ MHz, $\Theta = 5.6$ μs) with S-shaped delay curve. The ends of the impulse response are not sudden, resulting as they do from convolution of the signal emitted by the dispersive transducer with the rectangular impulse response of the receiver transducer (width 0.17 μs). The 3 dB width of the compressed pulse is 0.3 μs and sidelobes are at −25 dB. The first sidelobe is smaller than those following it, in agreement with theory (Problem 4.10).

For long pulses ($\Theta > 15$ μs), the number of electrode fingers can be reduced by sampling the transducer.

Many structures exist, based on the idea of dispersive transducers. Their characteristics can be found in manufacturers' catalogues. Generally speaking, they operate in the following conditions:

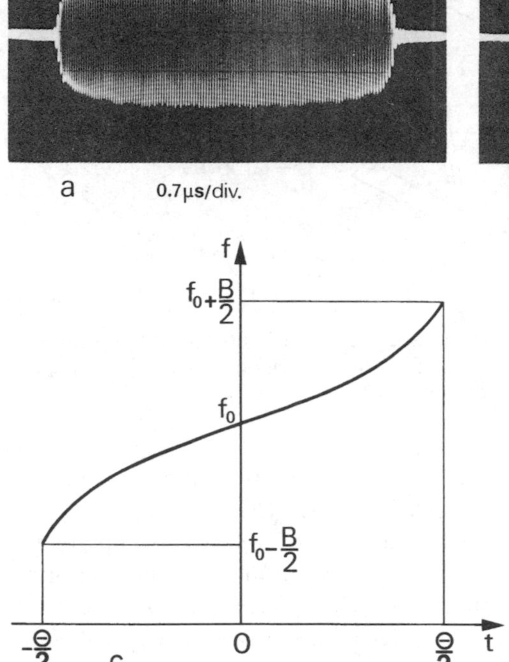

Fig. 4.53. Response of a Rayleigh wave transmission line with a single dispersive transducer ($f_0 = 22$ MHz, $B = 5$ MHz, $\Theta = 5.6$ μs) to: (**a**) a short pulse; (**b**) the signal to which it is matched and whose frequency varies with time as shown in (**c**). 3 dB width 0.3 μs. Sidelobe level −25 dB

Centre frequency	$10\ \text{MHz} < f_0 < 1\,000\ \text{MHz}$
Relative frequency sweep	$B/f_0 < 50\%$
Signal duration	$1\ \mu\text{s} < \Theta < 50\ \mu\text{s}$
Product $R = B\Theta$	$10 < R < 1\,000$
Loss (α-quartz)	30–50 dB
Sidelobe level	−15 to −40 dB
Temperature coefficient $\mathrm{d}\tau/\tau\mathrm{d}\theta$	Y cut: $-2.4 \times 10^{-5}\ \text{K}^{-1}$
(X-propagating α-quartz)	ST cut: 0

Although giving smaller losses, lithium niobate is less often used than quartz because of its high temperature coefficient (8.5×10^{-5}) and spurious effects due to its high coupling coefficient.

4.5.3.2 Dispersive Reflection Grating Filters. Elastic waves, like all waves, reflect from discontinuities. An array of discontinuities distributed in some well defined way can be made to produce a selective effect. The first convincing use of such effects was achieved by Martin, T.A., [4.65], in the form of a TH wave dispersive filter. These transverse horizontal waves are guided in a steel strip of thickness less than half a wavelength in order to prevent any other mode from propagating. The dispersive effect does not therefore

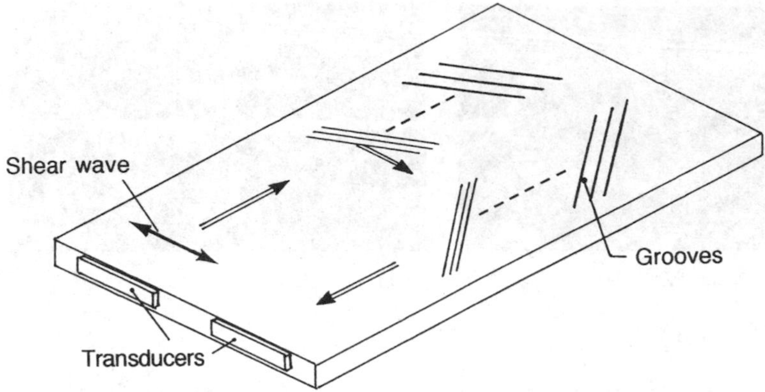

Fig. 4.54. Groove reflection grating. If the grating pitch is not constant, different spectral components in the acoustic signal (transverse horizontal waves), propagating near the surface, are reflected at different places

arise from the guide (Sect. 5.4, Vol. I), but is produced by two oblique groove gratings ruled on one of the two surfaces of the strip (Fig. 4.54).

Waves launched by a transducer fixed onto the edge face of the plate are reflected perpendicularly by the first grating, tilted at 45°, and then reflected by the second grating. In this way, they return along a parallel path towards the receiver transducer, fixed on the same edge as the emitter transducer. The grating pitch in each grating varies along the direction of propagation. Each wave is reflected by a zone which depends on its frequency. Dispersion arises because different paths are imposed on the various spectral components of the signal. Given the relative thinness of the strip, energy is concentrated for the main part close to the surface, whilst the frequency is necessarily low. Filters of this type operate with long pulses (> 100 μs) which have a rather narrow spectrum ($f_0 \leq 30$ MHz).

The principle of variable-pitch reflection gratings has been effectively transposed to surface waves by Williamson and Smith [4.66]. Dispersive gratings are arranged in a similar way to the transverse wave filter, but there is no longer any restriction on guide thickness. A piezoelectric substrate is used. The filter contains two non-dispersive interdigital transducers (Fig. 4.55). Waves launched by the emitter are reflected at right angles by the first oblique grating, from the zone in which the grating period equals their wavelength. Since grating period increases across the grating, high-frequency waves travel a shorter path than low-frequency waves. The second oblique grating plays the same role as the first, directing waves towards the receiver transducer.

The depth of grooves is a mere fraction (10^{-2}) of the wavelength and their reflection coefficient is a function of this depth. Consequently, amplitude weighting, and thereby sidelobe reduction in the compressed pulse, can be achieved by varying the groove depths.

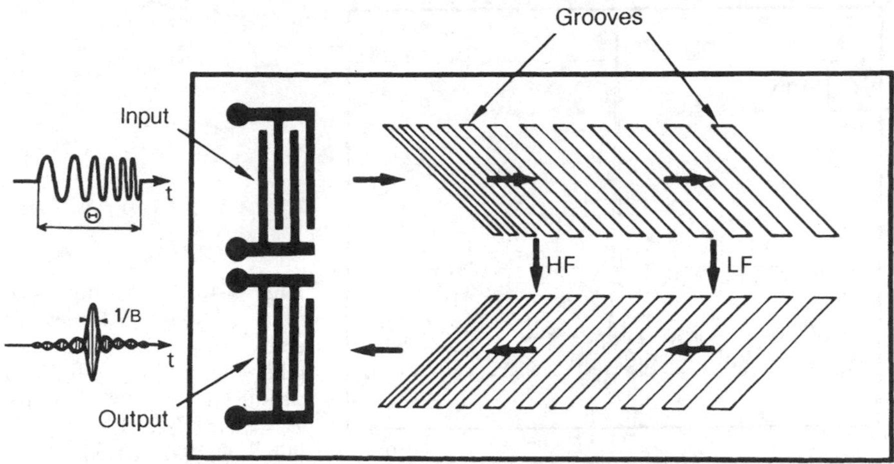

Fig. 4.55. Rayleigh wave filter with dispersive reflection gratings. Waves are reflected at right angles by each grating, from the zone in which the grating pitch is equal to (or close to) the wavelength. A metallic film deposited between the two gratings slightly modifies (to varying degrees, depending on its width) wave speeds

Wave speeds, and hence their phases, can be (slightly) altered by depositing a metallic film (e.g., aluminium) between the two reflection gratings. Such a modification is selective: it affects only those spectral components reflected by the first grating, which actually pass through it. The change in phase increases with film width.

The piezoelectric substrate is anisotropic. Wave vectors are related by

$$\boldsymbol{k}_{\mathrm{I}} + \boldsymbol{k}_{\mathrm{G}} = \boldsymbol{k}_{\mathrm{D}} , \tag{4.44}$$

where $k_{\mathrm{I}} = \omega/V_{\mathrm{I}}$ and $k_{\mathrm{D}} = \omega/V_{\mathrm{D}}$. Vector $\boldsymbol{k}_{\mathrm{G}} = 2\pi\boldsymbol{n}/p$ is the characteristic vector of the grating, where $p = \lambda/2$ is the grating pitch and $\boldsymbol{n}$ a unit vector perpendicular to the grooves. Consequently, the angle α at which the grating is inclined differs from 45° if speeds V_{I}, V_{D} are not equal:

$$\tan\alpha = \frac{V_{\mathrm{I}}}{V_{\mathrm{D}}} . \tag{4.45}$$

It is interesting to compare operating principles for these arrangements based on groove reflection gratings with those of ordinary systems using dispersive transducers.

- Technology for reflection grating filters is more sophisticated. Several thousand grooves are ruled using an ion gun. Their depths depend on time of exposure to the ion beam.
- The condition on the angle between gratings must be satisfied within the range of temperatures for which the filter is intended. However, speeds V_{I} and V_{D} general vary to different degrees with changes in temperature. Y-cut lithium niobate is more suitable than quartz, precisely because speed

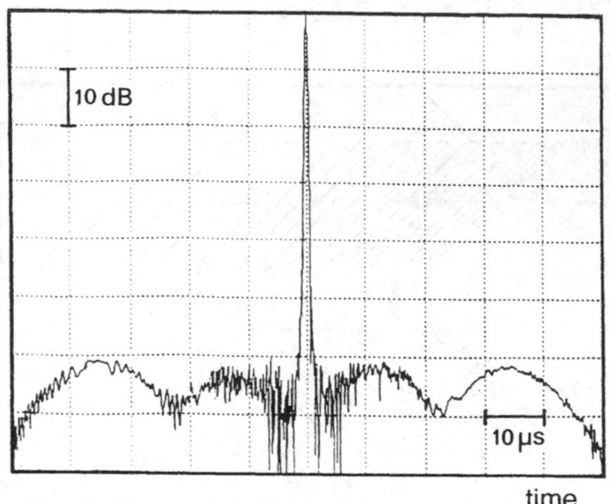

Fig. 4.56. Compressed pulse obtained with metallic strip dispersive gratings. The long pulse is digitally produced (Θ = 60 μs, B = 5 MHz). (Figure 8 in [4.68]. ©1992 IEEE)

variation coefficients are very close for this material. If two transmission lines are used, one as signal generator, and the other as filter matched to the signal, no temperature control is required.

- The radiation resistance of reflection grating filters is practically independent of their length, i.e., the number of grooves. This is not so for arrangements using dispersive transducers. In that case, radiation resistance falls (i.e., losses increase, Sect. 4.5.3.1) when the duration of the signal increases. This is because the capacitive part, in parallel with the active part, becomes more significant. Grating filters accept much longer pulses (up to 100 μs) than ordinary systems, and provide significantly bigger compression ratios (10 000 as opposed to only 1 000).
- More parameters are available to characterise reflection filters. Apart from the grating pitch, there is also groove depth and dimensions of the metallic film between the reflectors, which serves to correct phase errors.
- Bulk waves which may be excited by the emitter transducer have no consequence, being insensitive to effects of the gratings.

To sum up, filters built with dispersive groove reflection gratings, which can be analysed by the coupled mode method (see Sect. 2.5.1 and [4.67]), require much greater accuracy in their design and construction. This is due to their 2-dimensionality. They are considerably more expensive, but perform significantly better than simple 1-dimensional filters based on dispersive transducers.

An intermediate solution consists in replacing grooves by metallic strips. Figure 4.56 shows a compressed pulse [4.68] obtained using this technique [4.69]. The substrate is an ST-cut quartz crystal. Sidelobe level is extremely low (−60 dB). The long pulse ($\Theta = 60$ μs, $B = 5$ MHz) is a digital signal.

Figure 4.57 shows what this kind of filter looks like.

Fig. 4.57. Photograph of a dispersive array filter ($f_0 = 31.5$ MHz, $B = 1.5$ MHz, $\Theta = 100$ µs). Kindly communicated by B. King (Andersen)

4.5.3.3 Filter Matched to Discrete Codes. As we have already said, frequency modulation is an example of coding, widely used in radar. Discrete codes are used in telecommunications. The simplest is the binary code, a sequence of 0s and 1s (or 1, −1) produced at regular time intervals. Binary coding and decoding are easily achieved using Rayleigh waves, whether the data is carried by amplitude or phase. The principle is illustrated in Figs. 4.58 and 4.59 for the two cases. Applying a short pulse (shorter than the tran-

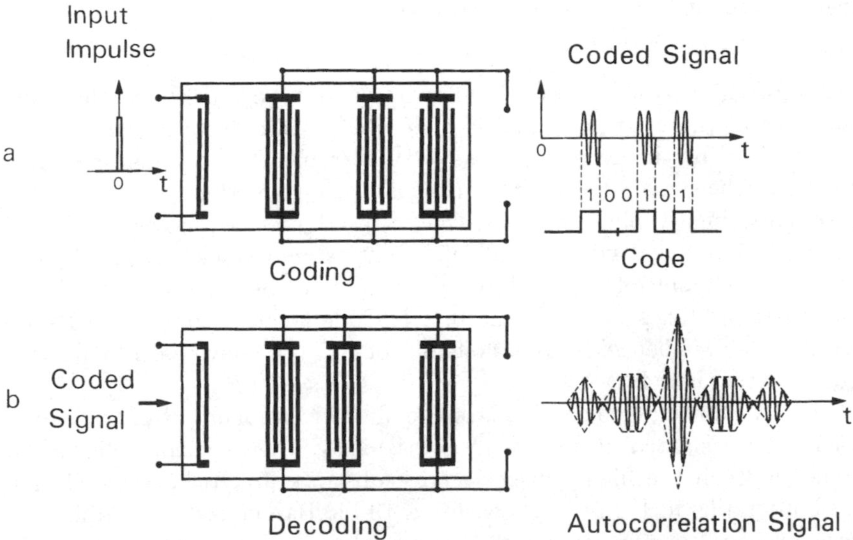

Fig. 4.58. Binary coding by amplitude modulation. (**a**) Signal generated by applying a short pulse at the input to the Rayleigh wave line. (**b**) Reception by the decoder, a filter matched to the signal. In fact, between coder output and decoder input, the signal propagates and suffers slight distortion due to noise (not shown). However, noise level is not affected by the decoder, whereas the signal level is increased. Signal-to-noise ratio is improved, as in the case of coding by frequency modulation

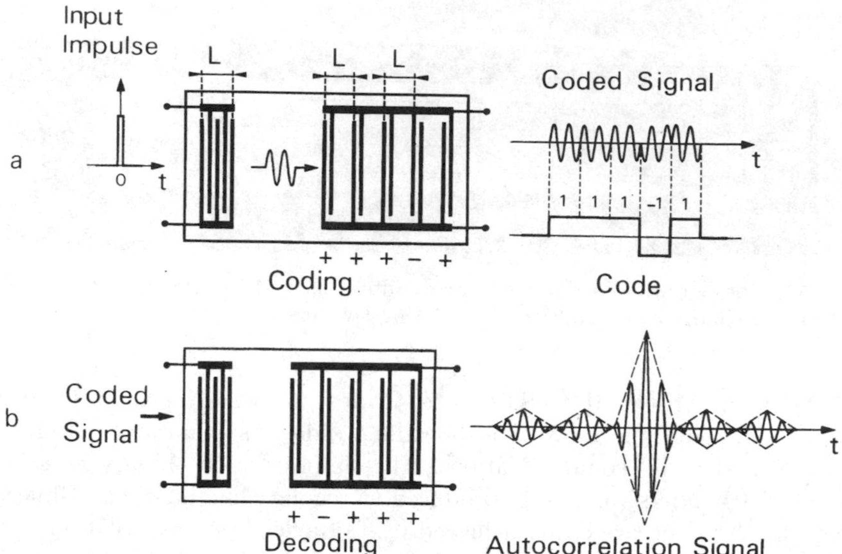

Fig. 4.59. Binary coding by phase modulation. 5-bit Barker sequence. (**a**) Generation of the signal. Phase opposition is produced by opposite connection of fingers in transducers receiving Rayleigh waves. (**b**) Reception by matched filter (decoder). Signal-to-noise ratio (not shown) is increased

sit time of the elastic wave between two consecutive fingers) to the input transducer of the coder, a signal is generated. After propagating, the signal is received by the decoder, whose output transducer, a transversal filter matched to the signal, delivers the autocorrelation function.

Amongst binary phase codings, the N-bit Barker sequences provide an autocorrelation signal with a peak of height N and sidelobes of height at most 1. At each multiple of the chip (duration of one bit), the autocorrelation signal therefore takes values N and 0, ± 1. Code length is limited to 13 bits when N is odd. When N is even, no codes of this type have been found with length greater than 4.

It is often advantageous to replace the aligned arrangement of transducer elements by a slanted arrangement, particularly when spurious effects due to reflections on the highly piezoelectric substrate are to be avoided. The two photographs in Fig. 4.60 show, for a 13-bit Barker code, the difference between responses with lithium niobate as substrate.

Elements of the coder output transducer may be distributed so as to produce a polyphased coding, i.e., the phase varies by $2\pi/n$ and not simply by π. In addition, modifying the active length of elements, sidelobes can be reduced. The same substrate can carry independent lines with complementary codes (e.g., Golay code). The resultant signal of autocorrelation signals leaving each system exhibits no sidelobes, as will be shown in Problem 4.11.

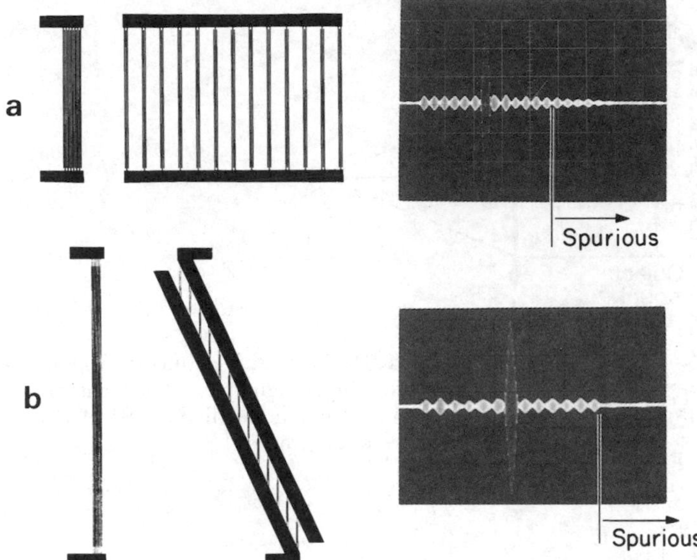

Fig. 4.60. Comparison of autocorrelation signals produced by a 13-bit Barker code, generated by Rayleigh waves on a lithium niobate substrate, for (**a**) aligned, (**b**) slanted transducer elements. (Picture kindly communicated by R.H. Tancrell, Raytheon-Waltham, Mass., USA)

4.5.4 Filters Based on Elasto-optic Interaction

Pattern recognition in optics appeals to the idea of the matched filter. The reference object is placed on the path of a light beam focussed by a lens. The various objects to be examined pass in front of the reference, like images in a film (Fig. 4.61). When the sought object passes, a maximum of light is produced at the focal point of the lens.

Since information must first be registered in some form, results of the analysis are delayed. However, using light diffraction by elastic waves, electrical signals can be processed in real time, as shown in Fig. 4.62. The signal, with fixed frequency and coded amplitude, is immediately transformed into an elastic wave beam (let us say, longitudinal waves). Light diffracted at the Bragg angle corresponding to the carrier frequency is concentrated at a point on the focal plane. An amplitude mask, a replica of the code, spatially modulates the beam of incident light. The detected light intensity, proportional to the square of the autocorrelation function of the signal, goes through a maximum when all elements of the wave train coincide with openings on the reference object. When the phase is coded, the amplitude mask is replaced by a phase plate.

Figure 4.63 shows the result of an experiment carried out by Torguet, R., Bauza, J.M., using a signal with binary coded phase (255 bits of length 20 ns, frequency 175 MHz), propagating along [001] in a lead molybdate crystal in

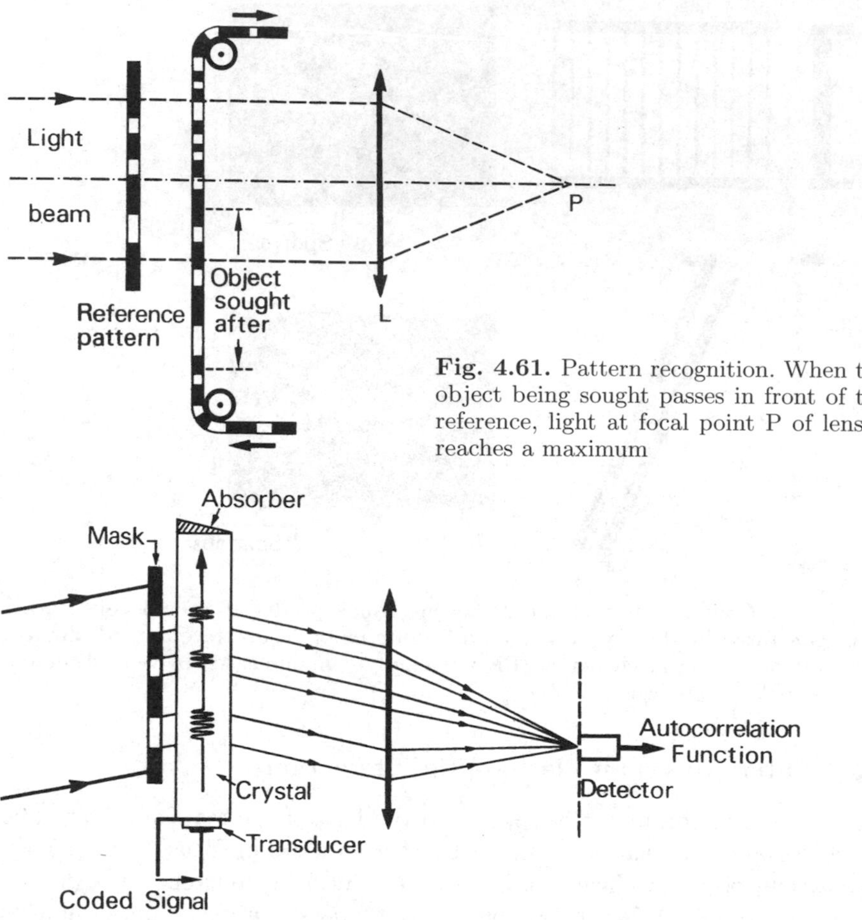

Fig. 4.61. Pattern recognition. When the object being sought passes in front of the reference, light at focal point P of lens L reaches a maximum

Fig. 4.62. Optical interaction filter. The intensity of the beam diffracted by the signal (converted into elastic waves) and concentrated at a point in the focal plane of the lens, goes through a maximum when the various elements of the wave train are located in front of openings in the mask (a replica of the code)

the form of longitudinal waves (speed 3 630 m/s). The signal diffracts a laser beam (6 328 Å) which has crossed a silica phase plate ($\pi \to 0.69$ μm). The 20 ns width of the correlation peak agrees with the theoretical prediction of 255 for the compression ratio.

The device is simpler when signal frequency is linearly modulated. Then the mask carrying a replica of the code becomes unnecessary. Indeed, the light beam is deflected through an angle θ proportional to elastic wave frequency F. This angle,

$$\theta \approx \frac{\lambda_0}{\Lambda} = \frac{\lambda_0}{V} F \,, \tag{4.46}$$

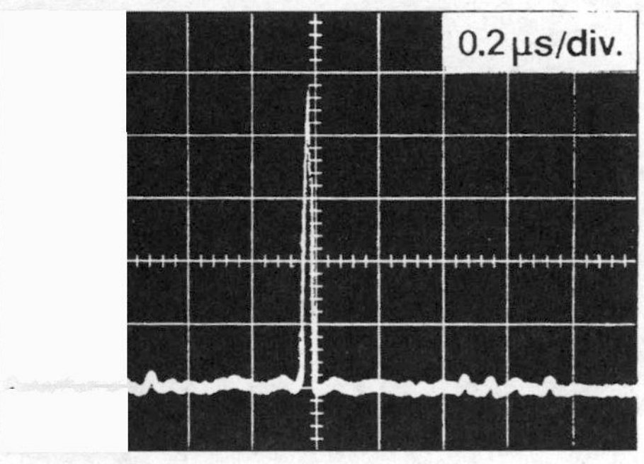

Fig. 4.63. Pulse compressed by an elasto-optic interaction filter matched to a signal with binary coded phase modulation (255 bits of length 20 ns)

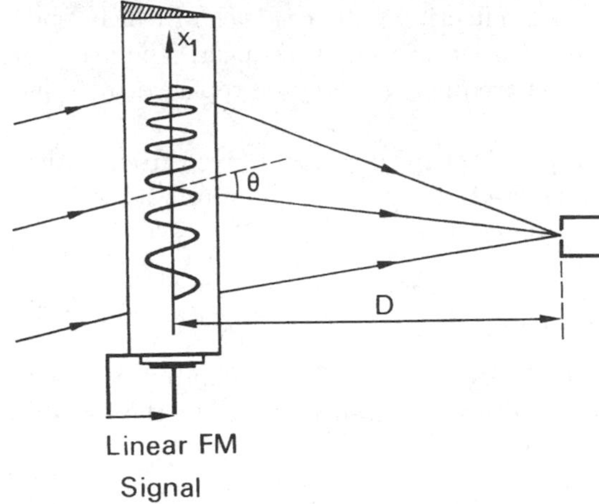

Fig. 4.64. Compression by Bragg diffraction of a pulse with linearly modulated frequency. Deflection angle θ varies with frequency, so that the diffracted beam converges at distance $D = \Theta V^2/B\lambda_0$

thus varies through the system according to the formula

$$\theta \approx \frac{\lambda_0}{V}F_0 + \frac{\lambda_0}{V}\frac{B}{\Theta}\left(t - \frac{x_1}{V}\right) ,$$

which takes frequency modulation into account [equation (4.24)], as well as propagation of the acoustic signal along x_1. The diffracted beam converges in mean direction $\theta_0 = \lambda_0 F_0/V$, at distance D given by (Fig. 4.64)

$$\frac{1}{D} = -\frac{\mathrm{d}\theta}{\mathrm{d}x_1} = \frac{\lambda_0}{V^2}\frac{B}{\Theta} . \tag{4.47}$$

When $B = 50$ MHz, $\Theta = 5$ μs, $\lambda_0 = 0.6328$ μm, $V = 3\,630$ m/s ($PbMoO_4$), this implies $D = 2.1$ m. The distance can be reduced by simple optical means.

Time variation of light intensity at the convergence point is proportional to the square of the autocorrelation function of the signal, whose frequency is linearly modulated. Correlation levels of several hundred have been obtained.

Needless to say, interest in optical calculators has inspired the study of devices capable of carrying out arithmetical operations. Fixed masks, replicas of amplitude or phase codes, are replaced by other Bragg modulators.

4.6 Spectrum Analysers and Fourier Transform Processors

Filters described above exploit the fact that an interdigital transducer generates (or detects) waves of frequency determined by its interdigital interval; and also that these waves reflect from metallic strip or groove gratings with period tuned to the frequency. Spectrometers can therefore be made using a transducer or a combination of transducers and reflection gratings. In both cases, the signal to be analysed excites a broadband transducer which converts it into a wave train. In the first case, this wave train propagates under several transducers, each of which extracts a particular spectral component. In the second, it passes under several reflection gratings, each of which sends the waves it reflects at 90° towards an associated transducer. The array of reflection gratings, tuned to different frequencies, may be replaced by a dispersive array.

In practice, it is in another type of setup, directly derived from optics, that dispersive array devices are inserted. They act as 'temporal lenses'.

The Fourier transform $X(f)$ of signal $x(t)$ is

$$X(f) = \int_{-\infty}^{\infty} x(\tau)\mathrm{e}^{-2\pi\mathrm{i}f\tau}\mathrm{d}\tau \,. \tag{4.48}$$

Dispersive devices have impulse responses of form $D_{\pm} = \exp(\pm\mathrm{i}\pi\beta t^2)$ [formula (4.27)]. In order to carry out the Fourier transform, make the variable change $f = \beta t$ and the transformation

$$-2t\tau = -t^2 - \tau^2 + (t-\tau)^2 \,.$$

Equation (4.48) becomes

$$X(\beta t) = \mathrm{e}^{-\pi\mathrm{i}\beta t^2} \int_{-\infty}^{\infty} \left[x(\tau)\mathrm{e}^{-\pi\mathrm{i}\beta\tau^2}\right] \mathrm{e}^{\pi\mathrm{i}\beta(t-\tau)^2}\mathrm{d}\tau \,,$$

or symbolically

$$X = D_- \bullet [(x \bullet D_-) \otimes D_+] \,. \tag{4.49}$$

The factor in square brackets is the output signal from a filter with impulse response $D_+(t)$ when the signal $x(t) \bullet D_-(t)$ is applied at its input.

A further decomposition of $X(\beta t)$ results from

$$\sqrt{\beta}\mathrm{e}^{-\pi\mathrm{i}/4} \int_{-\infty}^{\infty} \mathrm{e}^{\pi\mathrm{i}\beta(t-\tau)^2}\mathrm{d}\tau = 1 \,. \tag{4.50}$$

This follows from the standard formula $\int_{-\infty}^{\infty} \mathrm{e}^{-\pi x^2}\mathrm{d}x = 1$. Multiplying both sides by $X(\beta t)$, we find that

$$X(\beta t) = \sqrt{\beta}\mathrm{e}^{-\pi\mathrm{i}/4}\mathrm{e}^{\pi\mathrm{i}\beta t^2} X(\beta t) \int_{-\infty}^{\infty} \mathrm{e}^{\pi\mathrm{i}\beta\tau^2}\mathrm{e}^{-2\pi\mathrm{i}\beta t\tau}\mathrm{d}\tau \,.$$

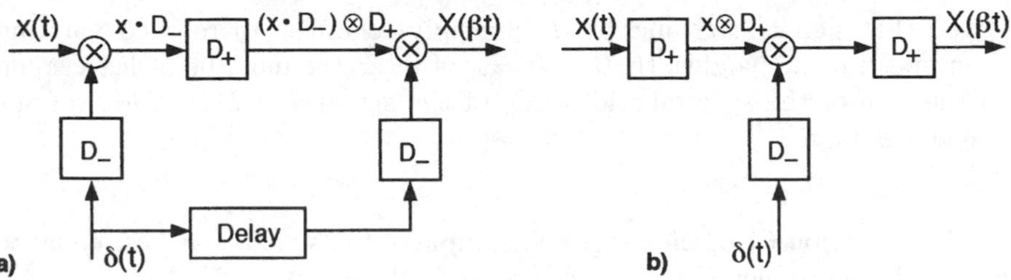

Fig. 4.65. Fourier transform processors. (**a**) MCM setup (Multiplication Convolution Multiplication) comprising two dispersive devices with impulse response $D_- = \exp(-\mathrm{i}\pi\beta t^2)$ and one with impulse response $D_+ = \exp(+\mathrm{i}\pi\beta t^2)$. (**b**) CMC setup (Convolution Multiplication Convolution) comprising two D_+ and one D_- dispersive devices

The integral is just the Fourier transform of $D_+(t)$ for $f = \beta t$. The product of Fourier transforms of $x(t)$ and $D_+(t)$ is the Fourier transform of the convolution product $x \otimes D_+$:

$$X(\beta t) = \sqrt{\beta}\mathrm{e}^{-\pi\mathrm{i}/4}\mathrm{e}^{\pi\mathrm{i}\beta t^2}\int_{-\infty}^{\infty}[x(\tau)\otimes D_+(\tau)]\mathrm{e}^{-2\pi\mathrm{i}\beta t\tau}\mathrm{d}\tau \,.$$

Hence,

$$X(\beta t) = \sqrt{\beta}\mathrm{e}^{-\pi\mathrm{i}/4}\int_{-\infty}^{\infty}\mathrm{e}^{\pi\mathrm{i}\beta(t-\tau)^2}\mathrm{e}^{-\pi\mathrm{i}\beta\tau^2}[x(\tau)\otimes D_+(\tau)]\,\mathrm{d}\tau \,.$$

The integral is a convolution product, so that $X(\beta t)$ can be written symbolically:

$$X(\beta t) = \sqrt{\beta}\mathrm{e}^{-\pi\mathrm{i}/4}D_+\otimes[D_-\bullet(x\otimes D_+)] \,. \tag{4.51}$$

Note that this expression can be deduced from (4.49), up to a factor, by permuting convolutions with multiplications, and swapping D_+ for D_-. Figure 4.65 shows the setups corresponding to these two situations.

In practice, signals D_+ and D_- are impulse responses of Rayleigh wave dispersive devices. These arrangements for real time spectral analysis are made possible by the existence of reflective array devices with large $B\Theta$ (several thousand), supplying signals with well defined amplitude and phase. These devices play an analogous role to lenses in optical arrangements. Their characteristics determine those of the analysers they make up. In the circuit of Fig. 4.65a, the finite duration Θ_- of the impulse response of the first device divides signal analysis into time intervals (e.g., 20 μs). As a result, if the wave applied at input is sinusoidal with frequency f, the system produces a $\sin x/x$ curve of width inversely proportional to Θ_-, rather than a line at this frequency. It is this width which defines resolution. The level (−13 dB) of undesirable sidelobes can be diminished by weighting. However, as we have already noted in Sect. 4.5.3.1, and with the help of Problem 4.10, there are then adverse effects on resolution, due to broadening of the central peak.

Given that signals $x(t)$ and $D_-(t)$ multiply together, thereby convoluting their spectra, the bandwidth $B_+ = \beta\Theta_+$ of filter D_+ must be at least equal to the sum of the spectral widths B_s of the signal and $B_- = \beta\Theta_-$ of the dispersive filter:

$$\beta\Theta_+ = B_s + \beta\Theta_- \ .$$

The duration Θ_s of the processed sample of the signal is at best equal to Θ_-, and the product $B_s\Theta_s$ for the system is then

$$B_s\Theta_s = \beta(\Theta_+ - \Theta_-)\Theta_- \ .$$

It is maximal and equal to $\beta\Theta_+^2/4 = B_+\Theta_+/4$ when $\Theta_- = \Theta_+/2$. The spectral range which can be exploited is thus $\beta\Theta_+/2$ and the resolution is $1/\Theta_- = 2/\Theta_+$. For typical centre frequencies (a few hundred [MHz]) and dispersive filter lengths (< 10 cm), the spectral range extends to several tens of [MHz] and the resolution is of order a few tens of [kHz].

These Fourier transform processors also have applications outside spectral analysis [4.70]. For example, they can be used as a continuously variable delay line. They also replace a correlation integral by a simple product.

As spectral components appear successively in time, they can be modified or selected by means of a gate. The device then filters frequencies. An example of application in satellite communications is described in [4.71]. Fourier transform processors are used to demodulate signals with coded frequency.

4.7 Convolvers

Devices described up to now operate in linear conditions. Indeed, it has been implicitly assumed that strain amplitudes, that is, relative variation of distances in the solids, remain much smaller than the threshold ($S < 10^{-4}$) beyond which simplified expressions for strain and Hooke's law are no longer valid (Sect. 3.2.1, Vol. I). When stresses and strains (and, as the case may be, electric fields) are no longer proportional, the linear theory of the past few chapters can no longer be applied. Now, a convolver generates a signal $e_3(t)$ from two signals $e_1(t)$ and $e_2(t)$ in the following way:

$$e_3(t) = \int_{-\infty}^{\infty} e_1(\tau)e_2(t-\tau)\,\mathrm{d}\tau \ . \tag{4.52}$$

The product in the integrand brings in non-linear effects.

Analysis of non-linear effects requires extra elastic constants. A first approach consists in adding a term of next highest order to equations used for the linear theory. For example, the relation between stresses and strains becomes

$$T_{ij} = c_{ijkl}S_{kl} + \frac{1}{2}c_{ijklpq}S_{kl}S_{pq} \ .$$

For a piezoelectric material, extra terms in E_kE_l and $S_{kl}E_m$ are now put in.

Third order terms such as $c_{ijklpq}S_{ij}S_{kl}S_{pq}$ are adjoined to the elastic potential energy quadratic form [equation (3.54), Vol. I]. Third order elastic constants significantly complicate the equation for propagation of elastic waves. The equation can nevertheless be treated in sufficiently simple cases (longitudinal waves in an unbounded non-piezoelectric medium, for example). The results, qualitatively predictable without calculation, are as follows:

- a monochromatic wave generates harmonic frequency waves as it propagates;
- two waves of frequencies f_1 and f_2 which interact in the solid give rise to waves of frequencies $f_1 + f_2$ and $f_1 - f_2$.

The study of non-linear effects is complicated even further when boundary conditions are taken into account, particularly in the case of surface waves. Mathematical study of the non-linear theory goes beyond the scope of this book. However, its use in forming products is within our reach.

Consider two signals with modulated amplitude:

$$x_1(t) = e_1(t)\cos\omega_1 t\ , \quad x_2(t) = e_2(t)\cos\omega_2 t\ .$$

These are applied, with electrical powers P_1 and P_2, respectively, to two transducers. The latter generate two waves u_1, u_2 propagating in opposite directions at the same speed V:

$$u_1(t,\, x) \propto P_1^{1/2} e_1(t - x/V)\cos(\omega_1 t - k_1 x)\ ,$$
$$u_2(t,\, x) \propto P_2^{1/2} e_2(t + x/V)\cos(\omega_2 t + k_2 x)\ .$$

Their non-linear interaction generates two further waves:

$$u_{3\pm}(t,\, x) \propto (P_1P_2)^{1/2} e_1(t - x/V) e_2(t + x/V)\cos[(\omega_1 \pm \omega_2)t + (k_1 \mp k_2)x]\ .$$

A transducer of length L detecting the wave with wave number $k_1 - k_2$ produces a voltage of carrier frequency $\omega_1 + \omega_2$ and amplitude

$$e_3(t) \propto (P_1P_2)^{1/2} \int_{-L/2}^{L/2} e_1\left(t - \frac{x}{V}\right) e_2\left(t + \frac{x}{V}\right) \mathrm{d}x\ . \tag{4.53}$$

If the two signals are short compared with the transit time L/V under the transducer, the latter can be treated as infinitely long. Change of variable $\tau = t - x/V$ reveals $e_3(t)$ as the convolution of the envelopes of the two signals:

$$e_3(t) \propto (P_1P_2)^{1/2} \int_{-\infty}^{\infty} e_1(\tau) e_2(2t - \tau)\,\mathrm{d}\tau\ . \tag{4.54}$$

The power gathered is proportional to P_1P_2.

The *bilinearity factor*

$$B = 10\log\frac{P_3}{P_1P_2} \tag{4.55}$$

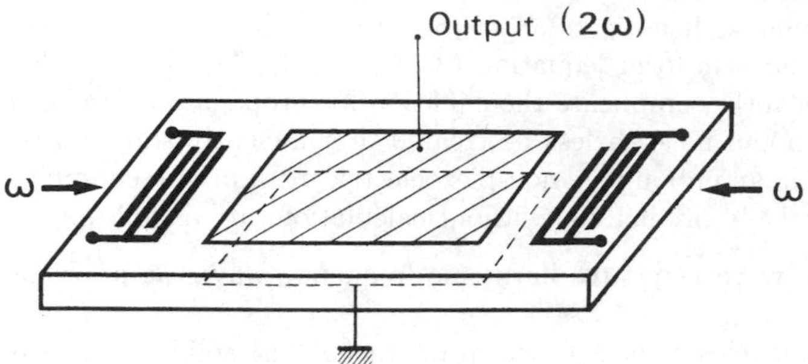

Fig. 4.66. Experiment to demonstrate convolution by non-linear effects (between Rayleigh waves) of two signals with the same frequency

characterises the elastic and/or piezoelectric non-linearity of the material for this configuration.

The fact that $e_3(t)$ is compressed by a factor of 2 relative to the ordinary convolution product occurs because the two waves carrying the two terms in the product propagate in opposite directions. Indeed, the relative speed of these two waves is $2V$. Equation (4.54) should be compared with the general formula (4.2), giving the output signal of a linear system as the convolution of input signal and impulse response. One of the input signals of the convolver plays the role of impulse response. But the fact that this response can be modified (up to certain restrictions on its spectrum and duration) shows the advantage of the convolver over a simple linear filter. If one of the signals, considered as reference, is the time reversal of the other signal, what is output is the autocorrelation function. The convolver thus operates as a filter which can be rapidly matched (the reference being controlled electronically) to different signals, coded in different ways.

Rayleigh waves propagating at the surface of a piezoelectric material are easily submitted to this principle. This is because they have high power density, due to the concentration of elastic energy close to the surface, over a depth of the same order as their wavelength. The convolution product is detected by an interdigital transducer with interdigital distance $d = \lambda_3/2$ fixed by the wavelength $\lambda_3 = 2\pi/|k_1 - k_2|$. When the two signals have the same frequency, λ_3 is infinite, and the transducer reduces to a plate. (This degenerate convolver has the advantage, compared with the non-degenerate interdigital transducer, of exciting almost no reflected bulk waves.) The associated electric field of frequency 2ω is now time independent. It can be detected by the voltage it produces between this plate and another electrode placed nearby (underneath, or even beside it). The idea is illustrated by the simple experiment described in Figs. 4.66 and 4.67.

One way of increasing the non-linear interaction is to increase power density by narrowing the beam. This is done by means of concave transducers or a beam compressor, i.e., a guide of variable cross-section or an asymmetric

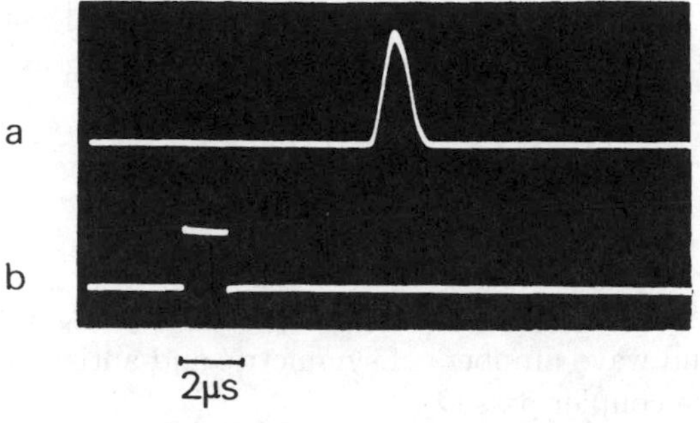

Fig. 4.67. (**a**) Autocorrelation of (**b**) a rectangular pulse. Carrier frequency 150 MHz. The base of the triangle is equal to the pulse width and not twice that value

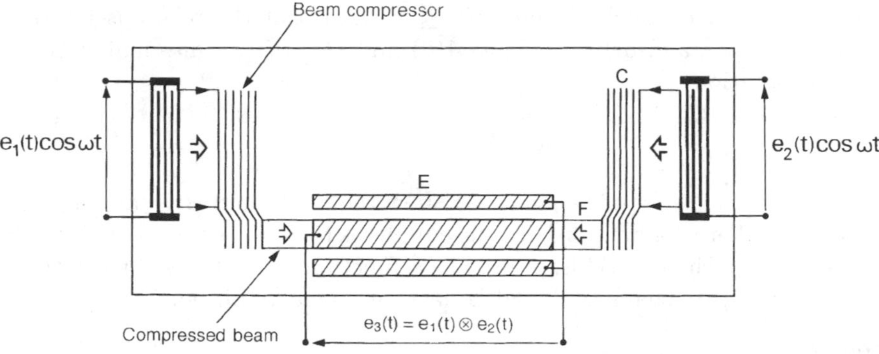

Fig. 4.68. Rayleigh wave convolver. The product of signals $e_1(t)$, $e_2(t)$ is carried out by non-linear interaction of the waves of angular frequency ω which carry them. Spatial integration is realised by means of long electrodes E. These detect the induced voltage of angular frequency 2ω. Wave beams F from each transducer are narrowed by a compressor C

coupler (Sect. 4.1). Figure 4.68 illustrates the last case. The convolution product is detected between a central electrode, which also acts as $\Delta V/V$ type wave guide, and two lateral ribbon-shaped electrodes, connected together electrically. This has the advantage of being a plane structure.

Characteristics of such a convolver are approximately [4.72]:

Material	Lithium niobate
Bilinearity factor	> -70 dB
Centre frequency	$f_0 = 150$ MHz
Spectral width of signals	40 MHz
Maximal input power	1 W
Process time	$\Theta = 12$ μs
Dynamic range	< 70 dB

A $ZnO/SiO_2/Si$ layered structure with Sezawa mode (vertical shear surface wave in a layer) has been studied at the University of Tohoku at Sendai, Japan [4.73]. This work led to design of a convolver, developed by Clarion ($f_0 = 210$ MHz, bandwidth 23 MHz, $\Theta = 9$ µs, output signal level ≈ -42 dBm, dynamic range 50 dB). It is the output signal level which explains its use in several systems [4.74, 4.75].

Problems

4.1 Consider the multistrip coupler of length L shown in Fig. 4.3a. Let M_{s}, k_{s} and M_{a}, k_{a} be the amplitudes and wave numbers of symmetric and antisymmetric modes with respect to the coupler axis Ox.

(a) Write expressions for complex amplitudes $A(x)$, $B(x)$ of the Rayleigh wave in coupler channels A and B ($0 \le x \le L$). Put the relations between $A(L)$ and $B(L)$, and between $A_0 = A(0)$ and $B_0 = B(0)$, into matrix form. Use the notation

$$k = \frac{k_{\mathrm{a}} + k_{\mathrm{s}}}{2} , \quad \phi = \frac{k_{\mathrm{a}} - k_{\mathrm{s}}}{2} L .$$

(b) The wave is incident in channel A. Show that $A(L)$ and $B(L)$ are in phase quadrature. For what length L_{t} of the coupler is the incident wave completely transmitted in channel B? Express ϕ as a function of L_{t}. For what length do outgoing waves in channels A and B have the same amplitude?

Solution.
(a) Expressions

$$\left. \begin{aligned} A(x) &= M_{\mathrm{s}} \mathrm{e}^{-\mathrm{i}k_{\mathrm{s}}x} + M_{\mathrm{a}} \mathrm{e}^{-\mathrm{i}k_{\mathrm{a}}x} \\ B(x) &= M_{\mathrm{s}} \mathrm{e}^{-\mathrm{i}k_{\mathrm{s}}x} - M_{\mathrm{a}} \mathrm{e}^{-\mathrm{i}k_{\mathrm{a}}x} \end{aligned} \right\} , \quad \text{where} \quad \left. \begin{aligned} M_{\mathrm{s}} &= (A_0 + B_0)/2 \\ M_{\mathrm{a}} &= (A_0 - B_0)/2 \end{aligned} \right\} ,$$

can be written

$$A(x) = \frac{1}{2} A_0 \left(\mathrm{e}^{-\mathrm{i}k_{\mathrm{s}}x} + \mathrm{e}^{-\mathrm{i}k_{\mathrm{a}}x} \right) + \frac{1}{2} B_0 \left(\mathrm{e}^{-\mathrm{i}k_{\mathrm{s}}x} - \mathrm{e}^{-\mathrm{i}k_{\mathrm{a}}x} \right) ,$$

$$B(x) = \frac{1}{2} A_0 \left(\mathrm{e}^{-\mathrm{i}k_{\mathrm{s}}x} - \mathrm{e}^{-\mathrm{i}k_{\mathrm{a}}x} \right) - \frac{1}{2} B_0 \left(\mathrm{e}^{-\mathrm{i}k_{\mathrm{s}}x} + \mathrm{e}^{-\mathrm{i}k_{\mathrm{a}}x} \right) .$$

In matrix form, this becomes

$$\begin{pmatrix} A(L) \\ B(L) \end{pmatrix} = \begin{pmatrix} \cos\phi & \mathrm{i}\sin\phi \\ \mathrm{i}\sin\phi & \cos\phi \end{pmatrix} \begin{pmatrix} A_0 \\ B_0 \end{pmatrix} \mathrm{e}^{-\mathrm{i}kL} . \tag{4.56}$$

(b) When $B_0 = 0$, amplitudes $A(L) = A_0 \cos\phi\, \mathrm{e}^{-\mathrm{i}kL}$, $B(L) = \mathrm{i} A_0 \sin\phi\, \mathrm{e}^{-\mathrm{i}kL}$ are $\pi/2$ out of phase. If $\phi = \pi/2$, the Rayleigh wave incident on channel A is transmitted entirely through channel B:

$$\phi = \frac{k_{\mathrm{a}} - k_{\mathrm{s}}}{2} L = \frac{\pi}{2} \frac{L}{L_{\mathrm{t}}} , \quad L_{\mathrm{t}} = \frac{\pi}{k_{\mathrm{a}} - k_{\mathrm{s}}} .$$

$A(L)$ and $B(L)$ have the same amplitude $A_0/\sqrt{2}$ if $\phi = \pi/4$, i.e., if $L = L_{\mathrm{t}}/2$ (coupler at 3 dB).

4.2 Consider a device consisting of two delay lines connected in parallel. The delay lines are identical except that they produce delays differing by τ. Show that the frequency response of the system has zeros and find the corresponding frequencies.

Solution. By the delay theorem, the frequency response corresponding to the impulse response $h(t) + h(t - \tau)$ of the system is

$$\mathcal{H}(f) = H(f) + H(f)\mathrm{e}^{-2\pi \mathrm{i} f\tau} = 2H(f)\mathrm{e}^{-\pi \mathrm{i} f\tau} \cos \pi f\tau \,.$$

This is zero at frequencies given by

$$f_n = \left(n + \frac{1}{2}\right)\frac{1}{\tau} \,.$$

Figure 4.69 shows the response of a filter of centre frequency 100 MHz. Zeros are spaced at 3.3 MHz intervals.

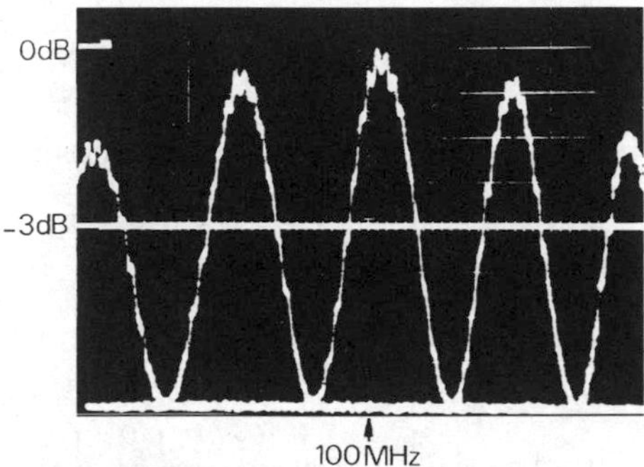

Fig. 4.69. Rejection filter. Adjacent zeros are separated by 3.3 MHz

4.3 A system is composed of an emitter transducer with N slanted fingers of constant length w and a straight receiver (Fig. 4.70). What is the frequency response of this system when sources are inclined at angles θ_n which vary linearly from $-\theta_0/2$ to $+\theta_0/2$ with varying abscissa $x_n = n\lambda_0/2$ (assuming θ_0 to be small)?

Solution. Decomposing each source into elements $\mathrm{d}z$, the signal from the nth source is

$$A_n = \frac{A_0}{w}\mathrm{e}^{-\mathrm{i}\omega l/V} \int_{-w/2}^{w/2} \mathrm{e}^{\mathrm{i}\omega x/V}\mathrm{d}z \,, \quad x = x_n + \theta_n z \,.$$

A_0 is the amplitude of the wave emitted by a straight source of length w. It follows that

$$A_n = A_0\mathrm{e}^{-\mathrm{i}\omega(l-x_n)/V} \frac{\sin[(w\theta_n/2V)\omega]}{(w\theta_n/2V)\omega} \,.$$

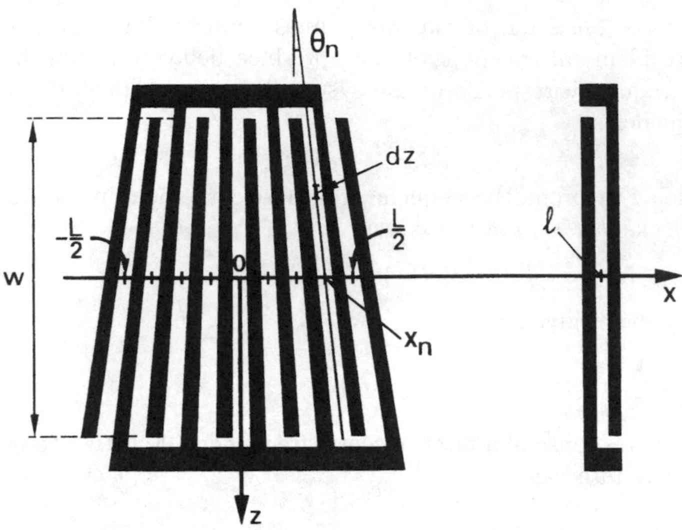

Fig. 4.70. Fan-shaped transducer

For large N, sources are distributed quasi-continuously and the frequency response

$$H(\omega) = \sum_n (-1)^n A_n = \sum_n A_n \cos \frac{\omega_0 x_n}{V}$$

is given by the integral

$$H(\omega) = \int_{-L/2}^{L/2} \frac{A(x)}{L} \cos \frac{\omega_0 x}{V} \, \mathrm{d}x \,, \quad A(x) = A_0 \mathrm{e}^{\mathrm{i}\omega x/V} \frac{\sin[(w\theta(x)/2V)\omega]}{(w\theta(x)/2V)\omega} \,,$$

where the mean delay $\tau = l/V$ has been omitted. For linear variation $\theta = \theta_0 x/L$, we find

$$H(\omega) = \int_{-L/2}^{L/2} \frac{A_0}{L} \frac{\sin[(w\theta_0/2LV)\omega x]}{(w\theta_0/2LV)\omega x} \cos \frac{\omega_0 x}{V} \mathrm{e}^{\mathrm{i}\omega x/V} \mathrm{d}x \,.$$

Extending integration limits to $\pm\infty$ ($N \gg 1$) and assuming angular frequency equal to ω_0 in the $\sin\alpha/\alpha$ factor (narrow bandwidth), the frequency response is the Fourier transform of a $\sin\alpha/\alpha$ function with carrier:

$$H(\omega) = \frac{A_0 V}{L} \int_{-\infty}^{\infty} \frac{\sin \pi B t}{\pi B t} \cos(2\pi f_0 t) \mathrm{e}^{2\pi \mathrm{i} f t} \mathrm{d}t \,, \quad B = \frac{w\theta_0 f_0}{L} \,.$$

This is a box function with centre frequency $\pm f_0$:

$$H(\omega) = \frac{A_0 \lambda_0}{2w\theta_0} \left[\Pi\left(\frac{f - f_0}{B}\right) + \Pi\left(\frac{f + f_0}{B}\right) \right] \,. \tag{4.57}$$

The relative bandwidth is $B/f_0 = w\theta_0/L$.

4.4 Consider a filter made from interdigital transducers with constant finger lengths. Show that a signal $h(t)$ with modulated amplitude can be treated as the sum of two signals of constant amplitude and modulated phase.

Solution. Given that

$$h(t) = e(t)\cos\omega_0 t = \cos\psi(t)\cos\omega_0 t ,$$

we can write

$$h(t) = \frac{1}{2}\cos[\omega_0 t + \psi(t)] + \frac{1}{2}\cos[\omega_0 t - \psi(t)] , \tag{4.58}$$

where $e(t) = \cos\psi(t)$, or $\psi(t) = \arccos e(t)$. The signal $h(t)$ with modulated amplitude is the sum of two signals of constant amplitude and modulated phase.

If $h(t)$ is an even function, i.e., $\psi(-t) = \psi(t)$, then denoting the first term on the right hand side of (4.58) by $g(t)$, we have

$$h(t) = g(t) + g(-t) . \tag{4.59}$$

The impulse response (4.59) has been produced [4.76]. The filter contains a central transducer T_0 implementing g and two lateral wideband transducers T_1, T_2. A short pulse applied to T_0 generates two waves propagating in opposite directions. One representing $g(t)$ excites T_1, the other representing $g(-t)$ excites T_2. Adding signals provided by T_1 and T_2 yields $x(t)$. Finger lengths are constant and the wave front is not distorted (Sect. 2.1.1).

4.5 What is the frequency response of a transducer comprising an infinite number of samples, spaced at intervals of $n\lambda_0$ and each containing 5 fingers?

Solution. The impulse response is a succession of pulses of width $2/f_0$, occurring at intervals $\tau = n/f_0$:

$$h(t) = \sum_{m=-\infty}^{\infty} \Pi\left[\frac{f_0(t - m\tau)}{2}\right] \cos 2\pi f_0 t .$$

The envelope has spectrum

$$E(f) = \frac{\sin(2\pi f/f_0)}{\pi f} \sum_{m=-\infty}^{\infty} \mathrm{e}^{-2\pi i m f\tau} .$$

Hence, using Poisson's formula

$$\sum_{m=-\infty}^{\infty} \mathrm{e}^{-2\pi i m x} = \sum_{p=-\infty}^{\infty} \delta(x - p) ,$$

this becomes

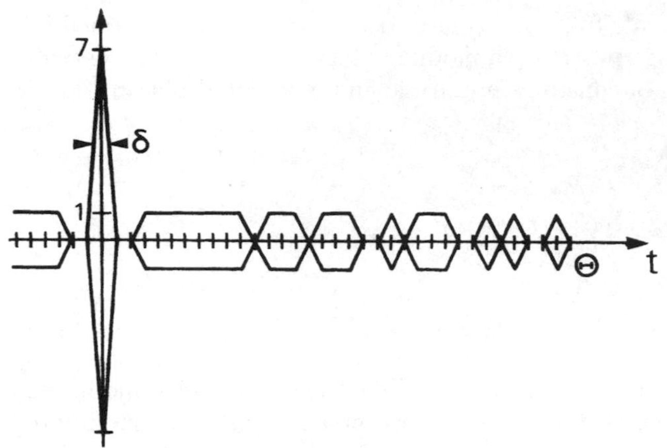

Fig. 4.71. Signal autocorrelation

$$E(f) = \frac{\sin(2\pi f/f_0)}{\pi f} \sum_{p=-\infty}^{\infty} \delta(f\tau - p) .$$

The frequency response

$$H(f) = \frac{1}{2}E(f - f_0) + \frac{1}{2}E(f + f_0)$$

is a succession of peaks at frequencies $f = \pm f_0 + pf_0/n$, with heights modulated by a $\sin x/x$ curve.

4.6 Find the response of the filter matched to the signal in Fig. 4.44b.

Solution. The autocorrelation of the signal, found graphically, is sketched in Fig. 4.71.

4.7 A conventional radar is equipped with a receiver filter whose frequency response is constant over the whole width B of the signal spectrum. The signal has duration Θ and constant amplitude A. Calculate the signal-to-noise ratio as defined in (4.19).

Solution. The output signal has amplitude $|y(t_0)| = |H(f)|A$, and the square of the noise has amplitude

$$b^2 = \int_0^\infty |H(f)|^2 N_0 \mathrm{d}f = BN_0|H(f)|^2 .$$

The signal-to-noise ratio can be written

$$\left(\frac{\mathrm{S}}{\mathrm{N}}\right)_{\mathrm{o}} = \frac{|y(t_0)|^2}{b^2} = \frac{A^2}{BN_0} = \frac{2E}{BN_0\Theta} , \tag{4.60}$$

where $E = \int_0^\infty x^2(t)\,\mathrm{d}t = A^2\Theta/2$ is the signal energy.

4.8 Calculate the spectrum of signal $x(t) = e(t)\cos\phi(t)$ with even envelope $e(t)$ and frequency symmetrically modulated about the centre frequency ω_0, i.e., $\phi(t) = \omega_0 t + \psi(t)$, where $\psi(-t) = \psi(t)$.

Solution. Expanding out the cosine, the spectrum of $x(t)$ can be written

$$X(\omega) = \frac{1}{2}\int_{-\infty}^{\infty} e(t)\exp \mathrm{i}[(\omega_0 - \omega)t + \psi(t)]\,\mathrm{d}t$$
$$+\frac{1}{2}\int_{-\infty}^{\infty} e(t)\exp -\mathrm{i}[(\omega_0 + \omega)t + \psi(t)]\,\mathrm{d}t\;.$$

The spectrum of $e(t)\mathrm{e}^{\mathrm{i}\psi(t)}$ is

$$E(\omega) = \int_{-\infty}^{\infty} e(t)\exp \mathrm{i}[\psi(t) - \omega t]\,\mathrm{d}t\;,$$

and $X(\omega)$ becomes

$$X(\omega) = \frac{1}{2}E(\omega - \omega_0) + \frac{1}{2}E^*(-\omega_0 - \omega)\;. \tag{4.61}$$

When $e(-t) = e(t)$ and $\psi(-t) = \psi(t)$, so that $E(-\omega) = E(\omega)$, this means that

$$X(\omega) = \frac{1}{2}E(\omega - \omega_0) + \frac{1}{2}E^*(\omega_0 + \omega)\;. \tag{4.62}$$

4.9 Using the stationary phase method (Appendix G), calculate the spectrum of the signal defined by (4.25) and (4.26).

Solution. Since envelope and frequency modulation of the signal $x(t)$ are symmetric, equation (4.62) of the previous problem applies with

$$E(\omega) = \int_{-\infty}^{\infty} \varPi\left(\frac{t}{\varTheta}\right)\mathrm{e}^{\mathrm{i}\alpha(t)}\mathrm{d}t\;, \quad \alpha(t) = \mu\frac{t^2}{2} - \omega t\;.$$

This integral has the same form as (G.1). At a given frequency ω_p, phase $\alpha(t)$ is stationary at time t_p such that

$$\frac{\mathrm{d}\alpha}{\mathrm{d}t} = 0 \Rightarrow t_\mathrm{p} = \frac{\omega_\mathrm{p}}{\mu}\;, \quad \alpha_\mathrm{p} = -\frac{\omega_\mathrm{p}^2}{2\mu} \quad \text{and} \quad \alpha''_\mathrm{p} = \left.\frac{\mathrm{d}^2\alpha}{\mathrm{d}t^2}\right|_{t_\mathrm{p}} = \mu\;.$$

Applying (G.5) in Appendix G yields

$$E(\omega) = \sqrt{\frac{2\pi}{\mu}}\varPi\left(\frac{\omega}{\mu\varTheta}\right)\exp\left[\mathrm{i}\left(\frac{\pi}{4} - \frac{\omega^2}{2\mu}\right)\right]\;, \quad \mu\varTheta = 2\pi B\;. \tag{4.63}$$

The spectrum

$$X(f) = \frac{1}{2}E(f - f_0) + \frac{1}{2}E^*(f + f_0)$$

is composed of two rectangles of width B, centred on frequencies $\pm f_0$. The phase varies parabolically with frequency. The full calculation shows that the stationary phase approximation becomes better as $B\Theta$ grows large compared with 1.

4.10 Using the stationary phase method, calculate the compressed pulse $y(t)$ output from the filter matched to a signal $x(t)$ with S-shaped frequency modulation given by

$$t = \Theta\left[\frac{f-f_0}{B} + \frac{a}{2\pi}\sin\left(2\pi\frac{f-f_0}{B}\right)\right], \quad a < 1,$$

for $-\Theta/2 < t < \Theta/2$, or $f_0 - B/2 < f < f_0 + B/2$.

Solution. The signal $y(t)$ can be deduced from its spectrum $Y(f) = H(f)X(f) = |X(f)|^2$:

$$y(t) = \int_{-\infty}^{\infty} |X(f)|^2 \mathrm{e}^{2\pi i f t}\mathrm{d}f .$$

Since $e[t(f)] = \Pi[(f-f_0)/B]$, equation (G.8) in Appendix G implies

$$|E(f-f_0)|^2 \approx \Pi\left(\frac{f-f_0}{B}\right)\frac{\Theta}{B}\left[1 + a\cos\left(2\pi\frac{f-f_0}{B}\right)\right].$$

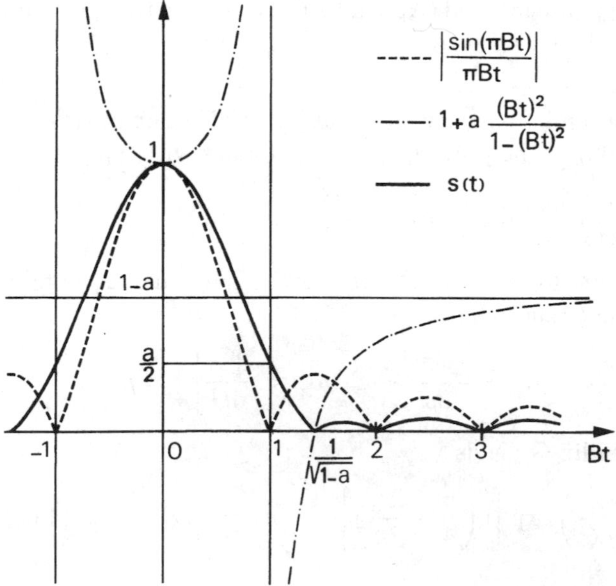

Fig. 4.72. Envelope of the compressed pulse delivered by a dispersive filter with S-shaped delay curve when $a = 1/2$ (*full curve*)

From (4.62),

$$|X(f)|^2 = \frac{1}{4}|E(f - f_0)|^2 + \frac{1}{4}|E(f + f_0)|^2 .$$

The envelope of the output signal $y(t) = s(t)\cos(2\pi f_0 t)$ is

$$s(t) = \frac{\Theta}{2B} \int_{-B/2}^{B/2} \left(1 + a\cos\frac{2\pi f}{B}\right) e^{2\pi i f t} df ,$$

or

$$s(t) = \frac{\Theta}{2} \frac{\sin \pi B t}{\pi B t} \left[1 + a\frac{(Bt)^2}{1-(Bt)^2}\right] . \tag{4.64}$$

Figure 4.72 shows that multiplying the term $\sin \pi Bt/\pi Bt$ by the function in square brackets, which vanishes when $Bt = 1/\sqrt{1-a}$, has the effect of broadening the central peak and reducing the first sidelobe. The latter is reduced much more than the other sidelobes, which are multiplied by $1 - a$.

4.11 Show that the sum of autocorrelation signals for each pair of signals in Fig. 4.73a has no sidelobes.

Solution. See Fig. 4.73b.

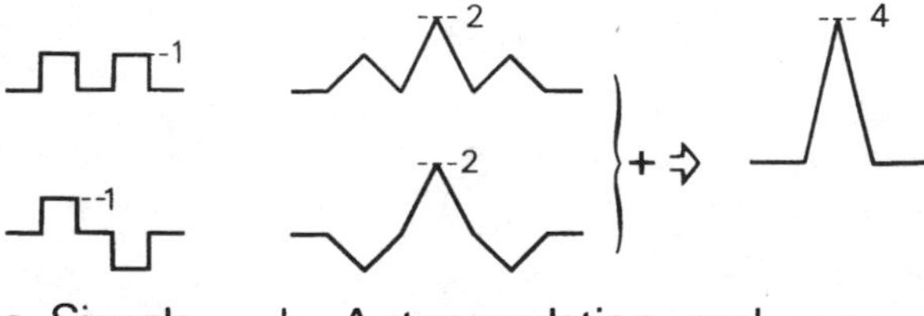

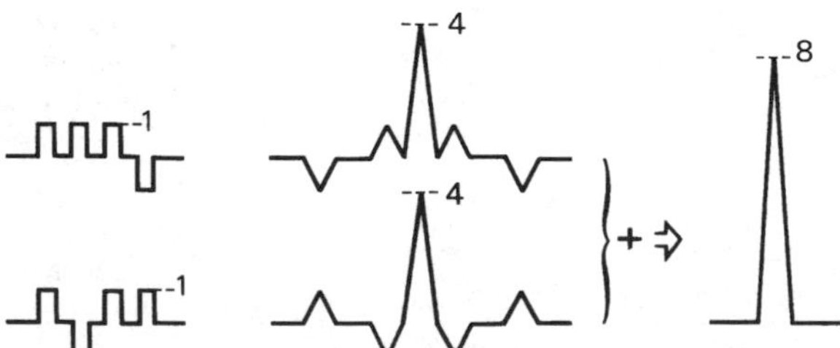

Fig. 4.73. (**a**) Golay complementary codes. (**b**) Autocorrelation

5. Sensors and Instrumentation

Scientific development of systems control, computing and instrumentation has produced a need for new sensors. These are devices transforming some physical quantity (e.g., thickness of a metallic film, liquid level in a tank, position of an object, a pressure, a temperature) into an electrical quantity (e.g., voltage, current intensity, wave frequency). An electrical signal is easier to handle, in the sense of processing (amplification) or transfer. Like any other wave, elastic waves can be used to design sensors, because their characteristics (i.e., speed, mode, amplitude, frequency) are altered when the propagation medium is perturbed. But some of their properties already mentioned explain why they play a different role to electromagnetic waves, such as light, in sensing and processing of signals. Let us review their most important properties:

- low speed ($1000 < V < 12\,000$ m/s), and hence comparable wavelength to light ($1\ \mu\text{m} < \lambda < 12\ \mu\text{m}$) for frequencies above $f = 1000$ MHz;
- propagation through opaque solids;
- generation and detection via the piezoelectric effect over a wide range of frequencies ($0.1\ \text{MHz} < f < 10\ \text{GHz}$), and using magnetorestriction at low frequencies (up to 1 MHz);
- modes with mechanical surface displacements accompanied by an electric field.

The aim of the present chapter is to show, through examples, how these properties are exploited in quite different areas.

The first part deals with sensors based on stationary elastic waves. Their main component may be a disk-shaped or rectangular quartz resonator vibrating in thickness shear mode (microbalance, pressure and temperature sensor); or it may be a strip vibrating in flexural mode (pressure sensor, accelerometer). The second part treats sensors based on guided waves. Examples discussed include: Rayleigh wave touch panels, remote sensors, Lamb wave coordinate sensors and liquid detectors, and cylindrical wave liquid level gauges. The third part is concerned with applications of acousto-optic components, such as modulators, tunable filters and deflectors, to printing, sorting, and radioastronomy. The subject of the fourth part is the ultrasonic motor. This is a new type of motor in which the moving (rotating or translating) part is driven by a mechanical wave. Several instruments (microscope, osteodensitometer) and methods (time reversal, pressure pulse inspection, parametric probes) are described in the final part.

5.1 Quartz Resonator Sensors

The frequency of a vibration is a remarkable physical quantity, in the sense that it is transferred unmodified from one fixed point to another [5.1]. This is why signals carried by frequency modulation suffer less distortion than signals carried by amplitude modulation. On the other hand, although the frequency of a free quartz resonator is defined to high accuracy ($Q > 10^5$, Sect. 4.4), it is altered by any form of action exerted on the monocrystal. Such effects vary according to crystallographic cut, vibrational mode and hence also the shape chosen (disk, strip, tuning fork, etc.). In this way, frequency shifts in a resonator of suitable cut are a means of detecting alterations on one of its faces (e.g., load, chemical effects), or changes in temperature, pressure or acceleration.

5.1.1 Thickness Shear Mode Sensors

We shall give two examples. The first is the microbalance, which has been in use for over 50 years now but remains topical. The second is a temperature and pressure sensor, which has existed for a few years now, and is remarkable in that it vibrates in two modes.

5.1.1.1 Microbalance. The thickness of a vacuum-deposited film (usually metallic) is determined by measuring the frequency change in a quartz resonator (AT-cut and vibrating in transverse mode) [5.2]. The basic relation here is

$$\frac{\Delta m}{m} = K\frac{\Delta T}{T}\,, \quad K = \text{Const.}\,, \tag{5.1}$$

where m is the quartz mass and $T = 1/f$ its oscillation period. Δm is the mass of deposited metal. The formula remains valid whilst $\Delta m/m$ is less than a few [%].

Another formula is found by treating the whole system of quartz plus metallic deposit as a composite resonator [5.3]:

$$\frac{\Delta m}{m} = -\frac{1}{\pi}\frac{Z_\mathrm{m} f}{Z f_\mathrm{c}} \arctan\left[\frac{Z}{Z_\mathrm{m}} \tan\left(\pi\frac{f_\mathrm{c}}{f}\right)\right]\,, \tag{5.2}$$

where f_c is the frequency of the loaded resonator, and Z_m and Z are acoustic impedances of the metal and the quartz (see Problem 1.11). This relation is empirically valid for values of $\Delta m/m$ up to 50%.

Film thicknesses measured in this way are often of order 1 µm. However, much smaller thicknesses can be determined. Let us mention one example. It concerns experiments to ascertain electronic structure in small clusters of

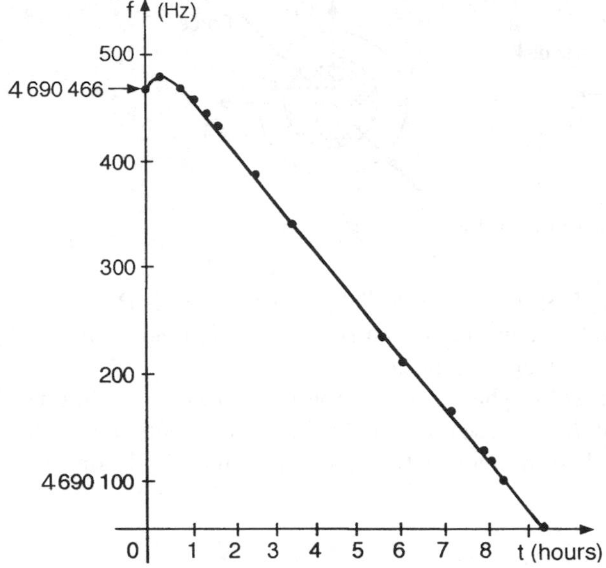

Fig. 5.1. Time dependence of microbalance frequency as iron atoms are deposited. Initial frequency 4 690 466 Hz

transition metal atoms (e.g., Fe, Ag) diluted in matrices of solid noble gases (e.g., Ar, Ne, Xe) at the temperature 4 K of liquid helium, on a transparent support (fluorine, CaF_2) [5.4]. The microbalance ($f = 4\,690\,466$ Hz) was calibrated using a steady jet of ion atoms supplied in the Knudsen regime by a temperature-controlled oven. The curve in Fig. 5.1 shows the time variation of the resonator frequency. After a stabilisation time of about 20 min., it is a linear dependence.

The simple formula $\Delta m/m \approx -\Delta f/f$ can be applied in this case. It yields $\Delta m \approx 2.8 \times 10^{-9}$ kg for a frequency change (over 8 hours) of 400 Hz ($m = 4 \times 10^{-5}$ kg). This value of Δm is confirmed by chemical methods (to within 15%). A 1 Hz frequency change thus corresponds to a mass change of 0.7×10^{-11} kg. This is comparable with the mass of an atomic monolayer on a deposition area of diameter 8 mm, in this case. The microbalance therefore allows accurate and very large dilutions of particles (volume concentration 10^{-3}) in a matrix. Such dilutions are required for the study, for example, by optical absorption, of electronic structure in isolated metallic clusters.

The microbalance has applications to chemistry. If its surface is coated by a suitable film, it can detect any product reacting with that film (e.g., detection of mercury by a gold electrode).

5.1.1.2 Dual-Mode Temperature and Pressure Sensor. The frequency of a quartz resonator, vibrating in a thickness transverse mode, also varies when forces are applied radially in its plane, over part or all of its periphery. The resonance frequency of an AT-cut crystal increases when it is compressed in the direction of its X crystallographic axis, whereas that of a BT-cut crys-

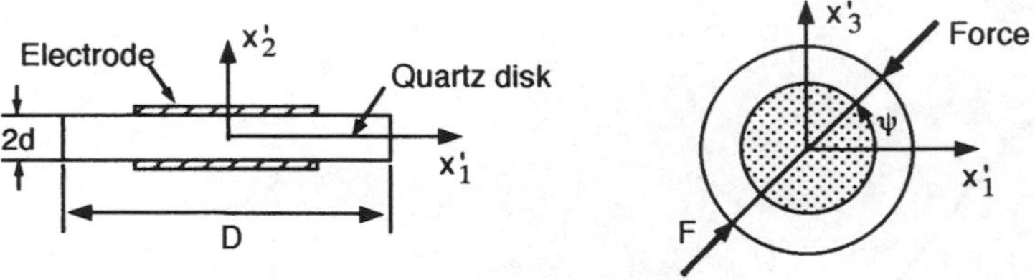

Fig. 5.2. Quartz disk subjected to a diametric force pair

tal decreases [5.5, 5.6, 5.7, 5.8]. (The article [5.8] by EerNisse, E.P., et al. is a historical review of developments in resonator sensor technology up until 1988 and contains a useful list of references.)

In the general case of a disk-shaped resonator (Fig. 5.2), we define the force–frequency coefficient K_f for given cut and vibrational mode [5.9]. This describes the sensitivity of its frequency to a pair of diametrically opposed forces F:

$$K_\mathrm{f} = \frac{\Delta f}{f}\frac{1}{F}\frac{2dD}{N}\,, \tag{5.3}$$

where $\Delta f/f$ is the fractional frequency shift of the resonator, $2d$ its thickness, D its diameter, and N is the frequency constant, equal to $V/2$, where V is the phase speed of the wave. When the resonator is subject to a uniform distribution of forces normal to its circumference (i.e., radial forces), it is characterised by a mean value of the above coefficient:

$$\langle K_\mathrm{f}\rangle = \frac{1}{\pi}\int_0^\pi K_\mathrm{f}(\psi)\,\mathrm{d}\psi\,. \tag{5.4}$$

In order to measure a pressure, the resonator must be mechanically and rigidly bonded onto an enclosure, upon which the external medium acts. Moreover, the temperature must be measured at (or close by) the location of the resonator; any temperature change will alter its frequency. The effect of simultaneous variations in the force (hence the pressure) and the temperature is given, for a thickness shear vibration, by the formula [5.10]

$$\frac{\Delta f}{f} = K'_\mathrm{f}(p-p_0) + A(\theta-\theta_0) + B(\theta-\theta_0)^2 + C(\theta-\theta_0)^3\,, \tag{5.5}$$

where

$$A = a\left[1 + (p-p_0)\left(K'_\mathrm{f} + \frac{1}{a}\frac{\mathrm{d}K'_\mathrm{f}}{\mathrm{d}\theta}\right)\right]\,, \tag{5.6a}$$

$$B = b\left[1 + (p-p_0)\left(K'_\mathrm{f} + \frac{a}{b}\frac{\mathrm{d}K'_\mathrm{f}}{\mathrm{d}\theta}\right)\right]\,, \tag{5.6b}$$

$$C = c\left[1 + (p-p_0)\left(K'_\mathrm{f} + \frac{b}{c}\frac{\mathrm{d}K'_\mathrm{f}}{\mathrm{d}\theta}\right)\right]\,. \tag{5.6c}$$

a, b, c are temperature coefficients of the first, second and third orders, measured at the reference pressure and temperature p_0, θ_0. The ratio $\mathrm{d}K'_\mathrm{f}/\mathrm{d}\theta$ ex-

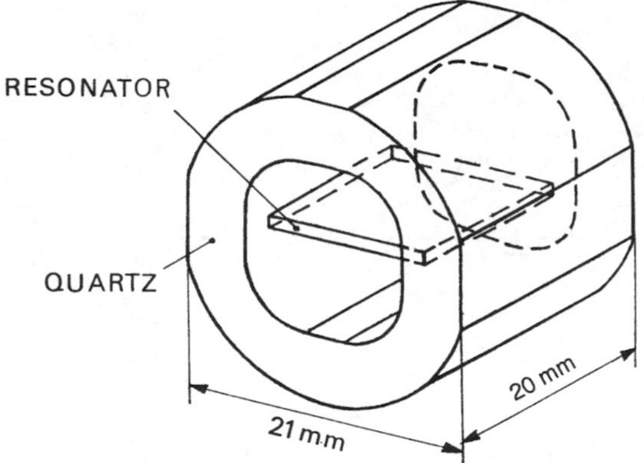

Fig. 5.3. Rectangular resonator cut from a quartz cylinder in such a way that it suffers forces in a particular direction. This cylindrical shell, on which pressure is exerted, is closed by two glass-sealed caps (not shown). The total length of the shell is roughly 38 mm, and its diameter is 25 mm

presses temperature sensitivity of the force–frequency coefficient. K_f' should be replaced by $\langle K_\mathrm{f}' \rangle$ when the resonator is subject to a uniform distribution of radial forces.

In order to measure the pressure, we must therefore measure the temperature and subtract off its contribution to the frequency shift. A first solution consists in using two resonators. One of them (with singly-rotated AT or BT cut) is affected by both temperature and pressure variations, whilst the other, placed nearby, is subjected only to the temperature variation. Such sensors have been commercialised for oil exploration, in particular, by Hewlett–Packard [5.11].

A second solution, suggested by Sinha, B.K., [5.12], uses a single doubly-rotated SBTC-cut quartz. This nomenclature indicates 'Stress compensated for the B mode and Temperature compensated for the C mode'. The crystal vibrates in two modes: temperature is measured by the B mode and pressure by the C mode (Sect. 4.4.1.1). In this way, the same resonator provides both frequency shifts (i.e., including the one related to its internal temperature) required to ascertain the pressure.

A further solution, related to the last, has been developed by Besson, R.J., et al. [5.13]. They have determined a crystal orientation (close to the SC cut) which can simultaneously support both B and C modes, but which is sensitive to uniaxial stresses. The rectangular resonator is cut out of a quartz cylinder by ultrasonic milling, as shown in Fig. 5.3. The length of the surrounding shell, closed by two sealed caps, is less than 40 mm, whilst its diameter is 25 mm. Frequency variations depending on the two parameters, pressure and temperature, are introduced into a calibration algorithm which simultaneously provides, the values of these two quantities, with period less than 1 s. Accuracy and resolution depend on the period chosen.

Several years' study were required to develop this sensor. Indeed, various unknown quantities had to be determined: the value of $\mathrm{d}K_\mathrm{f}/\mathrm{d}\theta$; the stress distribution in the quartz crystal (20 times more fragile under tension than

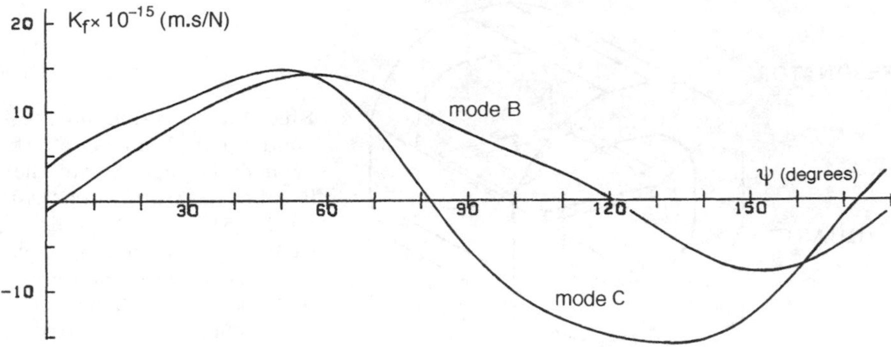

Fig. 5.4. Pressure sensitivity of modes B and C for different angles ψ

under compression); the shape required to trap the energy of the two modes; and responses to rapid changes in temperature and pressure.

Figure 5.4 shows coefficient K_f as a function of angle ψ. When $\psi = 120°$, mode B is insensitive to pressure, whilst mode C is extremely sensitive to it.

Figures 5.5a and b show frequency–temperature characteristics $f(\theta)$ for various pressure values. The mode B characteristic is independent of pressure. For a given value of pressure, the mode C characteristic varies only slightly with temperature. The main features of this pressure sensor are as follows:

Resonator frequency	5 MHz (partial 3)
Sensitivity	145 Hz/MPa
Resolution	≈ 70 Pa
Precision	≈ 7 kPa
Measurement range	0–100 MPa
Temperature range	-10 to 175°C
Aging	+7 kPa beyond 250 days

With dynamic corrections for temperature variations measured in situ, this sensor was developed for applications in the petrol industry (Schlumberger), in particular, for measuring pressures in oil fields. Knowledge of these pressures and their variations is all important in determining the size of oil deposits and conditions under which they may be extracted.

5.1.2 Vibrating Beam Sensors

The resonance frequency of a (laterally) vibrating beam, held at its ends, is sensitive to any tension or compression force. To first order, the fractional shift in this frequency is proportional to the applied force. It is given approximately, for the first mode, by [5.14]

$$\frac{\Delta f}{f} \approx \alpha s \left(\frac{l}{d}\right)^2 \frac{F}{wd}\,, \tag{5.7}$$

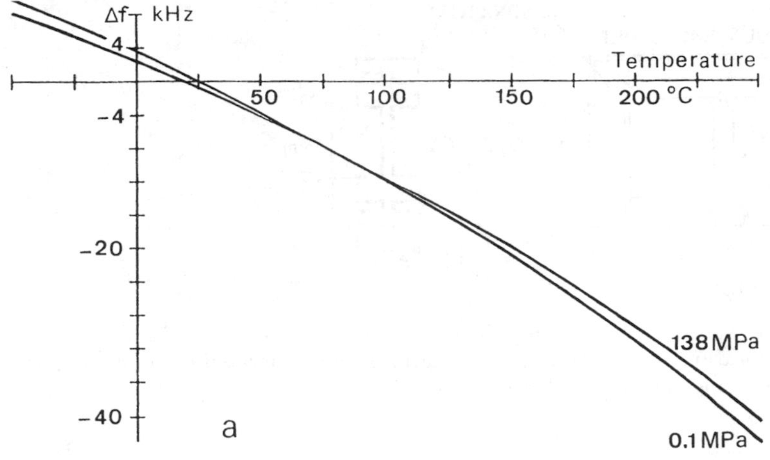

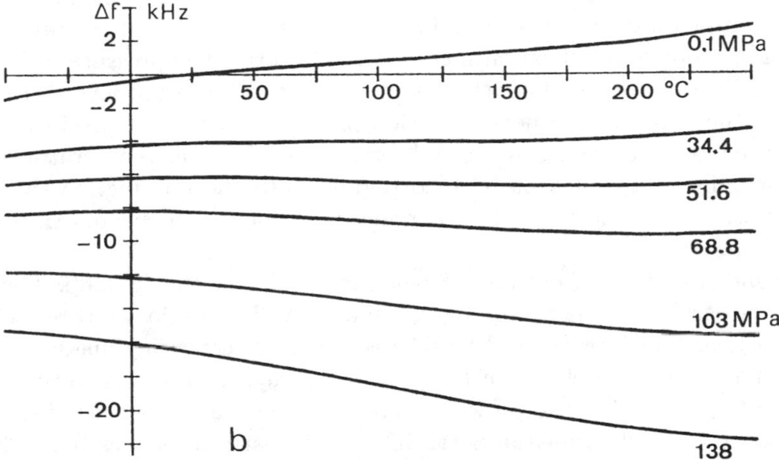

Fig. 5.5. Temperature dependence of frequency for different pressure values. (**a**) Mode B. (**b**) Mode C. Graphs kindly communicated by J.J. Boy

where l, w, d denote length, width and thickness of the beam, s is the flexibility constant (in the simplest case, $s = 1/E$, where E is Young's modulus), and α is a constant for the given orientation. The ratio F/wd represents the uniform stress over cross-section wd. For quartz, breaking stress is of order 10^8 Pa, $E \approx 8.7 \times 10^{10}$ Pa, $\alpha \approx 0.15$, so that, with $l/d = 25$, $\Delta f/f|_{\text{fracture}} \approx 10\%$. Consequently, a fractional frequency change of several [%] is acceptable [5.15].

This fact is exploited in force, pressure and acceleration measurements. One difficulty is to make mounting attachments through which only a very

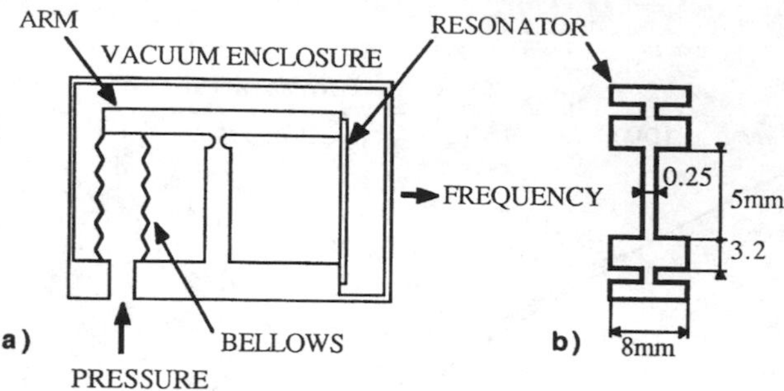

Fig. 5.6. Pressure sensor (Sextant, Type 51). (**a**) The frequency of the resonator, in flexural vibration, depends on the longitudinal force generated by pressure in the bellows. (**b**) Shape and size of the resonator

small proportion of vibrational energy is dissipated. Various beam designs and mountings have been developed [5.16].

5.1.2.1 Pressure Sensor. Figure 5.6a depicts the operating principle for a pressure sensor (Sextant, Type 51). The resonator, a single beam clamped at both ends, vibrates in a flexural mode at $f \approx 50$ kHz. The pressure to be measured generates a force along the beam axis, via bellows and lever arm, which alters the resonance frequency. The low pressure (coarse vacuum) inside the enclosure reduces damping by air and serves as a reference. Temperatures are measured by means of a probe incorporated into the housing, so that temperature variation effects on the resonance frequency of the beam can be subtracted off.

Shape and size of the Z-cut quartz beam are indicated in Fig. 5.6b. The beam axis is parallel to the crystallographic axis X. Energy losses through mounting attachments are reduced by means of two rotary spring masses.

In order to assess the effects of beam attachments, deformations of isolated resonators (without bellows) were recorded in our laboratory using the interferometric probe described in Sect. 3.2.2. The quartz beam was bonded onto a sensor mounting placed inside a vacuum enclosure. Figure 5.7 shows an example of the deformation. Experimental points are very close to theoretical predictions [5.17]. The displacement amplitude at the centre is 120 Å. The quality factor Q for this response is 15 000. Similar experiments carried out with resonators fixed in other ways have shown that the quality factor lies in the range 15 000 to 60 000.

Figure 5.8 shows schematically how the sensor works. f_θ is the frequency of the probe for a given temperature of the resonator mounting. f_p is the frequency of the resonator for the pressure p to be determined. The acquisition and computation system calculates the value of p from the values of f_p and f_θ, with the help of a model stored in memory.

The sensor operates over various ranges of pressure (low pressure, from 10^3 to 2.9×10^5 Pa, and high pressure from 0.05 to 35 bar). It has been developed

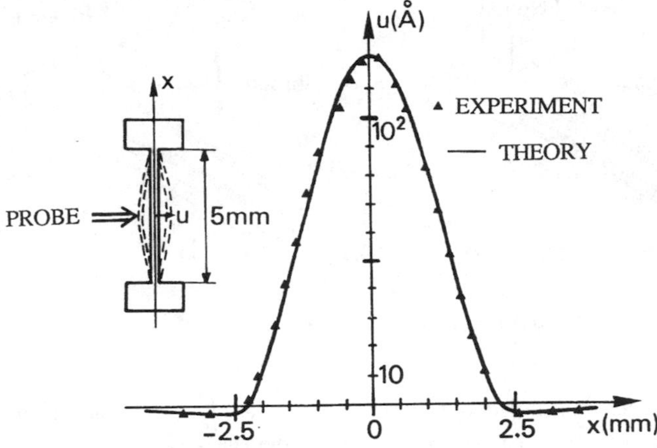

Fig. 5.7. Resonator deformation recorded by interferometric probe (Sect. 3.2.2). Mechanical displacement at the centre is 120 Å

industrially by Sextant Avionics. It is used for anemobarometric measurement on many civil and military aircraft (e.g., Airbus, Boeing, Rafale).

As an example, characteristics of sensor 51A16 over the temperature range 0–60°C are:

Measurement range	100–1 500 hPa
Error due to mobility	$< \pm 20$ Pa
Error due to hysteresis	$< \pm 10$ Pa
Time drift over six months	$< \pm 20$ Pa
Linear acceleration sensitivity	< 1.6 Pa/g
Angular acceleration sensitivity	< 1.6 Pa/rad s^{-2}

Flexural vibrations of this kind, at relatively low frequencies ($f <$ 100 kHz), are easily excited and detected optically [5.18]. The exciting light beam may be guided along a fibre. The vibration is detected by the knife-edge technique, or better, by measuring the rotation of polarisation produced by periodic birefringence, itself generated by stress variations. Optical excitation and detection, which require no mechanical contact with the resonator,

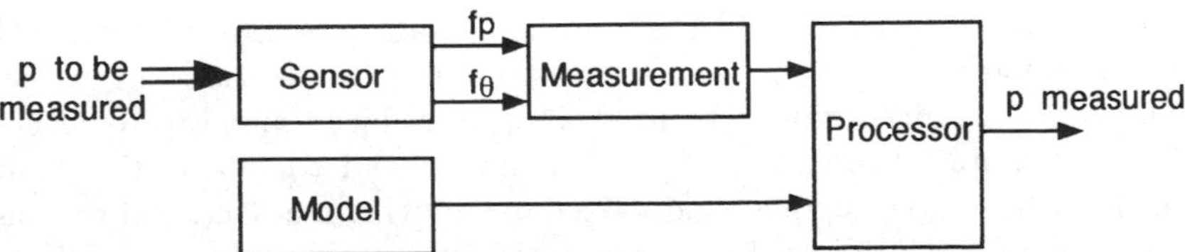

Fig. 5.8. Pressure sensor. Determination of pressure p from the pressure frequency f_p and temperature frequency f_θ, by means of a mathematical model stored in memory

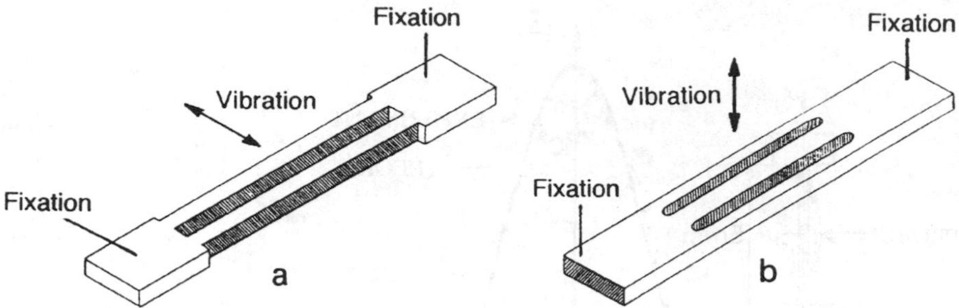

Fig. 5.9. Double (**a**) and triple (**b**) beam resonator force sensors

are not affected by electromagnetic perturbations induced in the electrical supply–wire–crystal circuit, required to excite a piezoelectric resonator. Moreover, the vibrating strip need no longer be piezoelectric and it is in principle possible to use a material which is itself insensitive to any electric field and provides a much higher quality factor than quartz (e.g., sapphire Al_2O_3, yttrium aluminium garnet YAG).

Double and Triple Beam Sensors. Apart from using a structure with spring masses near attachment points, another way of reducing leakage of vibrational energy through the mounting of the resonant element is as follows. Two beams vibrate in opposition in such a way that moments and forces created by one beam are cancelled by those of the other beam (Fig. 5.9a). Yet another way consists in using three beams vibrating as shown in Fig. 5.9b. The central beam is twice as wide as each of the other beams.

This kind of device has become much easier to produce with the development of photolithographic and chemical techniques for cutting crystals [4.50]. Small sensors, manufactured in large quantities [5.19], can seriously compete with stress gauge devices in markets eager to find simple and cheap sensors (e.g., the automobile industry). Generally speaking, characteristics of a double beam sensor are as follows: centre frequency in the range 30–100 kHz, linear variation range $\approx 5\%$, sensitivity $\approx 1\%/\mathrm{N}$, resolution 10^{-4} to 10^{-5}.

5.1.2.2 Accelerometers. A force sensor becomes an accelerometer when the force is exerted by a moving mass. In the case of a vibrating beam sensor, when the mass (often called the *proof mass*) is accelerated, it produces an axial tension or compression which alters the flexural vibration frequency of the beam [equation (5.7)]. In general, an accelerometer is made from two (double) strips arranged in a differential setup to reduce temperature effects [5.20]. For example, the two strips are mounted head to tail as shown in Fig. 5.10. Each (double) strip is rigidly fixed to a moving mass in the direction of its longitudinal axis. The value of the acceleration γ along this axis follows from the frequency difference:

$$f = f_0 + k_1\gamma + k_2\gamma^2 + k\theta \, , \tag{5.8a}$$

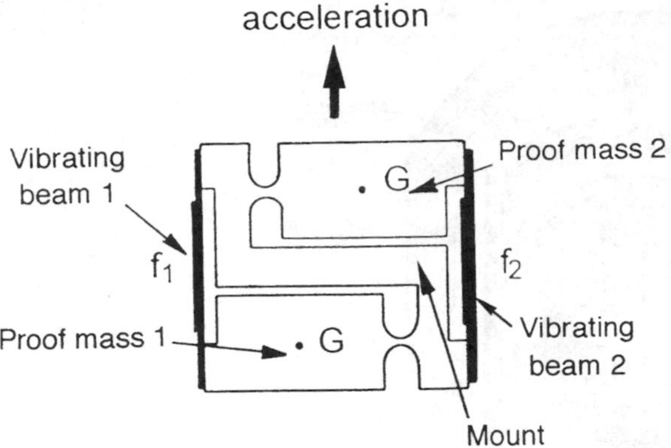

Fig. 5.10. Accelerometer with two beams mounted head to tail (SAGEM arrangement). Any acceleration along the longitudinal axis of the beams compresses one beam, whose frequency goes down, and extends the other, whose frequency goes up. The frequency difference yields the acceleration

$$f' = f_0 - k_1\gamma + k_2\gamma^2 + k\theta\ , \quad \gamma = \frac{f - f'}{2k_1}\ . \tag{5.8b}$$

Figure 5.11 shows another configuration in which the frequency variation of a strip is produced by motion of the proof mass perpendicular to its longitudinal axis. In fact this diagram represents one of the two (quartz) elements of a recently developed accelerometer [5.21]. This monolithic element includes an active part consisting of a proof mass and vibrating beam (thickness 30 μm, width 60 μm, length 2 mm), together with its mounting. One end of the strip, vibrating in the plane of the mounting, is fixed to an inactive mass (which nevertheless plays the important role of decoupling), whilst the other end is fixed to the proof mass. The latter is held by the inactive part through two flexible rods. The cross-section of these rods is chosen to reduce as far as possible any motion of the mass in the mounting plane. The central unit, consisting of the strip, proof mass and inactive mass, is mechanically and thermally decoupled from mounting points by means of a frame with opposite attachment points.

The accelerometer comprises two identical elements. They are fixed in the same small cylinder, at opposite ends. Each vibrating piezoelectric strip is an integral part of an oscillator. Electronic circuits are normally associated in a differential setup. The innovation here is in using two identical monolithic elements. Construction involves photolithographic and chemical etching techniques developed in the watch industry. It lends itself to batch production. For example, sixteen elements (of diameter 6 mm) are simultaneously cut from a quartz wafer of dimensions $38 \times 38 \times 0.4\ \text{mm}^3$, with Z crystallographic cut.

Characteristics of this accelerometer, developed by the Office National d'Etudes et de Recherches Aérospatiales (ONERA) in France, and designed

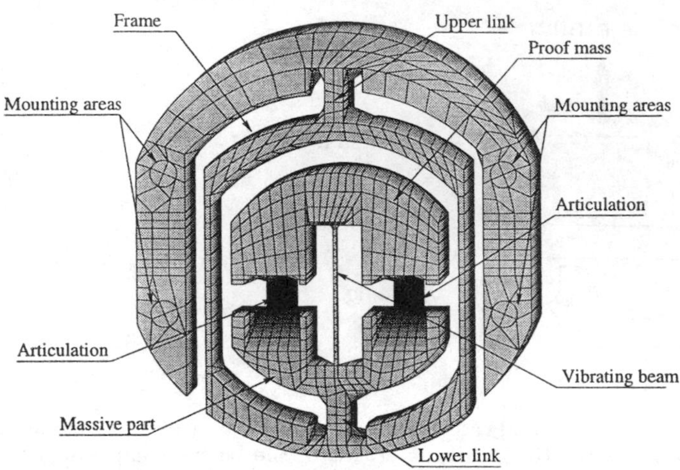

Fig. 5.11. Monolithic element in the ONERA accelerometer. The central part, containing the strip, proof mass and inactive mass, is mechanically and thermally isolated from mounting points by means of a frame. Taken from [5.21]

primarily for vehicle control and piloting, and inertial navigation of planes, helicopters and terrestrial vehicles, are as follows:

Flexural vibration frequency of one beam	60 kHz
Measurement range	± 100 g
Scale factor	24 Hz/g (12 Hz/g per beam)
Resolution	10 μg
Accuracy	3×10^{-4} g
Temperature span	-45 to $+90$°C
Volume without (with) electronics	2 cm^3 (10 cm^3)

5.1.3 Quartz Tuning Fork Thermometer

The temperature dependence of thickness shear vibration frequencies of resonators was soon flagged for potential applications. Resonators with appropriately sensitive cuts were therefore produced for temperature measurements. (First Y and then LC, a cut with almost linear $f(\theta)$ characteristic [5.22, 5.23], followed by other cuts [5.24, 5.25].) These thermometers can operate over temperatures from -80°C to 200°C, with accuracy and resolution of order one hundredth and one thousandth of a degree, respectively. However, although they perform remarkably, they are relatively bulky and expensive. Attempts were therefore made to produce miniature thermometers of more modest achievement. One idea in particular was to use a tuning fork structure in which the two tines vibrate in a torsional mode [5.26]. A cut close to the classic Z cut (tine length in the X direction), and also the dimensions, were determined in such a way as to produce an almost linear characteristic. The

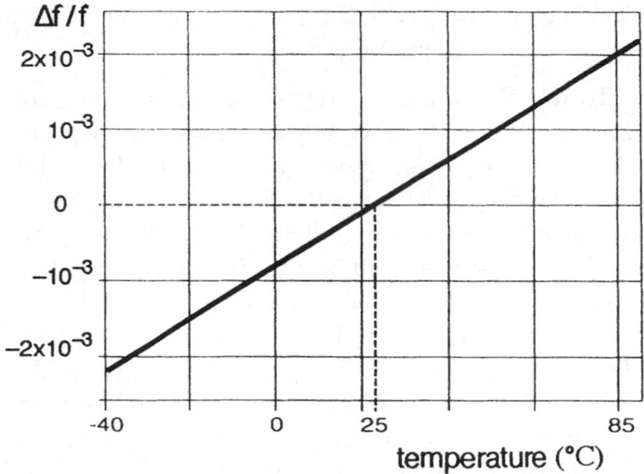

Fig. 5.12. Frequency shift as a function of temperature for a miniature quartz tuning fork thermometer (MX2T, Microcristal, Switzerland). Housing diameter 1.5 mm, length 5.5 mm

range of variations is limited by coupling between modes [5.27]. Figure 5.12 gives the characteristic of a sensor contained in a cylinder of diameter 1.5 mm and length 5.5 mm, vibrating at 262 kHz at 25°C. It has slope 33.5 ppm/°C.

Surface Wave Resonator Sensors. Sensors have been developed which use stationary surface waves. Their high quality factor makes them extremely sensitive. A Rayleigh wave sensor ($f = 500$ MHz) can detect vapours (from drugs or explosives) at the picogram level [5.28].

5.2 Guided Wave Sensors

The behaviour of a device is affected by any change in its environment, whether it operates in stationary or travelling wave conditions. The quantity causing perturbation is therefore detectable, not only with the kind of resonator described in the last section, by also by means of travelling wave devices. Detection is generally easier if waves are guided, for perturbations act on the guide surface, in contact with the waves.

5.2.1 Surface Wave Sensors

We shall describe two Rayleigh wave sensors with very different structure. The first is a touch screen. It contains a non-piezoelectric panel (with inclined reflectors) in which waves are generated by an associated electrical supply (and by a transducer comprising a bulk wave resonator mounted on a wedge). The second is a pressure sensor. It contains a passive piezoelectric device (with parallel reflectors), in which an interdigital transducer, equipped with antenna, is remotely excited by electromagnetic pulse.

5.2.1.1 Rayleigh Wave Touch Screen. The development of interactive equipment in railway stations, museums, amusement parks and other public places has brought a need for simple ways of selecting from a menu. The continuing success of the touch screen is clearly related to this need.

Various systems exist to identify the region indicated by a finger on the screen, based on resistance, capacitance or light waves. In the light system, for example, infrared beams are emitted by a series of diodes arranged along two adjacent sides of the screen. They are received by phototransistors placed along the opposite sides. The screen is thereby divided up by beams parallel to the x and y directions. When a finger touches the screen, it necessarily cuts some x_i beam and some y_j beam, there by revealing its location. Unfortunately, such a system may be triggered by something other than a finger, e.g., a fly. Resistive and capacitive systems also harbour certain disadvantages, because the conducting layers they require reduce screen transparency.

Adler, R., Desmares, P., have devised a system based on surface elastic waves [5.29]. The idea is to propagate surface waves from one side to the other, and also from top to bottom, across the screen surface (or rather, across the surface of a glass panel placed against the screen). These waves are absorbed by any finger touching the screen, because a finger behaves like a liquid (viz., water) with respect to such waves. The waves could be launched like the light waves mentioned earlier, by rows of emitter transducers arranged along adjacent edges of the screen, and detected by receiver transducers placed along the opposite edges. The finger position would then be given as the crossing point of the two absorbed wave beams x_i, y_j.

However, the authors had the idea of using fewer transducers by introducing reflectors. Two transducers, one emitting and the other receiving, are enough to determine each coordinate. The coordinate is deduced from a time measurement. Figure 5.13 shows elements needed to determine horizontal position (abscissa) of the finger: emitter transducer E, a first reflective array F_E with strips inclined at 45° to its axis, a second reflective array F_R, mirror image of the first in the horizontal axis of the screen, and a receiver transducer R.

An electrical impulse is applied to transducer E and it emits an elastic wave train which propagates along the axis of reflector F_E. A part of these waves is continuously reflected from the strips tilted at 45°, moving down towards the bottom of the screen. On reaching the second reflector, they are directed towards receiver transducer R. Of course, the electrical signal provided by output transducer R is much longer than the width of the pulse applied to input transducer E. If the waves encounter a finger in contact with the surface, as they cross the screen, part of their energy is absorbed. As a result, the amplitude of the signal delivered by output transducer R is reduced locally. The arrival time of the corresponding dip yields the abscissa

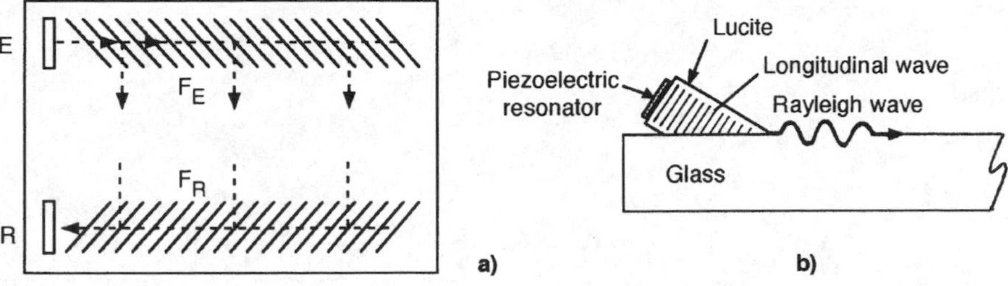

Fig. 5.13a,b. Surface wave touch screen using reflective arrays. Transducer E launches an elastic wave train towards reflector F_E opposite it. Waves travel along the axis of this reflector, but a fraction are continuously reflected from its strips, inclined at 45°, and cross the screen to be reflected once more, from strips of reflector F_R, mirror image of F_E in the screen axis, thus reaching transducer R. The electrical signal produced by output transducer R is much longer than the width of the pulse applied to input transducer E. If waves lose a small part of their energy in crossing the screen, absorbed by a finger tip in contact with the surface, the broad output pulse is tagged with a dip. The arrival time of this dip gives the abscissa of the contact point

of the contact point between finger and screen. The ordinate is found by the same method, with the help of elements placed along the two other edges of the screen.

Naturally, several problems remain to be solved before such a system can be put into practice.

- As the substrate is not piezoelectric, surface waves cannot be launched by a simple interdigital transducer. In this case, the transducer is made from two parts: a piezoceramic wafer and a wedge (Fig. 5.13b). The ceramic wafer launches longitudinal waves (of frequency around 5 MHz), and these are transformed into surface waves at the wedge–substrate interface, provided that angle β between wedge axis and surface is suitably chosen (for a Lucite wedge and glass substrate on which surface waves propagate at 3 150 m/s, $\beta = 33°$).
- The two reflective arrays must be easy and cheap to produce, given the intended applications. The interval between reflecting strips is relatively long ($f = 5$ MHz implies $\lambda = 0.63$ mm). The authors therefore used a mask technique with a high density glass powder emulsion, crystallising at 430°C. Strips are several [μm] thick.
- In the absence of perturbations, it is desirable that wave amplitudes, and hence power densities, should remain more or less constant over the screen (glass panel) surface. This requirement means that the reflection coefficient of an array should increase with distance from the transducer. For an array with constant reflection coefficient, and neglecting attenuation during propagation, if P_1 is the constant power density over the screen, power $P(x)$ decreases with distance from the transducer:

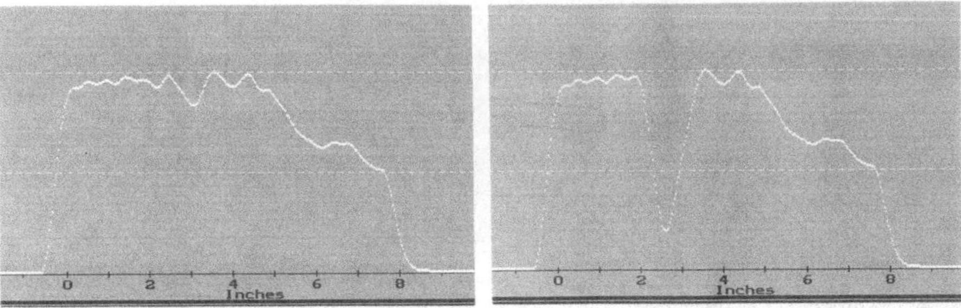

Fig. 5.14. Amplitude dip revealing the presence of a finger

$$P_1 \, \mathrm{d}x = -\frac{\mathrm{d}P}{\mathrm{d}x} \, \mathrm{d}x \quad \Rightarrow \quad P(x) = P_0 - P_1 x \, .$$

In order to compensate this reduction, the power reflection coefficient must vary as $1/P_1 x$ from the edge of the array. This can be achieved by arranging for a density of strips which increases with distance from the beginning of the array.

- If the surface of the cathode ray tube is not plane, reflective arrays are modified accordingly, allowing for curvature or using the guide effect provided by the reflective arrays [5.30].

Experiment (Fig. 5.14) shows that the device (developed under the name Intellitouch by Elo TouchSystems Raychem) detects not only the position at which a finger touches the screen (with resolution of order mm), but also the pressure it exerts. This sensitivity to pressure stems from the fact that the contact area between finger and substrate increases when the finger presses on the screen with greater force.

Dew Detector. Rayleigh waves are attenuated even more by liquid droplets on their propagating surface than by the presence of a finger tip. This fact is exploited in the Dewpoint sensor, Vaisala, Finland. When water droplets condense on a surface lying between the two transducers of a transmission line operating at 78 MHz, there is a sudden fall in output signal. This sensor is used to control humidity levels during the drying process in the paper industry. When in use, the test surface is regularly cleaned with an air jet.

Rayleigh waves, and also transverse waves, like Love waves, are extremely sensitive to any surface modifications. In principle, this means that they can detect gases (or vapours) if the surface is coated with some layer which selectively adsorbs the relevant gas. In order to attract the attention of chemists and facilitate their work, a number of subsystems have been developed: double transmission lines on ST-cut quartz, equipped with gas capsule, and excitation and detection electronics [5.31]. Sensitive layers have been studied in the laboratory. In particular, drug sensors have been developed.

5.2.1.2 Remote Controlled Pressure Sensor. This is an example of a sensor made from a passive echo delay line. Before describing it, let us note

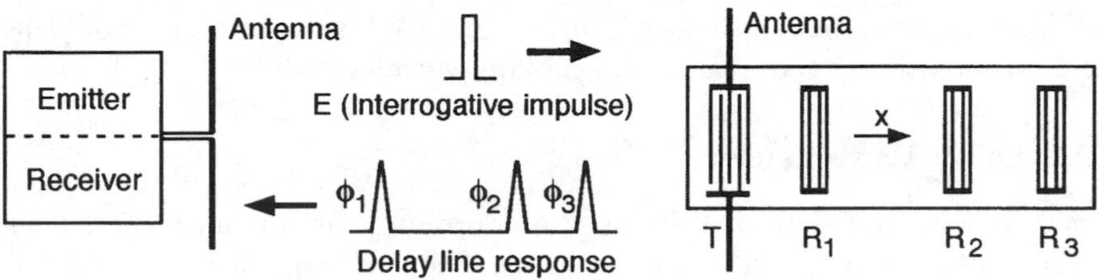

Fig. 5.15. Passive device responding to an interrogation pulse with a set of time-ordered pulses which is characteristic of its structure. Electromagnetic pulse E excites transducer T which launches a surface wave along x. Part of the wave is returned by each reflector R_i towards T and reconverted into a secondary pulse. These secondary pulses, identifiable by their arrival time, i.e., their phases ϕ_i, form a sequence which characterises the reflector arrangement. Pulses are transmitted with a carrier frequency which is not shown. The device becomes a sensor for any quantity altering the substrate and causing phase shifts $\Delta\phi_i$

that simple passive piezoelectric devices can serve as tags, to identify possibly mobile objects (including animals) [5.32]. An antenna is connected to the input transducer of the device and an electromagnetic pulse sensed by the antenna generates a surface wave (Fig. 5.15). The wave propagates towards reflectors, each of which returns a fraction of the wave to the emitter. The latter reconverts these reflected fractions. A time-ordered series of electromagnetic pulses is thereby radiated by the antenna.

This sequence, response of the device to the single initial (interrogation) pulse, is characteristic of the reflector configuration and hence the object carrying it. (There are variants with resonators or transducers instead of reflectors. These emit a pulse when the incident wave crosses them.) Such surface wave passive delay lines play a similar role in object identification to optically read bar codes. The difference is that their code can be read off even when they are not visible, for example, being located in some inaccessible place, and whatever their orientation, on a translating or even a rotating mount!

The code associated with the device is changed if the wave propagation is altered. Any quantity causing such a modification (temperature change, or other, distorting the wave path) can in principle be detected [5.33]. If τ_i is the delay of the pulse returned by reflector i, a substrate deformation causes a shift $\Delta\tau_i$ in this delay, and hence a phase shift $\Delta\phi_i$. The latter is a function of variations in device dimensions and wave speed, i.e., variations in elastic constants and density. Scherr, H., et al. have described a pressure sensor based on these ideas [5.34]. It comprises a quartz cavity closed by a Y-cut quartz membrane of thickness 500 μm. The transmission line ($f = 434$ MHz) is built on the membrane. Pressures are deduced from phase differences of three reflectors, arranged in such a way as to reduce temperature effects. The resulting signal varies linearly over the span 0–250 kPa, remaining insensitive to temperature in the range −20 to 100°C. These miniaturised manometers, weighing a few grams and which can be consulted at a distance, would be ideal for monitoring tyre pressure in moving vehicles [5.35].

5.2.2 Lamb Wave Sensors

In this section, we shall describe first a coordinate sensor and a liquid detector, then an interactive window display. The common feature of the first two devices is that they both use a symmetric propagation mode in a plate. However, they operate at different points on the dispersion curve. We shall therefore consider shapes of dispersion curves for propagation modes in a plate, and express the attenuation produced by a liquid in contact with one of its faces.

Let us review the results of Sect. 5.5, Vol. I. Waves propagating in an isotropic plate comprise a vertically polarised transverse component and a longitudinal component. The two components are coupled by reflection from the two free surfaces of the plate. The proportions of these transverse and longitudinal partial waves vary with the frequency, i.e., propagation is dispersive. There are two mode types, one symmetric (S) and the other antisymmetric (A). Figure 5.40, Vol. I shows dispersion curves for steel. Figure 5.16 shows the same for the first two modes S_0 and A_0 in glass.

At high frequencies, the two curves tend to the same limiting value, namely, the Rayleigh wave speed V_R. This mode is no longer a plate mode, vibration being localised close to the surface.

At low frequencies, the two modes behave differently. The longitudinal component dominates in the S_0 mode, whereas the transverse component does so in the A_0 mode. The S_0 mode phase speed is largely independent of plate thickness. It is equal to the *plate speed* V_P [equation (5.123), Vol. I], which is a function of longitudinal and transverse bulk wave speeds V_L and V_T in the material. Group speeds barely vary. On the other hand, the phase speed of mode A_0 starts out at zero. Near the origin ($kh \ll 1$), the dispersion relation is given approximately by (5.126), Vol. I. Continuous growth of the group speed up to V_R exhibits no particular features.

Phase speed and group speed of the S_0 mode evolve between the low frequency plateau V_P and the high frequency plateau $V_R < V_T$. In glass, for example,

$$V_L = 5\,960 \text{ m/s}, \qquad V_T = 3\,200 \text{ m/s},$$
$$V_P = 5\,400 \text{ m/s}, \qquad V_R = 2\,900 \text{ m/s}.$$

A minimum of the second curve occurs at the point of inflection of the first. At this point, we expect waves to be more sensitive to any surface modification, such as the presence of a liquid, or a finger contact (which is equivalent to the presence of water with regard to its elastic impedance). This observation is confirmed by amplitude variations in the two wave components within the plate. The curves of Fig. 5.17, which refer to a glass plate, correspond to points **x** on the dispersion curve of the S_0 mode in Fig. 5.16.

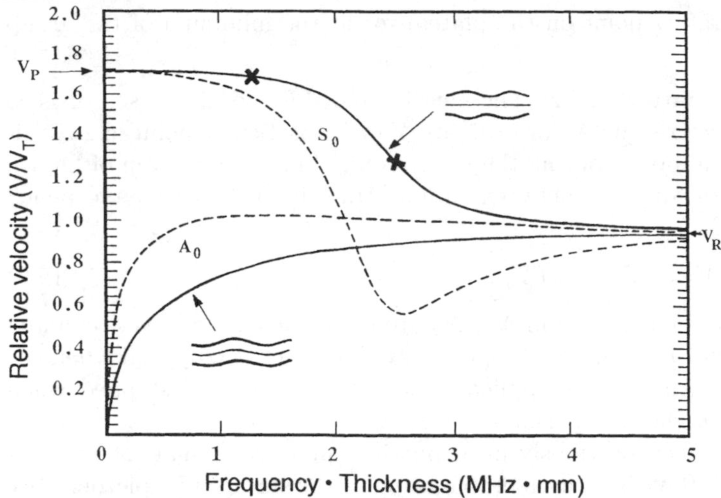

Fig. 5.16. Propagation modes in a glass plate ($V_L = 5\,960$ m/s, $V_T = 3\,200$ m/s). Phase speeds (*continuous curves*) and group speeds (*dotted curves*) are given for the first symmetric and antisymmetric modes S_0 and A_0 as a function of the product of frequency f [MHz] with plate thickness $2h$ [mm]. The S_0 mode group speed curve has a minimum

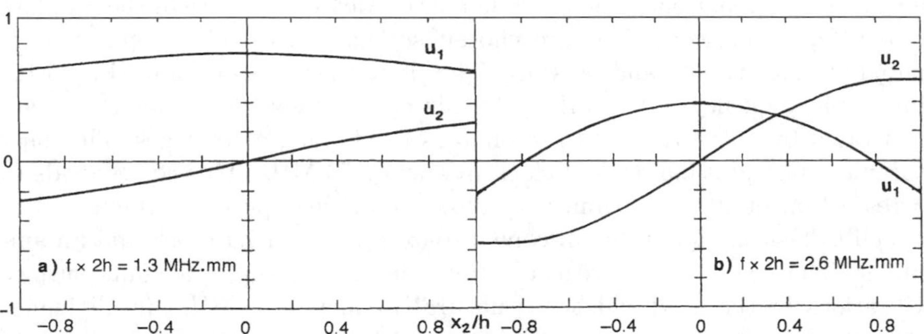

Fig. 5.17. Longitudinal and transverse components u_1, u_2 of mechanical displacement in a glass plate at points **x** on the symmetric mode dispersion curve in Fig. 5.16. (**a**) $f \times 2h = 1.3$ MHz mm. (**b**) $f \times 2h = 2.6$ MHz mm. If plate thickness $2h$ is 2 mm, frequencies are (**a**) 0.65 MHz and (**b**) 1.3 MHz. Note relative values of components at the surface: (**a**) $|u_1| \approx 2|u_2|$ and (**b**) $|u_2| \approx 3|u_1|$

At the point $f \times 2h = 1.3$ MHz mm, the amplitude u_1 of the longitudinal component of surface displacement is more than twice the value u_2 for the transverse component. At the point of inflection $f \times 2h = 2.6$ MHz mm, the transverse component u_2 is almost three times the value of u_1.

To sum up, depending on whether we require a sensor which is barely sensitive or sensitive to the presence of an object on the surface, we must choose the operating point on the plateau or at the minimum of the group speed curve.

5.2.2.1 Coordinate Sensor. The idea behind this coordinate sensor is to deduce the abscissa value X (or ordinate Y) of a particular point, e.g., indicated by a pencil, by measuring the time required for a wave train of known speed to cover the distance between emitter transducer X (or Y) and pencil tip:

$$X = N_x T V_x \,, \quad Y = N_y T V_y \,, \tag{5.9}$$

where T is the clock period, and N_x, N_y are numbers of successively counted periods for X and Y directions, respectively. The clock is triggered when the exciting electrical impulse is applied to the transducer, and stopped when the wave train reaches the pencil.

The principle had previously been put into practice to make a graphics tablet, first with Rayleigh waves propagating at the surface of a piezoelectric crystal [5.36], then with Lamb waves propagating in a glass plate [5.37]. The advantage in using a piezoelectric crystal is that elastic waves can be launched by comb electrodes. However, it limits the size of the tablet. The aim had been to record signatures and a 5×10 cm^2 lithium niobate crystal was used.

The Lamb wave tablet is a simple glass plate with usable surface area equivalent to about one page (24×30 cm^2). Waves propagate in the S_0 plate mode. Operating conditions are chosen so that wave train propagation is largely insensitive to hand pressure. The plate thickness of 2 mm (Fig. 5.16) implies a frequency of 650 kHz. For this carrier frequency, wave trains are generated by pulses applied to resonators fixed along plate edges. The measurement period is 0.5 ms and clock frequency 53 MHz. This corresponds to a resolution of 10 points/mm and allows a writing speed of 20 cm/s. The pencil is basically a duralumin cone, a piezoceramic (PZT) disk and an amplifier. The cone, coated with a Teflon film to aid slip, transmits surface vibrations to the disk, which has natural frequency (5 MHz) much higher than the transducer resonators. This type of tablet is in principle well suited to liquid crystal displays on pocket calculators. It was the starting point for the design of a coordinate sensor with a usable area of 1 m^2 (developed in collaboration between the authors' own laboratory and the Cimsa–Sintra division of Thomson–CSF) [5.38]. A diagram is given in Fig. 5.18. The aim was to measure coordinates of a point indicated on a map and to store them in memory. The map is placed on the sensor table, a duralumin plate to ensure solidity. A reticle is mounted on a tripod with a piezoelectric detector in each leg. The reticle is positioned over the selected point. Coordinates of the point are calculated by means of signals coming from the three detectors at the moment the tripod is lightly pressed onto the table. Dispersion and temperature effects are allowed for by the calculator which controls the device.

This coordinate sensor has the following features:

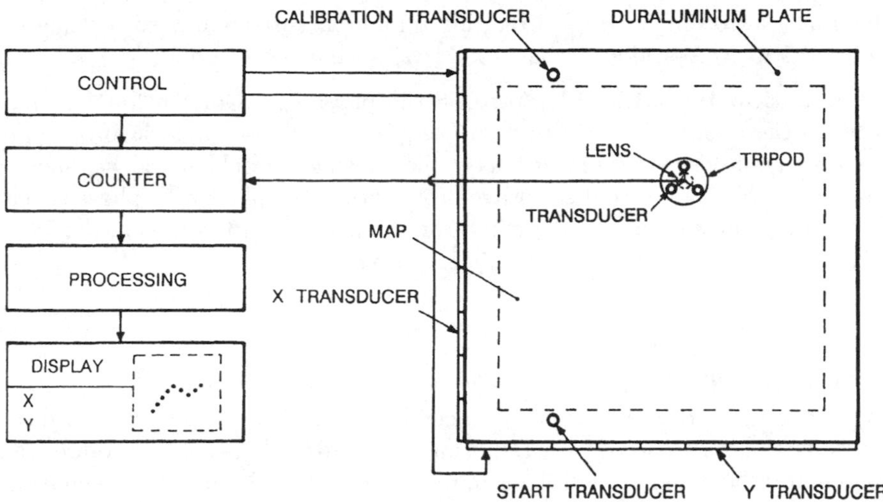

Fig. 5.18. Lamb wave coordinate sensor. The tripod, equipped with lens and reticle, is positioned on the map, above the selected point. Coordinates of the point are calculated from signals generated in the three transducers of the tripod by elastic waves periodically launched in the X and Y directions

- $1.2 \times 1.2\ \text{m}^2$ duralumin plate of thickness 1.5 mm and usable surface area $1 \times 1\ \text{m}^2$. The plate rests upon synthetic rubber blocks rigidly fixed onto a honeycomb structure.
- Resolution 0.1 mm.
- Accuracy 0.5 mm.
- Measurement rate 100 points/s.
- Transducers: 15 piezoelectric rods ($74 \times 3 \times 1.5\ \text{mm}^3$) on both sides. Electrical pulse applied to each row of transducers 200 V. Rise time 200 ns.
- Clock frequency 60 MHz.

The prototype passed mechanical, thermal and electromagnetic radiation tests imposed by military standards. Compared with other types of coordinate sensor, it has the advantage of radiating little and being insensitive to electromagnetic perturbations.

5.2.2.2 Detection of Liquids at Predetermined Levels. The liquid level in a tank is measured either continuously by means of a sensor known as a continuous gauge, or in discrete steps, using sensors known as point level gauges. A point level gauge merely indicates the presence or absence of a liquid at some predetermined height.

One way of building a non-intrusive point level gauge (i.e., one which is placed outside the tank, and hence also referred to as non-invasive) involves launching a Lamb wave in the tank wall at the chosen height. The device operates at the point of inflection on the symmetric mode dispersion curve, i.e., at the minimum group speed, so that attenuation due to the liquid is maximal. According to Fig. 5.16, this minimum corresponds to a value of about 2.5 for the product frequency [MHz] × thickness [mm].

Attenuation by a Liquid. Let c be the phase speed of the longitudinal wave in the liquid. The longitudinal wave is the only one able to propagate through a non-viscous liquid such as water, taken as model here. It was shown in Problem 5.7, Vol. I, that any wave propagating in a plate with phase speed V_ϕ greater than c radiates energy into the water (Fig. 5.53, Vol. I) in the form of plane waves moving in direction θ given by

$$\sin\theta = \frac{\lambda_\mathrm{L}}{\lambda} = \frac{k}{k_\mathrm{L}} = \frac{c}{V_\phi} \,. \tag{5.10}$$

After travelling distance x_1, the wave amplitude in the plate is multiplied by $\mathrm{e}^{-\alpha x_1}$, where α is the attenuation coefficient. The mean power carried by the wave thus varies as $\mathrm{e}^{-2\alpha x_1}$. Denoting $P_\mathrm{L} = -\mathrm{d}P/\mathrm{d}x_1$, the mean power radiated into the liquid per unit length, we find $\alpha = P_\mathrm{L}/2P$ and the attenuation coefficient over one wavelength is

$$\gamma = \alpha\lambda = \frac{\pi}{k}\frac{P_\mathrm{L}}{P} \,. \tag{5.11}$$

Power P_L is equal to the flux of the acoustic Poynting vector $-T_{ij}\dot{u}_j$ across the interface $x_2 = -h$. In the liquid, only the normal stress T_{22} is non-zero. It is equal and opposite to the overpressure $\delta p = Z\dot{u}$, where ρ is the density of the liquid, $Z = \rho c$ its acoustic impedance, and $u = u_2(h)/\cos\theta$ the longitudinal displacement in the liquid. In harmonic conditions and for a beamwidth w, it follows that

$$P_\mathrm{L} = \frac{w}{2}\mathrm{Re}\left[-T_{22}(h)\dot{u}_2^*(h)\right] = \frac{w}{2}Z\,\frac{|\dot{u}_2(h)|^2}{\cos\theta} \,. \tag{5.12}$$

The mean power P carried by the wave is given in terms of its group speed V_g and components $u_1(x_2)$, $u_2(x_2)$ of mechanical displacement in the plate (of density ρ_S) by equation (5.33), Vol. I:

$$P = \frac{w}{2}\rho_\mathrm{S} V_\mathrm{g}\omega^2 \int_{-h}^{h} \left[|u_1(x_2)|^2 + |u_2(x_2)|^2\right] \mathrm{d}x_2 \,. \tag{5.13}$$

Therefore, by (5.11),

$$\gamma = \frac{\pi\rho c}{\rho_\mathrm{S} V_\mathrm{g}\cos\theta}\,\frac{|u_2(h)|^2}{k\displaystyle\int_{-h}^{h}\left[|u_1|^2 + |u_2|^2\right]\mathrm{d}x_2} = \frac{\pi\rho c}{\rho_\mathrm{S} V_\mathrm{g}\cos\theta}\,|U_2(h)|^2 \,. \tag{5.14}$$

This formula, valid for any propagation mode in the plate, shows that attenuation over a distance of one wavelength is proportional to the square of the normal displacement at the interface, and inversely proportional to group speed. The normalised displacement $U_2(h)$ and coefficient γ calculated

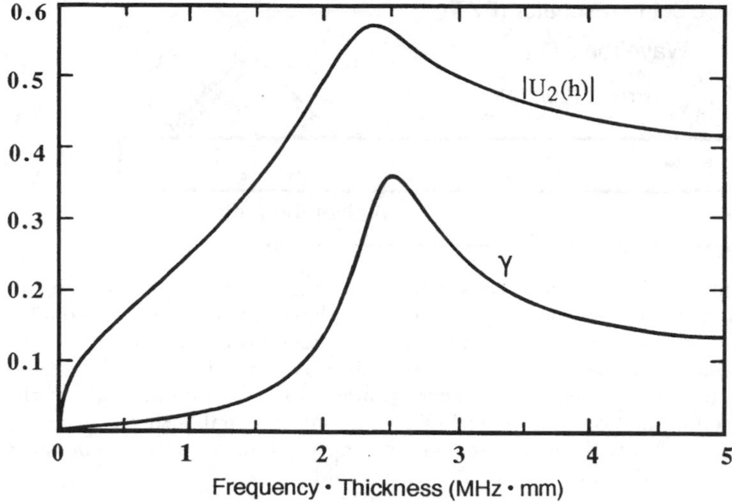

Fig. 5.19. Normalised displacement $|U_2(h)|$ and attenuation coefficient per wavelength γ when water is present on the glass plate, as a function of the frequency–thickness product $f \times 2h$. Coefficient γ is maximum when $f \times 2h = 2.5$ MHz mm

for symmetric mode S_0 in a glass plate are graphed against the frequency–thickness product in Fig. 5.19. The group speed (Fig. 5.16) and normalised displacement have a minimum and a maximum at very closely spaced values of frequency × thickness ($f \times 2h$), so that γ passes through a quite distinct maximum when $f \times 2h = 2.5$ MHz mm.

Structure. Figure 5.20 shows how the system operates. Waves are launched in the tank wall by a piezoelectric resonator fixed on a prism. The resonator vibrates in thickness and launches longitudinal waves propagating at speed V_B through the prism. At the prism–plate interface, these waves transform into a plate mode of wavelength λ if

$$\lambda \cos\beta = \lambda_B \,, \quad \lambda = \frac{V_\phi}{f} \,, \quad \lambda_B = \frac{V_B}{f} \,. \tag{5.15}$$

Angle β is thus defined by $\beta = \arccos(V_B/V_\phi)$. The wave speed V_B in the prism must be less than the wave speed V_ϕ in the wall.

If there is no liquid between the transducers, the output transducer detects any signal of frequency f applied to the input transducer. If there is a liquid, it detects a much weaker signal. Waves propagating in the wall are transformed into longitudinal waves which leak into the liquid. The whole setup constitutes a delay line in which the output signal disappears almost as soon as the liquid reaches the wall between the transducers, and reappears when the liquid goes down to a lower level. This all-or-nothing effect is accentuated if the delay line is inserted in cascade with an amplifier in such a way as to form a loop. Amplifier gain compensates for conversion losses in the transducers, so that

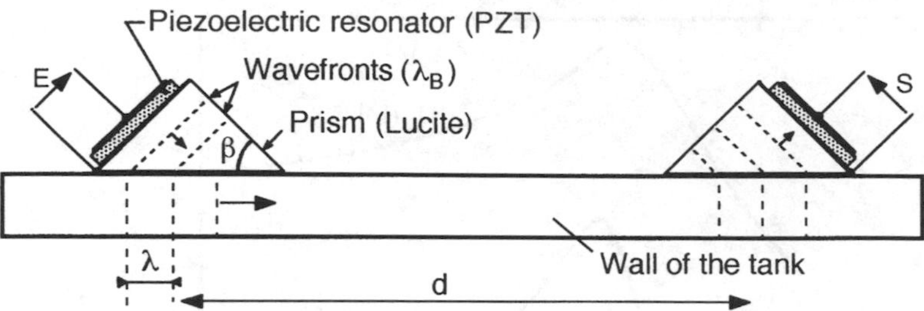

Fig. 5.20. Point liquid level sensor consisting of the tank wall region between two prisms. The input prism transforms bulk waves generated by the emitter resonator into Lamb waves which propagate through the wall when no liquid is present. The output prism transforms the Lamb waves into bulk waves which are then detected by the receiver resonator. Frequency f corresponds to the minimum value of the group speed, itself a function of wall thickness. Any input signal, either continuous at frequency f or pulsed with carrier frequency f, appears at the output unless the wall between the transducers is in contact with liquid

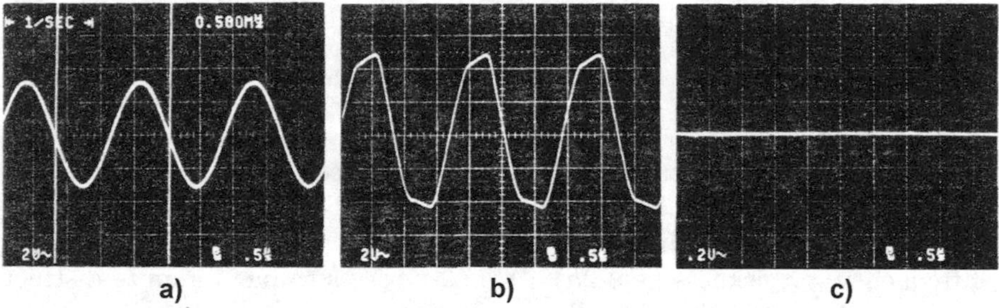

Fig. 5.21. Level sensor oscillator in which the delay line is a steel wall of thickness 4 mm and prisms are 150 mm apart. Oscillations in the absence of water, for amplifier gain (**a**) equal to, (**b**) greater than the oscillation threshold gain. (**c**) When water reaches the prism level, the loop no longer oscillates, whatever the gain

the loop oscillates if the line surface is free (no liquid) and stops oscillating as soon as it is wetted. A phase variation is thereby added to the amplitude variation produced by the liquid. Moreover, sensitivity can be adjusted via the amplifier gain.

Photographs in Fig. 5.21a,b refer to a steel wall of thickness 4 mm. The plexiglass prisms are 15 cm apart and oscillation frequency is 580 kHz. They show the signals from the (horizontal) delay line–amplifier loop when the gain is (**a**) equal to, and (**b**) greater than the oscillation threshold gain. Whatever the gain, when a liquid is present (water in this case), oscillation ceases. This type of point level sensor, used industrially, is suitable for both thin-walled (1 mm) and thick-walled tanks (10 cm). The frequency decreases with increasing thickness, so that the dimensions of the various elements, in particular, the prism spacing, increase.

Another non-invasive sensor, also made from two transducers fixed to the tank wall, has been developed by Liu and Lynnworth, and manufactured by Panametrics [5.39]. The wave used is different, being an A_0 flexural mode with frequency chosen so that its phase speed is less than the wave speed in the liquid. No energy radiates into the liquid as a result of propagation. When a liquid is present between the transducers, it is detected by the consequent change in wave speed, and hence phase. The very low frequency waves are launched and detected by magnetostrictive transducers, rigidly bonded to the wall by means of an epoxy resin.

5.2.2.3 Interactive Window Displays. The idea behind such a display is to provide visitors or customers with more information than is presented visibly. In a bookshop, for example, a number of books are on show in the window (Fig. 5.22). When the visitor indicates the region of the window corresponding to one of these books, information appears on a screen. He or she can choose a page, a drawing, a photo of the author, and so on. The zone is indicated by an impact of some kind, e.g., the impact of a pointer like a car key, or a ballpoint pen, because the coordinate measuring principle involves wave propagation [5.40]. This impact thus produces a wave train radiating outwards to the edges of the window. Experiment shows that these waves, generated by a normal impact on the plane guide constituted by the window, propagate for the main part in the first antisymmetric plate mode A_0. The wave train has a broad spectrum (20 Hz–200 kHz). It depends on the nature and shape of the pointer, as well as the motion given to it by the operator. The waves are sensed close to the edges of the window by a pair of transducers, one on the front face and one on the back face. Transducers are mounted in such a way as to suppress the symmetric part of the signal, which is small amplitude but arrives before the antisymmetric part.

Coordinates of the impact point are deduced from the known positions of the transducers and differences in arrival times of the waves under x and y tranducers, having previously measured their propagation speed through the glass of the window. Any impact (resolution ≈ 1 mm, accuracy ≈ 1 cm)

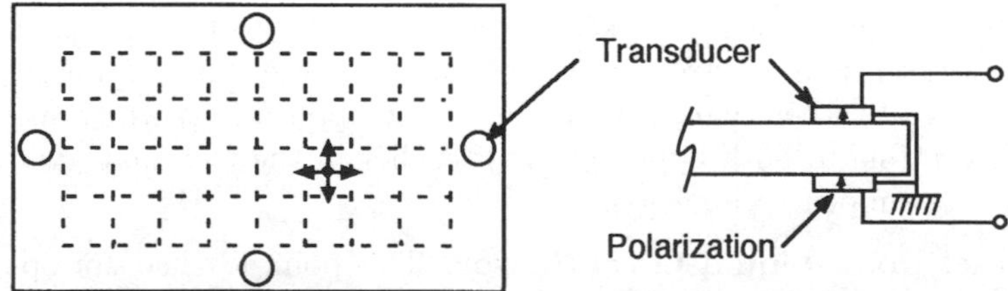

Fig. 5.22. Interactive window display. A small impact, effected by a ballpoint pen or car key, causes information to appear on a screen. The impact position determines which information will appear. Coordinates of this point are deduced from times at which the wave train generated by the impact arrives at the transducers

with coordinates lying inside the rectangle defining a zone triggers the menu associated with that zone.

This idea has been put into practice in laboratory conditions, using an ordinary glass window of area greater than 2 m^2 and thickness of order [cm], equipped with piezoceramic transducers (diameter $\approx$ 20 mm, thickness $\approx$ 0.5 mm, radial mode resonance frequency $\approx$ 100 kHz) bonded onto it with an epoxy resin. A reasonable impact with the metallic end of an object such as a key produces a voltage of about 50 mV across the terminals of a pair of transducers, loaded with a resistance of 1000 Ω and positioned 2 m away from the impact point. The device has recently been set up in a bookshop window with dimensions $3 \times 1.60 \times 0.008$ m^3, where it is at present under assessment.

5.2.3 Cylindrical Guide Sensors

Propagation of vibrations is more complicated to analyse in an infinitely long cylinder than in an infinitely wide plate, because of the extra dimension. Three types of motion can propagate: compression, torsion and flexion (see Sect. 5.6, Vol. I). A family of modes corresponds to each type. Their phase speeds depend on the product frequency × cylinder radius. However, considering only compressional motion ($n = 0$), in which matter is displaced essentially in the direction of the cylinder axis, a similar dispersion curve to the plate mode S_0 is found for the first mode $L(0, 1)$. At low frequencies, lateral motions can be neglected. The speed $V = V_b = \sqrt{E/\rho}$ is determined by Young's modulus E and density ρ. At high frequencies, motion is concentrated close to the surface and the speed tends to the Rayleigh wave speed. Curves in Fig. 5.44, Vol. I, are valid for typical materials with Poisson ratio around 0.3. A curve of the same shape is also obtained for the fundamental compression mode $L(0, 1)$ propagating in a tube (Fig. 5.46, Vol. I). At low frequencies, this mode causes longitudinal displacements and its phase speed is V_b. At high frequencies, it tends to the flexural mode A_0 of the shell. Consequently, results obtained for attenuation in a plate are also valid when dealing with these modes in solid cylindrical or tubular guides. We can therefore choose a frequency which makes the cylinder or tube of given radius either more or less sensitive to stresses exerted at its surface. The advantage in using a cylindrical rather than a plane guide, for applications which are in principle susceptible to both approaches, often lies in the fact that axial symmetry involves simpler technology.

5.2.3.1 Contact Point Liquid Level Sensor. This point level sensor operates like an echo device. However, the cylindrical guide is composed of two parts of different diameter. One acts as a mounting, the other as a probe. Diameters are chosen so that for the relevant frequency, the mounting is rather insensitive to stresses exerted at its surface and the probe is highly sensitive to contact with the liquid. Curves in Fig. 5.23 refer to two steel cylinders and

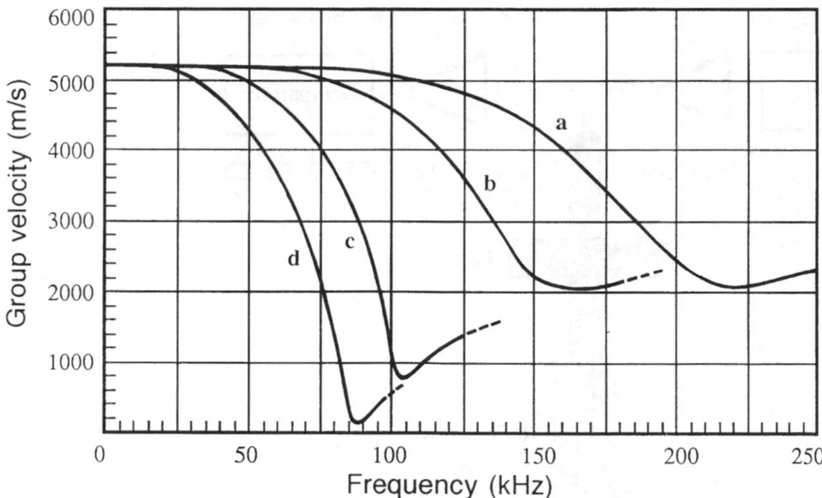

Fig. 5.23. Group speed against frequency for cylinders of diameter (*a*) 14 mm and (*b*) 20 mm, and tubes of internal and external diameters (*c*) 12 mm, 20 mm and (*d*) 16 mm, 20 mm

two steel tubes. They show that a guide made from a mounting cylinder of diameter 14 mm and probe cylinder of diameter 20 mm operates at around 150 kHz. Likewise, a guide consisting of a mounting cylinder of diameter 14 mm and a probe tube with internal diameter 12 mm and external diameter 20 mm operates at around 100 kHz. The second structure is a priori more sensitive. For a given external diameter, reducing tube thickness increases dispersion and hence guide sensitivity (Fig. 5.23d).

These predictions are confirmed by experiment. Figure 5.24 is an example of a solid cylinder point level gauge. The mounting part (12 mm) carries the ceramic emitter transducer and receiver, together with the attachment clamp (diameter 100 mm, thickness 20 mm). The probe has diameter 20 mm and length 130 mm. The inductance introduces a surge at emission and reception so that, given the low output impedance of the amplifier, electronic gates are not required. Figure 5.25a shows the signal detected in response to an electrical pulse applied to the transducer, whilst water has not yet reached the probe. Note the echo caused by reflection at the join between the two sections and also the principal end echo. Figure 5.25b is the detected signal when the probe is immersed to a depth of 50 mm in water. The echo from the join remains (serving as a reference). The principal echo is attenuated. Since the phenomenon is repeatable, different liquid levels can be distinguished, e.g., four levels at 15 mm intervals.

When immersed to the same depth, a similar sensor but with tubular probe provides almost twice the echo attenuation.

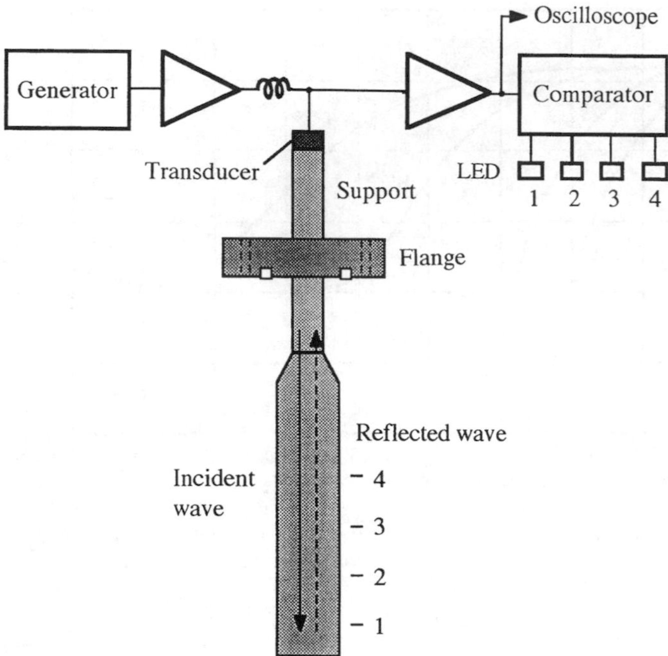

Fig. 5.24. Echo point level sensor. The cylindrical guide has a mounting part (diameter 12 mm) which carries the piezoelectric transducer, the attachment clamp and a probe cylinder (diameter 20 mm). The wave train is reflected from the end of the probe cylinder when no liquid is present. As the liquid rises to levels 1, 2, 3 and 4, successive electroluminescent diodes light up

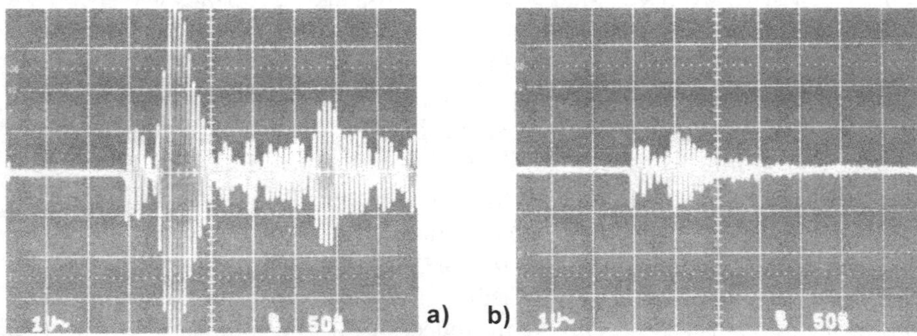

Fig. 5.25. Echos produced by the point level sensor in Fig. 5.24 when the probe is (**a**) out of water and (**b**) in water (≈ 5 mm)

The high sensitivity of these probes, requiring no impedance matching layers, stems from the fact that waves are guided. They reflect from guide edges as they travel, and the solid–liquid interaction extends over several wavelengths. Other point level gauge devices based on the same double cross-section principle are feasible. For example, the echo delay line can be replaced by a transmission line.

5.2.3.2 Continuous Level Gauge. This is a device which measures liquid levels on an almost continuous basis. A wide range of such gauges exists [5.41]. Among elastic wave devices, we may cite the radar-type gauge comprising basically a transducer located in the upper part of the tank. The liquid level is determined by measuring the time of flight of a pulse, launched and then detected by the transducer after reflecting from the liquid surface. In order to circumvent atmospheric disturbances, which change the speed of sound, it is best to guide the wave train. For instance, the guide may be a tube, with the echo produced by the guide–liquid interface. However, the range of the device is then limited by tube length, and in addition, the tube must somehow be mechanically held. A second solution which is useful for large storage tanks derives from the principle of the point level sensor described in the last section [5.42]. It consists in choosing as guide, insensitive to liquid presence, a wire wound around a drum, equipped with some sensitive part at its extremity. This probe controls the unwinding of the wire as long as it is not in contact with the liquid, stopping it as soon as the end is wetted. In principle the depth of liquid can be found by measuring the length of wire unwound (the number of turns of the drum). However, the echo method is very simple here. Use of a magnetostrictive wire simplifies generation and detection of the wave train. Indeed, waves are launched and detected by means of one, or better, two small coils around the wire (Fig. 5.26). The drum is merely a means of storage. The probe is a tube whose diameter is chosen, as in the point level sensor, to ensure maximal sensitivity. The connection between wire and probe is made through a conical part. Join A in the air still gives an echo, whereas the end of the probe ceases to do so when immersed.

Let t_0 be the time when the wave train is launched, V its propagation speed, and t_1 the arrival time of the echo produced by join A. The liquid level H is given by

$$H = V\frac{t_1 - t_0}{2} + d_2 + \frac{d_1}{2} \,. \tag{5.16}$$

The experiment was carried out in laboratory conditions using water, a nickel–iron wire of diameter 0.5 mm, and a stainless steel tube with closed end as probe ($l = 10$ cm, $r = 3.75$ mm, $e = 1.5$ mm). The cone ($l = 6$ cm) was also made of steel. The frequency and duration of the wave train were 260 kHz and 100 μs, respectively. The amplitude of the signal from B is reduced by a factor of 4 when the probe enters the water to a depth of 8 mm. This is sufficient to switch off a motor. The echo from junction A, used to measure the level, does not change.

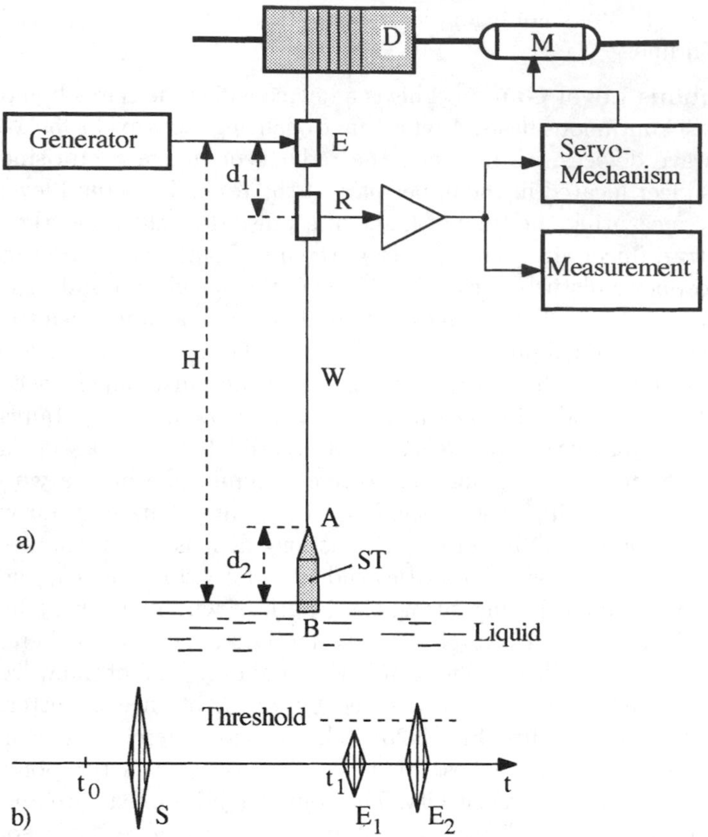

Fig. 5.26. Echo liquid level sensor. (**a**) The magnetostrictive wire is unwound from the drum and wave trains are launched periodically by coil E. Waves reflect off end B of the probe and induce a signal in coil R. The liquid level is found by measuring the arrival time of the echo from junction A. (**b**) Signals detected by coil R whilst the probe is out of the liquid

5.3 Acousto-optic Components

Effects of an elastic wave beam on a light wave beam were investigated in Sect. 3.1. The main results discussed there were modifications in light beam parameters (amplitude, direction, frequency and spectrum) when amplitude and frequency of the elastic waves are adjusted. However, applications were mentioned only briefly, apart from a frequency modulator inserted in an optical probe for measuring subnanometre displacements (Sect. 3.2.2.2). The aim here is to discuss some further examples of acousto-optic modulators, tunable filters and deflectors.

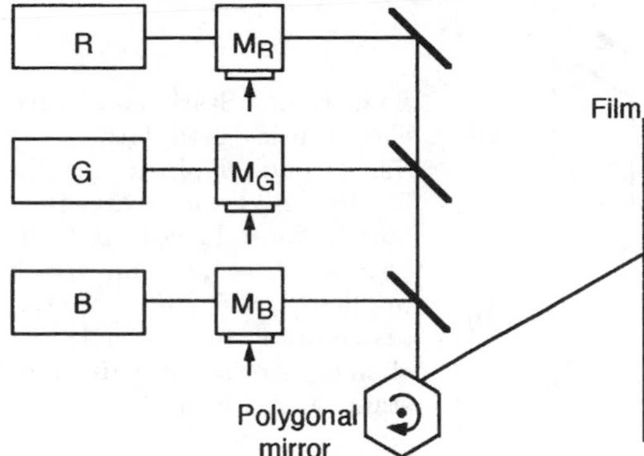

Fig. 5.27. Triple laser printer with colours red (R), green (G) and blue (B). The intensity of each laser beam is chosen from n values (where $n = 256$, for example) by means of an acousto-optic modulator. The beam composed by bringing together these three beams of different colours is directed by a rotating mirror onto a photosensitive film which it scans line by line

5.3.1 Modulator for Laser Printer

The laser printers discussed here are those intended for professional use, since these alone contain an acousto-optic modulator (abbreviated to AOM). There exist several types, mainly depending on the number of lasers (1 to 3) and the nature and shape of the photosensitive surface (e.g., drum, plane film). However, in each type, the printer contains at least one modulator and a deflection device which is often a polygonal mirror (rotating ring with polished faces). Figure 5.27 shows schematically how the setup works, without the optics, in a three-laser printer (red, green, blue). The intensity of each laser beam is controlled by an AO modulator before the three beams are brought together into a single beam. The latter hits the rotating polygonal mirror. As it rotates, each element of the mirror causes the beam to scan a line on a plane photosensitive film. Synchronised movement of the film ensures the correct line-by-line recording of the document (the latter has of course been entered digitally into a memory, which controls the modulators).

This type of modulator has the following characteristics (AA optoelectronic catalogue). Crystal TeO_2. Centre frequency 200 MHz. Rise time (equal to the time needed for the elastic wave beam to cross the light beams, which are generally focussed) 8–15 ns. Maximum control power applied to the transducer 1 W. Efficiency 85% at 633 nm. Number of grey levels ≥ 256. The power of each laser is expressed in mW. The length of a recorded line is of order 1 m. The diameter of the spot lies between 10 and 50 μm.

With regard to intensity modulators, note that travelling and stationary wave modulators are often inserted into laser cavities where they act as simple switches (Q switches) or mode lockers.

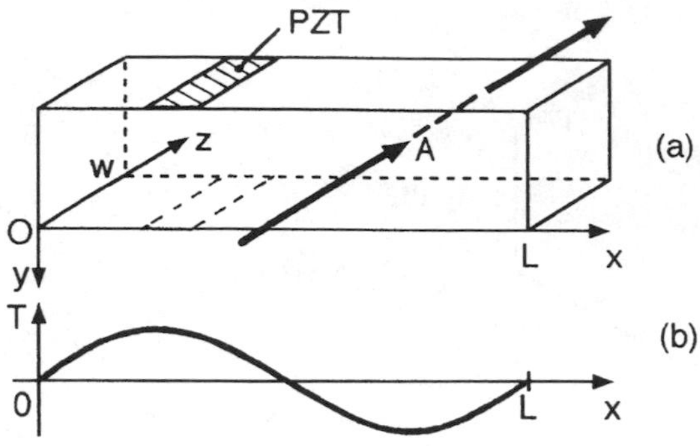

Fig. 5.28. Stationary wave elasto-optic modulator. (**a**) Silica parallelipiped ($f = 50\,\mathrm{kHz}$). PZT wafers are stuck onto its lateral faces. (**b**) λ vibration ($L = 11.4\,\mathrm{cm}$, $q = 2$ implies $f = 50\,\mathrm{kHz}$). Spatial distribution of internal mechanical stresses at a given instant. Taken from [5.43]

5.3.2 Stationary Wave Frequency Modulator

It is often necessary to modulate light beams in order to extract the information they carry at a frequency accessible to classic measurement equipment and with better signal-to-noise ratio. We shall describe a modulator operating with low frequency stationary waves [5.43], illustrated in Fig. 5.28. It consists of a silica parallelipiped with a (PZT ceramic plate) transducer stuck onto two of its lateral faces. These plates vibrate along the main axis of the parallelipiped, exciting longitudinal waves at its (λ or $\lambda/2$) resonance frequency. The alternating stress T thereby produced in the parallelipiped renders it birefringent:

$$T = T_{\mathrm{m}} \sin \frac{q\pi x}{L} \sin 2\pi f t \,,$$

where T_{m} is the maximal stress value, and q an integer defining the resonance mode. Figure 5.28 shows (for $q = 2$) the distribution of T at a given time. It also represents the spatial variation of the birefringence $n_x - n_y$. In the case $q = 2$, and for a length $L = 11.4$ cm, the resonance frequency is 50 kHz.

A phase shift is imposed upon the light beam crossing the width w of modulator at A. It is due partly to the change in refractive index $2\pi w \Delta n/\lambda$ and partly to the change in width $2\pi n \Delta w/\lambda$ caused by the Poisson effect. If waves are linearly polarised along x or y, only their phase is modified as they cross. If wave polarisation lies along a line bisecting the xy quadrant, i.e., if it has two components propagating at different speeds, it will be modified by time variation of the birefringence $n_x - n_y$ (the phase shift due to Δw is the same for both components and therefore plays no role). Hence, when the maximal stress generated by the stationary elastic wave induces a relative phase shift of $\pm\pi/2$, the light wave polarisation alternates from right circular to left circular. Between the two states, it passes through intermediate states of elliptical and then linear polarisation. Arranging an analyser perpendicular to the incident light polarisation, the flux obtained is modulated at a frequency equal to twice the elastic wave frequency.

This type of modulator differs from the electro-optic type of modulator by its very low control voltage (≈ 10 V), its suitability for processing wide-

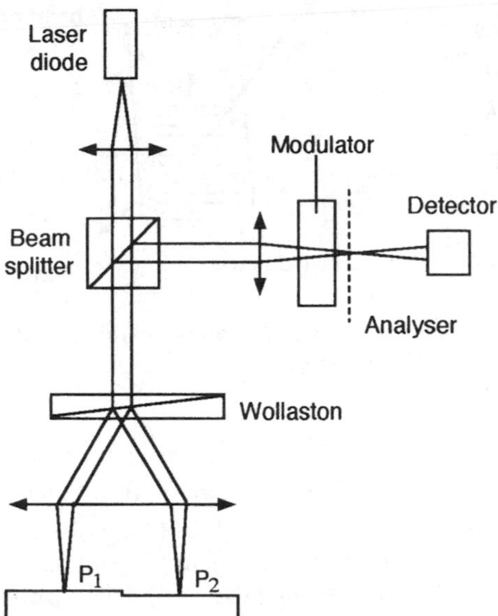

Fig. 5.29. Differential profilometer. The difference in elevation of points P_1, P_2 is measured with the help of an elasto-optic modulator. From [5.46]

angle beams (30°) of large cross-section (diameter > 1 cm), and the fact that it does not require precise alignment. However, the operating frequency is fixed, somewhere between a few hundred Hz and a few hundred kHz. Different materials are used depending on the light wavelength. Silica is suitable for the range 0.23–1.4 μm, and zinc selenide (ZnSe) for infrared wavelengths. The resonator can be excited by a transducer placed at one of its ends [5.44].

This modulator is built into several instruments. One is a dichrograph, used in biology for detection and dosing of molecules exhibiting dichroism or rotatory power. Another is an ellipsometer [5.45], used to study films and doping of integrated circuits in microelectronics. Figure 5.29 shows how the modulator is used in a laboratory profilometer [5.46]. The idea of the measurement is to separate a beam from a laser diode into two orthogonally polarised parts (with the help of a Wollaston prism). Each beam hits a point on the sample. After reflecting, waves from the two beams have a path length difference equal to twice the height difference between the points. The two beams are recombined and the resulting beam crosses the modulator. It superposes a periodic phase shift on the static phase difference related to the surface state. After passing through the analyser, the two beams interfere. The resulting periodic signal gives the difference in level. The smallest difference yet measurable is of picometric order.

5.3.3 Spectral Line Selection by Tunable Filter

The acousto-optic filter has inspired new techniques in the selection of spectral lines, and hence in the reading of light spectra. The idea was explained in

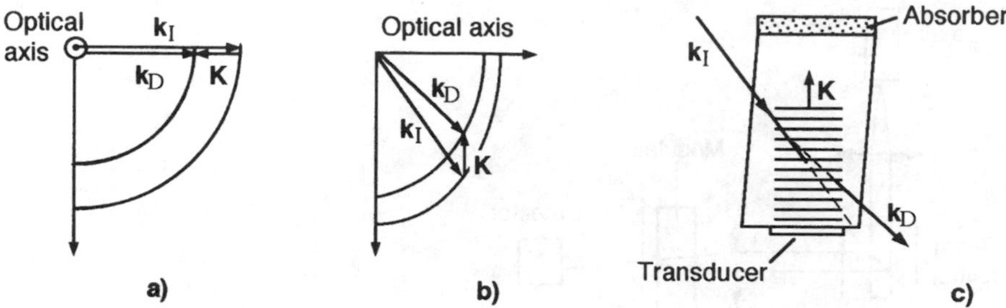

Fig. 5.30. Wave vector diagram in (**a**) collinear and (**b**) non-collinear interaction. Elastic waves may be generated by several ($LiNbO_3$) transducers

Sect. 3.1.5.3. It rests on the fact that a diffracted light wave of predetermined frequency is produced when a light wave and elastic wave of given frequencies enter the filter. By varying the radio frequency of the signal applied to the transducer, the various lines contained in the spectrum of any light beam illuminating the filter are successively diffracted. Considering that the filter must in general be largely insensitive to any angular shift in the wave vector $\boldsymbol{k}_\mathrm{I}$ of the incident light, the interaction medium is a birefringent crystal. If for example the incident vector $\boldsymbol{k}_\mathrm{I}$ has the extraordinary polarisation, then the diffracted vector $\boldsymbol{k}_\mathrm{D}$ has the ordinary one. The wave vector $\boldsymbol{K}$ of the elastic wave is chosen so that tangent planes to optical wave vectors are parallel, thus allowing a much wider incident beam. Figure 5.30 shows the two situations which may arise in the case of a uniaxial crystal. In the first (**a**), wave vectors are collinear. The two optical beams must be separated by polarisers (not shown). In the second (**b** and **c**), optical beams separate naturally on leaving the crystal. Inside, rays are collinear since energy vectors are, the tangent planes to the two index surfaces being parallel. The type of crystal is partly determined by the spectral lines to be selected. Tellurium oxide TeO_2 has a high acousto-optic coefficient but is too absorbent shortward of $\lambda = 350$ nm. It is suitable for visible and infrared light, and is indeed the material most often used. Quartz has a small coefficient but is transparent to shorter wavelengths, hence suitable for ultraviolet light.

Acousto-optic filters have various strong points. A wide band (one octave) can be scanned in a very short time, and they have large aperture ($> 20°$) and high resolution (of order nm). Combined with their compactness, light weight and reliability, it is not surprising to find them built into such a range of different devices. Spectra of light sources (laser or other) can be read in a fraction of a second by varying the control radio frequency at more than 50 MHz. It is relatively easy to follow the variation of a spectrum in time or with the change of some other parameter. Kurtz et al. [5.47] have been able to make a complete record of the spectrum of a match flame from the moment it was struck. Let us now mention three applications to very different areas.

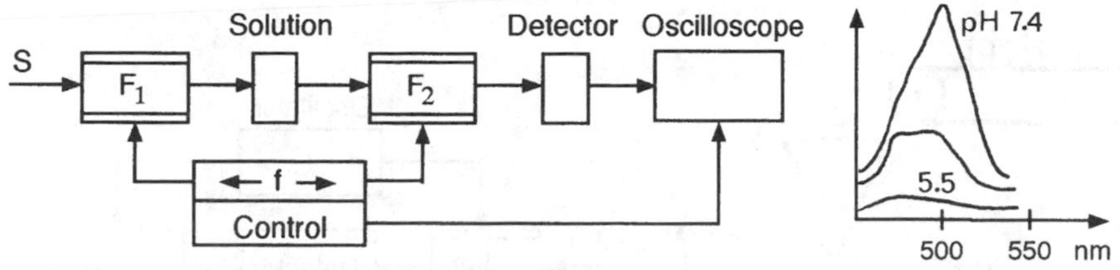

Fig. 5.31. Excitation of a dye solution by one or several lines chosen by means of an acousto-optic tunable filter F_1, together with a readout of its fluorescence spectrum made by a second tunable filter F_2. Several emission spectra, for different pH values, can be plotted in a matter of seconds. From [5.47]

Reading a Fluorescence Spectrum. Such a spectrum is normally read off by means of a single acousto-optic filter. This filter is often included in an instrument, such as a confocal microscope. However, we shall describe a laboratory experiment, due to Kurtz et al., involving two filters. It will demonstrate the ease with which these tunable filters can be used.

Figure 5.31 shows the setup for this experiment (without focussing lenses). It consists in exciting a dye solution, whose characteristic spectrum is sensitive to pH variations, and then recording its emission. The excitation line comes from source S (in this case, a mercury–xenon arc lamp). It is selected with the help of a first tunable (TeO_2) filter F_1, by choosing the radio frequency of the signal applied to the transducer. The beam is focussed in the solution. The light spectrum it emits is read off by a second tunable filter F_2 (also TeO_2) over a wavelength span determined by its radio frequency sweep. Of course, it is just as easy to adjust filter F_2 to a particular fixed wavelength and then excite the solution using a beam of broad spectrum, determined by radio frequency variations of F_1.

The designers of this experiment have excited a dye solution with a spectrum ranging over 452–518 nm and recorded the intensity emitted at 528 nm. They have also excited the same solution with a 498 nm line and observed the spectrum emitted in the 498–548 nm range (time needed < 20 ms). In less then 10 s, they plotted six emission spectra for different pH values of the solution (altered by acidification).

Sorting of Plastic Objects. It is not easy to separate plastic materials by measuring physical quantities such as density. Investigation of structural and molecular properties is required. Eisenreich, N., et al. [5.48] have suggested a selection method based on near-infrared absorption spectra, which can be scanned using acousto-optic tunable filters. Indeed, such filters are ideal for the desired scanning rates ($\ll 1$ s) and automation, imposed by economic considerations. Plastic objects are thus sorted with a view to recycling, as shown schematically in Fig. 5.32.

Items pass before a halogen light source. Light transmitted (reflected) by each item is carried by optical fibre to the filter input. Spectral lines

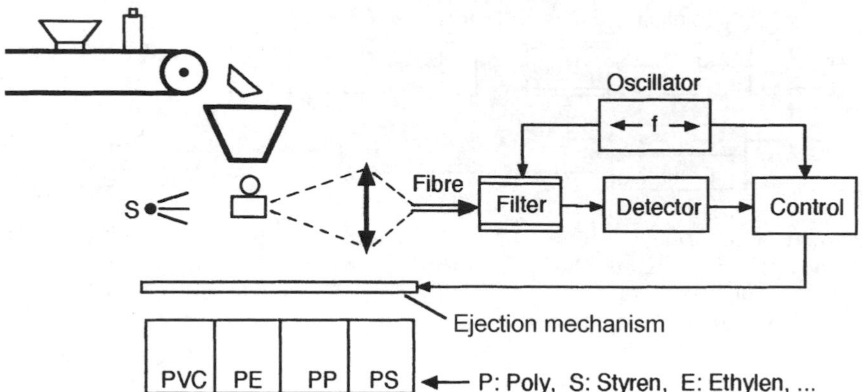

Fig. 5.32. Fast spectroscopic sorting of plastic materials, made possible by use of an acousto-optic tunable filter. Taken from [5.48]

of the light are successively diffracted by the filter whose radio frequency varies periodically over some defined range. A germanium or indium arsenide detector supplies the lines in electrical form to a computer which identifies them and controls ejection of the corresponding item into the appropriate bin. The radio frequency of the signal applied to the filter transducer is governed by an oscillator which is also controlled by the computer. There is a wide variety of plastic materials, depending on the various additives they may contain (e.g., dyes, pigments, stabilisers, hardeners, flame retarders). In order to identify them, a scanning range extending from 1000–1800 nm is required. Products from household consumption can be identified over a much narrower range, viz., 1600–1800 nm. Such a range is scanned in 1 ms by a TeO_2 tunable filter.

Son et Lumière Shows. These shows involve the projection of light rays of different colours over a large area, which may be indoors or outdoors. The surface may be the walls of a castle or a cloud! The project is basically a 'white' argon or krypton laser, together with an AO tunable filter and an oscillating (galvanometric) or rotating mirror. A different colour (spectral line) corresponds to each signal frequency applied to the filter transducer. In general, eight or twelve signals of different frequency (hence, colour) are available to the operator, who mixes them in the desired proportions. Operating conditions are roughly as follows: laser power 10 W, beam width 1.5 mm, TeO_2 (non-collinear interaction) filter, $LiNbO_3$ transducer, $80 < f < 150$ MHz, control power 0.1 W per line, efficiency 90% per line.

5.3.4 Radio Astronomy Spectrometer

Radio astronomers need spectrometers for a varied range of tasks: monitoring the ozone layer, close to Earth, and rapid plasma emissions from the Sun;

analysing the chemical composition of interstellar clouds, precursors for star birth, or detecting magnetic storms around Jupiter or other planets of the same type located in other stellar systems. Emission wavelengths of these diverse phenomena are equally varied (from a few μm to several metres). However, they can usually be brought within a fairly low frequency band (up to 1 GHz) by mixing two waves and studying the resulting beat. Acousto-optic deflectors have become an essential component in these spectrometers. It was Cole, T.W., in Australia [5.49], who first introduced acousto-optic interactions into radio astronomy when, in 1968, he carried out tests with a tank of water. However, the power sensed is extremely low, often less than a fraction of 1 aW (1 attowatt = 10^{-18} W). Very stable systems are needed to detect signals. Observation times are sometimes extremely long (several hours, but with a dynamic range of a few dB, in the case of some particular dilute molecule in a cold medium); and they are sometimes relatively short (of order [ms], with a dynamic range of 50–60 dB, in the case of a magnetic storm). This explains why, despite earlier studies in Australia [5.50] and Japan [5.51], it was not until the 1980s that the astronomical community came to recognise the importance of acousto-optic deflectors, through the work of Masson [5.52]. Experiments at the Owens Valley Observatory in California led to the construction of a first spectrometer with bandwidth 100 MHz and resolution 0.16 MHz, and a second with bandwidth 500 MHz and resolution 1 MHz.

Once it had been demonstrated that the AO deflector could compete with classic filter sets, which had been used for high bandwidths up until then, and digital autocorrelation techniques applied at low frequencies, several laboratories undertook research in the area. Deflectors were made and built into the main terrestrial observatories [5.53]. Materials used were (and still are, on the whole):

- tellurium dioxide TeO_2, with bandwidth B a few tens of MHz and product $B\Theta$ in the range 1000–2000 ;
- lead molybdate $PbMoO_4$, with bandwidth several hundred MHz and product $B\Theta$ less than 1000 ;
- gallium phosphide GaP, with bandwidth $B \approx 500$ MHz and resolution $R \approx 1$ MHz;
- lithium niobate $LiNbO_3$, with bandwidth $B \approx 1$ GHz and $R \approx 1$ MHz.

Transducers, generally $LiNbO_3$, usually generated longitudinal waves and interaction did not alter polarisation of the diffracted beam. The helium–neon laser (638 nm) was long used as the source. Since the 1990s, this source has been replaced by the laser diode (780 nm) and longitudinal elastic waves have given way to transverse waves. Interaction produces a diffracted light beam polarised perpendicularly to the incident beam (Sect. 3.1.5.3). This is a significant advantage since the useful signal is thereby easily distinguished from the zero order beam, a source of noise.

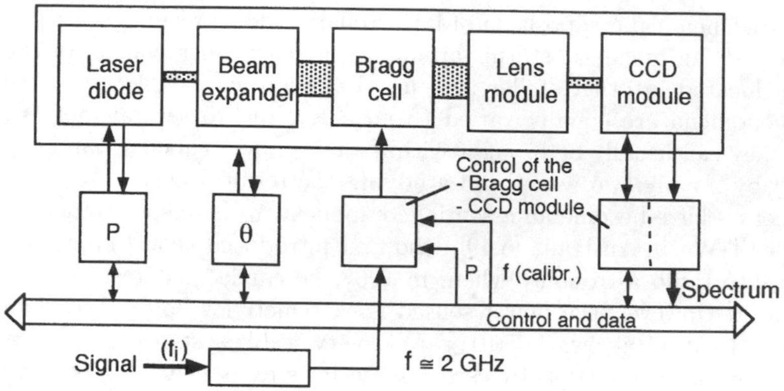

Fig. 5.33. Acousto-optic spectrometer. This comprises four elements mounted together: a laser diode with beam-widening facility, a Bragg cell, a focussing lens and a CCD detector (photodiode array with charge transfer readout). The absolute stability needed over periods of several minutes is guaranteed by servo systems controlling diode power P and temperature θ of the setup

Figure 5.33 depicts the four parts of the acousto-optic spectrometer: a laser diode with beam-widening optics (prisms and lenses), Bragg cell, lens and CCD detector (CCD stands for Charge Coupled Device, an array of photodiodes with sequential charge readout), and servo systems to satisfy stability requirements. The latter involve enslaving the power emitted by the diode; controlling the temperature of the whole setup and, in particular, of the mechanical mounting; and periodic recalibration of the deflector frequency response.

The intermediate frequency signal f_{i} (depending on the object under observation and lying in the range 10 to 1 200 MHz) is produced by interference between signals from the object and a local oscillator (a CO_2 laser or solid-state semiconductor source). The frequency of this signal may be modified (up to 2 GHz) so that it can command the deflector. For example, ozone observation through absorption of solar radiation requires, apart from a heliostat, a CO_2 laser local oscillator, a semi-transparent thin plate which superposes beams from the Sun and those from the laser, a ZnSe lens and a liquid nitrogen cooled HgCdTe mixer–detector, which supplies the signal applied at the spectrometer input.

Let us mention some features of these deflectors. At high frequencies (HF) $f \geq 2$ GHz [5.54], it is a lithium niobate crystal of length 5 mm, and at low frequencies (LF) $f \leq 40$ MHz, it is a paratellurite crystal of length 50 mm. Each crystal carries a lithium niobate transducer. HF elastic wave beams have dimensions around 0.3 mm × 0.3 mm, whereas LF beams are about 2 mm × 5 mm. The maximal power applied to each transducer is of order 100 mW. Light beams from a laser diode (20 mW, 780 nm) illuminate the crystal. The deflected light beam, which is polarised perpendicularly to the

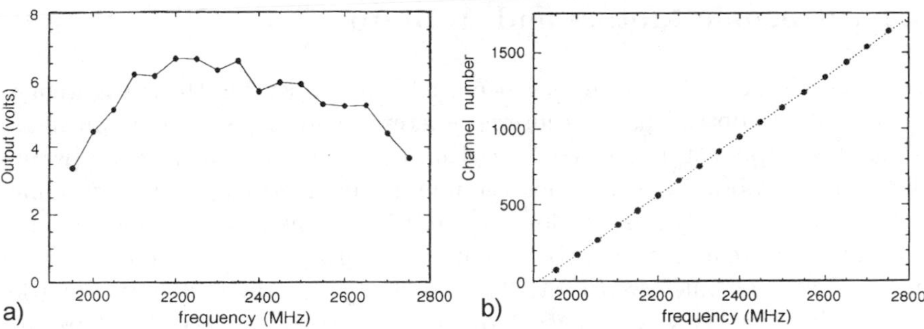

Fig. 5.34. Deflector with centre frequency 2.4 GHz. (**a**) Pass band. (**b**) Channel number–frequency correspondence (resolution 1.2 MHz). Adapted from Fig. 5 in [5.55]

incident beam and eliminates noise in the HF case, falls upon a CCD camera (with 1 750 elements).

Figure 5.34 refers to a deflector with relatively high centre frequency 2.4 GHz. The 3 dB bandwidth is 800 MHz. Figure 5.34b shows the correspondence between channel number and frequency (resolution 1.2 MHz, $B\Theta = 670$). An element of width 1.2 MHz receives power 0.15 mW. The signal is extracted by successive subtraction of two spectra: the first is the spectrum of the source under observation, the second comes from an 'empty field', so that a shift of less than 3 dB in the pass band has no consequence.

Progress over the last ten years has produced relatively light and compact instruments (total weight for the four parts of the spectrometer 1 kg, volume 1 dm^3). Several have been installed on satellites recently for imminent launch: SWAS (Submillimeter Wave Astronomical Satellite) and ODIN, a Franco-Swedish satellite [5.56]. Present studies aim to design spectrometers with 4 GHz bandwidth, by combining two or three Bragg cells, ready for the next generation of ESA (European Space Agency) satellites, e.g., FIRST (First InfraRed Space Telescope).

Coming back to Earth, we end this section by mentioning how introduction of AO deflectors into classical instruments and techniques generally simplifies measurement processes, whilst increasing their range. An example is triangulation measurement of an object's dimensions without mechanical contact (e.g., when the object is hot), using XY acousto-optic deflectors. Another is a radar experiment with an HF frequency modulator/changer. In order to increase the bandwidth of an electronic scanning antenna, it is interesting to use opto-electronic architecture, in particular to benefit from parallel optical processing of signals. The radar signal must be carried by an optical beam. Dolfi, D., et al. used an Nd:YAG laser beam, and an acousto-optic modulator whose transducer was excited by a radar signal of centre frequency 3 GHz [5.57]. They were thereby able to control optically the phase of 16 elements of an antenna operating between 2.5 and 3.5 GHz.

5.4 Ultrasonic Motors and Actuators

Piezoelectric actuators have long been used as adjustable thickness wedges to control the optical path in interferometers. They have become an indispensable component in a wide range of applications, from microscopy (where they shift the observation zone for scanning probe microscopes) to astronomy (where they optically adapt deformable mirrors). The idea of using these actuators to rotate a rotor disk, by placing them around its perimeter or on its free surface, was put forward over 25 years ago, when the first simple motors were built [5.58]. However, it was not until the 1980s that studies were undertaken on a large scale, in Japan, when Sashida, T., put together a novel travelling wave motor. In 1991, Canon built a motor of this type into their cameras for the automatic focussing mechanism. This fact naturally justified further studies outside Japan. The history of the subject is comprehensively covered in [5.59].

Ultrasonic motors work as follows: the fixed part (stator) is mechanically deformed by means of ceramic components subject to periodic electrical fields. This sets the moving part (rotor or slide block) into motion, by friction. Input electrical energy is thus partially converted into mechanical deformation energy (vibrations) via the piezoelectric effect. This energy is then partially transformed into translational or rotational energy by mechanical contact. The first transformation is usually quite efficient ($< 80\%$), whilst the second depends on the pressure with which the rotor is pushed against the stator.

Let us make a few general comments before describing examples of structures.

- Mechanical displacement produced by a piezoelectric actuator is very small. However, it is significantly greater for ceramics than for monocrystals. Hard PZT ceramics (Curie temperature $> 330°\mathrm{C}$) have values of d_{33} greater than 200 pm/V and d_{31} is of order -100 pm/V. These values should be compared with those for quartz and lithium niobate, found in Problems 3.9 and 3.10 of Vol. I. The (tangential) displacement is therefore of order μm per vibration cycle for ceramics. To generate a speed of a few cm/s, such displacements must be produced several times a second, and in such a way that they add together. Vibration frequencies of piezoelectric elements are therefore necessarily several tens of kHz and the operation point is chosen so as to benefit from a resonance. A speed of 6 cm/s requires a frequency close to 30 kHz. For a rotor diameter of 2 cm, this speed corresponds to 57 rpm. The speed of a piezoelectric motor is less than that of an electrical motor.
- Mechanical contact between two components can give rise to higher pressures than an electromagnetic link. The torque of a piezoelectric motor is therefore potentially greater than that of an electric motor. At standstill, the torque is large, being a function of the force holding the rotor against

the stator. This can be an advantage or a disadvantage, depending on the application.

- The rotor (or slide block, in the case of a linear motor) is a single element driven by friction. Its inertia can easily be reduced, if need be, thereby also cutting the response time of the motor (from standstill to normal running and vice versa). Times as short as some ms are accessible.
- No magnetic field is required and the presence of intense fields does not disturb its operation.
- The frequency of the elastic deformations which cause motion is located outside the audible range, so that it can run in silence.
- Frictional forces cause wear, thereby limiting the lifespan of the motor. It is therefore better suited to intermittent use.

To sum up, ultrasonic motors are characterised by their small volume (determined in part by the ceramic units) and speeds of at most a few tens of cm/s. Torques are relatively large (several N cm) considering the size of these motors, and direct connection with the load is possible, without reduction gears. A natural area of application is in microrobotics [5.60].

Several structures have been investigated. Figure 5.35 shows a Sashida-type travelling wave motor (developed by Shinsei). Both rotor and stator are rings. A flexural wave propagates on the toothed edge of the stator. This wave is accompanied by an elliptical motion, similar to that of a Rayleigh wave (although this would have too long a wavelength for the frequencies involved here: $f = 30$ kHz, $\lambda = 10$ cm, and the flexural wave is more easily excited). The normal component of mechanical displacement periodically raises the rotor which is maintained in contact with the stator by means of a spring. It is driven by the tangential component. The ellipse is followed in retrograde fashion by particles on the surface of the stator, and this means that the rotor is displaced in the opposite direction to the wave vector. In practice, the travelling wave is generated from two stationary waves temporally and spatially excited with phase difference $\pi/2$ $(\lambda/4)$, in conformity with the relation

$$\cos(\omega t - kx) = \cos \omega t \cos kx + \sin \omega t \sin kx \ .$$

Suitably polarised piezoelectric elements giving rise to these two waves are bonded onto the lower face of the stator, deforming it as shown in Fig. 5.35b. The teeth on the stator play a double role. They amplify tangential mechanical displacement and provide evacuation spaces for dust produced by friction between rotor and stator surface. The direction of rotation of the rotor is reversed, in agreement with the above formula, by a change of π in the phase of the voltage applied to one of the two series of ceramic elements which generate one of the two stationary waves (so that $\sin \omega t \to -\sin \omega t$). The principle is clearly valid for a concentric disk structure, as well as for a linear structure. In the latter case, the wave must be absorbed at the end of the

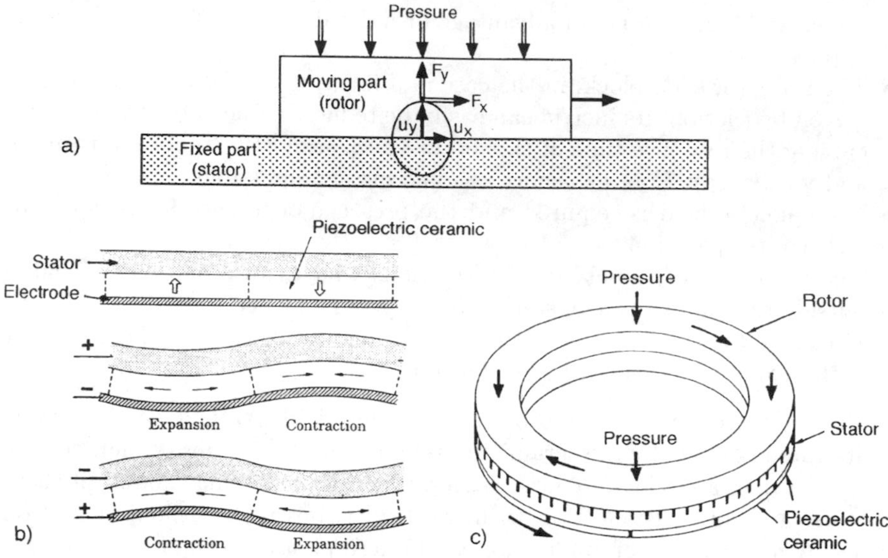

Fig. 5.35. Sashida travelling wave motor. (**a**) The idea is to produce an elliptical displacement at the surface of the stator, whose tangential component can drive the rotor. (**b**) The flexural wave generating this elliptical displacement is excited by ceramic elements bonded onto the lower surface of the stator and operating in extension–contraction. (**c**) The rotor (ring), pressed against the teeth of the stator by a spring (umbrella) which is not shown, is displaced in the opposite direction to propagation of the flexural wave in the stator, during each vibrational cycle. Taken from [5.59]. With the kind permission of Oxford University Press

wave guide. The annular structure described is useful when a light beam must be allowed to pass through the motor.

This kind of motor is difficult to study because of the frictional phenomena. Description has been achieved in part using equivalent circuits, introducing diodes to represent friction, and also finite element methods.

Figure 5.36a represents the second motor structure, in which the rotor is once again driven by an elliptical motion of the stator. The latter is made from four cylindrical units mounted on an axle, one end of which is threaded so that axial pressure can be exerted between stator and rotor with the help of a screw nut, spring and bearing. The behaviour of the motor depends on the force exerted by the rotor on the stator. This determines the contact area and hence also the elastic behaviour of the interface. Two of the four stator units are metallic. They act as loads for the other two units, which are piezoelectric, composed of stacked ceramic disks. The normal component of the elliptical displacement is produced by longitudinal vibration, along the axis, of one of the two ceramic stacks, each disk being polarised in its thickness direction. The tangential component is produced by torsional vibration of the other stack, each disk being polarised along its circumference. Simul-

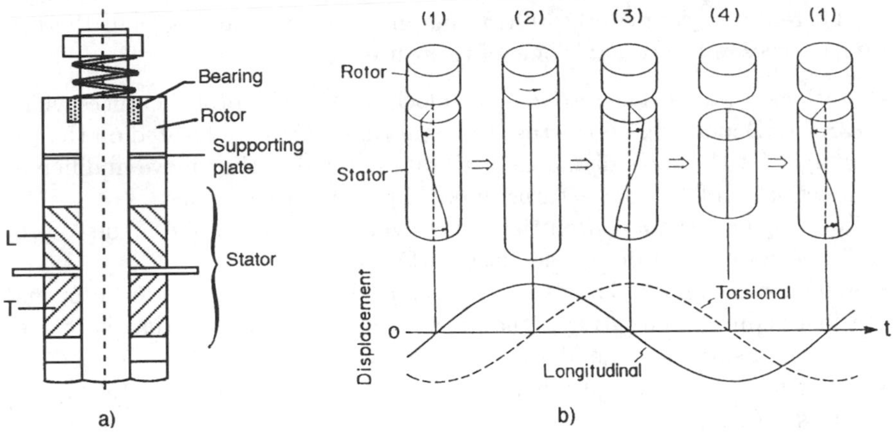

Fig. 5.36. Motor designed by Ohnishi, O., et al., consisting of two resonators. One resonator L vibrates in length, the other T in torsion. (**a**) Resonator L comprises 12 ceramic disks of thickness 0.5 mm, polarised in their thickness direction. Resonator T comprises 8 disks of thickness 1 mm, polarised around their circumference. (**b**) Stator deformation during one vibration period ($f = 30.5$ kHz). The rotor is driven during one half period. Taken from [5.61]. ©1993 IEEE

taneous excitation of the two vibrations, with appropriate phase difference, produces the sequence shown for one period in Fig. 5.36b. The turning force is transmitted from stator to rotor during the first half period. The direction of rotation is reversed when the phase difference between the two vibrations is changed by π. This motor has been studied by Ohnishi et al. [5.61], using the finite element method. The aim was to understand the effect of the force between rotor and stator on the torque, and conditions for equality between resonance frequencies of the two modes. They have shown that the frequency of the torsional motion, unlike that of the longitudinal motion, is more or less insensitive to force variations. They thus succeeded in adjusting the frequencies of the two modes to the same value, by applying a force of 550 N.

The motor constructed in this way has remarkable characteristics: centre frequency 30.5 kHz, efficiency (ratio of output mechanical power to input electrical power) 40%, with a torque of 30 N cm and at 400 rpm, startup time 2.5 ms, stopping time 0.5 ms, voltage applied across ceramic elements 70 V, diameter 20 mm, length 77 mm, rotor thickness 8 mm. The surface temperature of the rotor was measured as 80°C. This motor has better characteristics than the one described previously.

Let us mention in passing that a system made from two piezoelectric rings (or cylinders) and two rings (or cylinders) mounted tightly on an axle is often called a Langevin transducer. Its interest lies in the fact that ceramics have greater resistance to compression than to tension.

There are of course other ultrasonic motor configurations than the two we have chosen to present. Some of these use:

- Stationary waves. The rotor or slide block is moved by protuberances which carry it. Depending on whether these protuberances are located on the left or right (at the same distance) of vibrational nodes, they move in different directions and the motor turns in one direction or the other.
- Coupling between a concentric transverse mode and a radial mode to produce elliptical motion on the outer surface of a tube.
- A torsional coupler. This is an element positioned between the rotor and a longitudinally vibrating actuator, shaped in such a way that it rotates through a small angle under axial force from the actuator.
- A combination of longitudinally vibrating rods, capable of rotating two rotors [5.62].
- A ceramic arrangement able to excite a quasi-travelling wave along a given path across a rectangular plate [5.63].

To end this section, let us also mention a very different type of piezomotor, although they do not make use of elastic waves. These are linear motors, known as Inchworm motors. They are made from three piezoelectric tubes, disposed on a shaft. The two end tubes are clamps which are polarised so that they can be commanded to grip the shaft. The central tube is the actuator. It is polarised so that it extends or contracts lengthwise depending on the sign of the voltage applied to it. The reader can easily imagine the various stages of a motion similar to the advance of a worm! Maximal extension of the actuator is 2 μm. The maximal speed, depending on the opening and closing frequency of the clamps, is of order 2 mm/s. The range of motion of this linear motor is greater than 20 cm, with resolution 4 nm, and it can operate in vacuum conditions.

5.5 Instruments and Methods

This section is divided into five parts. The first two are devoted to two instruments: one, the acoustic microscope, is a relatively early development, but has recently been applied to non-destructive testing, whilst the other, the osteodensitometer, is newer and is still undergoing improvements. The last three parts describe novel ways of exploiting ultrasonic waves.

5.5.1 Acoustic Microscope

Figure 5.37 shows schematically an acoustic microscope operating by reflection [5.64]. It consists of a crystal in contact with a liquid. The plane end of the crystal carries a transducer, whilst the other end has a spherical cavity acting as a lens. The spherical surface is coated with a layer matching liquid

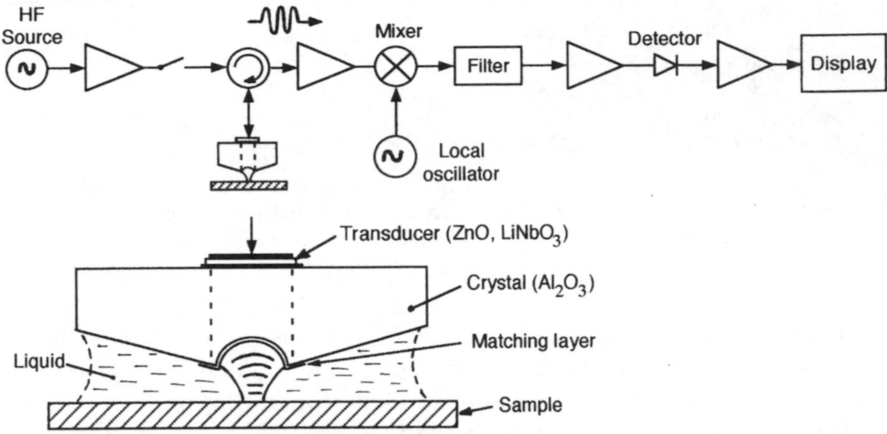

Fig. 5.37. Acoustic microscope operating by reflection and an example of the related electronic circuitry. The intensity of the elastic wave beam reflected from a point of the sample, and hence the voltage it induces across the transducer terminals, depends on the mechanical state of that point

and crystal impedances. The plane wave train launched by the transducer and transformed into spherical waves returns towards the transducer after reflecting pointwise off the sample surface or some discontinuity under the surface. A survey of the elastic state of the sample at given depth is obtained by mechanical translation of the sample.

The crystal may be aluminium oxide (or sapphire Al_2O_3, longitudinal waves propagating along A_3), the transducer a zinc oxide film (or lithium niobate $LiNbO_3$ monocrystal) between gold electrodes. The liquid is water and the impedance matching layer made from a glass. The sample is mounted on a support capable of motion in three dimensions. The diameter of the focal point is of the same order as the elastic wavelength in water or in the sample, depending on whether focussing occurs at the surface of or inside the material. It therefore measures a fraction of μm for a frequency of a few GHz.

This microscope only works in impulse conditions. It needs the corresponding electronics, able to supply excitation impulses of given width and carrier frequency and capable of selecting the relevant reflected pulses. The upper part of Fig. 5.37 shows an example of such a circuit.

An apparently simpler version of the acoustic microscope is made from two like crystals arranged symmetrically on either side of the sample. The plane face of each crystal carries a transducer, and the other ends are equipped with a lens. Elastic waves launched by the emitter transducer converge at a point inside the sample, then diverge to excite the receiver transducer. The voltage induced across its electrodes depends on the mechanical state of the sample point located at the common focus of the two lenses. The sample, thin enough not to cause too much attenuation, is held on a membrane mounted

Fig. 5.38. Image of an integrated circuit with bipolar transistor, taken using waves of frequency 4.4 GHz and a power of 25 dBm. Taken from [5.65]

on an xyz mobile support. As the sample moves, it is scanned line by line to produce a map of its mechanical characteristics in a plane. The instrument can work in continuous or impulse conditions. Operating by transmission, it requires accurate superposition of the two lens foci, a delicate matter in practice, as well as precise alignment of crystal axes. For this reason, the reflection microscope is preferred. The sample thickness is less important and it is easy to carry.

A single lens suffices to focus acoustic waves correctly. This is because their propagation speed in sapphire (11 100 m/s along the A_3 axis) is quite different from their propagation speed in water (1500 m/s). The refractive index for waves passing from sapphire to water is very small, viz., $n = 0.135$. Aberration (i.e., non-focussing of rays far off the axis) is therefore reduced, in fact, falling as n^2. The acoustic microscope is useful because it can penetrate optically opaque media and observe with high resolution. Many studies have emphasized this aspect, particularly those carried out with liquid gases, which attenuate less than water, at the extremely low temperatures required (e.g., argon, nitrogen, helium). Indeed a resolution of 200 Å has been achieved with helium and waves of frequency 8 GHz. However, water remains the liquid most commonly used in practice. It has been shown that non-linear effects occur which improve resolution. Figure 5.38 has become a classic image. It was taken in the following conditions: lens of focal length 15 μm and diameter 22 μm, pulses of width 3 ns, carrier frequency 4.4 GHz.

One disadvantage for imaging, at least outside medical and biological applications, lies in the demands made upon the coupling fluid. Although its mechanical impedance can be matched to that of the crystal, it is more difficult to match it to the impedance of the sample. Another problem is that mechanical scanning requires several seconds to produce an image.

Fortunately, acoustic microscopes exhibit a property which was rather unexpected during first studies, namely, excitation of surface waves. This has given rise to a new use for these instruments: they can measure elastic quantities locally near the surface. Surface waves, in particular, Rayleigh waves were first implicated when Weglein [5.66] observed extremely sensitive changes and even contrast reversals in the image as the distance from lens to object surface was reduced (defocussing). Contrast variations cause oscil-

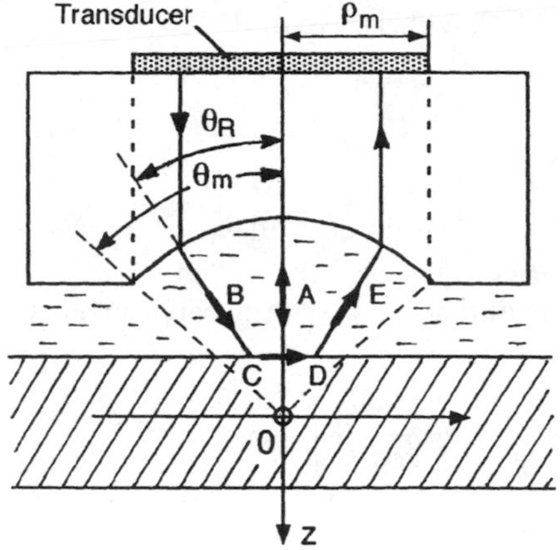

Fig. 5.39. Ray model. Rays B arriving at angle θ_R generate Rayleigh waves, of which a fraction are radiated into the liquid in the form of bulk waves. These radiated waves E return towards the lens and interfere with the normally reflected waves A

lations in the output signal $\mathcal{V}$ of the transducer. If z is the elevation of the sample surface relative to the lens focal point, the curve $\mathcal{V}(z)$ is a characteristic of the material, often called the *acoustic signature* of the material. [$\mathcal{V}(z)$ has become the standard notation, not to be confused with V denoting speed.] The shape of this curve (Fig. 5.40a) is understood by separating it into two parts:

$$\mathcal{V}(z) = \mathcal{V}_L(z) + \mathcal{V}_M(z) \, . \tag{5.17}$$

The first term $\mathcal{V}_L(z)$ is the normal response of the lens and transducer when illumination of a sample (of selected type) does not produce surface waves. Its amplitude, maximal for $z = 0$, decreases rapidly and continuously as z decreases and the surface moves away from the focal point to approach the lens (Fig. 5.40b). The second term $\mathcal{V}_M(z)$ results from interference caused by generation of surface waves in the observed object (Fig. 5.40c). This interference can be explained, in the context of the ray model, by considering Fig. 5.39. A family of incident rays A arrives perpendicularly at the surface at elevation $-z$. The family of rays B arrives at an angle θ_R such that

$$\sin \theta_R = \frac{c}{V_R} \, , \tag{5.18}$$

where c is the phase speed of waves in the liquid and V_R the phase speed of Rayleigh waves in the sample. Waves in the first family A are simply reflected, whereas those in the other family B give rise to Rayleigh waves which cross the surface from C to D, radiating bulk waves which return to the lens. The signal $\mathcal{V}_M(z)$ arises from interference between specularly reflected waves and these waves E radiated by the Rayleigh waves.

Let us denote the phase difference produced in these waves for a displacement $-z$ of the sample towards the lens, away from the focal plane, by $\phi_0(z)$ for normal incidence waves and $\phi_1(z)$ for waves at incidence θ_R. From the path lengths and taking into account the phase shift of π introduced by reflection at the Rayleigh angle (Sect. 4.4.2.2, Vol. I), we find

$$\phi_0(z) = -2kz\ ,$$
$$\phi_1(z) = -\frac{2kz}{\cos\theta_R} + 2k_R z \tan\theta_R - \pi\ ,$$

where $k = 2\pi/\lambda$ is the bulk wave number in the liquid and $k_R = 2\pi/\lambda_R$ is the Rayleigh wave number in the solid object.

A phase difference of 2π between $\phi_0(z)$ and $\phi_1(z)$ gives the separation Δz between two minima of $\mathcal{V}_M(z)$:

$$\phi_0(z) - \phi_1(z) = -2z\left[k\left(1 - \frac{1}{\cos\theta_R}\right) + k_R\frac{\sin\theta_R}{\cos\theta_R}\right] + \pi\ .$$

Since $k_R = k\sin\theta_R$,

$$\phi_0(z) - \phi_1(z) = -2kz(1 - \cos\theta_R) + \pi\ , \tag{5.19}$$

and the separation Δz between the two minima of $\mathcal{V}_M(z)$ is

$$\boxed{\Delta z = \frac{c}{2f}\,\frac{1}{1 - \cos\theta_R} = \frac{\lambda}{2(1 - \cos\theta_R)}}\ . \tag{5.20}$$

From this relation and (5.18), we deduce the Rayleigh wave speed

$$V_R = c\left[1 - \left(1 - \frac{c}{2f\Delta z}\right)^2\right]^{-1/2}\ . \tag{5.21}$$

Attenuation α_R of the Rayleigh waves is found from the decrease in amplitude of the oscillations. Indeed, the same approach shows that their amplitude varies as $e^{-\alpha z}$ with

$$\alpha = 2\left(\frac{\alpha_l}{\cos\theta_R} - \alpha_R\tan\theta_R\right)\ ,$$

where α_l is the attenuation coefficient in the liquid. Finally,

$$\boxed{\alpha_R = \frac{\alpha_l/\cos\theta_R - \alpha/2}{\tan\theta_R}}\ . \tag{5.22}$$

In fact, the measured speed V_R is the Rayleigh wave speed in the presence of a liquid. This differs somewhat from the speed they would have on a free surface. In addition, the above argument holds for all types of surface or pseudo-surface wave which can be generated in the solid by rays hitting it at the appropriate angles. To separate the various modes, of different frequencies, a Fourier transform is used.

If the solid is a crystal, its signature depends on the direction in which surface waves are excited. The spherical lens must then be replaced by a cylindrical lens, following the method of Kushibiki and Chubachi [5.67]. This lens concentrates waves from a rectangular transducer along a line. The generated surface waves propagate perpendicularly to this line focus. Curves in

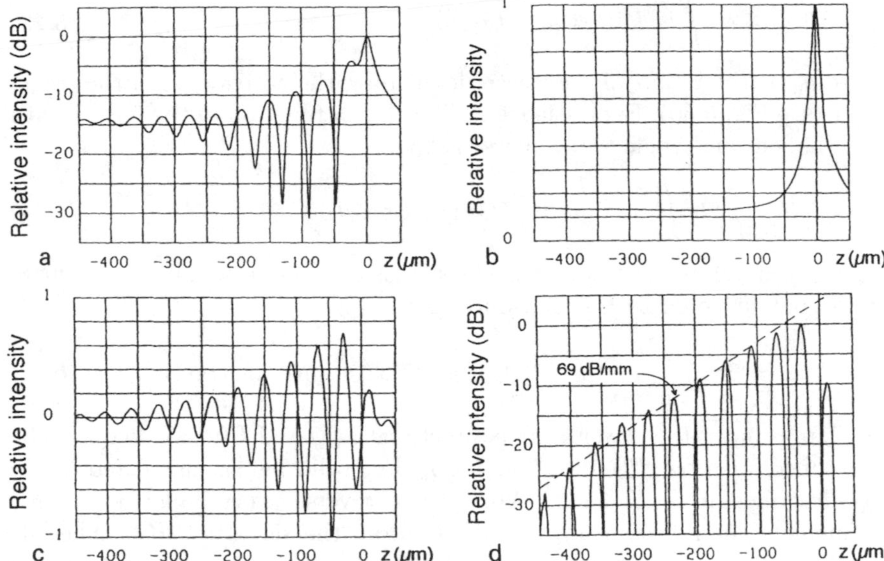

Fig. 5.40. Example of results obtained by Kushibiki, J., Chubachi, N. with their cylindrical lens microscope. (**a**) Signature $\mathcal{V}(z)$ for Y-cut α-quartz, with Rayleigh waves propagating in the Z direction. (**b**) Lens response $\mathcal{V}_{\mathrm{L}}(z)$ recorded for lead. (**c**, **d**) Curves $\mathcal{V}_{\mathrm{M}}(z)$ giving Rayleigh wave speed V_{R} and attenuation α_{R}. Taken from Fig. 14 of [5.67]. ©1985 IEEE

Fig. 5.40 were recorded by the above authors with a Y-cut quartz crystal, surface waves propagating in the Z direction. They extracted the Rayleigh wave contribution $\mathcal{V}_{\mathrm{M}}(z)$ to the curve $\mathcal{V}(z)$ by taking the lens response to be the curve $\mathcal{V}_{\mathrm{L}}(z)$ recorded for polycrystalline lead, which exhibits no oscillation. The values deduced from their measurements (at 225 MHz, cylindrical lens radius 1 mm, $\theta_{\mathrm{m}} = 60°$) were $V_{\mathrm{R}} = 3\,820$ m/s and normalised attenuation coefficient $\alpha_{\mathrm{N}} = \alpha_{\mathrm{R}}\lambda_{\mathrm{R}}/2\pi = 3.36 \times 10^{-2}$, in agreement with values calculated from physical constants.

The above description, referred to as the ray model, provides a good explanation for oscillations in the curve $\mathcal{V}(z)$, but analyses it into two parts: one describes the behaviour of the system, the other the sample reaction. This distinction is no longer necessary when Fourier acoustic are used to deal with the problem. Let us indicate the main steps in this global analysis, without going into detail. Let $T(\theta)$ be the acoustic field, at distance h, arising from the piston transducer (Sect. 1.3.2.2, Vol. I), vibrating with unit amplitude. Let $P(\theta)$ be the pupil function of the lens, including the transmission function of the impedance matching layer, and $r(\theta)$ the reflectance function of the object in the presence of the liquid. The field reflecting from the object surface at elevation $z = 0$, i.e., at the focus, and reaching the transducer is $P^2(\theta)r(\theta)T^2(\theta)$. It generates a signal

$$\mathcal{V}(0) = 2\pi \int_0^{\rho_m} P^2(\theta) r(\theta) T^2(\theta)\, \rho\, d\rho \,, \tag{5.23}$$

where ρ_m is the radius of the electrode defining the active part of the transducer (see Fig. 5.39). Expressing $\rho = F \sin\theta$ as a function of the focal length F of the lens and angle θ, it follows that

$$\mathcal{V}(0) = 2\pi F \int_0^{\theta_m} P^2(\theta) r(\theta) T^2(\theta) \sin\theta \cos\theta\, d\theta \,. \tag{5.24}$$

Shifting the sample $-z$ towards the lens leads to a phase advance $-2kz\cos\theta$. The signal delivered by the transducer is then:

$$\mathcal{V}(z) = 2\pi F \int_0^{\theta_m} P^2(\theta) r(\theta) T^2(\theta) e^{-2kiz\cos\theta} \sin\theta \cos\theta\, d\theta \,. \tag{5.25}$$

For a transducer–lens–layer system, functions $T(\theta)$ and $P(\theta)$ are known. The reflectance function $r(\theta)$ for a liquid–isotropic solid pair is calculated in the way described in Sect. 4.4.2.3, Vol. I. The reverse process can be applied to calculate the reflectance function $r(\theta)$ from the signature $\mathcal{V}(z)$. The latter arises as the Fourier transform of $\mathcal{V}(z)$, when the appropriate change of variable is made.

These spherical, cylindrical (or even conical) lens acoustic microscopes are often referred to by the acronym SAM, meaning Scanning Acoustic Microscope. Applications [5.68, 5.69] exist in medicine (eye and skin examinations), biology (in vivo cell studies, without dyes), material characterisation (measurement of elastic constants, observation of granular and layered structures, and stress distributions), and in non-destructive testing (adherence of films on substrates in microelectronics, detection of micro-inhomogeneities in metallurgy).

Scanning Laser Acoustic Microscopy SLAM. The principle behind these microscopes is illustrated in Fig. 5.41 [5.70]. The sample under test is immersed in a liquid (water) and illuminated by an ultrasonic beam from a transducer. Ultrasonic waves which have passed through the object generate a displacement field at the liquid surface, depending on its internal structure. The distribution of these displacements is read off by means of a laser beam which scans the liquid surface, line by line.

The scanning rate (30 images per second) is considerably higher than for a SAM. Resolution is determined here by the diameter of the light spot. Dimensions of the ultrasonic beam fix those of the object.

5.5.2 Ultrasonic Osteodensitometer

Elastic waves have several applications in medicine. The birth of any child, for example, is followed by a series of obstetric examinations based on ultrasonic imaging of the foetus (simple echography $f \approx 5$ MHz). Powerful ultrasonic

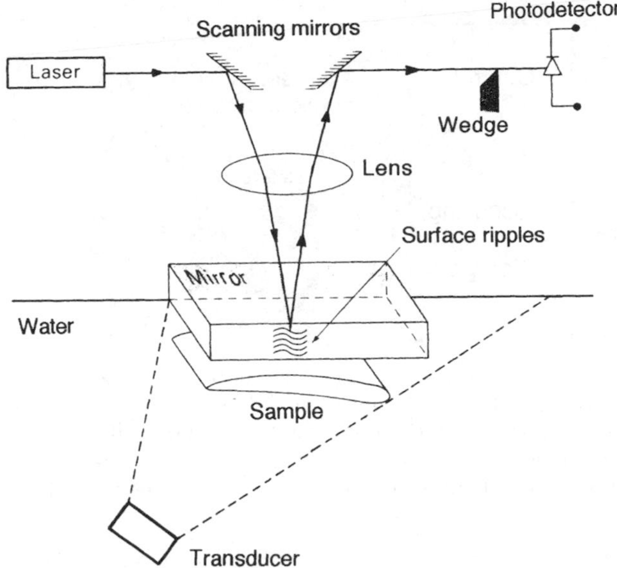

Fig. 5.41. Scanning laser acoustic microscope. The distribution of surface displacements of the liquid, created by waves which have passed through the object, is read off by a light beam scanning the surface

pulses (overpressure $\sim 10^8$ Pa) can destroy kidney stones (lithotripsy). Tumors can be reduced by ultrasonic hyperthermia. Cardiovascular observation is commonly made using velocimeters based on Doppler echography (f_0 in the range 2–10 MHz). These velocimeters are now available in all hospitals and some are even installed in space capsules and shuttles to monitor astronauts' cardiovascular systems [5.71].

We shall describe a new contribution made by elastic wave technology to progress in medical techniques. It involves the relatively recent development of a piece of equipment which assists in the diagnosis of osteoporosis. Several thousand of these are now in use throughout the world. Osteoporosis is a condition characterised by decrease in bone density and deteriorations in bone microstructure. The bone becomes fragile and more open to fracture. In order to anticipate such risks, it is crucial to ascertain the bone density. The classic approach is to measure absorption of X rays by either the lumbar vertebrae or the femoral neck. The elastic wave technique consists in measuring attenuation and speed of waves passing through the calcaneus (a bone in the heel). This bone is known to be a good observation site. However, it is a solid with rather complex shape, as well as being porous and heterogeneous. In fact, it is a biphase medium: the fluid phase is the marrow, and the solid phase is an interconnected network of plates and small columns. It is essential to identify the region when making measurements, so that any comparison with earlier recorded results, or other patients, is truly reliable.

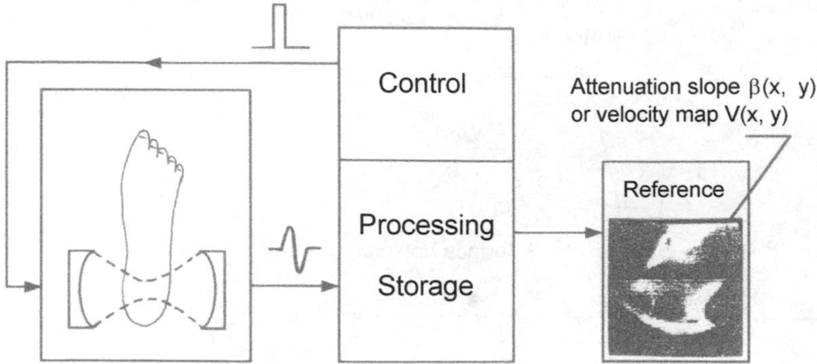

Fig. 5.42. Elastic wave examination of the calcaneus, great bone of the heel, for assessment of osteoporosis. Attenuation is measured for different frequencies (0.2–0.6 MHz), after waves have passed through the bone. At a given point, it is almost a straight line variation. The slope β of this line is a function of the local density of the bone. The scanner automatically maps out $\beta(x, y)$. It also produces the speed map $V(x, y)$. Taken from [5.72]

This scanner (Fig. 5.42) comprises a tank of water with two piezoceramic transducers (diameter 29 mm, focal length 50 mm, $f_0 = 0.5$ MHz) immersed in it. The patient's foot rests on the bottom of the tank, appropriately shaped, and the two rigidly connected transducers move parallel to the median plane of the foot in steps of 1 mm. At each step, a pulse of width 1 μs is applied to the input transducer, which launches a focussed wave beam. The spectrum of the signal transmitted through the bone to the output sensor is compared with the spectrum of the signal transmitted directly through the water. As ultrasound attenuation in water is virtually negligible here, the frequency dependence (over the range 0.2–0.6 MHz) of attenuation in the bone is simply deduced from the ratio of the two spectra. This dependence is practically linear at each point. The slope β of the corresponding straight line, a function of bone density at the point in question, is calculated and stored in memory. The map $\beta(x, y)$ is thus plotted, point by point, over the whole region scanned (60×60 mm^2). The attenuation slope varies from 20 to 100 dB/MHz. The speed map is recorded in the same way. Speeds vary from 1450 to 1650 m/s. This kind of examination takes about 1 min. Figure 5.43 gives an example of the results.

This method has been as successful as X-ray absorptiometry in predicting the likelihood of fracture. However, it has the advantage of not involving irradiation. In addition, the scanner is cheap to produce and easy to transport. Furthermore, it would appear that the elastic nature of the waves can uncover bone properties which X-ray absorptiometry does not reveal. The ultrasonic osteodensitometer seems well suited to monitoring child skeletal development, and also to studying osteoporosis induced by long space flights. Contact devices (with gel coupling) have been devised according to the same

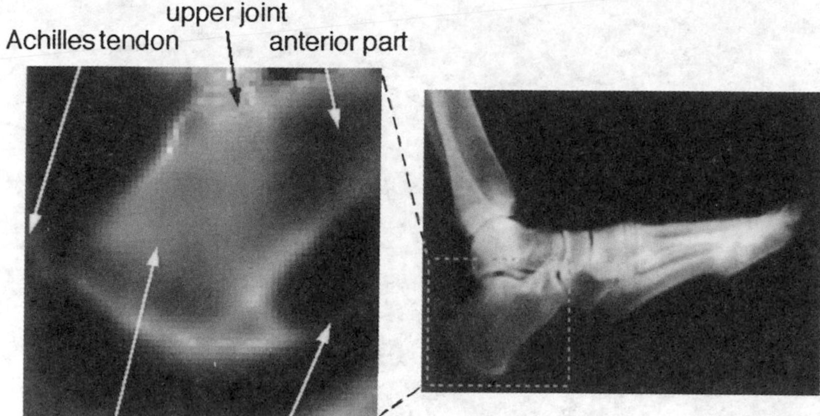

Fig. 5.43. (**a**) Map of the slope $\beta(x, y)$ of elastic wave attenuation in the calcaneus as a function of frequency. *White regions* correspond to high attenuation (100 dB/MHz), *black regions* to lower attenuation (20 dB/MHz). (**b**) X ray of the foot, showing the region scanned by the ultrasonic beam. Kindly communicated by P. Laugier [5.72]

principle. They do not provide images but serve to examine the phalanxes and also the calcaneus. Finally, let us mention, among the many areas of medical research in progress, ultrasonic acceleration of molecular transport through the skin (sonophoresis, $f \approx 1$ MHz) and acoustic microscopy studies ($20 < f < 200$ MHz) of the cartilage, the skin, the back of the eye and artery walls.

5.5.3 Time Reversal of Elastic Waves

The aim is to improve measurements made using these waves in heterogeneous media. The medium may be, for instance, the human body, a piece of metal or an undersea region. The idea behind time reversal of elastic waves, put forward by Fink, M. [5.73], is ideally illustrated in Fig. 5.44. It represents a cavity enclosing the heterogeneous material, bounded by a surface lined with piezoelectric transducers. Any pulse emitted by an internal source generates a pressure field $p(\boldsymbol{r}, t)$ which distorts as it propagates. The moment it reaches the walls, it is detected by transducers, which transmit their signals to a memory. After reading them in reverse chronological order, they are then restored in that reverse order. The pressure field $p(\boldsymbol{r}, \tau - t)$ generated by the transducers is thus the image of the incident field, but propagating in the opposite direction. τ is a delay. The field passes through the (stationary) medium in the opposite direction to reconstitute the original acoustic impulse at the source point. This process is reminiscent of a time reversed film of an exploding object which subsequently implodes. The only proviso is that

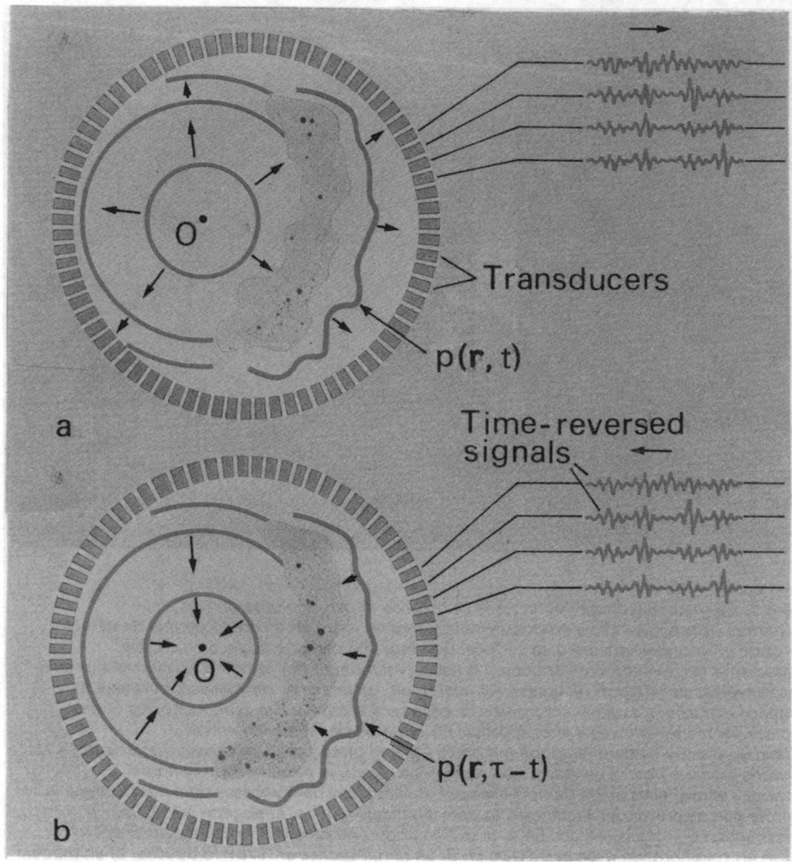

Fig. 5.44. Time reversal of ultrasonic fields. (**a**) A source located inside the cavity emits an acoustic impulse whose field is detected by sensors. These produce signals which are recorded in memory (not shown) and processed. (**b**) The memory restores image signals (the time reversal of recorded signals), which are sent back to the sensors, in reverse chronological order to their arrival. The resulting image field propagates in the opposite direction to reconstitute the original pulse. Taken from Fig. 2 of [5.74]

we neglect attenuation in the medium. (This introduces a derivative term into the propagation equation, thereby breaking the symmetry.) Hence, by hypothesis, the effects described here occur at relatively low frequencies.

If the aim is to reconstitute an object rather than a point source, then the object is first illuminated by a beam produced by a few of the transducers lining the cavity wall. In practice, the principle is more easily applied in the manner shown in Fig. 5.45, using a (Time Reversing TR) mirror M which occupies only a fraction of the useful region. An amplifier, analog-to-digital converter, memory and programmable generator are associated with each transducer. The last of these synthesises the time reversal of recorded signals.

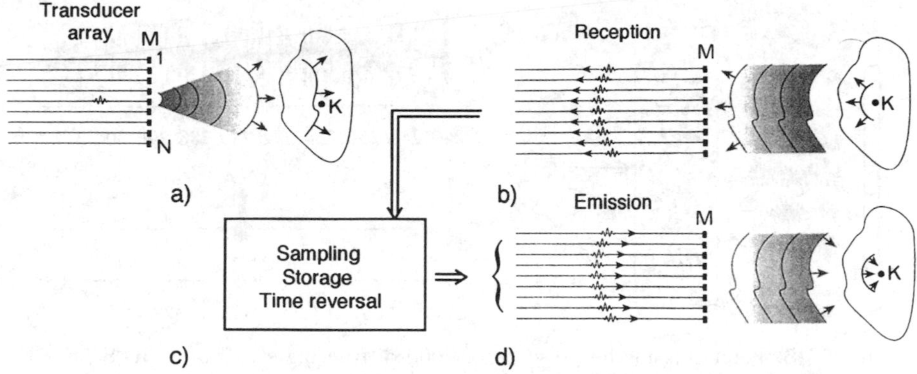

Fig. 5.45. Time reversal using a mirror. There are four stages. (**a**) Illumination of target. (**b**) Detection of field. (**c**) Signal processing. (**d**) Emission of time reversed field. The result is a concentration of energy on the target, despite the heterogeneity of the medium the waves pass through. Taken from Fig. 3 of [5.74]

There are four stages in the process which aims to concentrate energy on the target when the latter is located inside a heterogeneous medium.

- One of the N transducers in the mirror M emits a wave beam which illuminates target K for a given time.
- All N transducers in the mirror sense the pressure field $p(\boldsymbol{r}, t)$ returned by the target, until the field dies out after a time T.
- Signals delivered by the N transducers are processed (sampling, memorisation, reversal). Time required T_0.
- The image field $p(\boldsymbol{r}, T + T_0 - t)$ is emitted.

The effectiveness of the method has been demonstrated by several experiments carried out relatively recently. We shall review three of these, which illustrate the various approaches.

The first [5.75] involves randomly arranging about 2000 steel bars of diameter 0.8 mm, with mean spacing 2.6 mm, between a source S and TR mirror (Fig. 5.46a). The mirror is a linear array (40 mm) of transducers (T_k, $1 \leq k \leq 96$). Figure 5.46b is the signal r_i received by transducer T_i after exciting the source by a 1 μs pulse, i.e., after emitting a wave composed of 3 oscillations ($f_0 = 3.5$ MHz, $V = 1\,500$ m/s, so that 1 μs corresponds to 3λ). The duration of signal r_i, longer than 200 μs, stems from the fact that the original wave, and diffracted waves it generates when encountering the metal rods, follow an exceedingly indirect path. Figure 5.46c is the signal g detected by the source when it becomes detector (in the absence of electrical excitation), after signals r_i have been processed (20 MHz sampling with 8 bit converters) and chronologically reversed emission. Signal g has almost the same duration as the wave emitted by the source, and it is concentrated at the point where it was originally located. Time reversal thus gives rise to a compression equivalent to that of a filter matched to the medium.

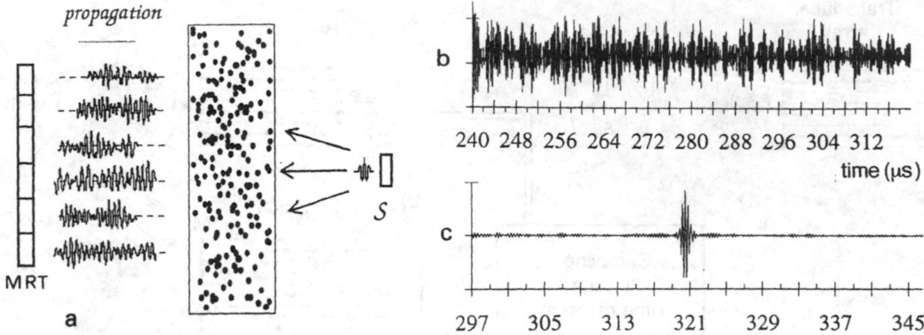

Fig. 5.46. Focussing in the presence of 2000 diffracting steel rods. (**a**) Schematic view of the experiment. (**b**) Very long signal r_i received by one of the 96 transducers in the mirror following a brief excitation (1 µs) of the source. (**c**) Pulse localised at the source point, resulting from time reversal of signals r_i

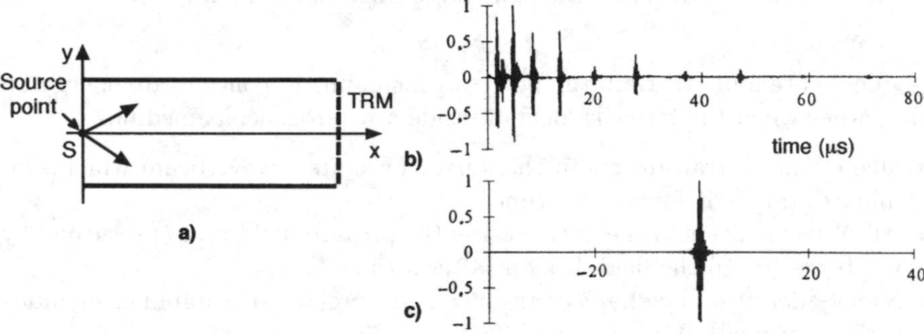

Fig. 5.47. Guide formed by two parallel immersed surfaces. (**a**) Experimental setup. (**b**) Sequence of echoes detected by each transducer in the TRM after emission of an acoustic pulse from S. (**c**) Energy of time reversed echoes is concentrated at S

This matching of the signal to an inhomogeneous medium can be applied in different situations, and the medium's response to the primary impulse may be discrete. Figure 5.47a refers to a wave guide formed by the free surface of water in a tank and a steel plate, parallel to the surface, immersed in the water [5.76]. A point source is placed at S, at one end of the guide, and a TR mirror at the other end, 80 cm away.

Figure 5.47b shows the sequence of echoes received by one of the 99 transducers in the mirror, in response to a pulse emitted by S. Figure 5.47c shows the single pulse reaching a detector located at S when the mirror returns the time reversed sequence of echoes. This guide is a model of a channel whose walls are an ocean surface and ocean bed. The experiment has been successfully carried out on a much larger scale (guide length 7000 m, depth of marine channel 120 m) [5.77].

The third experiment concerns non-destructive testing. The classic approach to locating defects in metal objects, immersed in a tank, by means of (longitudinal) wave beams, often encounters an interface problem. The difference of impedance between water and metal produces reflected waves, whilst the arrival of the beam at certain angles of incidence can generate transverse and/or surface waves in the object. Furthermore, granular structures may lead to echographic noise. Time reversal can simplify these investigations, through slice-by-slice scanning of the object. As speeds are constant, there is a correspondence between time and object size in the wave propagation direction. Each slice is determined by the same temporal segment of all signals r_i. By applying only the time reversal of this segment of the signals to the transducers, focussing occurs in the associated slice. This has been demonstrated in an experiment to locate a defect in a titanium block (Fig. 5.48a, a cylinder of diameter 25 cm and length 100 cm) [5.78]. Figure 5.48b is the block response to acoustic illumination by the central element of the mirror (128 elements, $f_0 = 5$ MHz). Peaks stem from specular reflection at incoming and outgoing interfaces of the cylinder. No anomaly is visible in the interval. And yet, after time reversal and reduction to the window indicated by the two vertical lines, an echo appears (Fig. 5.48c) which reveals a defect (a flat-bottomed calibration hole of diameter 0.4 mm, located at 140 mm from the incoming interface).

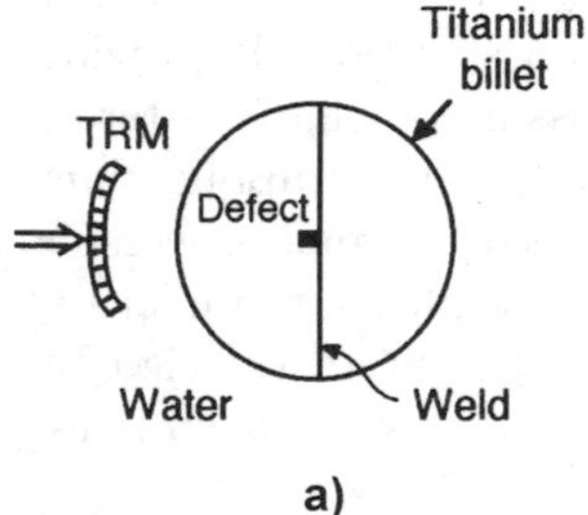

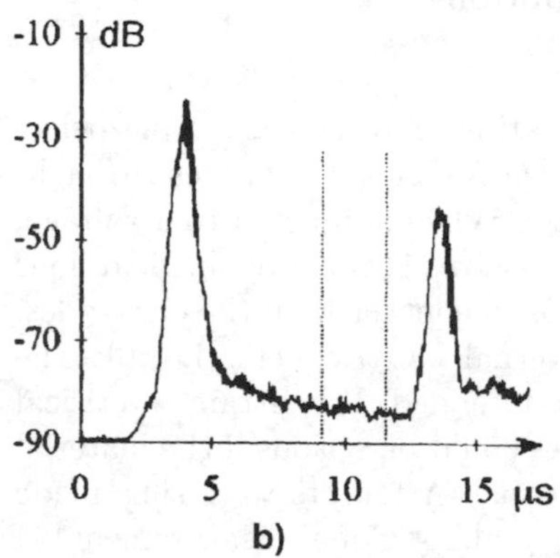

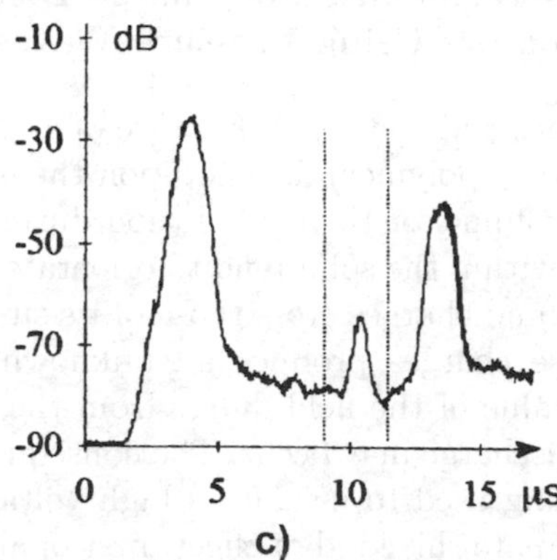

Fig. 5.48. Detection of a defect in a titanium cylinder. (**a**) Experimental setup. (**b**) Cylinder response to a pulse emitted by the central transducer of the mirror. (**c**) Response to the part of the time reversed field corresponding to a slice of material containing a defect (flat-bottomed calibration hole, diameter 0.4 mm, located 140 mm from the incoming interface)

If there are several scatterers in the same slice of material, the problem is more complicated. In fact, several successive reversals automatically yield the brightest scatterer. However, a method has been devised which can selectively target a single scatterer among others. The idea is worth mentioning [5.79]. Since each element of the TR mirror successively acts as emitter and receiver, it is useful to define an impulse response for each pair of elements. Let $h_{lm}(t)$ be the signal received by transducer l when transducer m is excited by an impulse $\delta(t)$. Then the signal received by the same transducer l when a signal $e_m(t)$, $1 < m < N$, is applied to each of the other transducers, is given by

$$r_l(t) = \sum_{m=1}^{N} h_{lm}(t) \otimes e_m(t) . \tag{5.26}$$

In the frequency domain, and in matrix form, this becomes

$$\boldsymbol{R}(\omega) = K(\omega)\boldsymbol{E}(\omega) ,$$

where $\boldsymbol{R}(\omega)$ and $\boldsymbol{E}(\omega)$ are the received and emitted signal vectors, and $K(\omega)$ is the transfer matrix for the system. The latter is symmetric: according to the reciprocity principle, swapping roles of elements l and m leaves the response unchanged, so that $h_{lm}(t) = h_{ml}(t)$. The time reversal operation $t \to -t$ amounts to a phase conjugation in the frequency domain. Hence, if the first signal emitted is E_0, the signal emitted after two time reversal operations is $E_2 = K^*KE_0$. Time reversal is defined by the operator K^*K. Since matrix K is symmetric, this operator is Hermitian and has positive eigenvalues. It can be shown that each eigenvector corresponds to a scatterer. It provides the amplitude and phase functions which allow focussing on that scatterer.

This time reversal method is therefore potentially rich in applications. Progress in computing, and in particular increased speed of signal processing, have made it all the easier to implement. Among other studies now underway, two worth mentioning are access to various parts of the brain through the skull, and scattering of acoustic waves by turbulent effects such as vortices.

5.5.4 Determination of Charge Distributions in Insulators Using Pressure Waves

Properties and hence performance of an insulating material (e.g., a piezoelectric solid or polymer) depend upon the existence of electric charges throughout its volume, or their generation during use. When subjected to a voltage, dipoles within the solid tend to orientate themselves parallel to the field, ions migrate and charges cross those of its surfaces in contact with the electrodes. All these charges produce an unknown internal electric field. Locally the actual value of the field differs from the one intended. If it attains a critical value, discharge may occur. The consequences could be serious if the material were being used to insulate a high voltage cable. Attempts were thus made to anticipate this kind of effect, first of all by making global measurements of

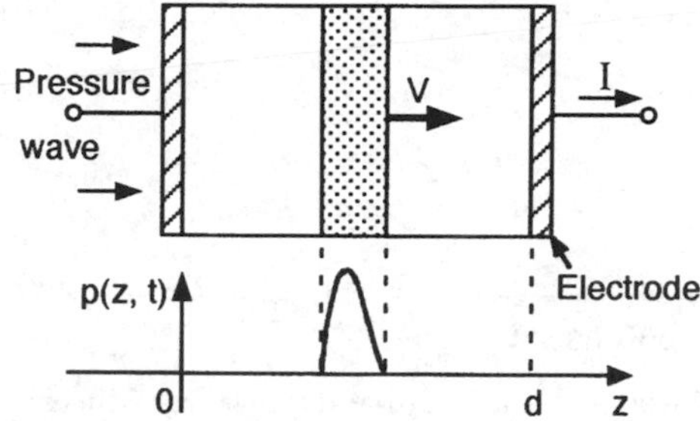

Fig. 5.49. Determining the spatial distribution of charges in a dielectric by means of a pressure pulse. The pulse, produced by suddenly applying a perturbation to one electrode, locally modifies the charge distribution as it moves through the solid. Resulting variations in current density I in the external circuit connecting the electrodes (not shown) provide an image of the spatial distribution

quantities such as conductivity, permittivity, or current intensity generated by depolarisation of the heated material. A new approach involved probing within the material by means of a heat pulse or electron beam, and finally, a pressure pulse (wave). This last type of inspection has the advantage of being quicker. The first trials were carried out with the pulse produced by a shock wave [5.80]. Other experiments followed: the pressure pulse was generated by capacitor discharge, light beam impact [5.81], or excitation of a piezoelectric transducer. The last two are the more commonly applied techniques.

The basic idea behind these measurements using pressure waves is depicted in Fig. 5.49 for the simple case of a plate shaped material placed between two electrodes. One of the electrodes is subjected to a mechanical perturbation which is extremely short compared with the transit time (the time needed for the perturbation to cross the plate). As it propagates, this elastic pressure pulse locally displaces charges rigidly attached to the structure, thereby modifying the dipole density. When the insulator is subjected to a fixed voltage, it induces a current in the external circuit connecting the electrodes, and variations in this current constitute an image of the spatial distribution of charges.

It is also possible to monitor changes in the charge distribution, caused by variations in the voltage across the insulator. This is done by emitting pulses at a suitable rate. The duration of the measurement, determined by the transit time of each pulse, must be short enough to justify treating the voltage as constant. For example, if the voltage varies at 50 Hz, the measurement duration can be 1 μs. The propagation speed of the pulse through the insulator fixes the sample thickness (2 mm for a polyethylene sample). The pulse can be generated by laser beam impact. Indeed a (Nd YAG 1.06 μm) laser supplies pulses of width a few ns and energy 300 mJ, which can produce a pressure of order 10 MPa on an absorbent surface (Sect. 3.3). Each pulse gives rise to an observable current intensity (1 mV/50 Ω). The technique has the advan-

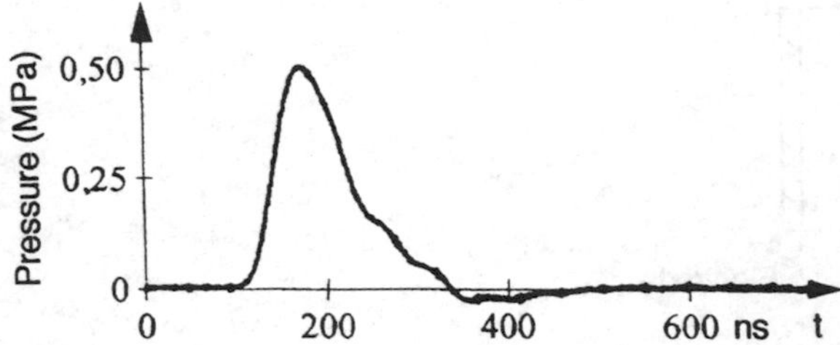

Fig. 5.50. Monopolar pressure pulse generated by a high bandwidth transducer

tage of involving no mechanical contact between sample and pressure wave generator. However, when the dielectric undergoes rapid voltage changes, so that the measurement rate has to be increased, the absorbent surface heats up and deteriorates. Results are no longer repeatable. Another method for generating the pressure wave consists in exciting a piezoelectric transducer. This involves mechanical contact with the sample, but transmission of a longitudinal displacement (through a gel) raises no real difficulties. The problem is to generate a pulse of suitable width and shape. One solution is to insert a piezoelectric disk between two blocks, one of which acts as load and the other as guide [5.82]. If the materials are chosen with the right impedances, the pressure is proportional to the derivative of the voltage applied to the disk.

Figure 5.50 is the response of a generator comprising a brass load (diameter 14 mm, length 20 mm), a silica guide and a piezoceramic disk of thickness 200 μm, when subjected to a 180 V step. This pressure pulse, with height of order 0.5 MPa and risetime 40 ns, can be applied to a sample at a rate of several kHz.

Figure 5.51 gives an example of the results. The dielectric is a polyethylene sheet of thickness 1 mm subject to a voltage of 55 kV for 2 s. Curves represent the charge distribution 0.1 s before and 0.1 s after the voltage is applied; then 0.1 s before and 0.2 s after it is removed. Before applying the voltage, the current intensity is zero, implying that there are no charges in the insulator. Just after the voltage is applied, two peaks appear at the beginning and end of the pressure pulse. These are due to charges on the electrodes. A small inflection between the peaks reveals the presence of highly mobile charges. The measurement just before cutoff shows that the positive charges (which have moved in the direction of the field) have gathered close to the end of the sample. When these charges appear, they increase the field locally by about 25%. They remain in part after cutoff (often for several minutes).

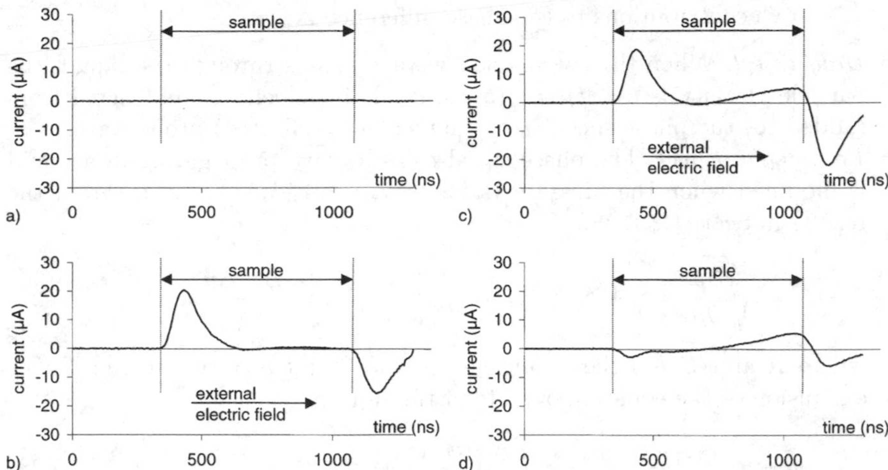

Fig. 5.51. Charge distribution at different times within a polyethylene sheet of thickness 1 mm. The sheet is submitted to a voltage of 55 kV for 2 s. (**a**) 0.1 s before and (**b**) 0.1 s after applying the voltage. (**c**) 0.1 s before and (**d**) 0.2 s after removing the voltage. Graphs kindly communicated by S. Holé

5.5.5 Parametric Interaction Probe

The motion of a vibrating surface changes the phase of any beam reflected from that surface. The optical probes described in Sect. 3.2.2 for the measurement of nanometric surface displacements are based on this effect. The idea of measuring a displacement by observing phase shifts can also be applied to an ultrasonic probe beam, provided that two conditions are taken into account. Firstly, the ultrasonic wave frequency must be higher than the vibrational frequency of the surface. And in addition, the medium (for example, a liquid) necessarily interposed between vibrating surface and ultrasonic wave source must be taken into consideration. Indeed, the phase shift $\Delta\phi$ due to the change in phase speed occurring in the liquid must be added to the phase shift $\Delta\phi_{\mathrm{d}} = 2k_0 u_{\mathrm{s}} = 4\pi u_{\mathrm{s}}/\lambda_0$ produced by surface displacement u_{s} (Doppler effect). The pressure variation is created by the surface here, assumed plane, vibrating at speed v_{s} (Fig. 5.52).

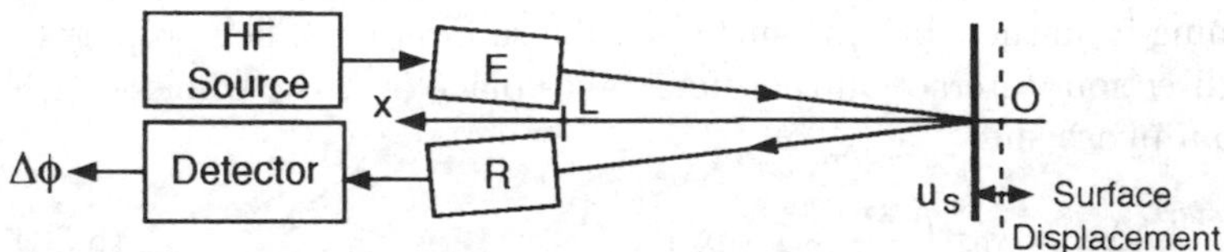

Fig. 5.52. The HF ultrasonic beam emitted by transducer E is received by transducer R after reflecting from the vibrating surface located at distance L. The phase shift $\Delta\phi$ in the HF signal, caused by the LF vibration, is extracted by a similar electronic circuit to the one in Fig. 3.25

The two contributions to the phase difference $\Delta\phi$ are:

- *Drag Effect.* When the low frequency wave passes through the liquid, the particle velocity is $v(x, t) = v_s(t - x/c)$. This velocity is subtracted from (added to) the phase speed c_0 of the incident (reflected) probe wave.
- *Parametric Effect.* The phase speed c_0 of a wave propagating in a liquid is modified when the pressure varies by $\Delta p = \rho_0 c_0 v_s(t - x/c)$. From the results in Sect. 1.2.1, Vol. I,

$$c = \sqrt{\frac{\partial p}{\partial \rho}} = c_0 + \Delta c\,, \quad c_0 = \sqrt{\frac{A}{\rho_0}}\,, \quad \Delta c = \frac{B}{2A}\frac{\Delta p}{\rho_0 c_0}\,, \tag{5.27}$$

where A and B are the coefficients of the first two terms in the Taylor expansion of the equation of state of the liquid:

$$p - p_0 = A\frac{\rho - \rho_0}{\rho_0} + \frac{B}{2}\left(\frac{\rho - \rho_0}{\rho_0}\right)^2 + \dots\,. \tag{5.28}$$

The total shift in phase speed of the incident wave is

$$\Delta c_i = \left(\frac{B}{2A} - 1\right) v_s(x, t)\,, \tag{5.29}$$

and for the reflected wave,

$$\Delta c_r = \left(\frac{B}{2A} + 1\right) v_s(x, t)\,. \tag{5.30}$$

The phase shift $\Delta\phi_i$ produced when the incident probe beam crosses the impulse emitted by the vibrating surface (distance L) is given by

$$\Delta\phi_i = \int_L^0 -\frac{\omega_0}{c_0^2}\Delta c_i\, dx = \int_L^0 \frac{\omega_0}{c_0}\left(\frac{B}{2A} - 1\right)\frac{\partial u}{\partial x}\, dx\,. \tag{5.31}$$

If the surface vibration lasts a short time Θ compared with the transit time $\tau = L/c_0$, then $u_s(L, t) = 0$ and

$$\Delta\phi_i = \frac{2\pi}{\lambda_0}\left(\frac{B}{2A} - 1\right) u_s\,. \tag{5.32}$$

This phase shift $\Delta\phi_i$, like $\Delta\phi_d$, is proportional to the displacement u_s of the surface.

As the probe beam is continuously emitted, the reflected probe wave and the wave coming from the vibrating surface will also interact. They propagate in the same direction. Interaction therefore takes place over the whole length L, leading to a phase shift

$$\Delta\phi_r = \frac{\omega_0 L}{c_0^2}\Delta c_r = k_0\tau\left(\frac{B}{2A} + 1\right) v_s\,. \tag{5.33}$$

$\Delta\phi_r$ is proportional to the velocity v_s of the surface.

The total phase difference at time t, $\Delta\phi = \Delta\phi_d + \Delta\phi_i + \Delta\phi_r$, is given by

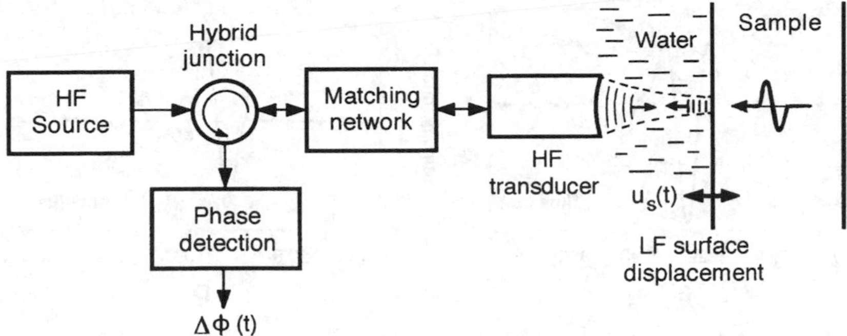

Fig. 5.53. Experimental setup. The same (concave) HF transducer emits the probe beam and receives it after reflection from the vibrating surface

$$\Delta\phi(t) = \frac{2\pi}{\lambda_0}\left(\frac{B}{2A}+1\right)\left[u_s(t-\tau)+\tau v_s(t-\tau)\right] . \tag{5.34}$$

It depends on the displacement u_s and the velocity of vibration v_s of the surface at the earlier time $t-\tau$.

It is interesting to compare the sensitivity of this ultrasonic probe with that of the optical probe. In the latter, surface vibrations generate only a Doppler phase shift equal to $4\pi u_s/\Lambda_0$. This Doppler phase shift is much smaller here, since $\lambda_0 \gg \Lambda_0$ (for a He-Ne laser, $\Lambda_0 = 0.633$ μm, whereas for a 30 MHz ultrasonic beam, $\lambda_0 = 50$ μm $\approx 79\Lambda_0$). It is the phase shift generated by parametric interaction on the reflected probe beam, proportional to velocity v_s, which prevails. Indeed, let $u_s = U\sin\omega t$ and $G = \Delta\phi_r/\Delta\phi_d$. Then

$$G = \frac{\omega\tau}{2}\left(\frac{B}{2A}+1\right) . \tag{5.35}$$

In water, $B/A = 5$, $c_0 = 1\,500$ m/s. With $L = 18$ mm, $\tau = 12$ μs and $f = 1$ MHz, $G = 132$.

Predominance of the $\Delta\phi_r$ term, and also its linear dependence on τ, have been confirmed with the help of the setup in Fig. 5.53 [5.83].

A concave transducer (focal length $L = 18$ mm), with impedance matched to that of the source, emits a probe beam of frequency $f_0 = 30$ MHz. The same transducer receives the probe beam after it has reflected from the surface set in motion by an LF transducer. The phase shift $\Delta\phi$ is extracted by means of an electronic circuit, similar to the one in Fig. 3.25. This circuit has two branches, one of which is traversed by the information-carrying signal and the other by a reference signal produced from the first by $\pi/2$ phase shift and suppression of spectral lines corresponding to $\Delta\phi(t)$. The component of frequency $2f_0$ at the mixer output is eliminated by the low frequency filter which delivers a signal $s(t)$ proportional to $\Delta\phi(t)$.

Figure 5.54 refers to vibrations of a duralumin plate of thickness 6 mm, ensonified by a 4 MHz transducer. Figure 5.54a represents the mechanical

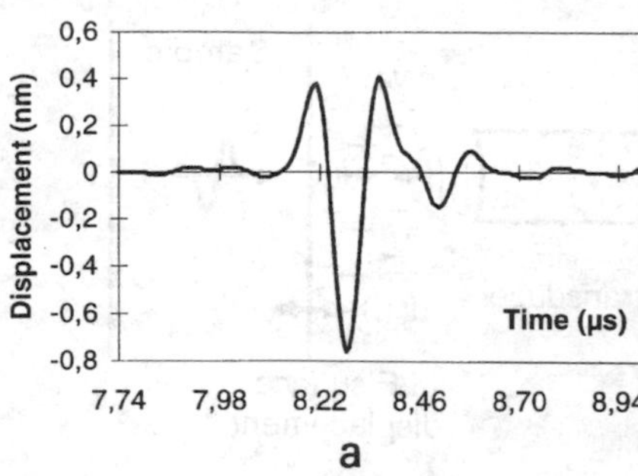

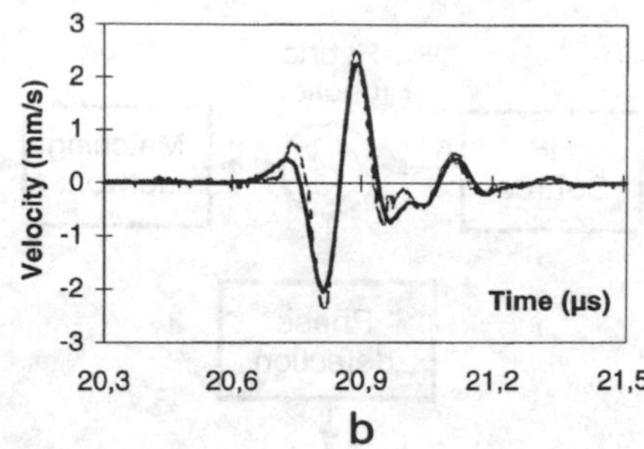

Fig. 5.54. Vibration of a duralumin plate ensonified in impulse regime by an LF transducer. (**a**) Displacement measured with an optical probe. (**b**) Velocity deduced from this displacement by differentiation (*dashed curve*) and phase difference measured by the ultrasonic probe (*continuous curve*), converted into a velocity with the help of (5.35)

displacement measured with an optical probe. Figure 5.54b shows (dashed lines) the derivative of the displacement (i.e., the velocity) and (continuous lines) the phase shift $\Delta\phi(t)$ given by the ultrasonic probe. The scale is marked off in [mm/s] with the help of (5.34), using the values cited above ($\tau = 12$ µs, $\lambda_0 = 50$ µm and $B/A = 5$). Similarity between the two curves shows that the main effect is indeed due to parametric interaction, proportional to the particle velocity.

Lateral resolution is enhanced by using a concave HF transducer. The parametric interaction is a bulk effect and this ultrasonic probe has spatial resolution, of order 0.5 mm, less than the optical probe. The spatial resolution of an optical probe is limited only by the diameter of the light spot on the surface. The minimal detectable velocity is less than 0.04 mm/s with bandwidth 5 MHz. The dynamic range is greater than 70 dB.

Appendix A. Spatial Distribution of Electrical Quantities

This can be calculated analytically for simple transducers with two comb-shaped electrodes having fingers of the same dimensions, separated by constant intervals. In general, however, numerical techniques are required.

A.1 Simple Transducer

Figure A.1 shows the geometry of an interdigital transducer together with the reference axes we shall use. The combs have an infinite number of fingers, long enough to ensure that no quantity depends on the x_3 coordinate. The thickness of the metallic film is small compared with interdigital distance d. Fingers have width a and are a distance $d - a$ apart. One electrode is at potential $U/2$ and the other at $-U/2$. To simplify calculation of the potential and electrostatic charge distribution, the piezoelectric medium is modelled by a dielectric with permittivities ε_{11} and ε_{22}, whilst ε_{12} is assumed zero.

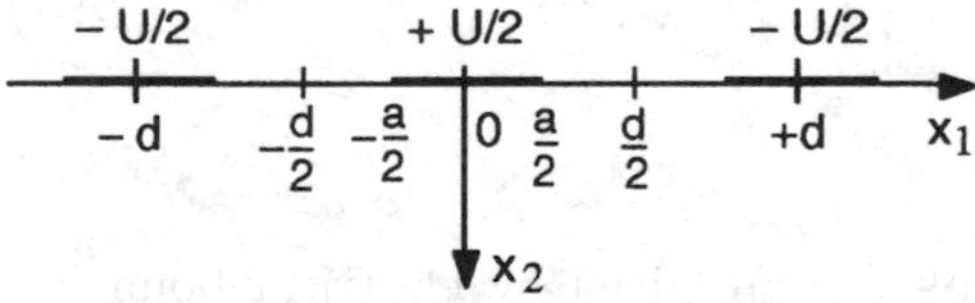

Fig. A.1. Elementary cell of a transducer with metallisation ratio $\eta = a/d$

The electric potential Φ satisfies Laplace's equation,

$$\varepsilon_{11}\frac{\partial^2\Phi}{\partial x_1^2} + \varepsilon_{22}\frac{\partial^2\Phi}{\partial x_2^2} = 0\,, \quad \varepsilon_{11} = \varepsilon_{22} = \varepsilon_0 \quad \text{in vacuum.} \tag{A.1}$$

The solution is an even, periodic function of x_1, of period $2d$, such that

$$\Phi(x_1 + d,\, x_2) = -\Phi(x_1,\, x_2) \qquad \forall\, x_1,\, x_2\,. \tag{A.2}$$

Its Fourier series expansion contains only odd harmonics of the $2d$ period:

$$\Phi(x_1,\, x_2) = \sum_{m=0}^{\infty} f_m(x_2)\cos(\chi_m x_1)\,, \quad \chi_m = (2m+1)\frac{\pi}{d}\,. \tag{A.3}$$

Functions $f_m(x_2)$ satisfy differential equations

$$\frac{\mathrm{d}^2 f_m}{\mathrm{d}x_2^2} - \frac{\varepsilon_{11}}{\varepsilon_{22}}\chi_m^2 f_m = 0\,, \quad \text{and} \quad \frac{\mathrm{d}^2 f_m}{\mathrm{d}x_2^2} - \chi_m^2 f_m = 0 \text{ in vacuum.}$$

The solution going to zero at infinity is

$$f_m(x_2) = \frac{UF_m}{(2m+1)\pi}\exp(-r\chi_m x_2)\,, \quad r = \begin{cases} \varepsilon_{11}/\varepsilon_{22} & \text{for } x_2 > 0\,, \\ -1 & \text{for } x_2 < 0\,. \end{cases}$$

The electric potential is found by substituting into (A.3),

$$\Phi(x_1, x_2) = \sum_{m=0}^{\infty} \frac{UF_m}{(2m+1)\pi}\exp(-r\chi_m x_2)\cos(\chi_m x_1)\,. \tag{A.4}$$

Introducing the function

$$F(\phi) = \sum_{m=0}^{\infty} F_m \exp \mathrm{i}(2m+1)\phi \tag{A.5}$$

of the dimensionless variable $\phi = \pi x_1/d$, the tangential component of the electric field on the free surface is

$$E_1(x_1, 0) = -\left.\frac{\partial \Phi}{\partial x_1}\right|_{x_2=0} = \frac{U}{d}\sum_{m=0}^{\infty} F_m \sin(\chi_m x_1) = \frac{U}{d}\mathrm{Im}[F(\phi)]\,. \tag{A.6}$$

The surface charge density

$$\sigma(x_1) = D_2(x_1, 0_+) - D_2(x_1, 0_-) = \varepsilon_0 \left.\frac{\partial \Phi}{\partial x_2}\right|_{x_2=0_-} - \varepsilon_{22}\left.\frac{\partial \Phi}{\partial x_2}\right|_{x_2=0_+}$$

becomes

$$\begin{aligned} \sigma(x_1) &= \left[\varepsilon_0 + (\varepsilon_{11}\varepsilon_{22})^{1/2}\right]\frac{U}{d}\sum_{m=0}^{\infty} F_m \cos(\chi_m x_1) \\ &= (\varepsilon_0 + \varepsilon_\mathrm{p})\frac{U}{d}\mathrm{Re}[F(\phi)]\,, \end{aligned} \tag{A.7}$$

where $\varepsilon_\mathrm{p} = (\varepsilon_{11}\varepsilon_{22})^{1/2}$. Coefficients F_m are given by electrical boundary conditions:

$$\begin{aligned} E_1(x_1, 0) &= 0 \quad \text{on electrodes, i.e.,} \quad |x_1| < a/2 = \eta d/2\,, \\ \sigma(x_1) &= 0 \quad \text{between electrodes, i.e.,} \quad a/2 < |x_1| < d/2\,. \end{aligned} \tag{A.8}$$

Then, from (A.6) and (A.7), putting $\theta = \eta\pi/2$,

$$\mathrm{Im}[F(\phi)] = 0 \quad \text{for} \quad |\phi| < \theta\,, \quad \mathrm{Re}[F(\phi)] = 0 \quad \text{for} \quad \theta < |\phi| < \pi/2\,.$$

Let us consider the generating function for Legendre polynomials (Appendix C),

$$w^{1/2}\sum_{m=0}^{\infty} w^m P_m(z) = (w - 2z + w^{-1})^{-1/2}\,,$$

for variables $w = \mathrm{e}^{2\mathrm{i}\phi}$ and $z = \cos 2\theta$. The function

$$G(\phi) = \sum_{m=0}^{\infty} P_m(\cos 2\theta)\mathrm{e}^{\mathrm{i}(2m+1)\phi} = 2^{-1/2}(\cos 2\phi - \cos 2\theta)^{-1/2} \tag{A.9}$$

is real

$$G(\phi) = \frac{1}{2}(\sin^2\theta - \sin^2\phi)^{-1/2} \quad \text{for} \quad |\phi| < \theta\,, \tag{A.10}$$

and imaginary

$$G(\phi) = \frac{\mathrm{i}}{2}(\cos^2\theta - \cos^2\phi)^{-1/2} \quad \text{for} \quad \theta < |\phi| < \pi/2\,. \tag{A.11}$$

The function $G(\phi)$ satisfies (A.8) and can be identified with $F(\phi)$ up to a constant: $F(\phi) = AG(\phi)$. The constant A can be found from the potential difference U applied across the electrodes, since $\Phi(x_1 = \pm d/2) = 0$:

$$U = 2\int_{a/2}^{d/2} E_1(x_1, 0)\,\mathrm{d}x_1 = \frac{2}{\pi}AU\int_{\theta}^{\pi/2} \mathrm{Im}[G(\phi)]\,\mathrm{d}\phi\,.$$

The charge per unit length Q on each finger is

$$Q = \int_{-a/2}^{a/2} \sigma(x_1)\,\mathrm{d}x_1 = \frac{2}{\pi}AU(\varepsilon_0 + \varepsilon_\mathrm{p})\int_0^{\theta} \mathrm{Re}[G(\phi)]\,\mathrm{d}\phi\,.$$

These involve complete elliptic integrals of the first kind with complementary variables $\sin\theta$ and $\cos\theta$:

$$K(\cos\theta) = \int_0^{\pi/2} (\cos^2\theta - \cos^2\phi)^{-1/2}\mathrm{d}\phi\,,$$

$$K(\sin\theta) = K' = \int_0^{\theta} (\sin^2\theta - \sin^2\phi)^{-1/2}\mathrm{d}\phi\,.$$

Hence,

$$A = \frac{\pi}{K(\cos\eta\pi/2)} \quad \text{and} \quad C_1 = \frac{Q}{U} = (\varepsilon_0 + \varepsilon_\mathrm{p})\frac{K(\sin\eta\pi/2)}{K(\cos\eta\pi/2)}\,. \tag{A.12}$$

The expression for capacitance per unit length C_1 of a finger pair is particularly simple when the metallisation ratio η is 1/2, i.e., $\theta = \pi/4$, since $\cos\theta = \sin\theta = 1/\sqrt{2} \Rightarrow K = K'$:

$$\boxed{C_1 = \varepsilon_0 + \varepsilon_\mathrm{p}} \quad \text{if} \quad \eta = \frac{1}{2}\,. \tag{A.13}$$

According to (A.10), harmonic coefficients of the electric field are proportional to Legendre polynomials of the variable $\cos\eta\pi$:

$$F_m = AP_m(\cos\eta\pi) \Rightarrow F_m = \pi\frac{P_m(\cos\eta\pi)}{K(\cos\eta\pi/2)}\,. \tag{A.14}$$

Figure 2.9 shows how they vary with metallisation ratio η. Substituting into (A.4), (A.6), (A.7), and taking (A.10) and (A.11) into account yields the potential, tangential component of electric field and electric charge density:

$$\Phi(x_1, x_2) = \frac{U}{K(\cos\theta)} \sum_{m=0}^{\infty} \frac{P_m(\cos 2\theta)}{2m+1} \mathrm{e}^{-r\chi_m x_2} \cos(\chi_m x_1) , \tag{A.15a}$$

$$E_1(x_1, 0) = \frac{U\chi_0}{2K(\cos\theta)} [\cos^2\theta - \cos^2(\chi_0 x_1)]^{-1/2} , \quad \eta\frac{d}{2} < |x_1| < \frac{d}{2} , \tag{A.15b}$$

$$\sigma(x_1) = \frac{\chi_0(\varepsilon_0 + \varepsilon_\mathrm{p})U}{2K(\cos\theta)} [\sin^2\theta - \sin^2(\chi_0 x_1)]^{-1/2} , \quad |x_1| < \eta\frac{d}{2} , \tag{A.15c}$$

$$\theta = \eta\frac{\pi}{2} , \quad \chi_m = (2m+1)\frac{\pi}{d} , \quad r = \begin{cases} \varepsilon_{11}/\varepsilon_{22} & \text{for } x_2 > 0 \\ -1 & \text{for } x_2 < 0 \end{cases} . \tag{A.15d}$$

These functions are plotted in Fig. 2.8 for the case $\eta = 1/2$, $\theta = \pi/4$. Since $K(\sqrt{2}/2) = 1.854$, the field at the centre of the interdigital interval and the charge density at the centre of a finger are given by

$$E_1(d/2) = \frac{\pi U}{dK(\sqrt{2}/2)\sqrt{2}} \approx 1.20\frac{U}{d} ,$$

$$\sigma(0) = \frac{\pi Q}{dK(\sqrt{2}/2)\sqrt{2}} \approx 1.20(\varepsilon_0 + \varepsilon_\mathrm{p})\frac{U}{d} .$$

Coefficients F_m are strongly dependent on the metallisation ratio η through the Legendre polynomials.

It is instructive to calculate the stored electrostatic energy per finger pair:

$$W_\mathrm{el} = \frac{1}{2}\int_{-d}^{+d} \mathrm{d}x_1 \int_{-\infty}^{\infty} (E_1 D_1 + E_2 D_2)\,\mathrm{d}x_2 .$$

Putting $\phi = \chi x_1$, $\psi = \chi x_2$ and separating dielectric ($x_2 < 0$) from piezoelectric solid ($x_2 > 0$),

$$W_\mathrm{el} = \frac{1}{2\chi^2}\int_{-\pi}^{+\pi} \mathrm{d}\phi \left[\int_{-\infty}^{0} \varepsilon_0(E_1^2 + E_2^2)\,\mathrm{d}\psi + \int_0^{\infty} (\varepsilon_{11}E_1^2 + \varepsilon_{22}E_2^2)\,\mathrm{d}\psi\right] ,$$

where $\chi = \pi/d$. Using expansion (A.4) for the potential, electric field components can be put into the form:

$$E_1 = -\frac{\partial\Phi}{\partial x_1} = \frac{U}{d}\sum_{m=0}^{\infty} F_m \mathrm{e}^{-r(2m+1)\psi} \sin(2m+1)\phi , \tag{A.16}$$

$$E_2 = -\frac{\partial\Phi}{\partial x_2} = r\frac{U}{d}\sum_{m=0}^{\infty} F_m \mathrm{e}^{-r(2m+1)\psi} \cos(2m+1)\phi . \tag{A.17}$$

Since

$$\int_{-\pi}^{\pi} \begin{Bmatrix} \sin(2m+1)\phi \\ \cos(2m+1)\phi \end{Bmatrix} \cdot \begin{Bmatrix} \sin(2n+1)\phi \\ \cos(2n+1)\phi \end{Bmatrix} \mathrm{d}\phi = \pi\delta_{nm} ,$$

the two field components contribute equally, both in the solid and in vacuum:

$$W_\mathrm{el} = \frac{\pi U^2}{(\chi d)^2}\sum_{m=0}^{\infty} F_m^2 \left[\varepsilon_0 \int_{-\infty}^{0} \mathrm{e}^{2(2m+1)\psi}\mathrm{d}\psi + \varepsilon_{11}\int_0^{\infty} \mathrm{e}^{-2r(2m+1)\psi}\mathrm{d}\psi\right] .$$

Then, since $\chi d = \pi$,

$$W_{\text{el}} = \frac{\pi}{2}U^2 \left[\varepsilon_0 + (\varepsilon_{11}\varepsilon_{22})^{1/2}\right] \sum_{m=0}^{\infty} \frac{(F_m/\pi)^2}{2m+1} \,.$$

Using (A.12) for the capacitance, $C_1 = (\varepsilon_0 + \varepsilon_{\text{p}})K/K'$, (A.14) for coefficients F_m, and the fact that $P_0(x) = 1$, $\forall x$, it follows that

$$W_{\text{el}} = \frac{\pi}{KK'} \left[1 + \sum_{m=1}^{\infty} \frac{P_m^2(\cos 2\theta)}{2m+1}\right] \frac{1}{2}C_1U^2 \,. \tag{A.18}$$

This formula reveals the relative importance of the harmonics. For example, for 50% metallisation ($\cos 2\theta = 0 \Rightarrow P_1 = P_3 = 0$, $P_2 = -0.5$, $P_4 = 3/8$, $K = K' = 1.854$), the fundamental term represents 91.4% of the stored energy and the fifth harmonic ($m = 2$) only 5%.

A.2 Single-Element Charge Distribution

The central electrode ($n = 0$) of the periodic array of period p in Fig. 2.19a is assumed to be held at potential $U_0 = 1$ V, the others being earthed (defining the zero potential). This distribution, $U_n = 1$ for $n = 0$ and $U_n = 0$ for $n \neq 0$, can be represented by the superposition of an infinite number of unit amplitude harmonic excitations which are spatially phase shifted:

$$U_n = \int_0^1 \mathrm{e}^{-2\pi \mathrm{i} ns} \mathrm{d}s \,, \quad 0 \leq s \leq 1 \,, \quad n \text{ integer.}$$

For given spatial frequency s, the charge distribution $\sigma_1(x, s)$ produced by the harmonic sequence of potential $\mathrm{e}^{-2\pi \mathrm{i} ns}$ is given by [A.1]

$$\sigma_1(x, s) = \frac{2}{p}(\varepsilon_0 + \varepsilon_{\text{p}}) \frac{\sin \pi s}{P_{-s}(-\cos \eta\pi)} \sum_{m=-\infty}^{\infty} P_m(\cos \eta\pi) \exp\left[-\mathrm{i}(s+m)\frac{2\pi}{p}x\right] . \tag{A.19}$$

Note that formula (A.7) for the alternating distribution in Fig. A.1 corresponds to the particular value $s = 1/2$ [$\Rightarrow U_n = (-1)^n$]. The Fourier transform of $\sigma_1(x, s)$,

$$\overline{\sigma}_1(k, s) = \int_{-\infty}^{\infty} \sigma_1(x, s)\mathrm{e}^{\mathrm{i}kx}\mathrm{d}x \,,$$

is a sequence of delta functions. Indeed, introducing wave number $S = kp/2\pi$ relative to that of the array $G = 2\pi/p$, and using

$$\int_{-\infty}^{\infty} \exp\left\{\mathrm{i}\frac{2\pi}{p}[S - (s+m)]x\right\} \mathrm{d}x = \delta\left[\frac{S - (s+m)}{p}\right] = p\delta[S - (s+m)] \,,$$

it follows that

$$\overline{\sigma}_1(k,\, s) = 2(\varepsilon_0 + \varepsilon_\mathrm{p})\frac{\sin \pi s}{P_{-s}(-\cos \eta\pi)} \sum_{m=-\infty}^{\infty} P_m(\cos \eta\pi)\delta(S - s - m)\,. \tag{A.20}$$

Deltas are centred on values $S = s + m$ for which m is a whole number, i.e., for k, f values such that

$$k = \frac{2\pi}{p}(s+m) = \frac{2\pi f}{V_\mathrm{R}} \quad \Rightarrow \quad f = 2f_0(s+m)\,,$$

where $f_0 = V_\mathrm{R}/2p$ is the synchronism frequency.

The Fourier transform of the single-element charge distribution is obtained by summing $\overline{\sigma}_1(k,\, s)$ over s: $\overline{\sigma}_{1\mathrm{e}}(k) = \int_0^1 \overline{\sigma}_1(k,\, s)\,\mathrm{d}s$. Only the polynomial $P_m(\cos\eta\pi)$ of degree m such that $s = S - m$ lies between 0 and 1 survives integration, yielding

$$\overline{\sigma}_{1\mathrm{e}}(k) = 2(\varepsilon_0 + \varepsilon_\mathrm{p})\frac{\sin \pi s}{P_{-s}(-\cos \eta\pi)} P_m(\cos \eta\pi)\,. \tag{A.21}$$

The single-element charge distribution depends on frequency through the variable

$$s = \frac{kp}{2\pi} - m = \frac{f}{2f_0} - m\,, \quad 0 \le s \le 1\,. \tag{A.22}$$

Its variations are plotted in Fig. 2.20 for different values of metallisation ratio η.

Electrostatic Contribution to Harmonic Admittance. Let us now calculate the incoming current at the nth electrode:

$$I_n(s) = \mathrm{i}\omega w \int_{np-a/2}^{np+a/2} \sigma_1(x,\, s)\mathrm{d}x = \mathrm{i}\omega \frac{wp}{2\pi} \int_{-\eta\pi}^{\eta\pi} \sigma_1(\theta,\, s)\,\mathrm{d}\theta\,,$$

where $x = np + p\theta/2\pi$. Substituting the expression given in (A.19) for the charge density produced by a harmonic excitation $U_n(s) = \mathrm{e}^{-2\pi \mathrm{i} ns}$, we find

$$\begin{aligned} I_n(s) \;=\; & \frac{\mathrm{i}\omega}{\pi}(\varepsilon_0 + \varepsilon_\mathrm{p})w\frac{\sin \pi s}{P_{-s}(-\cos \eta\pi)}\mathrm{e}^{-2\pi \mathrm{i} ns} \\ & \left[\int_{-\eta\pi}^{\eta\pi} \sum_{m=-\infty}^{\infty} P_m(\cos\eta\pi)\mathrm{e}^{-\mathrm{i}(s+m)\theta}\mathrm{d}\theta\right]\,. \end{aligned}$$

The harmonic admittance $Y(s)$ is defined, as in Sect. 2.6.3, by the ratio I_n/U_n which is independent of electrode position in the array. Using (C.8) with $\nu = -s$ and $\varDelta = \eta\pi$, the integral is equal to $2\pi P_{-s}(\cos\eta\pi)$ and

$$Y(s) = 2\mathrm{i}\omega(\varepsilon_0 + \varepsilon_\mathrm{p})w \sin \pi s \frac{P_{-s}(\cos \eta\pi)}{P_{-s}(-\cos \eta\pi)}\,.$$

In the special case $\eta = 1/2$, $\cos\eta\pi = 0$ and $(\varepsilon_0 + \varepsilon_\mathrm{p})w = C_1$, the capacitance per finger pair, the harmonic admittance takes the simple form

$$Y(s) = 2\mathrm{i}\omega C_1 \sin \pi s\,.$$

A.3 General Case. Charge Density. Current Intensity

The static charge density σ is given in terms of the potential Φ at the surface by the Green function $g_0(x)$ of the electrostatic problem [see (2.53)], the two quantities satisfying electrical boundary conditions (A.8).

A first method is to write the potential as a sum [A.2]

$$\Phi(x_i) = \sum_{j=1}^{P} A_{ij}\sigma(x_j) ,$$

where x_j is a point on an electrode separated from its neighbour x_i by distance Δx, and P is the number of points x_j. The matrix (A_{ij}) is symmetric and its elements are functions of $\Delta x g_0(x_i - x_j)$. The sum is only over points on the electrodes. Inverting the last equality, we find the following correspondence:

$$\Phi(x_j) = [g_0(x) \otimes \sigma(x)]_{x_j} \quad \Leftrightarrow \quad \sigma(x_i) = \sum_{j=1}^{P} B_{ij}\Phi(x_j) . \tag{A.23}$$

Elements of matrix (B_{ij}), inverse of (A_{ij}), are given in units [F/m^2].

Another way of finding the charge σ is to apply the superposition principle. We imagine all electrodes to be earthed (i.e., held at zero potential) except one, the nth electrode, which is held at a constant potential of 1 V. Let $p_n(x)$ be the polarity function, such that $p_n(x) = 1$ or 0 depending on whether x is a point of electrode n or not. The charge at a general point x_i of an electrode is denoted $c_n(x_i)$. It results from the potential at all other points x_j and is given by

$$c_n(x_i) = \sum_{j=1}^{P} B_{ij}p_n(x_j) . \tag{A.24}$$

If each of the M electrodes is held at a different potential U_n, the charge density is given by linear combination:

$$\sigma(x) = \sum_{n=1}^{M} U_n c_n(x) . \tag{A.25}$$

Let us apply this to the case of a transducer with constant finger overlap length w. Some of the M electrodes are held at potential U and the rest are earthed (Fig. A.2). The charge density $\sigma_1(x)$ corresponding to unit applied voltage, introduced in Sect. 2.3.1, is equal to

$$\sigma_1(x) = \sum_{n=1}^{M} P_n c_n(x) , \tag{A.26}$$

where the component P_n of the polarity vector is 1 if the nth electrode is at potential U and 0 if this electrode is earthed. The capacitance C_T is just

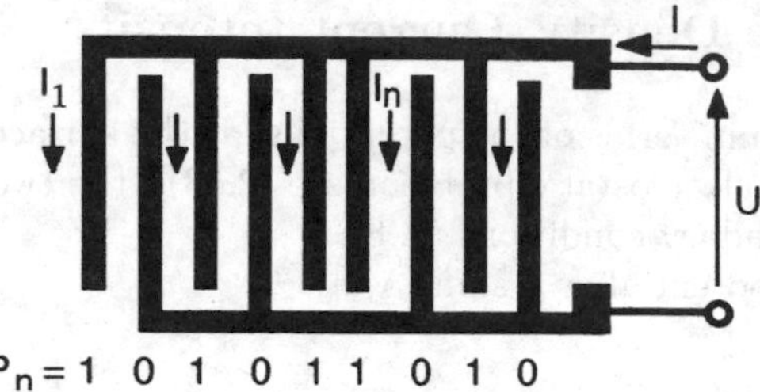

Fig. A.2. Component P_n of the polarity vector is equal to 1 if the nth electrode is at potential U and 0 if it is earthed

the sum of the charges, in phase with voltage U, carried by the fingers of an electrode when $U = 1$ V, viz.,

$$C_\mathrm{T} = w \sum_{n=1}^{M} P_n \int_n c(x)\,\mathrm{d}x \,, \tag{A.27}$$

where the integral extends over electrode n.

The current intensity through one electrode is the time derivative of the charge on that electrode. According to (2.67),

$$I_n = \mathrm{i}\omega w \int_n [\sigma_0(x,\,\omega) + \sigma_\mathrm{R}(x,\,\omega)]\,\mathrm{d}x = I_{0n} + I_{\mathrm{R}n} \,. \tag{A.28}$$

The term I_{0n} can be obtained if σ_0 is known. The second term $I_{\mathrm{R}n}$ is the contribution from Rayleigh waves propagating under the transducer. It arises only from points x_j on electrodes. Expressing the integral over the nth electrode in the form of a discrete sum over P points and using the distribution $p_n(x_j)$ as before, it becomes

$$I_{\mathrm{R}n} = \mathrm{i}\omega w \sum_{j=1}^{P} p_n(x_j)\sigma_\mathrm{R}(x_j,\,\omega)\Delta x \,. \tag{A.29}$$

The total incoming current intensity through the wire connected at potential U is

$$I = \sum_{n=1}^{M} P_n I_n \,.$$

Appendix B. Coupled Mode Theory

An exact solution for the propagation equations in a periodic medium can be found using the Bloch wave formalism, based on Floquet's theorem [B.1]. Spatially coupled modes were first studied in the context of optics, to describe coupling between waves propagating in a periodic array of dielectric layers, for example [B.2]. This approximate theory is justified provided that the periodic perturbation which causes mode coupling remains small [B.3]. We shall adopt a formalism applicable to both acoustics and optics.

B.1 Formulation. Coupled Mode Equations

To simplify the analysis, assume that the elastic wave can be represented by a scalar quantity a. It satisfies the propagation equation

$$c_{33}\frac{\partial^2 a}{\partial x_3^2} + c_{22}\frac{\partial^2 a}{\partial x_2^2} + c_{11}\frac{\partial^2 a}{\partial x_1^2} - \rho\frac{\partial^2 a}{\partial t^2} = 0\,, \quad \rho = \rho_0(x_2,\, x_3) + \Delta\rho\,. \tag{B.1}$$

In the unperturbed medium $\Delta\rho = 0$, this equation is invariant in the x_1 direction. In acoustics, a represents particle velocity v, ρ is density and $c_{\alpha\beta}$ are stiffnesses. In optics, these quantities are replaced by the electric field E, absolute permittivity $\varepsilon_0\varepsilon$ and reciprocal permeability $1/\mu$, respectively.

In the harmonic regime, the amplitude distribution $a_m(x_2,\, x_3)$ of a mode propagating in the x_1 direction through the unperturbed medium

$$a(x_1,\, x_2,\, x_3,\, t) = a_m(x_2,\, x_3)\mathrm{e}^{\mathrm{i}(\omega t - k_m x_1)} \tag{B.2}$$

satisfies

$$c_{33}\frac{\partial^2 a_m}{\partial x_3^2} + c_{22}\frac{\partial^2 a_m}{\partial x_2^2} + (\rho_0\omega^2 - c_{11}k_m^2)a_m = 0\,. \tag{B.3}$$

The index m which labels the mode varies continuously if the medium is unbounded or discretely if it is bounded (e.g., a waveguide). A general field distribution is represented by a linear combination of these modes:

$$a(x_i,\, t) = \sum_m A_m a_m(x_2,\, x_3)\mathrm{e}^{\mathrm{i}(\omega t - k_m x_1)}\,, \tag{B.4}$$

insofar as they are orthogonal and carry unit mean power $\langle P\rangle$ (normal modes). According to (5.33), Vol. I, and since $|k_m| = \omega/V_m$,

$$|k_m| \int a_m a_m^* \mathrm{d}x_2 \mathrm{d}x_3 = \frac{2\omega}{\rho V_m^2} \quad \text{when} \quad \langle P \rangle = 1\,\mathrm{W}\,. \tag{B.5}$$

The *orthogonality condition* for normal modes is then

$$c_{11} \int a_n a_m^* \mathrm{d}x_2 \mathrm{d}x_3 = \frac{2\omega}{|k_n|} \delta_{nm}\,. \tag{B.6}$$

A similar condition is obtained for optical modes in Problem 3.1.

Since solutions a_m are not eigenmodes for the perturbed structure, we express the quantity a characterising the wave in the perturbed medium in the same form as (B.4) but with coefficients A_m which depend on x_1:

$$a(x_i,\,t) = \sum_m A_m(x_1) a_m(x_2,\,x_3) \mathrm{e}^{\mathrm{i}(\omega t - k_m x_1)}\,. \tag{B.7}$$

This is analogous to the method using variation of integration constants. Substituting

$$\frac{\partial^2 a}{\partial x_1^2} = \sum_m \left(\frac{\mathrm{d}^2 A_m}{\mathrm{d}x_1^2} - 2\mathrm{i}k_m \frac{\mathrm{d}A_m}{\mathrm{d}x_1} - k_m^2 A_m \right) a_m(x_2,\,x_3) \mathrm{e}^{\mathrm{i}(\omega t - k_m x_1)}$$

into the propagation equation (B.1), we obtain

$$\begin{aligned} 0 = &\sum_m \left[c_{33} \frac{\partial^2 a_m}{\partial x_3^2} + c_{22} \frac{\partial^2 a_m}{\partial x_2^2} + (\rho_0 \omega^2 - c_{11} k_m^2) a_m \right] A_m \mathrm{e}^{-\mathrm{i}k_m x_1} \qquad \text{(B.8)} \\ &+ \sum_m \left[c_{11} \left(\frac{\mathrm{d}^2 A_m}{\mathrm{d}x_1^2} - 2\mathrm{i}k_m \frac{\mathrm{d}A_m}{\mathrm{d}x_1} \right) + \omega^2 \Delta\rho A_m \right] a_m(x_2,\,x_3) \mathrm{e}^{-\mathrm{i}k_m x_1}\,. \end{aligned}$$

From (B.3), the first term is zero. In addition, if the perturbation is weak enough to ensure that A_m varies only slightly over a wavelength $\lambda_m = 2\pi/k_m$, the approximation

$$\left| \frac{\mathrm{d}^2 A_m}{\mathrm{d}x_1^2} \right| \ll \left| k_m \frac{\mathrm{d}A_m}{\mathrm{d}x_1} \right| \tag{B.9}$$

is valid. Equation (B.8) simplifies to

$$2\mathrm{i}c_{11} \sum_m k_m \frac{\mathrm{d}A_m}{\mathrm{d}x_1} a_m(x_2,\,x_3) \mathrm{e}^{-\mathrm{i}k_m x_1} = \omega^2 \Delta\rho \sum_m A_m a_m(x_2,\,x_3) \mathrm{e}^{-\mathrm{i}k_m x_1}\,. \tag{B.10}$$

Multiplying both sides by a_n^* and integrating over x_2, x_3, the orthogonality relation (B.6) implies

$$4\mathrm{i}\omega \frac{k_n}{|k_n|} \frac{\mathrm{d}A_n}{\mathrm{d}x_1} \mathrm{e}^{-\mathrm{i}k_n x_1} = \omega^2 \sum_m \left(\int a_n^* \Delta\rho a_m \mathrm{d}x_2 \mathrm{d}x_3 \right) A_m(x_1) \mathrm{e}^{-\mathrm{i}k_m x_1}\,,$$

and hence

$$\frac{\mathrm{d}A_n}{\mathrm{d}x_1} = -\mathrm{i} \frac{k_n}{|k_n|} \frac{\omega}{4} \sum_m \left(\int a_n^* \Delta\rho a_m \mathrm{d}x_2 \mathrm{d}x_3 \right) A_m(x_1) \mathrm{e}^{\mathrm{i}(k_n - k_m)x_1}\,. \tag{B.11}$$

In the case of a periodic perturbation in the x_1 direction, of period p, the Fourier series expansion

$$\Delta\rho(x_i) = \sum_q \rho_q(x_2, x_3)\mathrm{e}^{-\mathrm{i}qGx_1} , \quad G = \frac{2\pi}{p} , \tag{B.12}$$

yields the linear system of first order differential equations ($x_1 = x$)

$$\frac{\mathrm{d}A_n}{\mathrm{d}x} = -\mathrm{i}\frac{k_n}{|k_n|}\sum_m\sum_q C_{nm}^{(q)} A_m(x)\mathrm{e}^{\mathrm{i}(k_n - k_m - qG)x} , \tag{B.13}$$

where

$$C_{nm}^{(q)} = \frac{\omega}{4}\int a_n^*\rho_q a_m \mathrm{d}x_2\mathrm{d}x_3 . \tag{B.14}$$

The coefficient $C_{mn}^{(q)}$ measures the coupling strength between modes m and n due to the qth Fourier component of the perturbation. Another important factor in the interaction is

$$\Delta k = k_n - k_m - qG ,$$

which characterises the *detuning* between wave vectors. Indeed, the variation in amplitude of mode m due to interaction with mode n, over a large distance L relative to the wavelength, is given by

$$\Delta A_n = -\mathrm{i}\frac{k_n}{|k_n|}\int_L C_{nm}^{(q)} A_m(x)\mathrm{e}^{\mathrm{i}\Delta kx}\mathrm{d}x .$$

Because of the oscillating exponential, the integral will be small if the phase detuning ΔkL is greater than π (see Appendix G on the Stationary Phase Method). The phase tuning condition

$$\Delta k = 0 \quad \Rightarrow \quad k_n - k_m = qG , \quad q \text{ integer} , \tag{B.15}$$

also called the Bragg condition, ensures maximal coupling between the two modes, whatever the interaction length L.

To sum up, slow variations in amplitude (relative to the wavelength) of modes propagating through a medium subject to a weak periodic perturbation are governed by the system of first order differential equations (B.13). A large interaction occurs between modes if the amplitude of the coupling coefficient is large and if, in addition, the tuning condition is satisfied between wave vectors and a reciprocal array vector.

Usually, only two modes (1 and 2) are efficiently coupled. The system (B.13) then reduces to two equations:

$$\begin{aligned} \frac{\mathrm{d}A_1}{\mathrm{d}x} &= -\mathrm{i}\frac{k_1}{|k_1|}C_{12}^{(q)} A_2(x)\mathrm{e}^{\mathrm{i}\Delta kx} , \\ \frac{\mathrm{d}A_2}{\mathrm{d}x} &= -\mathrm{i}\frac{k_2}{|k_2|}C_{21}^{(-q)} A_1(x)\mathrm{e}^{-\mathrm{i}\Delta kx} . \end{aligned} \tag{B.16}$$

Denote the coupling coefficient by

$$\kappa = C_{12}^{(q)} = \left[C_{21}^{(-q)}\right]^* , \tag{B.17}$$

and the detuning parameter by

$$\delta = \frac{\Delta k}{2} = \frac{k_1 - k_2}{2} - q\frac{G}{2} . \tag{B.18}$$

Putting

$$A_1(x) = R(x)\mathrm{e}^{\mathrm{i}\delta x} , \quad A_2(x) = S(x)\mathrm{e}^{-\mathrm{i}\delta x} , \tag{B.19}$$

the coupled equations become

$$\begin{aligned} \frac{\mathrm{d}R}{\mathrm{d}x} &= -\mathrm{i}\delta R(x) - \mathrm{i}\frac{k_1}{|k_1|}\kappa S(x) , \\ \frac{\mathrm{d}S}{\mathrm{d}x} &= \mathrm{i}\delta S(x) - \mathrm{i}\frac{k_2}{|k_2|}\kappa^* R(x) . \end{aligned} \tag{B.20}$$

$R(x)$ and $S(x)$ represent slow amplitude variations in the two coupled modes.

B.2 Solution

Solutions of this homogeneous linear system of two first order differential equations differ depending on whether modes propagate in the same or opposite directions.

B.2.1 Propagation in the Same Direction

Ratios $k_1/|k_1|$ and $k_2/|k_2|$ are equal to 1. In order to apply the results of Appendix F, we shall put (B.20) into a matrix form analogous to the equation of state which describes evolution of a linear dynamic system:

$$\frac{\mathrm{d}\boldsymbol{X}}{\mathrm{d}x} = \boldsymbol{F}\boldsymbol{X}(x) , \quad \boldsymbol{X} = \begin{pmatrix} R(x) \\ S(x) \end{pmatrix} , \quad \boldsymbol{F} = \begin{pmatrix} -\mathrm{i}\delta & -\mathrm{i}\kappa \\ -\mathrm{i}\kappa^* & \mathrm{i}\delta \end{pmatrix} . \tag{B.21}$$

Vector $\boldsymbol{X}$ describes the state of waves under the array for a given frequency.

The general solution is found using a particular matrix solution known as the transition matrix $\boldsymbol{M}(x_1, x_0)$ which gives wave amplitudes at a point x_1 in terms of their amplitudes at x_0, viz.,

$$\boldsymbol{X}(x_1) = \boldsymbol{M}(x_1, x_0)\boldsymbol{X}(x_0) , \quad \boldsymbol{M} = \begin{pmatrix} M_{11} & M_{12} \\ M_{21} & M_{22} \end{pmatrix} . \tag{B.22}$$

Uniform Array. For an x invariant perturbation, parameters δ and κ only depend on frequency and the matrix $\boldsymbol{F}$ is constant. The transition matrix $\boldsymbol{M}(x_1, x_0)$ of the stationary system of equations depends only on $x = x_1 - x_0$. It is the matrix exponential $\mathrm{e}^{\boldsymbol{F}x}$ (see Appendix F). Calculating the square of $\boldsymbol{F}$,

$$\boldsymbol{F}^2 = -\begin{pmatrix} \delta^2 + |\kappa|^2 & 0 \\ 0 & \delta^2 + |\kappa|^2 \end{pmatrix} = -\eta^2 \boldsymbol{I} \,, \quad \eta^2 = \delta^2 + |\kappa|^2 \,, \tag{B.23}$$

where $\boldsymbol{I}$ is the identity matrix, and other powers

$$\boldsymbol{F}^3 = -\eta^2 \boldsymbol{F} \,,\ \boldsymbol{F}^4 = \eta^4 \boldsymbol{I} \,,\ \boldsymbol{F}^5 = \eta^4 \boldsymbol{F} \,, \dots \,,$$

yields

$$\mathrm{e}^{\boldsymbol{F}x} = \left(1 - \frac{\eta^2 x^2}{2!} + \frac{\eta^4 x^4}{4!} - \dots\right)\boldsymbol{I} + \left(x - \frac{\eta^2 x^3}{3!} + \frac{\eta^4 x^5}{5!} - \dots\right)\boldsymbol{F} \,. \tag{B.24}$$

Hence,

$$\mathrm{e}^{\boldsymbol{F}x} = \boldsymbol{I}\cos\eta x + \frac{\boldsymbol{F}}{\eta}\sin\eta x \,. \tag{B.25}$$

The transition matrix for the array

$$\boldsymbol{M}(x) = \begin{pmatrix} \cos\eta x - \mathrm{i}\dfrac{\delta}{\eta}\sin\eta x & -\mathrm{i}\dfrac{\kappa}{\eta}\sin\eta x \\ -\mathrm{i}\dfrac{\kappa^*}{\eta}\sin\eta x & \cos\eta x + \mathrm{i}\dfrac{\delta}{\eta}\sin\eta x \end{pmatrix} \tag{B.26}$$

is unitary, so that $||\boldsymbol{M}(x)|| = 1$, because of power conservation.

Amplitudes $R(x)$ and $S(x)$ are given by

$$\begin{aligned} R(x) &= M_{11}(x)R(0) + M_{12}(x)S(0) \,, \\ S(x) &= M_{21}(x)R(0) + M_{22}(x)S(0) \,, \end{aligned} \tag{B.27}$$

where $R(0)$ and $S(0)$ are amplitudes at $x = 0$. The power fraction transferred from mode R to mode S, or vice versa, is

$$|M_{21}(x)|^2 = \frac{|\kappa|^2}{\eta^2}\sin^2\eta x = \frac{|\kappa|^2}{|\kappa|^2 + \delta^2}\sin^2\left(x\sqrt{|\kappa|^2 + \delta^2}\right) \,. \tag{B.28}$$

This varies periodically with interaction length. Energy conservation implies that the power carried by the other mode is $|M_{11}(x)|^2 = 1 - |M_{21}(x)|^2$. Complete transfer is only possible for perfect phase matching $\delta = 0$:

$$|M_{21}(x)|^2 = \sin^2|\kappa|x \,, \quad |M_{11}(x)|^2 = \cos^2|\kappa|x \,, \tag{B.29}$$

and after a path length L such that $|\kappa|L = \pi/2$.

Figure B.1 shows the power distribution for each of the modes in the two cases: $\delta = 0$ (phase tuning) and $\delta \neq 0$.

B.2.2 Propagation in Opposite Directions

In the coupled equations (B.20), ratios $k_1/|k_1|$ and $k_2/|k_2|$ are equal to 1 and -1. In matrix form,

$$\frac{\mathrm{d}}{\mathrm{d}x}\begin{pmatrix} R \\ S \end{pmatrix} = \begin{pmatrix} -\mathrm{i}\delta & -\mathrm{i}\kappa \\ \mathrm{i}\kappa^* & \mathrm{i}\delta \end{pmatrix}\begin{pmatrix} R(x) \\ S(x) \end{pmatrix} \,. \tag{B.30}$$

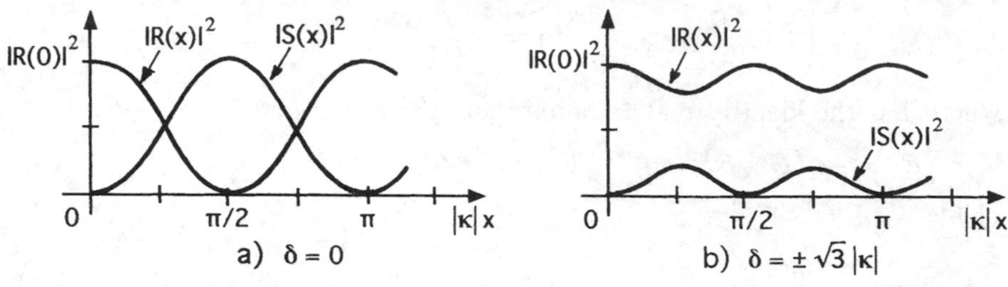

Fig. B.1. Power transfer between two coupled modes propagating in the same direction. (**a**) Perfect phase tuning. (**b**) Imperfect phase tuning $\delta = \pm|\kappa|\sqrt{3}$

The square of matrix $\boldsymbol{F}$ is proportional to the identity matrix:

$$\boldsymbol{F}^2 = \begin{pmatrix} |\kappa|^2 - \delta^2 & 0 \\ 0 & |\kappa|^2 - \delta^2 \end{pmatrix} = -\gamma^2 \boldsymbol{I}\,, \quad \gamma^2 = \delta^2 - |\kappa|^2\,. \tag{B.31}$$

The transition matrix is then given by (B.26) with κ^* replaced by $-\kappa^*$ and η replaced by γ:

$$\boldsymbol{M}(x) = \begin{pmatrix} \cos\gamma x - \mathrm{i}\dfrac{\delta}{\gamma}\sin\gamma x & -\mathrm{i}\dfrac{\kappa}{\gamma}\sin\gamma x \\ \mathrm{i}\dfrac{\kappa^*}{\gamma}\sin\gamma x & \cos\gamma x + \mathrm{i}\dfrac{\delta}{\gamma}\sin\gamma x \end{pmatrix}. \tag{B.32}$$

γ is real or pure imaginary depending on whether $|\delta|$ is greater or less than $|\kappa|$. When γ is imaginary, trigonometric functions become hyperbolic functions.

Solutions $R(x)$ and $S(x)$ of the homogeneous system (B.30) (eigenmodes) have form $\mathrm{e}^{\mathrm{i}\gamma x}$. From (B.19), coupled mode amplitudes $A_1(x)$, $A_2(x)$ vary as $\mathrm{e}^{\mathrm{i}(\gamma+\delta)x}$ and $\mathrm{e}^{\mathrm{i}(\gamma-\delta)x}$, so that the wave given in (B.7) is the sum of two terms:

$$a(x,\,t) = a_1 \mathrm{e}^{\mathrm{i}[\omega t - (k_1 - \gamma - \delta)x]} + a_2 \mathrm{e}^{\mathrm{i}[\omega t - (k_2 - \gamma + \delta)x]}\,.$$

From (B.18), corresponding wave numbers are

$$\beta_1 = k_1 - \gamma - \delta = \frac{k_1 + k_2}{2} + q\frac{G}{2} - \gamma\,,$$

$$\beta_2 = k_2 - \gamma + \delta = \frac{k_1 + k_2}{2} - q\frac{G}{2} - \gamma\,.$$

As the two modes propagate in opposite directions, at the same frequency: $k_1 = -k_2 = \omega/V$, where $q = 1$ and $G = 2\pi/p$, it follows that

$$\beta_1 = \frac{\pi}{p} \pm \sqrt{\delta^2 - |\kappa|^2}\,, \quad \beta_2 = -\frac{\pi}{p} \pm \sqrt{\delta^2 - |\kappa|^2}\,, \tag{B.33}$$

where

$$\delta = \frac{\omega}{V} - \frac{G}{2} = \frac{\omega - \omega_0}{V}\,, \quad \omega_0 = \frac{G}{2}V = \pi\frac{V}{p}\,. \tag{B.34}$$

In the frequency band such that $|\delta| < |\kappa|$, called the forbidden band or *stop band*, wave numbers β_1 and β_2 are complex:

$$\left|\frac{\omega-\omega_0}{V}\right| < |\kappa| \quad \Rightarrow \quad \beta_1 = \frac{\pi}{p} \pm \mathrm{i}\sqrt{|\kappa|^2 - \left(\frac{\omega-\omega_0}{V}\right)^2}\,. \tag{B.35}$$

The wave no longer propagates without attenuation in the x direction. The centre frequency of this stop band is determined by the array period: $\omega_0 = \pi V/p \Rightarrow \beta_0 = \pi/p$. It corresponds to the Bragg condition for constructive interference between waves reflected from each obstacle in the array. It has width $\Delta\omega = 2|\kappa|V$.

Figure B.2 shows variation of $\mathrm{Re}(\beta)$ and $\mathrm{Im}(\beta)$ with ω. The imaginary part is maximal for $\omega = \omega_0$ and equals $|\kappa|$. For frequencies close to ω_0, the array behaves as a reflector. Reflection and transmission coefficients are calculated in Sect. 2.5.1.

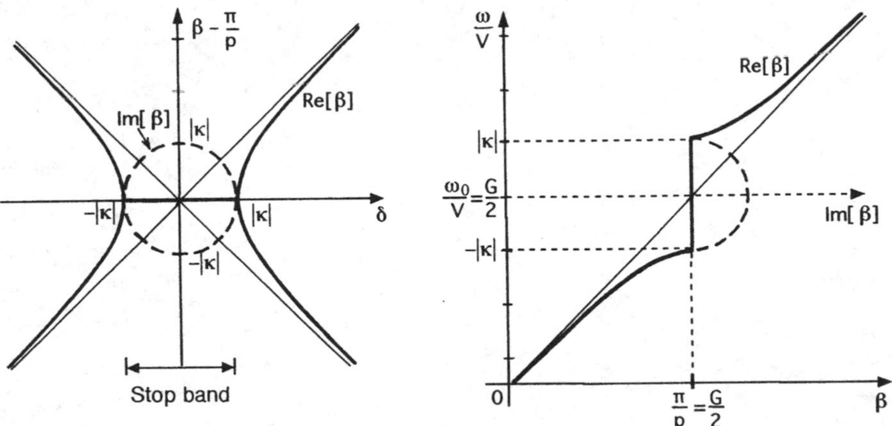

Fig. B.2. Coupled mode approximation. Dispersion curves $\omega(\beta)$ for a periodic array. *Full curve*: $\mathrm{Re}(\beta)$. *Dashed curve*: $\mathrm{Im}(\beta)$. In the stop band, the real part of β does not depend on frequency, whereas the imaginary part is maximal in the centre of the stop band

Appendix C. Legendre Functions and Polynomials

The Legendre function of degree ν is defined by the series

$$P_\nu = \sum_{m=0}^{\infty} a_m(x) , \quad a_0 = 1 , \quad |x| \le 1 , \tag{C.1}$$

$$a_m(x) = \frac{(m-1-\nu)(m+\nu)}{2m^2}(1-x)a_{m-1}(x) , \quad m \ge 1 .$$

As a function of ν, it is symmetric in the value $\nu = -1/2$ (see Fig. C.1):

$$P_{-\nu}(x) = P_{\nu-1}(x) , \tag{C.2}$$

and in particular

$$P_{-1}(x) = P_0(x) = 1 , \quad \forall x \quad \text{and} \quad P_\nu(1) = 1 , \quad \forall \nu .$$

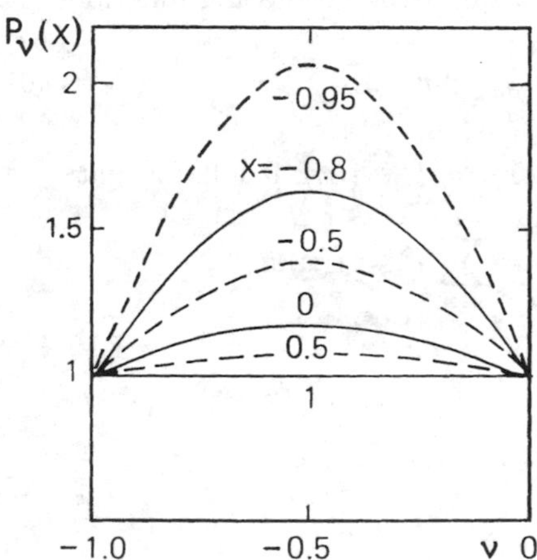

Fig. C.1. Legendre function $P_\nu(x)$ as a function of ν in the range $-1 \le \nu \le 0$ for several values of x. If ν is not an integer, $P_\nu(-1) = \infty$

Legendre functions satisfy the recurrence relation

$$\nu P_\nu(x) = (2\nu - 1)xP_{\nu-1}(x) - (\nu - 1)P_{\nu-2}(x) . \tag{C.3}$$

Putting $x = \cos\Delta$, the Mehler–Dirichlet formula is

$$P_\nu(\cos\Delta) = \frac{1}{\pi\sqrt{2}}\int_{-\Delta}^{\Delta} \frac{\exp \mathrm{i}(\nu+1/2)\theta}{\sqrt{\cos\theta-\cos\Delta}}\,\mathrm{d}\theta$$
$$= \frac{\sqrt{2}}{\pi}\int_{0}^{\Delta} \frac{\cos(\nu+1/2)\theta}{\sqrt{\cos\theta-\cos\Delta}}\,\mathrm{d}\theta\,, \quad 0<\Delta<\pi\,. \tag{C.4}$$

When $\nu = -1/2$,

$$P_{-1/2}(\cos\Delta) = \frac{\sqrt{2}}{\pi}\int_{-\Delta}^{\Delta} \frac{\mathrm{d}\theta}{\sqrt{\cos\theta-\cos\Delta}} = \frac{2}{\pi}K[\sin(\Delta/2)]\,,$$

where K is the elliptic integral of the first kind, defined by

$$P_{-1/2}(-\cos\Delta) = \frac{2}{\pi}K[\cos(\Delta/2)] \quad \text{or}$$
$$P_{-1/2}(x) = \frac{2}{\pi}K[\sqrt{(1-x)/2}]\,, \quad |x|<1\,. \tag{C.5}$$

When ν is an integer n, the series (C.1) stops at $m = 1+n$ and the Legendre function is then a polynomial $P_n(x)$ of degree n:

$$P_0(x) = 1\,, \quad P_1(x) = x\,, \quad P_2(x) = \frac{3x^2-1}{2}\,, \quad \ldots\,.$$

Further polynomials then follow by the recurrence relation (C.3). They are orthogonal in the interval $[-1, +1]$ and satisfy the symmetry

$$P_n(-x) = (-1)^n P_n(-x)\,.$$

Legendre polynomials occur in the expansion of the generating function

$$G(w) = \frac{1}{(1-2wx+w^2)^{1/2}} = \sum_{m=0}^{\infty} P_m(x)w^m\,. \tag{C.6}$$

Changing w to w^{-1} and using (C.2), $P_n(x) = P_{-n-1}(x)$, it follows that

$$G(w^{-1}) = wG(w) = \sum_{n=0}^{\infty} P_n(x)w^{-n} = \sum_{n=0}^{\infty} P_{-n-1}(x)w^{-n}\,.$$

Then with $m = -n-1$,

$$G(w) = \sum_{m=-1}^{-\infty} P_m(x)w^m\,,$$

and hence

$$\sum_{m=-\infty}^{\infty} P_m(x)w^m = 2G(w) = \frac{2w^{-1/2}}{(w^{-1}-2x+w)^{1/2}}\,.$$

The change of variable $x = \cos\Delta$ and $w = \mathrm{e}^{-\mathrm{i}\theta}$ yields

$$\sum_{m=-\infty}^{\infty} P_m(\cos\Delta)\mathrm{e}^{-im\theta} = \frac{\sqrt{2}\mathrm{e}^{\mathrm{i}\theta/2}}{\sqrt{\cos\Delta-\cos\theta}}\,, \quad -\Delta<\theta<\Delta<\pi\,. \tag{C.7}$$

Multiplying both sides by $\mathrm{e}^{\mathrm{i}\nu\theta}$ and applying the Mehler–Dirichlet formula (C.4),

$$\int_{-\Delta}^{\Delta} \sum_{m=-\infty}^{\infty} P_m(\cos\Delta)\mathrm{e}^{-\mathrm{i}(m-\nu)\theta}\mathrm{d}\theta = 2\pi P_\nu(\cos\Delta)\,, \quad 0 < \Delta < \pi\,. \tag{C.8}$$

Appendix D. Scattering Matrix and Mixed Matrix

This appendix has two parts. In the first, relations between scattering matrix and transmission matrix are established for a quadripole, and conditions under which modes propagate through a passive array of N cascaded elementary cells are investigated. This array represents a transducer with shorted electrodes.

In the second part, the aim is to describe a transducer globally by means of a mixed matrix (adjoining electromechanical coupling and electrical admittance terms to the scattering matrix in order to model the electrical port). Reciprocity and energy conservation are applied by introducing a generalised admittance matrix. The number of independent (complex) parameters in the mixed matrix is reduced to five. General conclusions about directionality of acoustoelectric devices can then be drawn. The interaction between two transducer cells, needed for the study of harmonic admittance (Sect. 2.6.2), is calculated by transforming to the transmission eigenmode basis. We shall therefore examine the effect of basis change on the mixed matrix.

D.1 Quadripole. Scattering Matrix

In the low frequency range, it is natural to express the linear relation between input and output voltages and currents of an electrical quadripole by an impedance or admittance matrix, $\boldsymbol{Z}$ or $\boldsymbol{Y}$, respectively. At higher frequencies, when circuit dimensions are comparable with the wavelength, it is better to deal with transmission line sections and formulate the input-output transformation in terms of waves. At each port 1 and 2 of the quadripole, voltage and current intensity are then replaced by incident and reflected waves.

Let U_n^+ (U_n^-) be the complex amplitude of the voltage accompanying the incident (reflected) wave at port n (see Fig. D.1). It is convenient to normalise amplitudes a_n, b_n of incident and reflected waves so that

$$a_n = \frac{U_n^+}{\sqrt{Z_{cn}}}, \quad b_n = \frac{U_n^-}{\sqrt{Z_{cn}}}, \tag{D.1}$$

where Z_{cn} is the characteristic impedance of the line section located at port n.

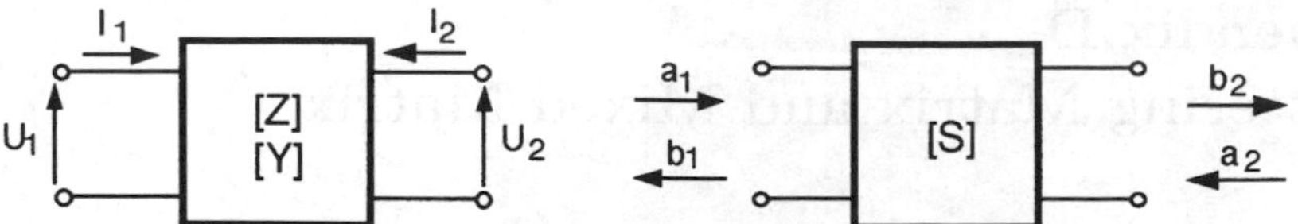

Fig. D.1. Representing an electrical circuit by an impedance (admittance) or scattering matrix

In the reference plane of port n, voltage and current intensity are related to wave amplitudes by

$$U_n = U_n^+ + U_n^- = \sqrt{Z_{cn}}(a_n + b_n) , \tag{D.2a}$$

$$I_n = \frac{1}{Z_{cn}}(U_n^+ - U_n^-) = \frac{1}{\sqrt{Z_{cn}}}(a_n - b_n) . \tag{D.2b}$$

The mean incoming power at port n,

$$P_n = \frac{1}{2}\mathrm{Re}\,(U_n I_n^*) = \frac{1}{2}\mathrm{Re}\,[(a_n a_n^* - b_n b_n^*) + (b_n a_n^* - b_n^* a_n)] ,$$

is equal to the difference between power carried by the incident wave and power carried by the reflected wave:

$$P_n = \frac{1}{2}(a_n a_n^* - b_n b_n^*) . \tag{D.3}$$

The *scattering matrix* $\boldsymbol{S}$ relates amplitudes of reflected and incident waves:

$$\boldsymbol{b} = \boldsymbol{S}\boldsymbol{a} , \quad \begin{pmatrix} b_1 \\ b_2 \end{pmatrix} = \begin{pmatrix} s_{11} & s_{12} \\ s_{21} & s_{22} \end{pmatrix} \begin{pmatrix} a_1 \\ a_2 \end{pmatrix} . \tag{D.4}$$

If $a_2 = 0$, i.e., when port 2 is matched,

$$b_1 = s_{11} a_1 , \quad b_2 = s_{21} a_1 .$$

s_{11} and s_{21} are reflection and transmission coefficients, respectively, (in amplitude and in phase) when this matching condition holds.

In the case where characteristic impedances Z_{c1} and Z_{c2} are equal, relations (D.2a, b) are given in matrix form by

$$\boldsymbol{I}_c = \frac{1}{\sqrt{Z_c}}(\boldsymbol{a} - \boldsymbol{b}) = \frac{1}{\sqrt{Z_c}}(\boldsymbol{I} - \boldsymbol{S})\boldsymbol{a} , \quad \boldsymbol{U} = \sqrt{Z_c}(\boldsymbol{I} + \boldsymbol{S})\boldsymbol{a} ,$$

where $\boldsymbol{I}$ is the identity matrix. The admittance matrix $\boldsymbol{Y}$, defined by $\boldsymbol{I}_c = \boldsymbol{Y}\boldsymbol{U}$, is given in terms of the scattering matrix by

$$\boldsymbol{Y} = \frac{1}{Z_c}(\boldsymbol{I} - \boldsymbol{S})(\boldsymbol{I} + \boldsymbol{S})^{-1} . \tag{D.5}$$

For a circuit satisfying the reciprocity principle, the matrix $\boldsymbol{Y}$ is symmetric and so is the matrix $\boldsymbol{S}$, i.e., $s_{21} = s_{12}$.

In a lossless circuit,

$$\sum_{n=1}^{2} b_n b_n^* = \sum_{n=1}^{2} a_n a_n^* \,, \quad \boldsymbol{b}^{\mathrm{t}}\boldsymbol{b}^* = \boldsymbol{a}^{\mathrm{t}}\boldsymbol{a}^* \,,$$

where $\boldsymbol{b}^{\mathrm{t}}$ is the transpose of $\boldsymbol{b}$. The scattering matrix is unitary:

$$\boldsymbol{S}^{\mathrm{t}}\boldsymbol{S}^* = \boldsymbol{I} \quad \Rightarrow \quad \boldsymbol{S}^{-1} = \boldsymbol{S}^{*\mathrm{t}} \,. \tag{D.6}$$

Its components satisfy

$$\sum_{n=1}^{2} s_{ni} s_{nj}^* = \delta_{ij} \quad \text{for} \quad i, j = 1, 2 \,. \tag{D.7}$$

In detail,

$$|s_{11}|^2 + |s_{12}|^2 = 1 \,, \tag{D.8a}$$

$$|s_{12}|^2 + |s_{22}|^2 = 1 \,, \tag{D.8b}$$

$$s_{11}s_{12}^* + s_{12}s_{22}^* = 0 \,. \tag{D.8c}$$

Introducing moduli and arguments of the matrix elements $s_{jk} = \sigma_{jk}\mathrm{e}^{\mathrm{i}\theta_{jk}}$, the first two equations imply

$$\sigma_{11}^2 = \sigma_{22}^2 = 1 - \sigma_{12}^2 \,. \tag{D.9}$$

The third equation implies

$$\mathrm{e}^{\mathrm{i}(\theta_{11}-\theta_{12})} + \mathrm{e}^{\mathrm{i}(\theta_{12}-\theta_{22})} = 0 \quad \Rightarrow \quad \mathrm{e}^{\mathrm{i}(\theta_{11}+\theta_{22}-2\theta_{12})} = -1 \,,$$

and hence,

$$\frac{\theta_{11}+\theta_{22}}{2} = \theta_{12} \pm \frac{\pi}{2} \,.$$

At a given frequency, the zero phase can always be chosen so that the quadripole is symmetric:

$$s_{22} = s_{11} \quad \Rightarrow \quad \theta_{22} = \theta_{11} = \theta_{12} \pm \frac{\pi}{2} \,. \tag{D.10}$$

Reflection and transmission coefficients $r = s_{11}$, $t = s_{12}$ are in phase quadrature. Putting $\theta_{12} = -\varphi$ and using (D.8a), we find

$$s_{21} = s_{12} = t = \cos\Delta\,\mathrm{e}^{-\mathrm{i}\varphi} \,,$$

$$s_{22} = s_{11} = r = -\mathrm{i}\sin\Delta\,\mathrm{e}^{-\mathrm{i}\varphi} \,, \quad \Delta \text{ real} \,. \tag{D.11}$$

The scattering matrix then depends on only one parameter:

$$\boldsymbol{S} = \begin{pmatrix} r & t \\ t & r \end{pmatrix} \,, \quad \text{since} \quad t^2 - r^2 = \mathrm{e}^{-2\mathrm{i}\varphi} \,. \tag{D.12}$$

Transmission Matrix. Array of Identical Cells. In the case where several quadripoles are connected in cascade, we introduce the transfer (or transmission) matrix which relates output waves at 2 to input waves at 1:

$$\begin{pmatrix} b_2 \\ a_2 \end{pmatrix} = \begin{pmatrix} t_{11} & t_{12} \\ t_{21} & t_{22} \end{pmatrix} \begin{pmatrix} a_1 \\ b_1 \end{pmatrix} \,. \tag{D.13}$$

Indeed, the vector $(b_2,\ a_2)^{\mathrm{t}}$ at the output of the first quadripole represents complex amplitudes at the input to the second; the transfer matrix $\boldsymbol{T}$ resulting from the combination is the product of those for the two quadripoles, i.e., $\boldsymbol{T} = \boldsymbol{t}_2\boldsymbol{t}_1$.

According to (D.4),

$$a_2 = \frac{b_1 - s_{11}a_1}{s_{12}} , \quad b_2 = s_{12}a_1 + s_{22}\frac{b_1 - s_{11}a_1}{s_{12}} .$$

Therefore, transmission matrix entries are related to those of $\boldsymbol{S}$ by

$$\begin{aligned} t_{11} &= s_{21} - \frac{s_{11}s_{22}}{s_{12}} , \quad & t_{12} &= \frac{s_{22}}{s_{12}} . \\ t_{21} &= -\frac{s_{11}}{s_{12}} , & t_{22} &= \frac{1}{s_{12}} . \end{aligned} \tag{D.14}$$

In the case of a reciprocal quadripole, its determinant is $||\boldsymbol{t}|| = s_{21}/s_{12} = 1$. Given the form (D.11) of matrix $\boldsymbol{S}$ for a symmetric quadripole, transfer matrix entries are:

$$t_{11} = \frac{\mathrm{e}^{-2\mathrm{i}\varphi}}{t} , \quad t_{12} = -t_{21} = \frac{r}{t} , \quad t_{22} = \frac{1}{t} . \tag{D.15}$$

The transfer matrix for an array containing a large number N of identical cells is $\boldsymbol{T} = \boldsymbol{t}^N$, viz.,

$$\begin{pmatrix} b_{n+1} \\ a_{n+1} \end{pmatrix} = \begin{pmatrix} t_{11} & t_{12} \\ t_{21} & t_{22} \end{pmatrix} \begin{pmatrix} a_n \\ b_n \end{pmatrix} \Rightarrow \begin{pmatrix} b_{N+1} \\ a_{N+1} \end{pmatrix} = \begin{pmatrix} T_{11} & T_{12} \\ T_{21} & T_{22} \end{pmatrix} \begin{pmatrix} a_1 \\ b_1 \end{pmatrix} . \tag{D.16}$$

Far from the edges, the solution is a wave and, for a cell of length p, we can write

$$b_{n+1} = \mathrm{e}^{-\mathrm{i}\beta p}a_n , \quad a_{n+1} = \mathrm{e}^{-\mathrm{i}\beta p}b_n . \tag{D.17}$$

The homogeneous linear system obtained by substituting into (D.16),

$$\begin{aligned} 0 &= \left(t_{11} - \mathrm{e}^{-\mathrm{i}\beta p}\right) a_n + t_{12}b_n , \\ 0 &= t_{21}a_n + \left(t_{22} - \mathrm{e}^{-\mathrm{i}\beta p}\right) b_n , \end{aligned}$$

has a non-trivial solution if the determinant of amplitudes a_n, b_n is zero:

$$t_{11}t_{22} - t_{12}t_{21} - (t_{11} + t_{22})\mathrm{e}^{-\mathrm{i}\beta p} + \mathrm{e}^{-2\mathrm{i}\beta p} = 0 .$$

By the reciprocity principle, $||\boldsymbol{t}|| = 1$ and the dispersion relation of the array can be expressed in the form

$$2\cos\beta p = t_{11} + t_{22} = \mathrm{Tr}(\boldsymbol{t}) . \tag{D.18}$$

If the trace of $\boldsymbol{t}$ is less than 2, the wave number β is real (propagating mode). Propagation then introduces a phase shift without attenuation. If the trace is greater than 2, β is complex and the incident wave is attenuated (evanescent mode). It is in fact reflected by the array since we have assumed it to be lossless.

The zero phase was chosen so that the quadripole was symmetric. From the expression for the transmission coefficient, $t = \cos\Delta\, \mathrm{e}^{-\mathrm{i}\varphi}$, and formula (D.15) for the transfer matrix entries, equation (D.18) becomes

$$\cos\beta p = \frac{\cos\varphi}{\cos\Delta} \,. \tag{D.19}$$

Modes propagate outside the *stop band*, defined by

$$\pi - \Delta \le \varphi \le \pi + \Delta \,, \quad \text{where} \quad \varphi = \frac{\omega}{V} p \tag{D.20}$$

is the phase shift of the wave propagating at phase speed V over the length p of the cell. In general, the modulus $\sin\Delta$ of the reflection coefficient is small ($\Delta \ll 1$ rad). Modes are evanescent near the stop angular frequency $\omega_0 = \pi V/p$, where φ is close to π, i.e., $\varphi = \pi \pm \varepsilon$, with $\varepsilon < \Delta$. The dispersion relation

$$\cos\beta p \approx -\frac{1 - \varepsilon^2/2}{1 - \Delta^2/2} \approx -1 + \frac{1}{2}(\varepsilon^2 - \Delta^2) \Rightarrow \beta p \approx \pm\pi \pm \mathrm{i}\sqrt{\Delta^2 - \varepsilon^2} \tag{D.21}$$

is exactly the same as the one obtained by coupled mode theory in (B.35), using the correspondence $|\Delta| \approx |r| = \kappa p$ and $\varepsilon = p\delta = (\omega - \omega_0)p/V$.

The matrix $\boldsymbol{T}$ for the array (see [2.25], p. 541) is given in terms of the matrix $\boldsymbol{t}$ for one element by

$$\boldsymbol{T} = \boldsymbol{t}\frac{\sin N\beta p}{\sin\beta p} - \boldsymbol{I}\frac{\sin(N-1)\beta p}{\sin\beta p} \,. \tag{D.22}$$

This is useful for calculating the reflection coefficient of the array.

D.2 Hexapole. Mixed Matrix. Directionality

A surface elastic wave transducer is an acoustoelectric device with two acoustic ports and one electrical port. It is represented, as in Sect. 2.4.2 by a 3×3 *mixed matrix* relating wave amplitudes at acoustic ports to electrical quantities (voltage U and current intensity I). Introducing vectors $\boldsymbol{a}$ and $\boldsymbol{b}$ to represent incoming and outgoing waves, respectively, the mixed matrix is, in condensed form,

$$\begin{pmatrix} \boldsymbol{b} \\ I \end{pmatrix} = \begin{pmatrix} \boldsymbol{S} & \boldsymbol{\alpha} \\ \boldsymbol{\beta} & Y \end{pmatrix} \begin{pmatrix} \boldsymbol{a} \\ U \end{pmatrix} \,, \quad \boldsymbol{a} = \begin{pmatrix} a_1 \\ a_2 \end{pmatrix} \,, \quad \boldsymbol{b} = \begin{pmatrix} b_1 \\ b_2 \end{pmatrix} \,. \tag{D.23}$$

$\boldsymbol{S}$ is the acoustic scattering matrix, Y the electrical admittance, and vectors $\boldsymbol{\alpha} = (\alpha_1, \alpha_2)^{\mathrm{t}}$, $\boldsymbol{\beta} = (\beta_1, \beta_2)$ characterise electroacoustic and acoustoelectric couplings.

The reciprocity principle and energy conservation imply relations between parameters of the P-matrix. They can be simply expressed through the *admittance matrix* $\boldsymbol{Y}$ relating velocity and generalised force vectors:

$$\begin{pmatrix} \boldsymbol{v} \\ I \end{pmatrix} = \boldsymbol{Y} \begin{pmatrix} \boldsymbol{F} \\ U \end{pmatrix} , \quad \boldsymbol{v} = \begin{pmatrix} v_1 \\ v_2 \end{pmatrix} , \quad \boldsymbol{F} = \begin{pmatrix} F_1 \\ F_2 \end{pmatrix} . \tag{D.24}$$

For symmetry reasons, it is best to choose rms values to represent forces and velocities, as well as voltage U and current intensity I, rather than amplitudes. Relations (2.106) of Sect. 2.4.2 become

$$\boldsymbol{F} = \sqrt{\frac{Z_{\mathrm{a}}}{2}}(\boldsymbol{a}+\boldsymbol{b}) , \quad \boldsymbol{v} = \frac{1}{\sqrt{2Z_{\mathrm{a}}}}(\boldsymbol{a}-\boldsymbol{b}) . \tag{D.25}$$

Replacing $\boldsymbol{b}$ by $\boldsymbol{S}\boldsymbol{a}+\boldsymbol{\alpha}U$,

$$\boldsymbol{F} = \sqrt{\frac{Z_{\mathrm{a}}}{2}}\left[(\boldsymbol{I}+\boldsymbol{S})\boldsymbol{a}+\boldsymbol{\alpha}U\right] \Rightarrow \boldsymbol{a} = \sqrt{\frac{2}{Z_{\mathrm{a}}}}(\boldsymbol{I}+\boldsymbol{S})^{-1}\boldsymbol{F} - (\boldsymbol{I}+\boldsymbol{S})^{-1}\boldsymbol{\alpha}U , \tag{D.26}$$

$$\boldsymbol{v} = \frac{1}{\sqrt{2Z_{\mathrm{a}}}}\left[(\boldsymbol{I}-\boldsymbol{S})\boldsymbol{a}-\boldsymbol{\alpha}U\right] , \tag{D.27}$$

and substituting the first of these into the second, and into the relation $I = \boldsymbol{\beta}\boldsymbol{a}+YU$, we find

$$\boldsymbol{v} = \frac{1}{Z_{\mathrm{a}}}(\boldsymbol{I}-\boldsymbol{S})(\boldsymbol{I}+\boldsymbol{S})^{-1}\boldsymbol{F} - \frac{1}{\sqrt{2Z_{\mathrm{a}}}}\left[(\boldsymbol{I}-\boldsymbol{S})(\boldsymbol{I}+\boldsymbol{S})^{-1}+\boldsymbol{I}\right]\boldsymbol{\alpha}U ,$$

and for the current intensity,

$$I = \sqrt{\frac{2}{Z_{\mathrm{a}}}}\boldsymbol{\beta}(\boldsymbol{I}+\boldsymbol{S})^{-1}\boldsymbol{F} + \left[Y-\boldsymbol{\beta}(\boldsymbol{I}+\boldsymbol{S})^{-1}\boldsymbol{\alpha}\right]U .$$

Using the identity $\left[(\boldsymbol{I}-\boldsymbol{S})(\boldsymbol{I}+\boldsymbol{S})^{-1}+\boldsymbol{I}\right](\boldsymbol{I}+\boldsymbol{S}) = 2\boldsymbol{I}$, the admittance matrix is

$$\begin{pmatrix} \boldsymbol{v} \\ I \end{pmatrix} = \begin{pmatrix} \frac{1}{Z_{\mathrm{a}}}(\boldsymbol{I}-\boldsymbol{S})(\boldsymbol{I}+\boldsymbol{S})^{-1} & -\sqrt{\frac{2}{Z_{\mathrm{a}}}}(\boldsymbol{I}+\boldsymbol{S})^{-1}\boldsymbol{\alpha} \\ \sqrt{\frac{2}{Z_{\mathrm{a}}}}\boldsymbol{\beta}(\boldsymbol{I}+\boldsymbol{S})^{-1} & Y-\boldsymbol{\beta}(\boldsymbol{I}+\boldsymbol{S})^{-1}\boldsymbol{\alpha} \end{pmatrix} \begin{pmatrix} \boldsymbol{F} \\ U \end{pmatrix} . \tag{D.28}$$

The reciprocity principle implies that it is symmetric and therefore,

$$\boldsymbol{S} = \boldsymbol{S}^{\mathrm{t}} , \quad \boldsymbol{\beta} = -\boldsymbol{\alpha}^{\mathrm{t}} , \quad \text{hence} \quad \beta_1 = -\alpha_1 , \quad \beta_2 = -\alpha_2 . \tag{D.29}$$

Entries of matrix $\boldsymbol{P}$ defined in Sect. 2.4.2 satisfy relations:

$$P_{21} = P_{12} , \; P_{31} = \beta_1\sqrt{2} , \; \alpha_1 = P_{13}\sqrt{2} \;\Rightarrow\; P_{31} = -2P_{13} , \; P_{32} = -2P_{23} .$$

The admittance matrix for a lossless device is pure imaginary. For the diagonal term,

$$(\boldsymbol{I}-\boldsymbol{S})(\boldsymbol{I}+\boldsymbol{S})^{-1} = 2\boldsymbol{A}-\boldsymbol{I} , \quad \text{where} \quad \boldsymbol{A} = (\boldsymbol{I}+\boldsymbol{S})^{-1} ,$$

this implies

$$2\boldsymbol{A}-\boldsymbol{I}+2\boldsymbol{A}^*-\boldsymbol{I} = 0 \Rightarrow \boldsymbol{A}+\boldsymbol{A}^* = \boldsymbol{I} .$$

Multiplying on the left by $\boldsymbol{A}^{-1} = \boldsymbol{I} + \boldsymbol{S}$, it follows that $(\boldsymbol{I} + \boldsymbol{S})\boldsymbol{A}^* = \boldsymbol{S}$. Then, multiplying on the right by $(\boldsymbol{A}^*)^{-1} = \boldsymbol{I} + \boldsymbol{S}^*$,

$$\boldsymbol{I} + \boldsymbol{S} = \boldsymbol{S} + \boldsymbol{S}\boldsymbol{S}^* \quad \Rightarrow \quad \boxed{\boldsymbol{S}\boldsymbol{S}^* = \boldsymbol{I}} \,. \tag{D.30}$$

The term $(\boldsymbol{I} + \boldsymbol{S})^{-1}\boldsymbol{\alpha} = \boldsymbol{A}\boldsymbol{\alpha}$ is imaginary if

$$\boldsymbol{A}\boldsymbol{\alpha} + \boldsymbol{A}^*\boldsymbol{\alpha}^* = 0 \quad \Rightarrow \quad \boldsymbol{A}\boldsymbol{\alpha} + (\boldsymbol{I} - \boldsymbol{A})\boldsymbol{\alpha}^* = 0\,.$$

Multiplying on the left by $\boldsymbol{A}^{-1} = \boldsymbol{I} + \boldsymbol{S}$,

$$\boxed{\boldsymbol{S}\boldsymbol{\alpha}^* + \boldsymbol{\alpha} = 0} \,. \tag{D.31}$$

Likewise,

$$\boxed{\boldsymbol{\beta}^*\boldsymbol{S} + \boldsymbol{\beta} = 0} \,. \tag{D.32}$$

Putting $Y = G + \mathrm{i}B$, the last diagonal entry in the admittance matrix is imaginary if

$$2G = \boldsymbol{\beta}\boldsymbol{A}\boldsymbol{\alpha} + \boldsymbol{\beta}^*\boldsymbol{A}^*\boldsymbol{\alpha}^* = (\boldsymbol{\beta}^* - \boldsymbol{\beta})\,\boldsymbol{A}^*\boldsymbol{\alpha}^*\,.$$

Since $\boldsymbol{\beta}^* - \boldsymbol{\beta} = -\boldsymbol{\beta}(\boldsymbol{S}^* + \boldsymbol{I}) = -\boldsymbol{\beta}(\boldsymbol{A}^*)^{-1}$, the electrical conductance G is given by

$$2G = -\boldsymbol{\beta}\boldsymbol{\alpha}^* = \boldsymbol{\alpha}^{\mathrm{t}}\boldsymbol{\alpha}^* \quad \Rightarrow \quad \boxed{2G = |\alpha_1|^2 + |\alpha_2|^2} \,. \tag{D.33}$$

Making a suitable choice for the zero phase, the scattering matrix can be put into the form (D.12) and the P-matrix involves 5 independent parameters:

$$\begin{pmatrix} b_1 \\ b_2 \\ I \end{pmatrix} = \begin{pmatrix} r & t & \alpha_1 \\ t & r & \alpha_2 \\ \beta_1 & \beta_2 & Y \end{pmatrix} \begin{pmatrix} a_1 \\ a_2 \\ U \end{pmatrix}\,, \tag{D.34}$$

with $\beta_1 = -\alpha_1$, $\beta_2 = -\alpha_2$ and $Y = G + \mathrm{i}B$. r and t are reflection and transmission coefficients in the shorted device. Since $t + r = \mathrm{e}^{-\mathrm{i}(\varphi+\Delta)}$ and $t - r = \mathrm{e}^{-\mathrm{i}(\varphi-\Delta)}$, condition (D.31) implies

$$\left.\begin{array}{l} r\alpha_1^* + t\alpha_2^* + \alpha_1 = 0 \\ t\alpha_1^* + r\alpha_2^* + \alpha_2 = 0 \end{array}\right\} \Rightarrow \begin{array}{l} (\alpha_1^* + \alpha_2^*)\mathrm{e}^{-\mathrm{i}(\varphi+\Delta)} + (\alpha_1 + \alpha_2) = 0 \\ (\alpha_1^* - \alpha_2^*)\mathrm{e}^{-\mathrm{i}(\varphi-\Delta)} - (\alpha_1 - \alpha_2) = 0 \end{array}.$$

Introducing generation parameters α_{S} and α_{A} for modes which are symmetric and antisymmetric relative to acoustic ports, it follows that

$$\begin{aligned} \alpha_{\mathrm{S}} &= \frac{\alpha_1 + \alpha_2}{2} = \mathrm{i}\sqrt{G_{\mathrm{S}}}\exp -\mathrm{i}\frac{\varphi + \Delta}{2}\,, \\ \alpha_{\mathrm{A}} &= \frac{\alpha_1 - \alpha_2}{2} = \sqrt{G_{\mathrm{A}}}\exp -\mathrm{i}\frac{\varphi - \Delta}{2}\,. \end{aligned} \tag{D.35}$$

These modes are orthogonal and hence the conductance, deduced from (D.33), is the sum of contributions from each separately:

$$G = G_{\mathrm{S}} + G_{\mathrm{A}}\,, \quad G_{\mathrm{S}} = G\cos^2\delta\,, \quad G_{\mathrm{A}} = G\sin^2\delta\,, \tag{D.36}$$

where δ is an angle characterising the symmetry and directionality of the transducer.

Directionality. A transducer is acoustoelectrically *symmetric* if angle δ is zero:

$$\alpha_1 = \alpha_2 = \alpha = \mathrm{i}\sqrt{G}\exp -\mathrm{i}\frac{\varphi + \Delta}{2} \quad \Rightarrow \quad G = -\frac{\alpha^2}{r+t} = \frac{\alpha\beta}{r+t} \,. \tag{D.37}$$

If $\delta = \pm\pi/2$, the transducer is *antisymmetric* and

$$\alpha_1 = -\alpha_2 = \alpha = \mathrm{i}\sqrt{G}\exp -\mathrm{i}\frac{\varphi - \Delta}{2} \,. \tag{D.38}$$

In both cases, the transducer is bidirectional. Complex amplitudes of waves emitted at each acoustic port are equal and opposite. The general expressions

$$\alpha_{2,1} = \mathrm{i}\sqrt{G}\mathrm{e}^{-\mathrm{i}\varphi/2}\left[\mathrm{e}^{-\mathrm{i}\Delta/2}\cos\delta \pm \mathrm{i}\mathrm{e}^{\mathrm{i}\Delta/2}\sin\delta\right] \,, \tag{D.39}$$

show that the *directionality factor* defined by (2.173),

$$d = \frac{|\alpha_1|^2 - |\alpha_2|^2}{|\alpha_1|^2 + |\alpha_2|^2} = \sin 2\delta \, \sin\Delta \Rightarrow d_{\mathrm{max}} = \sin\Delta = |r| \,, \tag{D.40}$$

has an extremum when $\delta = \pm\pi/4$. Its absolute value is at most equal to the short-circuit reflection coefficient. Perfect directionality $d = \pm 1$ is only possible if $\Delta = \pm\pi/2$. Generation parameters take the form

$$\begin{aligned} \alpha_1 &= \mathrm{i}\sqrt{G(1+d)}\mathrm{e}^{-\mathrm{i}\varphi_1} \,, \qquad & \varphi_1 &= \frac{\varphi}{2} + \phi + \psi \,, \\ \alpha_2 &= \mathrm{i}\sqrt{G(1-d)}\mathrm{e}^{-\mathrm{i}\varphi_2} \,, \qquad & \varphi_2 &= \frac{\varphi}{2} + \phi - \psi \,, \end{aligned} \tag{D.41}$$

which brings out their common phase ϕ and their phase difference 2ψ:

$$\tan 2\phi = \cos 2\delta \tan\Delta \,, \quad \tan 2\psi = \tan 2\delta \cos\Delta \,. \tag{D.42}$$

For a *unidirectional transducer* $\delta = \pm\pi/4$, we have $\phi = 0$ and $\psi = \pm\pi/4 + n\pi$. Given (D.11) for the reflection coefficient and the phase shift $\varphi = 2(\varphi/2)$ due to the return path of reflected waves, generation and reflection coefficients are $\pm\pi/4$ out of phase, i.e., generation and reflection centres are spatially separated by a distance $\pm\lambda/8 + n\lambda/2$. This result, a consequence of the reciprocity principle and energy conservation, applies to any transducer, whatever its structure [2.39]. It is more general than the identical phase shift condition proven locally for uniform transducers by the coupled mode method (Sect. 2.5.2.2).

Identical Cell Array. Change of Basis. A transducer made by repeating a large number of identical cells is analysed by cascading acoustic ports. Each cell is represented by an elementary P-matrix of the type given in (D.34). When waves emitted either downstream or upstream cross a given cell, effects are best described in terms of the transmission eigenmode basis in which the short-circuit reflection coefficient is zero. We shall therefore investigate the effects of a basis change on the mixed matrix.

The definition of the mixed matrix in Sect. 2.4.2 uses a reference acoustic impedance Z_{a} of arbitrary value. According to (D.25), choosing another impedance $\tilde{Z}_a$ amounts to changing amplitudes $\boldsymbol{a}$, $\boldsymbol{b}$ of incoming and outgoing waves. In this basis change, parameters of the P-matrix are modified. The new amplitude vectors $\tilde{\boldsymbol{a}}$ and $\tilde{\boldsymbol{b}}$ are defined by

$$\boldsymbol{F} = \sqrt{\frac{\tilde{Z}_a}{2}} \left(\tilde{\boldsymbol{a}} + \tilde{\boldsymbol{b}} \right) , \quad \boldsymbol{v} = \frac{1}{\sqrt{2\tilde{Z}_a}} \left(\tilde{\boldsymbol{a}} - \tilde{\boldsymbol{b}} \right) .$$

The basis change relations

$$\boldsymbol{a} + \boldsymbol{b} = \sqrt{\frac{\tilde{Z}_a}{Z_a}} \left(\tilde{\boldsymbol{a}} + \tilde{\boldsymbol{b}} \right) , \quad \boldsymbol{a} - \boldsymbol{b} = \sqrt{\frac{Z_a}{\tilde{Z}_a}} \left(\tilde{\boldsymbol{a}} - \tilde{\boldsymbol{b}} \right) ,$$

can be written

$$\boldsymbol{a} = \frac{\tilde{\boldsymbol{a}} - \xi \tilde{\boldsymbol{b}}}{\tau} , \quad \boldsymbol{b} = \frac{\tilde{\boldsymbol{b}} - \xi \tilde{\boldsymbol{a}}}{\tau} , \text{ where } \xi = \frac{Z_a - \tilde{Z}_a}{Z_a + \tilde{Z}_a} , \ \tau = \sqrt{1 - \xi^2} \tag{D.43}$$

are (amplitude) reflection and transmission coefficients at the interface between two media with acoustic impedances Z_a and $\tilde{Z}_a$. The inverse transformation follows from (D.43) by changing ξ to $-\xi$.

The half-sum and half-difference of the first two lines in the expansion (D.34) of the P-matrix involve symmetric and antisymmetric modes:

$$\begin{aligned} b_S &= (r+t)a_S + \alpha_S U , & b_S &= \frac{b_1 + b_2}{2} , & a_S &= \frac{a_1 + a_2}{2} , \\ b_A &= (r-t)a_A + \alpha_A U , & b_A &= \frac{b_1 - b_2}{2} , & a_A &= \frac{a_1 - a_2}{2} . \end{aligned} \tag{D.44}$$

In the new basis, as functions of reflection and transmission coefficients $\tilde{r}$, $\tilde{t}$, it follows that

$$\begin{aligned} \tilde{b}_S &= \left(\tilde{r} + \tilde{t} \right) \tilde{a}_S + \tilde{\alpha}_S U , \\ \tilde{b}_A &= \left(\tilde{r} - \tilde{t} \right) \tilde{a}_A + \tilde{\alpha}_A U . \end{aligned} \tag{D.45}$$

The change of basis applied to (D.44) leads to

$$\begin{aligned} \tilde{b}_S - \xi \tilde{a}_S &= (r+t) \left(\tilde{a}_S - \xi \tilde{b}_S \right) + \alpha_S \tau U , \\ \tilde{b}_A - \xi \tilde{a}_A &= (r-t) \left(\tilde{a}_A - \xi \tilde{b}_A \right) + \alpha_A \tau U . \end{aligned}$$

Identifying with (D.45), matrix entries of $\tilde{\boldsymbol{P}}$ are given in terms of matrix entries of $\boldsymbol{P}$ by

$$\tilde{r} + \tilde{t} = \frac{r + t + \xi}{1 + (r+t)\xi} , \quad \tilde{r} - \tilde{t} = \frac{r - t + \xi}{1 + (r-t)\xi} , \tag{D.46}$$

$$\tilde{\alpha}_S = \frac{\alpha_S \sqrt{1 - \xi^2}}{1 + (r+t)\xi} , \quad \tilde{\alpha}_A = \frac{\alpha_A \sqrt{1 - \xi^2}}{1 + (r-t)\xi} . \tag{D.47}$$

These expressions are inverted by changing ξ to $-\xi$. Since the transformation (D.43) is unitary, relations (D.29) still hold, viz., $\tilde{\beta}_1 = -\tilde{\alpha}_1$ and $\tilde{\beta}_2 = -\tilde{\alpha}_2$. Formulas for the conductance follow from (D.35):

$$G_{\mathrm{S}} = \frac{-\alpha_{\mathrm{S}}^2}{r+t} , \quad G_{\mathrm{A}} = \frac{-\alpha_{\mathrm{A}}^2}{r-t} . \tag{D.48}$$

Hence,

$$\tilde{G}_{\mathrm{S}} = \frac{-\tilde{\alpha}_{\mathrm{S}}^2}{\tilde{r}+\tilde{t}} = \frac{-\alpha_{\mathrm{S}}^2(1-\xi^2)}{[1+(r+t)\xi]^2} = G_{\mathrm{S}} \frac{(r+t)(1-\xi^2)}{(r+t+\xi)[1+(r+t)\xi]} ,$$

or

$$\begin{aligned} \tilde{G}_{\mathrm{S}} &= G_{\mathrm{S}} \frac{\xi^{-1}-\xi}{(r+t)^{-1}+(r+t)+\xi^{-1}+\xi} , \\ \tilde{G}_{\mathrm{A}} &= G_{\mathrm{A}} \frac{\xi^{-1}-\xi}{(r-t)^{-1}+(r-t)+\xi^{-1}+\xi} . \end{aligned} \tag{D.49}$$

A particularly useful change of basis is the one which reduces the short-circuit reflection coefficient $\tilde{r}$ to zero. By (D.46), the parameter ξ for the transformation to transmission eigenmodes satisfies

$$\frac{r+t+\xi}{1+(r+t)\xi} = -\frac{r-t+\xi}{1+(r-t)\xi} \quad \Rightarrow \quad r+\xi+r\xi^2 = (t^2-r^2)\xi .$$

Dividing by $r\xi$,

$$\xi^{-1}+\xi = \frac{t^2-r^2-1}{r} = \frac{\mathrm{e}^{-2\mathrm{i}\varphi}-1}{-\mathrm{i}\sin\Delta\,\mathrm{e}^{-\mathrm{i}\varphi}} = 2\frac{\sin\varphi}{\sin\Delta} . \tag{D.50}$$

If ξ is one solution, the other is ξ^{-1}:

$$\xi = \frac{\sin\varphi \pm (\cos^2\Delta - \cos^2\varphi)^{1/2}}{\sin\Delta} = \frac{\sin\varphi \pm \cos\Delta\sin\varphi_{\mathrm{sc}}}{\sin\Delta} ,$$

putting

$$\cos\varphi_{\mathrm{sc}} = \frac{\cos\varphi}{\cos\Delta} . \tag{D.51}$$

Since $(r+t)^{-1}+(r+t) = 2\cos(\varphi+\Delta)$, conductances (D.48) become

$$\tilde{G}_{\mathrm{S}} = G_{\mathrm{S}} \frac{\cos\Delta\sin\varphi_{\mathrm{sc}}}{\sin\Delta\cos(\varphi+\Delta)+\sin\varphi} = G_{\mathrm{S}} \frac{\sin\varphi_{\mathrm{sc}}}{\sin(\varphi+\Delta)} \tag{D.52}$$

and

$$\tilde{G}_{\mathrm{A}} = G_{\mathrm{A}} \frac{\sin\varphi_{\mathrm{sc}}}{\sin(\varphi-\Delta)} . \tag{D.53}$$

These show that in an infinite array the admittance of the symmetric eigenmode becomes infinite at the beginning of the stop band $\varphi = \pi - \Delta$. The admittance of the antisymmetric eigenmode is infinite at the end of the stop band $\varphi = \pi + \Delta$. The result is reversed if $\Delta < 0$.

Results established in this section are used to calculate the contribution of surface acoustic waves to the harmonic admittance of an electrode array (Sect. 2.6.2).

Appendix E. Matrix Representation of a Dioptric System

Phenomena arising from emission, detection and diffraction of elastic waves at the surface $x_2 = 0$ of a solid are formulated by means of a matrix $\boldsymbol{M}$. This matrix relates the amplitude vector $\boldsymbol{b}$ of reflected (emitted) bulk waves and mechanical displacements $\boldsymbol{u}$ at the surface, on the one hand, to the amplitude vector $\boldsymbol{a}$ of incident bulk waves and mechanical stress $\boldsymbol{t}$, on the other [E.1]. This matrix $\boldsymbol{M}$ is made up of 4 submatrices $\boldsymbol{D}$, $\boldsymbol{E}$, $\boldsymbol{R}$ and $\boldsymbol{G}$:

$$\begin{pmatrix} \boldsymbol{b} \\ \boldsymbol{u} \end{pmatrix} = \begin{pmatrix} \boldsymbol{D} & \boldsymbol{E} \\ \boldsymbol{R} & \boldsymbol{G} \end{pmatrix} = \begin{pmatrix} \boldsymbol{a} \\ \boldsymbol{t} \end{pmatrix} . \tag{E.1}$$

If the solid is not piezoelectric, vectors $\boldsymbol{a}$ and $\boldsymbol{b}$ generally have three components (one for each bulk wave). Vectors $\boldsymbol{u}$ and $\boldsymbol{t}$ also have three components:

$$\boldsymbol{u} = \begin{pmatrix} u_1 \\ u_2 \\ u_3 \end{pmatrix} , \qquad \boldsymbol{t} = \begin{pmatrix} -T_{12}/\mathrm{i}\omega \\ -T_{22}/\mathrm{i}\omega \\ -T_{32}/\mathrm{i}\omega \end{pmatrix} . \tag{E.2}$$

The submatrices are then square (3×3). $\boldsymbol{G}$ is a mechanical admittance matrix, whose entries are ratios of a pressure $-T_{i2}$ to the particle velocity given by $\boldsymbol{v} = \mathrm{i}\omega \boldsymbol{u}$.

If the solid is piezoelectric, electrical phenomena are taken into account by adjoining surface electric potential $u_4 = \Phi(0)$ to surface displacement, and normal component of electric displacement $t_4 = D_2(0_+)$ to mechanical stress. The rank of each matrix $\boldsymbol{E}$, $\boldsymbol{R}$ and $\boldsymbol{G}$ is thereby increased by one. Each matrix has a physical meaning.

- The diffraction matrix $\boldsymbol{D}$ gives reflected wave amplitudes in terms of incident wave amplitudes when the surface is free $\boldsymbol{t} = 0$ and in open circuit $D_2(0) = 0$.
- The emission matrix $\boldsymbol{E}$ gives, when there are no incident waves $\boldsymbol{a} = 0$, amplitudes of bulk waves produced by a line force or charge distribution on the surface.
- The reception matrix $\boldsymbol{R}$ is useful for calculating mechanical displacement and electric potential induced by bulk waves on a free surface $\boldsymbol{t} = 0$.
- The admittance matrix $\boldsymbol{G}$ gives, in the harmonic regime, displacements and electric potential produced by a line distribution of mechanical forces or electric charges. Components of its Fourier transform are Green functions

for the half-space of mechanical, electrical and electromechanical surface phenomena.

Consider an incident plane wave of amplitude a and polarisation given by unit vector p_l, with wave vector $k_i = s_i\omega$ contained in the (x_1, x_2)-plane:

$$u_l = p_l a \exp[-\mathrm{i}\omega(s_k x_k - t)] . \tag{E.3}$$

In a non-piezoelectric medium, stresses have the form

$$T_{ij} = c_{ijkl}\frac{\partial u_l}{\partial x_k} = -\mathrm{i}\omega c_{ijkl} s_k p_l a . \tag{E.4}$$

Hence, for normal stresses at the interface $x_2 = 0$, $T_{i2} = -\mathrm{i}\omega t_i$. Boundary conditions are

$$t_i = c_{i2kl} s_k p_l a = z_i a , \quad i = 1, 2, 3 . \tag{E.5}$$

z_i is the acoustic impedance vector for each of the three bulk waves. The mechanical tension is the sum of those produced by each incident wave (indexed by α),

$$t_i = \sum_{\alpha=1}^{3} z_i^\alpha a_\alpha \Rightarrow \boldsymbol{t} = \boldsymbol{A}\boldsymbol{a} . \tag{E.6}$$

$\boldsymbol{A} = (z_i^\alpha)$ is the 3×3 matrix obtained by juxtaposing the three columns z_i^α. Similar considerations for reflected waves, of amplitude $\boldsymbol{b}$, lead to a matrix $\boldsymbol{B}$. The total mechanical tension is given by the sum:

$$\boldsymbol{t} = \boldsymbol{A}\boldsymbol{a} + \boldsymbol{B}\boldsymbol{b} \Rightarrow \boldsymbol{b} = \boldsymbol{B}^{-1}\boldsymbol{t} - \boldsymbol{B}^{-1}\boldsymbol{A}\boldsymbol{a} . \tag{E.7}$$

By the Snell–Descartes law (Sect. 4.2, Vol. I), $s_3 = 0$ and the component s_1 is identical for all incident and reflected waves.

The mechanical displacement produced at the surface by incident waves is

$$u_i = \sum_{\alpha=1}^{3} p_i^\alpha a_\alpha \Rightarrow \boldsymbol{u} = \boldsymbol{P}\boldsymbol{a} , \tag{E.8}$$

where $\boldsymbol{P} = (p_i^\alpha)$ is the polarisation matrix for incident waves. Denoting the polarisation matrix for reflected waves by $\boldsymbol{Q}$, the total surface displacement is

$$\boldsymbol{u} = \boldsymbol{P}\boldsymbol{a} + \boldsymbol{Q}\boldsymbol{b} \Rightarrow \boldsymbol{u} = \left(\boldsymbol{P} - \boldsymbol{Q}\boldsymbol{B}^{-1}\boldsymbol{A}\right)\boldsymbol{a} + \boldsymbol{Q}\boldsymbol{B}^{-1}\boldsymbol{t} . \tag{E.9}$$

Using (E.7) and (E.9), submatrices of $\boldsymbol{M}$ are given by

$$\boldsymbol{D} = -\boldsymbol{B}^{-1}\boldsymbol{A} , \quad \boldsymbol{E} = \boldsymbol{B}^{-1} , \quad \boldsymbol{R} = \boldsymbol{P} - \boldsymbol{Q}\boldsymbol{B}^{-1}\boldsymbol{A} , \quad \boldsymbol{G} = \boldsymbol{Q}\boldsymbol{B}^{-1} . \tag{E.10}$$

Application: Isotropic Solid. Longitudinal and transverse waves polarised in the sagittal plane produce stresses:

$$T_{12} = c_{66}\left(\frac{\partial u_2}{\partial x_1} + \frac{\partial u_1}{\partial x_2}\right) = -\mathrm{i}\omega c_{66}(s_1 p_2 + s_2 p_1)a = -\mathrm{i}\omega t_1 \,,$$

$$T_{22} = c_{21}\frac{\partial u_1}{\partial x_1} + c_{22}\frac{\partial u_2}{\partial x_2} = -\mathrm{i}\omega(c_{21}s_1 p_1 + c_{22}s_2 p_2)a = -\mathrm{i}\omega t_2 \,.$$

For the longitudinal wave, of speed $V_{\mathrm{L}} = \sqrt{c_{11}/\rho}$, we have $p_1 = s_1 V_{\mathrm{L}}$, $p_2 = s_{2\mathrm{L}} V_{\mathrm{L}}$ and $c_{66} = \rho/S_{\mathrm{T}}^2$, implying

$$z_1^{\mathrm{L}} = Z_{\mathrm{L}}\frac{2 s_1 s_{2\mathrm{L}}}{S_{\mathrm{T}}^2} \,, \qquad Z_{\mathrm{L}} = \rho V_{\mathrm{L}} \,, \quad S_{\mathrm{T}} = 1/V_{\mathrm{T}} \,,$$

$$z_2^{\mathrm{L}} = \left[c_{11}\left(s_1^2 + s_{2\mathrm{L}}^2\right) + (c_{12} - c_{11})s_1^2\right] V_{\mathrm{L}} \,. \tag{E.11}$$

Using $c_{11} - c_{12} = 2\rho/S_{\mathrm{T}}^2$ and $s_1^2 + s_{2\mathrm{L}}^2 = S_{\mathrm{L}}^2 = \rho/c_{11}$,

$$z_2^{\mathrm{L}} = \rho V_{\mathrm{L}}\left(1 - 2\frac{s_1^2}{S_{\mathrm{T}}^2}\right) = Z_{\mathrm{L}}\frac{S_{\mathrm{T}}^2 - 2s_1^2}{S_{\mathrm{T}}^2} \,. \tag{E.12}$$

For the transverse wave, $p_1 = -s_2 V_{\mathrm{T}}$, $p_2 = s_1 V_{\mathrm{T}}$, since $s_1 p_1 + s_2 p_2 = 0$:

$$z_1^{\mathrm{T}} = \frac{\rho V_{\mathrm{T}}}{S_{\mathrm{T}}^2}(s_1^2 - s_2^2) = -Z_{\mathrm{T}}\frac{S_{\mathrm{T}}^2 - 2s_1^2}{S_{\mathrm{T}}^2} \,, \quad Z_{\mathrm{T}} = \rho V_{\mathrm{T}} \,, \tag{E.13}$$

$$z_2^{\mathrm{T}} = (c_{22} - c_{12})s_1 s_{2\mathrm{T}} V_{\mathrm{T}} = Z_{\mathrm{T}}\frac{2 s_1 s_{2\mathrm{T}}}{S_{\mathrm{T}}^2} \,. \tag{E.14}$$

The half-space impedance matrix is given by

$$A = \frac{1}{S_{\mathrm{T}}^2}\begin{pmatrix} 2Z_{\mathrm{L}} s_1 s_{2\mathrm{L}} & -Z_{\mathrm{T}}\left(S_{\mathrm{T}}^2 - 2s_1^2\right) \\ Z_{\mathrm{L}}\left(S_{\mathrm{T}}^2 - 2s_1^2\right) & 2Z_{\mathrm{L}} s_1 s_{2\mathrm{T}} \end{pmatrix} . \tag{E.15}$$

$s_{2\mathrm{L}}$ ($s_{2\mathrm{T}}$) is the component of the slowness vector of the longitudinal (transverse) incident wave. Matrix $\boldsymbol{B}$ follows from $\boldsymbol{A}$ on replacing these quantities by their values for reflected waves.

Solid–Vacuum Interface. As the surface is free $\boldsymbol{t} = 0$, reflected wave amplitudes are given by

$$\begin{pmatrix} b_{\mathrm{L}} \\ b_{\mathrm{T}} \end{pmatrix} = \begin{pmatrix} D_{11} & D_{12} \\ D_{21} & D_{22} \end{pmatrix}\begin{pmatrix} a_{\mathrm{L}} \\ a_{\mathrm{T}} \end{pmatrix} , \quad \boldsymbol{D} = -\boldsymbol{B}^{-1}\boldsymbol{A} \,. \tag{E.16}$$

A longitudinal (transverse) incident wave is reflected with coefficient r_{LL} (r_{TT}) and converted into a transverse (longitudinal) wave with coefficient r_{LT} (r_{TL}). These coefficients are equal to diffraction matrix entries

$$r_{\mathrm{LL}} = D_{11} \,, \quad r_{\mathrm{LT}} = D_{21} \,, \quad r_{\mathrm{TL}} = D_{12} \,, \quad r_{\mathrm{TT}} = D_{22} \,. \tag{E.17}$$

Calculation gives back (4.107–110) in Sect. 4.4.2.2, Vol. I.

Surface Modes. From (E.9), when there is no bulk wave, $\boldsymbol{a} = 0 \Rightarrow \boldsymbol{t} = \boldsymbol{G}^{-1}\boldsymbol{u}$, there is displacement on the free surface if the determinant of matrix $\boldsymbol{G}^{-1} = \boldsymbol{B}\boldsymbol{Q}^{-1}$ is zero. The condition for there to be a surface wave is therefore $||\boldsymbol{B}|| = 0$. Hence, by (E.15),

$$4s_1^2 s_{2\mathrm{L}}^{\mathrm{R}} s_{2\mathrm{T}}^{\mathrm{R}} + \left(S_{\mathrm{T}}^2 - 2s_1^2\right)^2 = 0 \,.$$

Replacing components $s_{2\mathrm{L}}^{\mathrm{R}}$ and $s_{2\mathrm{T}}^{\mathrm{R}}$ of reflected wave slownesses by

$$\left(s_{2\mathrm{L}}^{\mathrm{R}}\right)^2 = S_{\mathrm{L}}^2 - s_1^2 \Rightarrow s_{2\mathrm{L}}^{\mathrm{R}} = \mathrm{i}\left(s_1^2 - S_{\mathrm{L}}^2\right)^{1/2} \,,$$

$$\left(s_{2\mathrm{T}}^{\mathrm{R}}\right)^2 = S_{\mathrm{T}}^2 - s_1^2 \Rightarrow s_{2\mathrm{T}}^{\mathrm{R}} = \mathrm{i}\left(s_1^2 - S_{\mathrm{T}}^2\right)^{1/2} \,,$$

we obtain

$$R(s_1) = \left(S_{\mathrm{T}}^2 - 2s_1^2\right)^2 - 4s_1^2\left(s_1^2 - S_{\mathrm{T}}^2\right)^{1/2}\left(s_1^2 - S_{\mathrm{L}}^2\right)^{1/2} = 0 \,. \qquad \text{(E.18)}$$

This is the same as Rayleigh's equation (4.136) with $k_1 = \omega s_1$. Given that the diffraction matrix (E.16) can be written

$$\boldsymbol{D} = -\boldsymbol{B}^{-1}\boldsymbol{A} = -\frac{\mathrm{Adj}(\boldsymbol{B})}{||\boldsymbol{B}||}\boldsymbol{A} = -\frac{\mathrm{Adj}(\boldsymbol{B})}{R(s_1)}\boldsymbol{A} \,, \qquad \text{(E.19)}$$

the solution $s_1 = S_{\mathrm{R}} = 1/V_{\mathrm{R}}$ of Rayleigh's equation (E.18) corresponds to a pole for all bulk wave reflection and conversion coefficients at the half-space boundary.

Piezoelectric Solid. Green Functions. When there are no incident waves and in the harmonic case, mechanical displacement $\overline{\boldsymbol{u}}$ and electric potential $\overline{\Phi}$ at the surface $x_2 = 0$ are given by

$$\begin{pmatrix} \overline{\boldsymbol{u}}(s,\,0) \\ \overline{\Phi}(s,\,0) \end{pmatrix} = \overline{\boldsymbol{G}}(s)\begin{pmatrix} \overline{\boldsymbol{t}}(s,\,0_+) \\ \overline{D}_2(s,\,0_+) \end{pmatrix} \,, \qquad s = s_1 = k/\omega \,. \qquad \text{(E.20)}$$

Components $\overline{G}_{ij}$, i, $j = 1, 2, 3$, describe surface mechanical admittances. The electrical component $\overline{G}_{44}$ is directly related to the piezoelectric permittivity,

$$\overline{G}_{44}(k/\omega) = \left.\frac{\Phi(0)}{D_2(0_+)}\right|_{\boldsymbol{t}=0} = \frac{1}{|k|\overline{\varepsilon}(k/\omega)} \,, \qquad \text{(E.21)}$$

defined in (5.72), Vol. I, for a free surface.

In order to analyse generation by a one-dimensional array of sources distributed along the axis $x = x_1$ in the surface $x_2 = 0$, we return to the spatial domain by an inverse Fourier transform:

$$\begin{pmatrix} \boldsymbol{u}(x) \\ \Phi(x) \end{pmatrix} = \int_{-\infty}^{\infty} \boldsymbol{G}(x - x')\begin{pmatrix} \boldsymbol{t}(x') \\ \sigma(x') \end{pmatrix} \mathrm{d}x' \,, \qquad \text{(E.22)}$$

where $\boldsymbol{G}(x)$ is the Green function matrix

$$\boldsymbol{G}(x) = \frac{1}{2\pi}\int_{-\infty}^{\infty} \overline{\boldsymbol{G}}(k)\mathrm{e}^{-\mathrm{i}kx}\mathrm{d}k \,. \qquad \text{(E.23)}$$

$\sigma(x) = D_2(x,\,0_+)$ and $\boldsymbol{t}(x) = \boldsymbol{t}(x,\,0_+)$ are the respective Fourier transforms of $\overline{D}_2(k,\,0_+)$ and $\overline{\boldsymbol{t}}(k,\,0_+)$, representing distributions of charges and mechanical stresses at the surface.

Appendix F. Linear Systems of Differential Equations

Consider the linear system of first order differential equations

$$\frac{\mathrm{d}\boldsymbol{X}}{\mathrm{d}x} = \boldsymbol{F}(x)\boldsymbol{X}(x) + \boldsymbol{G}(x)U \,, \tag{F.1}$$

where $\boldsymbol{X}$, $\boldsymbol{G}$ are $n \times 1$ vectors, $\boldsymbol{F}$ an $n \times n$ matrix and U a (constant) scalar. Its solution is the sum of the general solution to the homogeneous equation and a particular solution to the full equation.

Solution of the Homogeneous Equation

Let $\boldsymbol{M}(x)$ be an n-column matrix solving

$$\frac{\mathrm{d}\boldsymbol{M}}{\mathrm{d}x} = \boldsymbol{F}(x)\boldsymbol{M}(x) \,, \quad \boldsymbol{M} = \left(\boldsymbol{M}^{(1)}, \boldsymbol{M}^{(2)}, \ldots, \boldsymbol{M}^{(n)}\right) . \tag{F.2}$$

By the rule for matrix multiplication, each column $\boldsymbol{M}^{(i)}$ of $\boldsymbol{M}$ must satisfy (F.2).

(**a**) Assume that columns of $\boldsymbol{M}$ are linearly dependent for some $x = x_0$, i.e., there is a constant vector $\boldsymbol{C}$ with $\boldsymbol{M}(x_0)\boldsymbol{C} = 0$. At any point $x \neq x_0$,

$$\frac{\mathrm{d}}{\mathrm{d}x}[\boldsymbol{M}(x)\boldsymbol{C}] = \frac{\mathrm{d}\boldsymbol{M}}{\mathrm{d}x}\boldsymbol{C} = \boldsymbol{F}[\boldsymbol{M}(x)\boldsymbol{C}] \,,$$

and the vector $\boldsymbol{M}(x)\boldsymbol{C}$ is a solution of the homogeneous equation with $\boldsymbol{M}(x_0)\boldsymbol{C} = 0$. It follows that $\boldsymbol{M}(x)\boldsymbol{C} = 0\,,\ \forall x$ and hence the columns of $\boldsymbol{M}(x)$ are dependent $\forall x$. Conversely, if the columns of $\boldsymbol{M}(x)$ are independent at some point x_0, then they are independent everywhere.

(**b**) Let $\boldsymbol{M}$ be a square ($n \times n$) matrix solution such that $||\boldsymbol{M}|| \neq 0$, and let $\boldsymbol{X}$ be a solution of the homogeneous equation (F.2). Making the change $\boldsymbol{X} = \boldsymbol{M}\tilde{\boldsymbol{X}}$,

$$\frac{\mathrm{d}\boldsymbol{X}}{\mathrm{d}x} = \frac{\mathrm{d}\boldsymbol{M}}{\mathrm{d}x}\tilde{\boldsymbol{X}} + \boldsymbol{M}\frac{\mathrm{d}\tilde{\boldsymbol{X}}}{\mathrm{d}x} = \boldsymbol{F}\boldsymbol{M}\tilde{\boldsymbol{X}} + \boldsymbol{M}\frac{\mathrm{d}\tilde{\boldsymbol{X}}}{\mathrm{d}x} \quad \Rightarrow \quad \boldsymbol{M}\frac{\mathrm{d}\tilde{\boldsymbol{X}}}{\mathrm{d}x} = 0 \,.$$

Since $||\boldsymbol{M}|| \neq 0$,

$$\frac{\mathrm{d}\tilde{\boldsymbol{X}}}{\mathrm{d}x} = 0 \quad \Rightarrow \quad \tilde{\boldsymbol{X}} = \boldsymbol{C} \quad \text{constant vector} \,.$$

The general solution of the homogeneous equation is obtained by taking arbitrary linear combinations of the columns of the solution matrix $\boldsymbol{M}$. These independent columns form a fundamental set of solutions.

(**c**) Let $\boldsymbol{M}(x, x_0)$ be a matrix solution with

$$\boldsymbol{M}(x_0, x_0) = \boldsymbol{I} . \tag{F.3}$$

As $||\boldsymbol{M}(x_0, x_0)|| = 1$, the columns of this matrix are independent at all x. They form a fundamental set of solutions. The linear combination of columns

$$\boldsymbol{X}(x) = \boldsymbol{M}(x, x_0)\boldsymbol{C} , \tag{F.4}$$

given by constant vector $\boldsymbol{C}$, is the solution of the homogeneous system such that

$$\boldsymbol{X}(x_0) = \boldsymbol{M}(x_0, x_0)\boldsymbol{C} = \boldsymbol{C} \Rightarrow \boxed{\boldsymbol{X}(x) = \boldsymbol{M}(x, x_0)\boldsymbol{X}(x_0)} . \tag{F.5}$$

This particular matrix solution is called the *transition matrix* and it yields the solution at x in terms of the solution at x_0.

Given that $\boldsymbol{M}(x_0, x_1)$ is a constant matrix, the two solution matrices $\boldsymbol{M}(x, x_0)\boldsymbol{M}(x_0, x_1)$ and $\boldsymbol{M}(x, x_1)$ satisfy homogeneous equation (F.2) with the same value at $x = x_0$, viz., $\boldsymbol{M}(x_0, x_1)$. They are therefore identical:

$$\boldsymbol{M}(x, x_0)\boldsymbol{M}(x_0, x_1) = \boldsymbol{M}(x, x_1) . \tag{F.6}$$

Putting $x_1 = x$, the formula $\boldsymbol{M}(x, x_0)\boldsymbol{M}(x_0, x) = \boldsymbol{I}$ shows that the transmission matrix is inverted by simply swapping variables:

$$\boldsymbol{M}(x_0, x) = \boldsymbol{M}^{-1}(x, x_0) . \tag{F.7}$$

Solution of the Full Equation

Let us now solve (F.1) by varying the coefficients $\boldsymbol{C}$ in (F.4). Substituting

$$\boldsymbol{X}(x) = \boldsymbol{M}(x, x_0)\boldsymbol{C}(x) \quad \Rightarrow \quad \boldsymbol{X}(x_0) = \boldsymbol{C}(x_0) , \tag{F.8}$$

into (F.1), we find that

$$\frac{\mathrm{d}\boldsymbol{M}(x, x_0)}{\mathrm{d}x}\boldsymbol{C}(x) + \boldsymbol{M}(x, x_0)\frac{\mathrm{d}\boldsymbol{C}(x)}{\mathrm{d}x} = \boldsymbol{F}\boldsymbol{M}(x, x_0)\boldsymbol{C}(x) + \boldsymbol{G}(x)U .$$

Then, using (F.2) and $\boldsymbol{M}^{-1}(x, x_0) = \boldsymbol{M}(x_0, x)$,

$$\frac{\mathrm{d}\boldsymbol{C}(x)}{\mathrm{d}x} = \boldsymbol{M}(x_0, x)\boldsymbol{G}(x)U .$$

Integrating from x_0 to x and substituting into (F.8) yields

$$\boldsymbol{C}(x) = \left[\int_{x_0}^{x} \boldsymbol{M}(x_0, x')\boldsymbol{G}(x')\,\mathrm{d}x'\right] U + \boldsymbol{C}(x_0) .$$

Using (F.6), the general solution can be written,

$$\boldsymbol{X}(x) = \left[\int_{x_0}^{x} \boldsymbol{M}(x, x')\boldsymbol{G}(x')\,\mathrm{d}x'\right] U + \boldsymbol{M}(x, x_0)\boldsymbol{X}(x_0) . \tag{F.9}$$

Stationary Case

When matrix $\boldsymbol{F}$ is constant, independent of x, the transition matrix $\boldsymbol{M}(x, x_0)$ depends only on the difference $x-x_0$, i.e., $\boldsymbol{M}(x, x_0) = \boldsymbol{M}(x-x_0)$. Expanding in powers of x,

$$\boldsymbol{M}(x) = \boldsymbol{I} + \boldsymbol{M}_1 x + \boldsymbol{M}_2 x^2 + \ldots + \boldsymbol{M}_k x^k + \ldots ,$$

which satisfies the homogeneous equation (F.2) if

$$\begin{aligned}\boldsymbol{M}_1 + 2\boldsymbol{M}_2 x + \ldots + k\boldsymbol{M}_k x^{k-1} + \ldots \\ = \boldsymbol{F} + \boldsymbol{F}\boldsymbol{M}_1 x + \ldots + \boldsymbol{F}\boldsymbol{M}_k x^{k-1} + \ldots .\end{aligned}$$

Identifying term by term, the condition is

$$\boldsymbol{M}_k = \frac{\boldsymbol{F}}{k}\boldsymbol{M}_{k-1} \quad \Rightarrow \quad \boldsymbol{M}_k = \frac{\boldsymbol{F}^k}{k!} .$$

The transition matrix is therefore the matrix exponential defined by

$$\boldsymbol{M}(x) = \boldsymbol{I} + \boldsymbol{F}x + \frac{\boldsymbol{F}^2}{2!}x^2 + \ldots + \frac{\boldsymbol{F}^k}{k!}x^k + \ldots = \mathrm{e}^{\boldsymbol{F}x} .$$

Then according to (F.9), $\boldsymbol{X}$ evolves according to

$$\boxed{\boldsymbol{X}(x) = \left[\int_{x_0}^{x} \mathrm{e}^{\boldsymbol{F}(x-x')}\boldsymbol{G}(x')\,\mathrm{d}x'\right] U + \mathrm{e}^{\boldsymbol{F}(x-x_0)}\boldsymbol{X}(x_0)} . \qquad \text{(F.10)}$$

Appendix G. Stationary Phase Method

This method is useful when evaluating integrals of type

$$I = \int_{-\infty}^{\infty} f(x) \mathrm{e}^{\mathrm{i}\alpha(x)} \mathrm{d}x \ . \tag{G.1}$$

x-dependent variations in the phase $\alpha(x)$ lead to very rapid oscillations in the exponential which tend to cancel each other out. The main contribution to integral I comes from the region (or regions) where the phase varies slowly (or is stationary). Let us calculate the contribution I_0 due to an interval around some point x_0 such that

$$\left.\frac{\mathrm{d}\alpha}{\mathrm{d}x}\right|_{x_0} = 0 \quad \Rightarrow \quad \alpha(x) \approx \alpha(x_0) + \frac{1}{2} \left.\frac{\mathrm{d}^2\alpha}{\mathrm{d}x^2}\right|_{x_0} (x - x_0)^2 \ . \tag{G.2}$$

Substituting the truncated expansion into (G.1) yields

$$I = \mathrm{e}^{\mathrm{i}\alpha_0} \int_{-\infty}^{\infty} f(x) \exp\left[\frac{\mathrm{i}}{2}\alpha_0''(x - x_0)^2\right] \mathrm{d}x \ , \quad \alpha_0 = \alpha(x_0) \ , \quad \alpha_0'' = \left.\frac{\mathrm{d}^2\alpha}{\mathrm{d}x^2}\right|_{x_0} \ .$$

If $f(x)$ varies slowly compared to the exponential then, making the change of variable

$$\alpha_0''(x - x_0)^2 = 2\pi y^2 \quad \Rightarrow \quad \mathrm{d}x = \sqrt{\frac{2\pi}{\alpha_0''}}\, \mathrm{d}y \ ,$$

it follows that

$$I = \mathrm{e}^{\mathrm{i}\alpha_0} f(x_0) \sqrt{\frac{2\pi}{\alpha_0''}} \int_{-\infty}^{\infty} \mathrm{e}^{\mathrm{i}\pi y^2} \mathrm{d}y \ . \tag{G.3}$$

The result

$$\int_{-\infty}^{\infty} \mathrm{e}^{\mathrm{i}\pi y^2} \mathrm{d}y = \mathrm{e}^{\mathrm{i}\pi/4} \tag{G.4}$$

leads to the approximation

$$I_0 = \sqrt{\frac{2\pi}{\alpha_0''}} f(x_0) \mathrm{e}^{\mathrm{i}(\alpha_0 + \pi/4)} \ , \quad \left.\frac{\mathrm{d}\alpha}{\mathrm{d}x}\right|_{x_0} = 0 \ , \quad \alpha_0'' = \left.\frac{\mathrm{d}^2\alpha}{\mathrm{d}x^2}\right|_{x_0} \ . \tag{G.5}$$

Application. Consider a signal $x(t) = e(t)\cos\phi(t)$ with modulated instantaneous frequency $\omega = \mathrm{d}\phi/\mathrm{d}t$, i.e., such that the phase $\phi(t)$ does not vary linearly with time: $\phi(t) = \omega_0 t + \psi(t)$. When the envelope $e(t)$ and modulation $\psi(t)$ are even functions, the spectrum $x(t)$ is given by [4.62]:

$$X(\omega) = \frac{1}{2}E(\omega - \omega_0) + \frac{1}{2}E^*(\omega + \omega_0) ,$$

$$E(\omega) = \int_{-\infty}^{\infty} e(t)\exp\mathrm{i}[\psi(t) - \omega t]\,\mathrm{d}t .$$

The integral giving $E(\omega)$ can be calculated to a good approximation by the stationary phase method. At frequency ω_p, the phase $\alpha(t) = \psi(t) - \omega_\mathrm{p} t$ is stationary at time t_p such that

$$\left.\frac{\mathrm{d}\alpha}{\mathrm{d}t}\right|_{t_\mathrm{p}} = 0 \quad\Rightarrow\quad \psi'(t_\mathrm{p}) = \left.\frac{\mathrm{d}\psi}{\mathrm{d}t}\right|_{t_\mathrm{p}} = \omega_\mathrm{p} , \tag{G.6}$$

and

$$E(\omega_\mathrm{p}) \approx \sqrt{\frac{2\pi}{\psi''(t_\mathrm{p})}}\, e(t_\mathrm{p})\exp\mathrm{i}\left[\frac{\pi}{4} + \psi(t_\mathrm{p}) - \omega_\mathrm{p} t_\mathrm{p}\right] . \tag{G.7}$$

Since $\psi(t) = \phi(t) - \omega_0 t$ and $\omega = \mathrm{d}\phi/\mathrm{d}t$, it follows that

$$\omega_\mathrm{p} = \left.\frac{\mathrm{d}\phi}{\mathrm{d}t}\right|_{t_\mathrm{p}} - \omega_0 = \omega(t_\mathrm{p}) - \omega_0 ,$$

$$\psi''(t_\mathrm{p}) = \left.\frac{\mathrm{d}\omega}{\mathrm{d}t}\right|_{t_\mathrm{p}} = \frac{1}{\dfrac{\mathrm{d}t}{\mathrm{d}\omega}[\omega(t_\mathrm{p})]} = \frac{2\pi}{t'[f(t_\mathrm{p})]} .$$

The modulus squared of the spectrum is given by

$$|E(f - f_0)|^2 \approx |t'(f)|e^2[t(f)] . \tag{G.8}$$

References

Chapter 1

1.1 Mason, W.P. (1948) Electromechanical Transducers and Wave Filters, 2nd edn. Van Nostrand–Reinhold, Princeton

1.2 Redwood, M. (1961) J. Acoust. Soc. Amer. **33**, 527–536

1.3 Krimholtz, R., Leedom, D.A., Matthaei, G.L. (1970) New equivalent circuit for elementary piezoelectric transducers, Electron. Lett. **6**, 398–399

1.4 Ristic, V.M. (1983) Principles of Acoustic Devices. John Wiley, New York

1.5 Holland, R., Nisse, E. (1969) IEEE Trans. Son. Ultrasonics **16**, 173

1.6 Trotel, J., Dieulesaint, E., Autin, B. (1968) Calculation of the bandshape factor of a piezoelectric thin-film transducer, Electron. Lett. **4**, 156–157

1.7 Smith, W.A., Auld, B.A. (1991) Modelling 1-3 composite piezoelectrics: thickness-mode oscillations, IEEE Trans. Ultrasonics, Ferro. and Freq. Control **38**, 40–47

1.8 Ohigashi, H., Koga, K., Suzuki, M., Nakanishi, T., Kimura, K., Hasimoto, N. (1984) Piezoelectric and ferroelectric properties of P(VDF-TrFE) copolymers and their application to ultrasonic transducers, Ferroelectrics **60**, 263–276

1.9 Brown, L.F. (1992) Ferroelectric polymers: current and future ultrasound applications, IEEE Ultrasonics Symp. Proc. 539–545

1.10 Foster, N.F., Rozgonyi, G.A. (1966) Zinc oxide film transducers, Appl. Phys. Lett. **8**, 221

1.11 Selfridge, A.R., Kino, G.S., Khuri-Yakub, B.T. (1980) Fundamental concepts in acoustic transducer array design, IEEE Ultrason. Symp. Proc. 989–993

1.12 Datta, S. (1986) Surface Acoustic Wave Devices, p. 37. Prentice Hall, Englewood Cliffs

General Reading

1.13 Kino, G.S. (1987) Acoustic Waves: Devices, Imaging and Analog Signal Processing. Prentice Hall, Englewood Cliffs

1.14 Ristic, V.M. (1983) Principles of Acoustic Devices. John Wiley, New York

1.15 Berlincourt, D., Kikuchi, Y., Meitzler, A.H. (1971) Ultrasonic transducer materials, Mattiat, O.E. (Ed.). Plenum Press, New York

1.16 Jaffe, B., Cook Jr., W.R., Jaffe, H. (1971) Piezoelectric Ceramics. Academic Press, London

1.17 Heising, R.A. (1946) Quartz Crystals for Electrical Circuits. Van Nostrand, Princeton, New Jersey

1.18 Auld, B. (1973) Acoustic Fields and Waves in Solids, Vol. 2. John Wiley, New York

Chapter 2

2.1 White, R.M., Voltmer, F.W. (1965) Direct piezoelectric coupling to surface elastic waves, Appl. Phys. Lett. **7**, 314–316

2.2 Tancrell, R.H., Williamson, R.C. (1971) Wavefront distortion of acoustic surface waves from apodised interdigital transducers, Appl. Phys. Lett. **19**, 456–459

2.3 Morgan, D.P. (1985) Surface Wave Devices for Signal Processing, 363–392. Elsevier, New York

2.4 Auld, B.A. (1973) Acoustic Fields and Waves in Solids, Vol. 2, Chap. 12. Wiley, New York

2.5 de Vries, A.J., Miller, R.L., Wojcik, T.J. (1972) Reflection of a surface wave from three types of ID transducers, IEEE Ultrason. Symp. Proc. 353–358

2.6 Li, R.C.M., Melngailis, J. (1975) The influence of stored energy at step discontinuities on the behavior of surface-wave gratings, IEEE Trans. Son. and Ultrason. **SU-22**, 189–198

2.7 Datta, S., Hunsinger, B.J. (1979) First order reflection of surface acoustic waves from thin-strip overlays, J. Appl. Phys. **50**, 5661–5665

2.8 Datta, S., Hunsinger, B.J. (1980) An analytical theory for the scattering of surface acoustic waves by a single electrode in a periodic array on a piezoelectric substrate, J. Appl. Phys. **51**, 4817–4823

2.9 Datta, S., Hunsinger, B.J. (1980) An analysis of energy storage effects on SAW propagation in periodic arrays, IEEE Trans. Son. and Ultrason. **SU-27**, 333–341

2.10 Engan, H. (1969) Excitation of elastic surface waves by spatial harmonics of interdigital transducers, IEEE Trans. Electron. Devices **ED-16**, 1014–1017

2.11 Tancrell, R.H., Holland, M.G. (1971) Acoustic surface wave filters, IEEE Proc. **59**, 393–409

2.12 Dieulesaint, E., Royer, D. (1990) Automatique appliquée, Vol. 2, Chap. 1. Masson, Paris

2.13 Kallman, H.E. (1940) Transversal Filters, Proc. IRE **28**, 302

2.14 Hartmann, C.S., Bell, D.T., Rosenfeld, R.C. (1973) Impulse model design of acoustic surface-wave filters, IEEE Trans. Microwave Theory and Tech. **MTT-21**, 162–175

2.15 Tancrell, R.H. (1977) Principle of surface wave filter design. In Matthews, H. (Ed.) Surface Wave Filters, 109–164. Wiley, New York

2.16 Smith, W.R., Gerard, H.M., Collins, J.H., Reeder, T.M., Shaw, H.J. (1969) Analysis of interdigital surface wave transducers by use of an equivalent circuit model, IEEE Trans. Microwave Theory and Tech. **MTT-17**, 856–864

2.17 Auld, B., Kino, G.S. (1971) Normal mode theory for acoustic waves and its application to the interdigital transducer, IEEE Trans. Electron. Devices **ED-18**, 898–908

2.18 Milsom, R.F., Reilly, N.H.C., Redwood, M. (1977) Analysis of generation and detection of surface and bulk acoustic waves by interdigital transducers, IEEE Trans. Son. and Ultrason. **SU-24**, 147–166

2.19 Ingebrigtsen, K.A. (1969) Surface waves in piezoelectrics, J. Appl. Phys. **40**, 2681–2686

2.20 Milsom, R.F., Redwood, M., Reilly, N.H.C. (1977) The interdigital transducer. In Matthews, H. (Ed.) Surface Wave Filters, 55–108. Wiley, New York

2.21 Morgan, D.P. (1985) Surface Wave Devices for Signal Processing, Chap. 3. Elsevier, New York

2.22 Datta, S., Hunsinger, B.J. (1980) Element factor for periodic transducers, IEEE Trans. Son. and Ultrason. **SU-27**, 42–44

2.23 Peach, R.C. (1981) A general approach to the electrostatic problem of the SAW interdigital transducer, IEEE Trans. Son. and Ultrason. **SU-28**, 96–105

2.24 Morgan, D.P. (1985) Surface Wave Devices for Signal Processing, Sect. 4.4. Elsevier, New York

2.25 Ramo, S., Whinnery, J.R., van Duzer, T. (1965) Fields and Waves in Communication Electronics, 535–550. Wiley, New York

2.26 Lewis, M.F. (1983) A different approach to the wave-scattering properties of interdigital transducers, IEEE Trans. Son. and Ultrason. **SU-30**, 55–57

2.27 Tobolka, G. (1979) Mixed matrix representation of SAW transducers, IEEE Trans. Son. and Ultrason. **SU-26**, 426–428

2.28 Hartmann, C.S., Wright, P.V., Kansy, R.J., Garber, E.M. (1982) An analysis of SAW interdigital transducers with internal reflections and the application to the design of single-phase unidirectional transducers, IEEE Ultrason. Symp. Proc. 40–45

2.29 Chen, D.P., Hauss, H.A. (1985) Analysis of metal-strip SAW gratings and transducers, IEEE Trans. Son. and Ultrason. **SU-32**, 395–408

2.30 Hartmann, C.S., Abbot, B.P. (1988) A generalised impulse response model for SAW transducers including effects of electrode reflections, IEEE Ultrason. Symp. Proc. 29–34

2.31 Wright, P.V. (1989) A new generalised modelling of SAW transducers and gratings, 43rd Annual Symp. on Frequency Control, 596–605

2.32 Hartmann, C.S., Abbot, B.P. (1989) Overview of design challenges for single phase unidirectional SAW filters, IEEE Ultrason. Symp. Proc. 78–79

2.33 Lewis, M.F. (1983) Low loss SAW devices employing single stage fabrication, IEEE Ultrason. Symp. Proc. 104–108

2.34 Yamanouchi, K., Furuyashiki, H. (1984) Low-loss SAW filters using internal reflection types of new single-phase unidirectional transducers, IEEE Ultrason. Symp. Proc. 68–71

2.35 Wright, P.V. (1985) The natural single-phase unidirectional transducer: a new low-loss SAW transducer, IEEE Ultrason. Symp. Proc. 58–63

2.36 Thorvaldsson, T. (1989) Analysis of the natural single phase unidirectional SAW transducer, IEEE Ultrason. Symp. Proc. 91–96

2.37 Kodoma, T., Kawabata, H., Yasuhara, Y., Sato, H. (1986) Design of low-loss SAW filters employing distributed acoustic reflection transducers, IEEE Ultrason. Symp. Proc. 59–64

2.38 Ventura, P., Solal, M., Dufilie, P., Hodé, J.M., Roux, F. (1994) A new concept in SPUDT design: the RSPUDT (Resonant SPUDT), IEEE Ultrason. Symp. Proc. 1–6

2.39 Hodé, J.M., Desbois, J., Dufilie, P., Solal, M., Ventura, P. (1995) SPUDT-based filters: design principles and optimisation, IEEE Ultrason. Symp. Proc. 39–50

2.40 Scholl, G., Christ, A., Ruile, W., Russer, P., Weigel, R. (1991) Efficient analysis tool for coupled-SAW-resonator filters, IEEE Trans. Ultrason. Ferroelec. and Freq. Control **38**, 243–251

2.41 Datta, S. (1983) Reflection and mode conversion of surface acoustic waves in layered media, IEEE Ultrason. Symp. Proc. 362–368

2.42 Wright, P.V. (1984) Modelling and experimental measurements of the reflection properties of SAW metallic gratings, IEEE Ultrason. Symp. Proc. 54–56

2.43 Morgan, D.P. (1996) Cascading formulas for identical transducer P-matrices, IEEE Trans. Ultrason. Ferroelec. and Freq. Control **43**, 985–987

2.44 Morgan, D.P. (1995) Reflective array modelling for SAW transducers, IEEE Ultrason. Symp. Proc. 215–220

2.45 Zang, Y., Desbois, J., Boyer, L. (1993) Characteristic parameters of surface acoustic waves in a periodic metal grating on a piezoelectric substrate, IEEE Trans. Ultrason. Ferroelec. and Freq. Control **40**, 183–192
2.46 Ventura, P., Hodé, J.M. (1997) A new accurate analysis of periodic IDTs built on unconventional orientation on quartz, IEEE Ultrason. Symp. Proc. 139–142
2.47 Ventura, P., Hodé, J.M., Lopes, B. (1995) Rigorous analysis of finite SAW devices with arbitrary electrode geometries, IEEE Ultrason. Symp. Proc. 257–262
2.48 Smith, W.R. (1981) Circuit model analysis and design of interdigital transducers for surface acoustic wave devices. In: Mason, W.P., Thurston, R.N. (Eds.) Physical Acoustics, Vol. XV, 99–189. Academic Press, New York
2.49 Kino, G.S. (1987) Acoustic Waves: Devices, Imaging and Analog Signal Processing. Prentice Hall, Englewood Cliffs
2.50 Thorvaldsson, T., Nyffeler, F.M. (1986) Rigorous derivation of the Mason equivalent circuit parameters from coupled mode theory, IEEE Ultrason. Symp. Proc. 91–96
2.51 Inagawa, K., Koshiba, M. (1994) Equivalent networks for SAW interdigital transducers, IEEE Trans. Ultrason. Ferroelec. and Freq. Control **41**, 402–411
2.52 Pereira da Cunha, M., Adler, E.L. (1995) High Velocity Pseudo-Surface Waves (HVPSAW), IEEE Trans. Ultrason. Ferroelec. and Freq. Control **42**, 840–844
2.53 Pereira da Cunha, M. (1996) High Velocity Pseudo-Surface Waves (HVPSAW): further insight, IEEE Ultrason. Symp. Proc. 97–106
2.54 Smith, H.I., Bachner, F.J., Efremow, N. (1971) A high-yield photolithographic technique for surface wave devices, Journal of the Electrochemical Society **118**, 821
2.55 Ruppel, C.C., Reindl, L., Berek, S., Knauer, U., Heide, P., Vossiek, M. (1996) Design, fabrication and application of precise delay lines at 2.45 GHz, IEEE Ultrason. Symp. Proc. 261–265
2.56 Broers, A.N., Lean, E.G., Hatzakis, M. (1969) Appl. Phys. Lett. **15**, 98
2.57 Cahen, O., Sigelle, R., Trotel, J. (1972) 141st Nat. Meeting, The Electrochemical Society. Houston, Texas
2.58 Lean, E.G., Broers, A.N. (1970) The Microwave Journal **13**, 97
2.59 Yamanouchi, K., Cho, Y., Meguro, T. (1988) SHF-range surface acoustic wave interdigital transducers using electron beam exposure, IEEE Ultrason. Symp. Proc. 115–118
2.60 Yamanouchi, K., Qureshi, J.A., Odagawa, H. (1997) 5–10 GHz range surface acoustic wave filters using electrode thickness difference type and new reflector bank type of unidirectional interdigital transducers, IEEE Ultrason. Symp. Proc. 61–64
2.61 Yuhara, A., Mizutani, T., Hosaka, N., Yamada, J., Kobayashi, S. (1989) Dry process technology for high frequency SAW devices, IEEE Ultrason. Symp. Proc. 343–349
2.62 Hickernell, F.S., Shulda, G.F., Brewer, J.W. (1970) ZnO and CdS overlay surface wave transducers, IEEE Ultrason. Symposium
2.63 Shiosaki, T., Ieki, E., Kawabata, A. (1976) 58 MHz surface acoustic wave video intermediate frequency using ZnO sputtered film, Appl. Phys. Lett. **28**, 475–476
2.64 Shiosaki, T. (1978) High-speed fabrication of high-quality sputtered ZnO thin films for bulk and surface wave applications, IEEE Ultrason. Symp. Proc. 100–110
2.65 Hata, T., Noda, E., Morimoto, O., Hada, T. (1979) High rate deposition of piezoelectric zinc oxide films using new reactive sputtering technique, IEEE Ultrason. Symp. Proc. 936–939

2.66 Hickernell, F.S. (1980) ZnO processing for bulk and surface wave devices, IEEE Ultrason. Symp. Proc. 785–794
2.67 Defranould, P. (1981) High deposition rate sputtered ZnO thin films for BAW and SAW applications, IEEE Ultrason. Symp. Proc. 483–488
2.68 Hickernell, F.S. (1996) Measurement techniques for evaluating piezoelectric thin films, IEEE Ultrason. Symp. Proc. 235–240
2.69 Kadota, M., Minakata, M. (1995) Piezoelectric properties of zinc oxide films on glass substrates deposited by RF-magnetron mode electron cyclotron resonance sputtering system, IEEE Trans. Ultrason. Ferroelec. and Freq. Control **42**, 345–350
2.70 Nakahata, H., Kitabayashi, H., Fujii, S., Higaki, K., Tanabe, K., Seki, Y., Shikata, S. (1996) Fabrication of 2.5 GHz SAW retiming filter with SiO_2/ZnO/diamond structure, IEEE Ultrason. Symp. Proc. 285–288
2.71 Yamanouchi, K., Sakurai, N., Satoh, T. (1989) SAW propagation characteristics and fabrication technology of piezoelectric thin film/diamond structure, IEEE Ultrason. Symp. Proc. 351–354
2.72 Foster, N.F. (1969) The deposition and piezoelectric characteristics of sputtered lithium niobate films, J. Appl. Phys. **40**, 420–421
2.73 Shibata, Y., Kuze, N., Kaya, K., Matsui, M. (1996) Piezoelectric $LiNbO_3$ and $LiTaO_3$ films for SAW device applications, IEEE Ultrason. Symp. Proc. 247–254

General Reading

2.74 Feldmann, M., Hénaff, J. (1989) Surface Acoustic Waves for Signal Processing. Artech House, Boston
2.75 Matthews, H. (Ed.) (1977) Surface Wave Filters. Wiley, New York
2.76 Kino, G.S. (1987) Acoustic Waves: Devices, Imaging and Analog Signal Processing. Prentice Hall, Englewood Cliffs
2.77 Ristic, V.M. (1983) Principles of Acoustic Devices. John Wiley, New York
2.78 Morgan, D.P. (1985) Surface Wave Devices for Signal Processing. Elsevier, New York
2.79 Campbell, C. (1989) Surface Acoustic Wave Devices and their Signal Processing Applications. Academic Press, Boston
2.80 Datta, S. (1986) Surface Acoustic Wave Devices. Prentice Hall, Englewood Cliffs
2.81 White, R.M. (1970) Surface elastic waves, Proc. IEEE **58**, 1238–1276
2.82 Ash, E.A., Paige, E.G.S. (Eds.) (1985) Rayleigh-Wave Theory and Applications. Springer, Berlin
2.83 Gulyaev, Y.V., Plessky, V. (1989) Propagation of acoustic surface waves in periodic structures, Sov. Phys. Usp. **32** (1), American Institute of Physics, 51–74
2.84 Ruppel, C.C., Ruile, W., Scholl, G., Wagner, K.C., Männer, O. (1994) Review of models for low-loss filter design and applications, IEEE Ultrason. Symp. Proc. 313–324
2.85 Haus, H.A., Huang, W. (1991) Coupled-mode theory, Proc. IEEE **79**, 1505–1518

Chapter 3

3.1 Brillouin, L. (1922) Diffusion de la lumière et des rayons X par un corps transparent homogène, influence de l'agitation thermique, Ann. Phys. **17**, 88–122
3.2 Lucas, R., Biquard, P. (1932) Propriétés optiques des milieux solides et liquides soumis aux vibrations élastiques ultrasonores, J. Physique **10**, 464–477
3.3 Debye, P., Sears, F.W. (1932) On the scattering of light by supersonic waves, Proc. Nat. Acad. Sci. **18**, 409–414
3.4 Raman, C.V., Nath, N.S.N. (1935) The Diffraction of Light by High Frequency Sound: Part I, Proc. Indian Acad. Sci. **2**, 406; (1935) Part II **2**, 413; (1936) Part III **3**, 75; (1936) Part IV **3**, 119; (1936) Part V **3**, 459; (1936) Generalised Theory **4**, 222
3.5 Adler, R. (1967) Interaction between light and sound, IEEE Spectrum **4**, 42–54
3.6 Huard, S. (1994) Polarisation de la lumière. Masson, Paris
3.7 Kittel, C. (1996) Introduction to Solid State Physics, 7th ed., 390. Wiley, New York
3.8 Primak, W., Post, D. (1959) J. Appl. Phys. **30**, 779
3.9 Pinnow, D.A., Dixon, R.W. (1968) Alpha-iodic acid: a solution-grown crystal with a high figure of merit for acousto-optic device applications, Appl. Phys. Lett. **13**, 156–158
3.10 Coquin, G.A., Pinnow, D.A., Warner, A.W. (1971) Physical properties of lead molybdate relevant to acousto-optic device applications, J. Appl. Phys. **42**, 2162–2168
3.11 Dixon, R.W. (1967) Photoelastic properties of selected materials and their relevance for applications to acoustic light scanners and modulators, J. Appl. Phys. **38**, 5149–5153
3.12 Uchida, N., Ohmachi, Y. (1969) Elastic and photoelastic properties of TeO_2 single crystals, J. Appl. Phys. **40**, 4692–4695
3.13 Pockels, F. (1889) Ann. Physik. Chem. **37**, 269–372
3.14 Nelson, D.F., Lax, M. (1970) Phys. Rev. Lett. **24**, 379
3.15 Nelson, D.F., Lazay, P.D. (1970) Phys. Rev. Lett. **25**, 1187
3.16 Nelson, D.F., Lazay, P.D., Lax, M. (1972) Phys. Rev. B **6**, 3109
3.17 Nelson, D.F., Lax, M. (1971) Theory of the photoelastic interaction, Phys. Rev. B **3**, 2778–2794
3.18 Klein, W.R., Cook, B.D. (1967) Unified approach to ultrasonic light diffraction, IEEE Trans. Son. Ultrason. **SU-14**, 123–134
3.19 Smith, T.M., Korpel, A. (1965) IEEE J. Quantum Electron. **QE-1**, 283
3.20 Pinnow, D.A. (1970) Guidelines for the selection of acousto-optic materials, J. Quantum Electron. **QE-6**, 223–238
3.21 Yano, T., Watanabe, A. (1974) Acousto-optic figure of merit of TeO_2 for circularly polarized light, J. Appl. Phys. **45**, 1243–1245
3.22 Uchida, N. (1972) Acoustic attenuation in TeO_2, J. Appl. Phys. **43**, 2915–17
3.23 Dixon, R.W. (1967) Acoustic diffraction of light in anisotropic media, IEEE J. Quantum Electron. **3**, 85–93
3.24 Harris, S.E., Nieh, S.T.K., Feigelson, R.S. (1970) $CaMoO_4$ electronically tunable optical filter, Appl. Phys. Lett. **17**, 223–225
3.25 Montgomery, R.M., Young Jr., E.H. (1971) J. Appl. Phys. **42**, 2585
3.26 Tsaï, C.S. (1992) IEEE Trans. Ultrason. Ferroelec. Freq. Control **39**, 529
3.27 Dutoit, M. (1973) IEEE Trans. Son. Ultrason. **SU-20**, 279
3.28 Spencer, E.G., Lenzo, P.V. (1967) J. Appl. Phys. **38**, 423
3.29 Wilkinson, C.D.W., Caddes, D.E. (1966) J. Acoust. Soc. Am. **40**, 498

3.30 Varasi, M., Signorazzi, M., Vannucci, A., Dunphy, J. (1996) A high-resolution integrated optical spectrometer with applications to fibre sensor signal processing, Meas. Sci. Technol. **7**, 173–178

3.31 Whitman, R.L., Korpel, A. (1969) Probing of acoustic surface perturbations by coherent light, Applied Optics **8**, 1567–76

3.32 Adler, R., Korpel, A., Desmares, P. (1968) An instrument for making surface waves visible, IEEE Trans. Son. Ultrason. **15**, 157

3.33 Kessler, L.W., Yuhas, D.E. (1979) Acoustic microscopy 1979, Proc. IEEE **67**, 526–536

3.34 Engan, H. (1978) Phase sensitive laser probe for high-frequency surface acoustic wave measurements, IEEE Trans. Son. Ultrason. **SU-25**, 372–377

3.35 Slobodnik, A.J. (1970) Microwave acoustic surface wave investigations using laser light deflection, Proc. IEEE **58**, 488–490

3.36 Rouvaen, J.M., Bridoux, E., Grémillet, N., Torguet, R., Hartemann, P. (1974) Electron. Lett. **10**, 297

3.37 Royer, D., Dieulesaint, E. (1989) Mesure optique de déplacements d'amplitude 10^{-4} à 10^{2} Å. Application aux ondes élastiques, Rev. Phys. Appl. **24**, 833–846

3.38 Kwaaitaal, T. (1974) Contribution to the interferometric measurement of subangström vibrations, Rev. Sci. Instr. **45**, 39–41

3.39 de la Rue, R.M., Humphryes, R.F., Mason, I.M., Ash, E.A. (1972) Acoustic surface wave amplitude and phase measurements using laser probes, Proc. IEEE **119**, 117–126

3.40 Royer, D., Dieulesaint, E. (1986) Optical detection of sub-angström transient mechanical displacements, IEEE Ultrason. Symp. Proc. 527–530

3.41 Cretin, B., Hauden, D. (1984) Thermoacoustic scanning microscope using a laser probe, IEEE Ultrason. Symp. Proc. 656–659

3.42 Royer, D., Dieulesaint, E., Martin, Y. (1985) Improved version of a polarised beam heterodyne interferometer, IEEE Ultrason. Symp. Proc. 432–435

3.43 Monchalin, J.P. (1986) Optical detection of ultrasound, IEEE Trans. Ultrason. Ferroelec. and Freq. Control **33**, 485–499

3.44 Padioleau, C., Bouchard, P., Héon, R., Monchalin, J.P., Chang, F.H., Drake, T.E., MacRea, K.I. (1993) Laser ultrasonic inspection of graphite epoxy laminates, Review of Progress in QNDE **12B**, 1345–1352

3.45 Royer, D., Casula, O. (1994) Quantitative imaging of transient acoustic fields by optical heterodyne interferometry, IEEE Ultrason. Symp. Proc. 1153–1163

3.46 Certon, D., Casula, O., Patat, F., Royer, D. (1997) Theoretical and experimental investigations of lateral modes in 1-3 piezocomposites, IEEE Trans. Ultrason. Ferroelec. and Freq. Control **44**, 643–651

3.47 Carslaw, H.S., Jaeger, J.C. (1959) Conduction of Heat in Solids. Clarendon Press, Oxford

3.48 Ready, J.F. (1971) Effect of High Power Radiation. Academic Press, New York

3.49 Aki, K., Richards, G. (1980) Quantitative Seismology Vol. I, Chap. 6. Freeman

3.50 Scruby, C.B., Drain, L.E. (1990) Laser Ultrasonics Techniques and Applications. Adam Hilger

3.51 Rose, L.R.F. (1984) Point-source representation for laser-generated ultrasound, J. Acoust. Soc. Am. **75**, 723–732

3.52 Dewhurst, R., Hutchins, D.A., Palmer, S.B., Scruby, C. (1982) Quantitative measurements of laser-generated acoustic waveforms, J. Appl. Phys. **53**, 4064–4071

3.53 Pekeris, C.L., Lifson, H. (1957) Motion of the surface of a uniform elastic half-space produced by a buried pulse, J. Acoust. Soc. Am. **29**, 1233–38

3.54 Hutchins, D.A. (1988) Ultrasonic Generation by Pulsed Laser. In: Mason, W.P., Thurston, R.N. (Eds.) Physical Acoustics **18**, Chap. 2. Academic Press, New York

3.55 Miller, G.F., Pursey, H. (1954) The field and radiation impedance of mechanical radiators on the free surface of a semi-infinite isotropic solid, Proc. Roy. Soc. London **A223**, 521–541

3.56 Berthelot, Y.H., Busch-Vishniac, I.J. (1987) Thermoacoustic radiation of sound by a moving laser source, J. Acoust. Soc. Am. **81**, 317–327

3.57 Noroy, M.H., Royer, D., Fink, M. (1993) The laser-generated ultrasonic phased array: analysis and experiments, J. Acoust. Soc. Am. **94**, 1934–43

3.58 White, R.M. (1963) Generation of elastic waves by transient surface heating, J. Appl. Phys. **34**, 3559–67

3.59 von Gutfeld, R.J., Melcher, R.L. (1977) 20 MHz acoustic waves from pulsed thermoelastic expansions of constrained surfaces, Appl. Phys. Lett. **30**, 257–259

3.60 Royer, D., Dieulesaint, E., Leclaire, P. (1989) Remote sensing of the thickness of hollow cylinders from optical excitation and detection of Lamb waves, IEEE Ultrason. Symp. Proc. 1163–66

3.61 Pouet, B. (1991) Modélisation physique par ultrasons laser. Application à la modélisation sismique. Thesis, University of Paris VII

3.62 Royer, D. (1996) Génération et détection optiques d'ondes élastiques. Techniques de l'Ingénieur, Electronics treatise E 4415

General Reading

3.63 Sapriel, J. (1976) L'acousto-optique. Masson, Paris

3.64 Korpel, A. (1996) Acousto-optics, 2nd edn. Marcel Dekker, New York

3.65 Yariv, A., Yeh, P. (1984) Optical Waves in Crystals, Chaps. 9 and 10. Wiley, New York

3.66 Das, P.K., Decusatis, C.M. (1991) Acousto-optic Signal Processing: Fundamentals and Applications. Artech House, Boston

3.67 Kino, G.S. (1987) Acoustic Waves: Devices, Imaging and Analog Signal Processing. Prentice Hall, Englewood Cliffs

3.68 Nelson, D.F. (1979) Electric, Optic and Acoustic Interactions in Dielectrics. Wiley, New York

3.69 Gordon, E.I. (1966) A review of acousto-optical deflection and modulation devices, Proc. IEEE **54**, 1391–1401

3.70 Spencer, E.G., Lenzo, P.V., Ballman, A.A. (1967) Dielectric materials for electrooptic, elastooptic and ultrasonic device applications, Proc. IEEE **55**, 2074

3.71 Dixon, R.W. (1970) Acoustooptic interactions and devices, IEEE Trans. Electron. Devices **ED-17**, 229

Chapter 4

4.1 McFee, J.H. (1960) Transmission and amplification of acoustic waves in piezoelectric semiconductors. In: Mason, W.P., Thurston, R.N. (Eds.) Physical Acoustics, Vol. 4A, Chap. 1. Academic Press, New York

4.2 Marshall, F.G., Paige, E.G.S. (1971) Novel acoustic surface wave directional coupler with diverse applications, Electron. Lett. **7**, 460–464

4.3 Marshall, F.G., Newton, C.O., Paige, E.G.S. (1973) Theory and design of the surface acoustic wave multistrip coupler, IEEE Trans. Microwave Theory Tech. **21**, 206–215
4.4 Max, J., Lacoume, J.L. (1996) Méthodes et techniques de traitement du signal et applications aux mesures physiques, Vol. 1. Masson, Paris
4.5 Marshall, F.G., Newton, C.O., Paige, E.G.S. (1973) Surface acoustic wave multistrip components and their applications, IEEE Trans. Microwave Theory Tech. **21**, 216–225
4.6 Feldmann, M., Hénaff, J. (1989) Surface Acoustic Waves for Signal Processing 114–128. Artech House, Boston
4.7 Morozumi, K., Kadota, M., Hayashi, S. (1996) Characteristics of BGS wave resonators using ceramic substrates and their applications, IEEE Ultrason. Symp. Proc. 81–86
4.8 Cahen, O., Dieulesaint, E., Torguet, R. (1968) Ligne à retard variable utilisant l'effet Brillouin, C.R. Acad. Sci. Paris **266**, 1009
4.9 Dieulesaint, E., Royer, D. (1987) Automatique appliquée Vol. 1, 70. Masson, Paris
4.10 Lewis, M. (1974) The surface acoustic wave oscillator – a natural and timely development of the quartz crystal oscillator, 28th Freq. Control Symp. Proc. 304–314
4.11 Smith, W.R., Parker, T.E. (1985) Precision oscillators. In: Gerber, A., Ballato, A. (Eds.) Precision Frequency Control Vol. 2, Chap. 8. Academic Press, Orlando
4.12 Feldmann, M. (1981) Théorie des réseaux et systèmes linéaires, Chap. 17, 275. Eyrolles, Paris
4.13 Dieulesaint, E., Royer, D. (1987) Automatique appliquée Vol. 1, 45. Masson, Paris
4.14 Campbell, C. (1989) Surface Acoustic Wave Devices and their Signal Processing Applications 177–192. Academic Press, Boston
4.15 McClellan, J.H., Parks, T.W., Rabiner, L.R. (1973) A computer program for designing optimum FIR linear phase digital filters, IEEE Trans. on Audio and Electroacoustics **AU-21**, 506–526
4.16 Hartemann, P., Dieulesaint, E. (1969) Acoustic surface wave filters, Electron. Letters **5**, 657
4.17 Hartmann, C.S. (1973) Weighting interdigital surface wave transducers by selective withdrawal of electrodes, IEEE Ultrason. Symp. Proc. 423–426
4.18 Lewis, M.F. (1985) Practical frequency source for use in agile radar, Electron. Letters **21**, 1017–18
4.19 Lewis, M.F. (1982) SAW filters employing interdigitated interdigital transducers IIDT, IEEE Ultrason. Symp. Proc. 12–17
4.20 Yatsuda, H. (1998) Design technique for nonlinear phase SAW filters using slanted finger interdigital transducers, IEEE Trans. Ultrason. Ferroelec. Freq. Control **45**, 41–47
4.21 Shiosaki, T., Ieki, E, Kawabata, A. (1976) 58-MHz surface acoustic wave video intermediate frequency using ZnO sputtered film, Appl. Phys. Lett. **28**, 475
4.22 Nakahata, H., Higaki, K., Fujii, S., Hachigo, A., Kitabayashi, H., Tanabe, K., Seki, Y., Shikata, S. (1995) SAW devices on diamond, IEEE Ultrason. Symp. Proc. 361–370
4.23 Marshall, F.G. (1972) New technique for the suppression of triple-transit signal in surface acoustic wave delay line, Electron. Lett. **8**, 311–312
4.24 Porter, W.A., Smilowitz, B. (1982) Frequency sidelobes generated by apodized transducer pairs, IEEE Ultrason. Symp. Proc. 35–39

4.25 Sudhakar, P., Battacharyya, A.B., Mathur, B. (1978) SAW bandpass filter with −50 dB sidelobes using unweighted IDTS, Electron. Lett. **14**, 437–439
4.26 Feldmann, M., Hénaff, J. (1989) Surface Acoustic Waves for Signal Processing 164. Artech House, Boston
4.27 Maerfeld, C., Farnell, G.W. (1973) Nonsymmetrical multistrip coupler as a surface wave beam compressor of large bandwidth, Electron. Lett. **9**, 432–434
4.28 Ash, E.A. (1970) Surface wave grating reflectors and resonators, IEEE International Microwave Symp. Proc. 385–386
4.29 Takeuchi, M., Yamanouchi, K. (1986) New types of SAW reflectors and resonators consisting of elements with positive and negative reflectivity, IEEE Trans. Ultrason. Ferroelec. Freq. Control **33**, 369–374
4.30 Adler, E.L. (1994) SAW and pseudo-SAW properties, using matrix methods, IEEE Trans. Ultrason. Ferroelec. Freq. Control **41**, 876–882
4.31 Pereira da Cunha, M., Adler, E.L. (1994) High velocity pseudo surface waves (HVPSAW), IEEE Ultrason. Symp. Proc. 281–286
4.32 Tiersten, H.F., Smythe, R.C. (1975) Guided acoustic surface wave filters, IEEE Ultrason. Symp. Proc. 293–294
4.33 Lambert, C., Kondratiev, S., Plessky, V., Thorvaldsson, T. (1997) Main approaches to design of low loss SAW filters for mobile communications, 11th European Frequency and Time Forum 123
4.34 Ikata, O., Miyashita, T., Matsuda, T., Nishihara, T., Satoh, Y. (1992) Development of low-loss band-pass filters using SAW resonators for portable telephones, IEEE Ultrason. Symp. Proc. 111–115
4.35 Machui, J., Bauregger, J., Riha, G., Schropp, I. (1995) SAW devices in cellular and cordless phones, IEEE Ultrason. Symp. Proc. 121–130
4.36 Corvisier, P. (1998) Les filtres à ondes de surface, L'Electronique **82**, 114–124
4.37 Besson, R.J., Boy, J.J., Mourey, M.M. (1995) BVA resonators and oscillators: a review. Relation with space requirements and quartz material characterization, Proc. 47th Symp. on Freq. Control 590–599
4.38 Gerber, E.A., Ballato, A. (1985) Precision Frequency Control Vols. 1, 2. Academic Press, New York
4.39 EerNisse, E.P. (1976) Calculations on the stress compensated (SC-cut) quartz resonators, Proc. 30th Symp. on Freq. Control 8–11
4.40 Aubry, J.P. (1996) Matériaux et composants piézoélectriques, Techniques de l'Ingénieur, E 1890, 1–14, E 1891, 1–15, E 2205, 1–24
4.41 Détaint, J., Capelle, B., Zarka, A. (1998) Whispering gallery like modes in quartz energy trapping resonators, 50th Symp. on Freq. Control
4.42 Besson, R.J. (1977) A new 'electrodeless' resonator design, Proc. 31st Symp. on Freq. Control 147–152
4.43 Besson, R.J. (1976) A new piezoelectric resonator design, Proc. 30th Symp. on Freq. Control 78–83
4.44 Aubry, J.P., Debaisieux, A. (1984) Further results on 5 MHz and 10 MHz resonators with BVA and QAS design, Proc. 38th Symp. on Freq. Control 190–200
4.45 Gagnepain, J.J., Besson, R.J. (1975) Nonlinear effects in piezoelectric quartz crystals. In: Mason, W.P., Thurston, R.N. (Eds.) Physical Acoustics, Vol. 11, 245–288. Academic Press, New York
4.46 Besson, R.J., Mourey, M. (1997) A miniature ultrastable quartz oscillator with extremely good performances, Proc. 11th European Freq. and Time Forum 227–229
4.47 Berté, M. (1977) Acoustic bulk wave resonators and filters operating in the fundamental mode at frequencies greater than 100 MHz, Proc. 31st Symp. on Freq. Control 122–125

4.48 Berté, M., Hartemann, P. (1978) Quartz resonators at fundamental frequencies greater than 100 MHz, IEEE Ultrason. Symp. Proc. 148–151
4.49 Heimann, R.B. (1982) Principle of chemical etching, the art and science of etching crystals, Crystals **8**, 214–224. Springer, Berlin
4.50 Staudte, J.H. (1973) Subminiature quartz tuning fork resonator, Proc. 27th Symp. on Freq. Control 50–54
4.51 Studer, B., Zingg, W. (1990) Technology and characteristics of chemically milled miniature quartz crystals, 4th European Freq. and Time Control Proc. 653–658
4.52 Danel, J.S., Delapierre, G. (1991) Quartz as a material for microdevices, J. Micromech. and Microeng. **1**, 187–198
4.53 Studer, B., Zingg, W., Dalla Piazza, S. (1996) Characteristics of high frequency fundamental rectangular quartz crystal resonators, 10th European Freq. and Time Control Proc. 14–20
4.54 Dick, G.J., Santigo, D.G., Wang, R.T. (1995) Temperature-compensated sapphire resonators for ultrastable oscillator capability at temperatures above 77 K, IEEE Trans. Ultrason. Ferroelec. Freq. Control **42**, 812–819
4.55 Dieulesaint, E., Mazerolle, D., Royer, D. (1987) YAG resonators, Electron. Lett. **23**, 581–582
4.56 Spencer, W.J. (1972) Monolithic Crystal Filters. In: Mason, W.P., Thurston, R.N. (Eds.) Physical Acoustics, Vol. 9, 167–220. Academic Press, New York
4.57 Détaint, J., Schwartzel, J. (1994) Materials for filtering and frequency control in the next generations of mobile communication systems, Suppl. C2, J. Phys. III **4**, 93–106
4.58 Lefèvre, R., Jenselme, L., Servajean, D. (1985) Laser processed miniature $LiTaO_3$ resonators and monolithic filters, Proc. 39th Symp. on Freq. Control 333–337
4.59 Woodward, P.M. (1953) Probability and Information Theory, with Application to Radar. Pergamon Press, Oxford
4.60 Carpentier, M. (1966) Radars. Concepts nouveaux, Chap. 2. Dunod, Paris
4.61 Cook, C.E., Bernfeld, M. (1967) Radar Signals: An Introduction to Theory and Application, Chap. 2. Academic Press, New York
4.62 Bruhat, G., Kastler, A. (1965) Cours de physique générale. Optique 202–209. Masson, Paris
4.63 Tortoli, P., Guidi, F., Atzeni, C. (1994) Digital vs SAW matched filter implementation for radar pulse compression, IEEE Ultrason. Symp. Proc. 199–202
4.64 Tournois, P. (1964) Annales de Radioélectricité **78**, 267
4.65 Martin, T.A. (1973) The IMCON pulse compression filter and its applications, IEEE Trans. Microwave Theory Tech. **21**, 186–194
4.66 Williamson, R.C., Smith, H.I. (1973) The use of surface elastic wave reflection gratings in large time-bandwidth pulse-compression filters, IEEE Trans. Microwave Theory Tech. **21**, 195–205
4.67 Wright, P.V., Haus, H.A. (1980) A closed-form analysis of reflective array gratings, IEEE Ultrason. Symp. Proc. 282–285
4.68 Dubouis, P., Gragnolati, J.P., Psaila, E., Solal, M. (1992) A −60 dB sidelobe pulse compression system for space application, IEEE Ultrason. Symp. Proc. 231–235
4.69 Riha, G. (1982) RAC filters with position weighted metallic strip arrays, IEEE Ultrason. Symp. Proc. 175–178
4.70 Jack, M.A., Grant, P.M., Collins, J.H. (1980) The theory, design and applications of surface acoustic wave Fourier transform processors, Proc. IEEE **68**, 450–458

4.71 Williamson, R.C., Dolat, V.S., Rhodes, R.R., Boroson, D.M. (1979) A satellite-borne SAW chirp-transform system for uplink demodulation of FSK communication signals, IEEE Ultrason Symp. Proc. 741–747
4.72 Defranould, P., Maerfeld, C. (1976) A SAW planar piezoelectric convolver, Proc. IEEE **64**, 748–751
4.73 Minagawa, S., Okamoto, T., Niitsuma, T., Mitsutsuka, S., Tsubouchi, K., Mikoshiba, N. (1984) Sezawa correlator using monolithic $ZnO/SiO_2/Si$ structure, IEEE Ultrason. Symp. Proc. 298–302
4.74 Tsubouchi, K., Mikoshiba, N. (1989) An asynchronous multi-channel spread spectrum transceiver using a SAW convolver, IEEE Ultrason. Symp. Proc. 165–172
4.75 Tsubouchi, K., Nakase, H., Namba, A., Mazu, K. (1994) Full duplex transmission operation of a 2.45 GHz asynchronous spread spectrum modem using a SAW convolver, IEEE Trans. Ultrason. Ferroelect. Freq. Control **41**, 478–482
4.76 Atzeni, C., Manes, G., Masotti, L. (1973) Synthesis of amplitude-modulated SAW filters with constant-length fingers, IEEE Ultrason. Symp. Proc. 414–418

General Reading

4.77 Campbell, C. (1989) Surface Acoustic Wave Devices and their Signal Processing Applications. Academic Press, Boston
4.78 Morgan, D.P. (1985) Surface Wave Devices for Signal Processing. Elsevier, New York
4.79 Feldmann, M., Hénaff, J. (1989) Surface Acoustic Waves for Signal Processing. Artech House, Boston
4.80 Kino, G.S. (1987) Acoustic Waves: Devices, Imaging and Analog Signal Processing. Prentice Hall, Englewood Cliffs
4.81 Matthews, H. (Ed.) (1977) Surface Wave Filters. Design, Construction and Use. Wiley, New York
4.82 Shibayama, K., Yamanouchi, K. (1992) Proceedings of the international symposium on surface acoustic wave devices for mobile communication, Sendai.
4.83 Gerber, E.A., Ballato, A. (1985) Precision Frequency Control Vols. 1, 2. Academic Press, New York
4.84 Datta, S. (1986) Surface Acoustic Wave Devices. Prentice Hall, Englewood Cliffs
4.85 Hickernell, F.S. (1988) High reliability SAW bandpass filters for space applications, IEEE Trans. Ultrason. Ferroelect. and Freq. Control **35**, 652–656
4.86 Ruppel, C.C.W., et al. (1993) SAW devices for consumer communication applications, IEEE Trans. Ultrason. Ferroelect. and Freq. Control **40**, 438–452
4.87 Gagnepain, J.J. (1985) Rayleigh wave resonators and oscillators 151–172; Maerfeld, C. (1985) Rayleigh wave nonlinear components 191–218. In: Ash, E.A., Paige, E.G.S. (Eds.) Rayleigh Wave Theory and Application. Springer, Berlin
4.88 Audoin, C., Bernard, M.Y., Besson, R., Gagnepain, J.J., Groslambert, J., Granveaud, M., Neau, J.C., Olivier, M., Rutman, J. (1991) La mesure de la fréquence des oscillateurs. (Chronos) Masson, Paris
4.89 Détaint, J. (1997) Résonateurs piézoélectriques à ondes de volume. Matériaux, modélisation, visualisation. Thesis, University of Paris 7

Chapter 5

5.1 Audoin, C., Bernard, M.Y., Besson, R., Gagnepain, J.J., Groslambert, J., Granveaud, M., Neau, J.C., Olivier, M., Rutman, J. (1991) La mesure de la fréquence des oscillateurs. Masson, Paris

5.2 Walls, F.L., Gagnepain, J.J. (1985) Special Applications. In: Gerber, A., Ballato, A. (Eds.) Precision Frequency Control Vol. 2, 287–296. Academic Press, Orlando

5.3 Lu, J.C., Lewis, O. (1972) Investigation of film-thickness determination by oscillating quartz resonators with large mass load, J. Appl. Phys. **43**, 4385–90

5.4 Rivoal, J.C., Grisolia, C., Lignieres, J., Kreisle, D., Fayet, P., Woste, L. (1989) Absorption spectroscopy of size-selected trimers and pentamers transition metal clusters isolated in rare gas solids: preliminary steps, Zeitschrift für Physik D **12**, 481

5.5 EerNisse, E.P. (1975) Quartz resonator frequency shifts arising from electrode stress, 29th Freq. Control Symp. Proc. 1–4

5.6 Ballato, A., EerNisse, E.P., Lukaszek, T. (1977) The force–frequency effect in doubly rotated quartz resonators, 31st Freq. Control Symp. Proc. 8–16

5.7 Janiaud, D., Nissim, L., Gagnepain, J.J. (1978) Analytical calculation of initial stress effects on anisotropic crystals: Application to quartz resonators, 32nd Freq. Control Symp. Proc. 169–179

5.8 EerNisse, E.P., Ward, R.W., Wiggins, R.B. (1988) Survey of quartz bulk resonator sensor technologies, IEEE Trans. Ultrason. Ferroelec. Freq. Control **35**, 323–330

5.9 Ratajski, J.M. (1968) Force–frequency coefficient of singly-rotated vibrating crystals, IBM J. Res. Dev. **12**, N° 1, 92–99

5.10 Valdois, M., Sinha, B.K., Boy, J.J. (1989) Experimental verification of stress compensation in the SBTC-cut, IEEE Trans. Ultrason. Ferroelec. Freq. Control **36**, 643–651

5.11 Karrer, H.E., Leach, J. (1969) A quartz resonator pressure transducer, IEEE Trans. Ind. Electron. Constr. Instrum. **16**, 44–50

5.12 Sinha, B.K. (1981) Stress compensated orientations for thickness-shear quartz resonators, 35th Freq. Control Symp. Proc. 213–221

5.13 Besson, R.J., Boy, J.J., Glotin, B., Jinzaki, Y., Sinha, B., Valdois, M. (1993) A dual-mode thickness-shear quartz pressure sensor, IEEE Trans. Ultrason. Ferroelec. Freq. Control **40**, 584–591

5.14 Albert, W.C. (1984) Force sensing using quartz crystal flexure resonators, 38th Freq. Control Symp. Proc. 233–239

5.15 Zingg, W. (1985) Résonateurs et capteurs à quartz subminiatures. Séminaire sur les étalons de fréquence, Besançon

5.16 Langdon, R.M. (1985) Resonator sensors – a review, J. Phys. E: Sci. Instrum. **18**, 103–115

5.17 de Sorbier, G. (1987) Capteur de pression à quartz vibrant type 51, 1st European Freq. and Time Forum 179–183

5.18 Dieulesaint, E., Royer, D., de Hond, P., Morbieu, B. (1989) Résonateur miniature à entrée et sortie optiques, 3rd European Freq. and Time Forum 293–295

5.19 Chuang, S.S. (1983) Force sensor using double-ended tuning fork quartz crystals, 37th Symp. on Freq. Control Proc. 248–254

5.20 Albert, W.C. (1982) Vibrating quartz crystal beam accelerometer, Proc. 28th Intern. Instrum. Symp. 33–34, Las Vegas

5.21 le Traon, O., Janiaud, D., Muller, S., Bouniol, P. (1998) The VIA vibrating beam accelerometer: concept and performances, Position, Location and Navigation Symp., Palm Springs

5.22 Wade, W.H., Slutsky, L.J. (1962) Quartz Crystal Thermometer, Rev. Sci. Instr. **33**, 212–213
5.23 Hammond, D.L., Adams, C.A., Schmidt, P. (1964) A linear quartz crystal temperature sensing element, 19th ISA Conf., New York
5.24 Kaitz, G. (1984) Extended pressure and temperature operation of BT-cut pressure transducers, 38th Freq. Control Symp. Proc. 245–250
5.25 Nakazawa, M., Yamaguchi, H., Ballato, A., Lukaszek, T. (1984) Stress compensated quartz resonators having ultra-linear frequency–temperature responses, 38th Freq. Control Symp. Proc. 240–245
5.26 Dinger, R.J. (1982) The torsional tuning fork as a temperature sensor, 36th Freq. Control Symp. Proc. 265–269
5.27 EerNisse, E.P., Wiggins, R.B. (1986) A resonator temperature transducer with no activity dips, 40th Freq. Control Symp. Proc. 216–223
5.28 Watson, G., Horton, W., Staples, E. (1992) Portable detection system for illicit materials based on SAW resonators, IEEE Ultrason. Symp. Proc. 269–273
5.29 Adler, R., Desmares, P. (1985) An economical touch panel using SAW absorption, IEEE Ultrason. Symp. Proc. 499–502
5.30 Adler, R., Desmares, P. (1986) SAW touch systems on spherically curved panels, IEEE Ultrason. Symp. Proc. 289–292
5.31 Vellekoop, M.J., Lubking, G.W., Venema, A. (1994) Acoustic-wave based monolithic microsensors, IEEE Ultrason. Symp. Proc. 565–574
5.32 Plessky, V.P., Kondratiev, S.N., Stierlin, R., Nyffeler, F. (1995) SAW tags: new ideas, IEEE Ultrason. Symp. Proc. 117–120
5.33 Reindl, L., Scholl, G., Ostertag, T., Ruppel, C.C.W., Bulst, W.E., Seifert, F. (1996) SAW devices as wireless passive sensors, IEEE Ultrason. Symp. Proc. 363–367
5.34 Scherr, H., Scholl, G., Seifert, F., Weigel, R. (1996) Quartz pressure sensor based on SAW reflective delay line, IEEE Ultrason. Symp. Proc. 347–350
5.35 Pohl, A., Ostermayer, G., Reindl, L., Seifert, F. (1997) Monitoring the tire pressure on cars using passive SAW sensors, IEEE Ultrason. Symp. Proc. 471–474
5.36 Dieulesaint, E., Royer, D., Ballegeer, J.C. (1976) SAW coordinate sensor, Electron. Lett. **12**, 586–587
5.37 Dieulesaint, E., Royer, D., Chaabi, A., Formery, B. (1987) Lamb wave graphic tablet, Electron. Lett. **23**, 982–984
5.38 Royer, D., Dieulesaint, E., Legras, O. (1992) Capteurs à ondes élastiques guidées, J. de Phys. III **2**, 145–168
5.39 Liu, Y., Lynnworth, L.C. (1993) Flexural wave sidewall sensor for noninvasive measurement of discrete liquid levels on large storage tanks, IEEE Ultrason. Symp. Proc. 385–390
5.40 Nikolovski, J.P., Fournier, D. (1996) Dispositif d'acquisition des coordonnées d'une source acoustique appliquée à une plaque, French patent 94/11954, international patent WO 96/11378
5.41 Lynnworth, L.C. (1989) Ultrasonic Measurements for Process Control. Theory, Techniques, Applications. Academic Press, Boston
5.42 Royer, D., Levin, L., Legras, O. (1993) A liquid level sensor using the absorption of guided acoustic waves, IEEE Trans. Ultrason. Ferroelect. Freq. Control **40**, 418–421
5.43 Yang, D., Canit, J.C., Gaignebet, E. (1995) Photoelastic modulator: polarization modulation and phase modulation, J. Optics **26**, 151–159
5.44 Kemp, J.C. (1969) Piezo-optical birefringence modulators: new use for a long-known effect, J. Opt. Soc. Am. **59**, 950

5.45 Drevillon, B., Perrin, J., Marbot, R., Violet, A., Dalby, J.L. (1982) First polarization modulated ellipsometer using a microprocessor system for digital Fourier transformation, Rev. Sci. Instrum. **53**, 969–977

5.46 Gleyzes, P., Boccara, A.C. (1994) Profilométrie picométrique par interférométrie de polarisation. 1. L'approche monodétecteur, J. Optics **25**, 207–224

5.47 Kurtz, I., Dwelle, R., Katzka, P. (1987) Rapid scanning fluorescence spectroscopy using an acousto-optic tunable filter, Rev. Sci. Instrum. **58**, 1996–2003

5.48 Eisenreich, N., Herz, J., Kull, H., Mayer, W., Rohe, T. (1996) Fast on-line identification of plastics by near-infrared spectroscopy for use in recycling processes, ANTEC 96, 3131–3135

5.49 Cole, T.W. (1968) Opt. Technol. **1**, 31

5.50 Milne, D.K., Cole, T.W. (1977) An acousto-optical spectrometer for radio astronomy, Pro. ASA **3**, 108

5.51 Kaifu, N., Ukita, N., Chikada, Y., Miyaji, T. (1977) Pub. Astron. Soc. Japan **29**, 429

5.52 Masson, C.R. (1982) A stable acousto-optical spectrometer for millimeter radio astronomy, Astron. and Astrophys. **114**, 270

5.53 Rosolen, C., Michet, D., le Fouiller, F., Lecacheux, A., Dierich, P., Marteaud, M., Vola, P. (1993) Acoustooptical spectrometers in the world Part 5. In: ESA Work Package 1422

5.54 Ponçot, J.C. (1991) Progress and perspectives of LiNbO3 wide band Bragg cells, IEEE Ultrason. Symp. Proc. 563–567

5.55 Dierich, P., Rosolen, C., Michet, D. (1996) 12 years of experience in using acousto-optic spectrometers for radioastronomical observations, Advances in Acousto-Optics, 10th Meeting European Optical Soc., Issy-les-Moulineaux

5.56 Lecacheux, A., Rosolen, C., Michet, D. (1996) Space qualified, wide band and ultra-wide band acousto-optical spectrometers for millimetre and submillimetre radioastronomy, 30th ESLAB Symp. Proc., Submillimetre and far-infrared space instrumentation, Noordwijk, Netherlands

5.57 Dolfi, D., Joffre, P., Antoine, J., Huignard, J.P., Philippet, D., Granger, P. (1996) Experimental demonstration of a phased-array antenna optically controlled with phase and time delays, Appl. Opt. **35**, 5293–5300

5.58 Barth, H.V. (1973) Ultrasonic driven motor, IBM Technical Disclosure Bulletin **16**, 2263

5.59 Sashida, T., Kenjo, T. (1993) An Introduction to Ultrasonic Motors. Clarendon Press, Oxford

5.60 Ferreira, A., Minotti, P., le Moal, P. (1995) New multi-degree of freedom piezoelectric micromotors for micromanipulator applications, IEEE Ultrason. Symp. Proc. 417–422

5.61 Ohnishi, O., Myohga, O., Uchikawa, T., Tamegai, M., Inoue, T., Takasashi, S. (1993) Piezoelectric ultrasonic motor using longitudinal–torsional composite resonance vibration, IEEE Trans. Ultrason. Ferroelec. Freq. Control **40**, 687–693

5.62 Petit, L., Briot, R., Lebrun, L., Gonnard, P. (1998) A piezomotor using longitudinal actuators, IEEE Trans. Ultrason. Ferroelec. Freq. Control **45**, 277–283

5.63 Manceau, J.F., Bastien, F. (1996) Linear motor using a quasi-travelling wave in a rectangular plate, Ultrasonics **34**, 257–260

5.64 Lemons, R.A., Quate, C.F. (1974) Acoustic Microscope-Scanning version, Appl. Phys. Lett. **24**, 163–165

5.65 Hadimioglu, B., Quate, C.F. (1983) Water acoustic microscopy at suboptical wavelengths, Appl. Phys. Lett. **43**, 1006–7

5.66 Weglein, R.D. (1979) A model for predicting acoustic material signatures, Appl. Phys. Lett. **34**, 179–181
5.67 Kushibiki, J., Chubachi, N. (1985) Material characterization by line-focus beam acoustic microscope, IEEE Trans. Son. Ultrason. **32**, 189–212
5.68 Briggs, A. (1992) Acoustic Microscopy. Clarendon Press, Oxford
5.69 Yu, Z., Boseck, S. (1995) Scanning acoustic microscopy and its applications to material characterization, Rev. Mod. Phys. **67**, 863–891
5.70 Kessler, L.W., Yuhas, D.E. (1979) Acoustic microscopy 1979, Proc. IEEE **67**, 526–536
5.71 Pourcelot, L., Savilov, A., Bystrov, V., Kakurin, L., Kotovskaya, A., Patat, F., Pottier, J., Zhernakov, A. (1983) Results of echocardiographic examination during 7 days flight on board Saliout VII, June 1982, The Physiologist **26** N° 6, Suppl.
5.72 Laugier, P., Fournier, B., Berger, G. (1996) Ultrasound parametric imaging of the calcaneous: in vivo results with a new device, Calcif. Tissue Int. **58**, 326–331
5.73 Fink, M. (1992) Time reversal of ultrasonic fields, IEEE Trans. Ultrason. Ferroelec. Freq. Cont. **39**, Part 1: Basic Principles 555–566; Wu, F., Thomas, J.L., Fink, M. (1992) Part 2: Experimental Results 567–578; Cassereau, D., Fink, M. (1992) Part 3: Theory of the closed time-reversal cavity 579–592
5.74 Fink, M. (1994) Le retournement temporel des ondes acoustiques, La Recherche **25**, N° 264, 392–400
5.75 Derode, A., Roux, P., Fink, M. (1995) Robust acoustic time reversal with high order multiple scattering, Phys. Rev. Lett. **75**, 4206–4209
5.76 Roux, P., Roman, B., Fink, M. (1997) Time-reversal in an ultrasonic wave guide, Appl. Phys. Lett. **70**, 1811–1813
5.77 Kuperman, W.A., Hodgkiss, W.S., Chun Song, H., Akal, T., Ferla, C., Jackson, D.R. (1998) Phase conjugation in the ocean: experimental demonstration of an acoustic time-reversal mirror, J. Acoust. Soc. Am. **103**, 25–40
5.78 Miette, V., Sandrin, L., Wu, F., Fink, M. (1996) Optimization of time reversal processing in titanium inspections, IEEE Ultrason. Symp. Proc. 643–647
5.79 Prada, C., Manneville, S., Spoliansky, D., Fink, M. (1996) Decomposition of the time reversal operator: detection and selective focusing on two scatterers, J. Acoust. Soc. Am. **99**, 2067–76
5.80 Laurenceau, P., Dreyfus, G., Lewiner, J. (1977) New principle for the determination of potential distributions in dielectrics, Phys. Rev. Lett. **38**, 46–49
5.81 Alquié, C., Dreyfus, G., Lewiner, J. (1977) Stress-wave probing of electric field distributions in dielectrics, Phys. Rev. Lett. **47**, 1483–1487
5.82 Holé, S., Alquié, C., Lewiner, J. (1997) Measurement of space-charge distributions in insulators under very rapidly varying voltage, IEEE Trans. on Dielectrics and Elect. Insulation **4**, 719–724
5.83 Casula, O., Royer, D. (1997) Transient surface velocity measurements in a liquid by an active ultrasonic probe, IEEE Trans. Ultrason. Ferroelec. Freq. Control **45**, 760–767

General Reading

5.84 Benes, E. (1984) Improved quartz crystal microbalance technique, J. Appl. Phys. **56**, 608–626
5.85 Studer, B., Zingg, W. (1990) Technology and characteristics of chemically milled miniature quartz crystals, 4th European Freq. and Time Forum Proc. 653–658

5.86 Martin, S.J., Frye, G.C., Spates, J.J., Butler, M.A. (1996) Gas sensing with acoustic devices, IEEE Ultrason. Symp. Proc. 423–434
5.87 Hartmann, C.S. (1985) Future high volume applications of SAW devices, IEEE Ultrason. Symp. Proc. 64–73
5.88 Ballantine, D.S., White, R.M., Martin, S.J., Ricco, A.J. Frye, G.C., Zellers, E.T., Wohltjen, H. (1997) Acoustic Wave Sensors: Theory, Design and Physico-chemical Applications. Academic Press, New York
5.89 Muller, R.S., Howe, R.T., Senturia, S.D., Smith, R.L., White, R.M. (1990) Microsensors. IEEE Press, New York
5.90 Special issue on acoustic sensors (1987) IEEE Trans. Ultrason. Ferroelec. Freq. Control **34** N° 2, 122–277. Special issue on sensors and actuators (1998) IEEE Trans. Ultrason. Ferroelec. Freq. Control **45**, N° 5, 1123–1427
5.91 Ueha, S. (1989) Present status of ultrasonic motors, IEEE Ultrason. Symp. Proc. 749–753
5.91 Lemons, R.A., Quate, C.F. (1979) Acoustic microscopy. In: Mason, W.P., Thurston, R.N. (Eds.) Physical Acoustics, Vol. 14, 1–92. Academic Press, New York
5.93 Special issue on acoustic microscopy (1985) IEEE Trans. Son. Ultrason. **32** N° 2, 130–378
5.94 Briggs, A. (1992) Acoustic Microscopy. Clarendon Press, Oxford
5.95 Briggs, A., Arnold, W. (Eds.) (1996) Advances in Acoustic Microscopy Vol. 2. Plenum Press, New York
5.96 Lynnworth, L.C. (1989) Ultrasonic Measurements for Process Control. Theory, Techniques, Applications. Academic Press, Boston
5.97 Kino, G.S. (1987) Acoustic Waves: Devices, Imaging and Analog Signal Processing. Prentice Hall, Englewood Cliffs

Appendix A

A.1 Peach, R.C., Dix, C. (1978) A low loss medium bandwidth filter on lithium niobate, IEEE Ultrason. Symp. Proc. 509–512
A.2 Morgan, D.P. (1985) Surface Wave Devices for Signal Processing, Chap. 4. Elsevier, New York

Appendix B

B.1 Elachi, C. (1976) Waves in active and passive structures: A review, IEEE Proc. **64**, 1666–1698
B.2 Yariv, A., Yeh, P. (1984) Optical Waves in Crystals. Propagation and Control of Laser Radiation, Chap. 6. Wiley, New York
B.3 Biryukov, S.V., Gulyaev, Y.V., Krylov, V.V., Plessky, V.P. (1995) Surface Acoustic Waves in Inhomogeneous Media. Springer, Berlin

Appendix E

E.1 Boyer, L. (1994) Etude des phénomènes de réflexion-réfraction des ondes planes acoustiques dans les milieux piézoélectriques, doctoral thesis, Université Paris 7

Index

Ablation 212
– regime 220–222
Absorption coefficient 213
Accelerometer 328–330
Acoustic microscope 362–368
Acousto-optic
– BAW filter 232, 351–354
– components 348–357
– deflector 355
– modulator 348–351
– SAW filter 195
Acousto-optic interaction 158–196
– bandwidth 183–186, 191, 232
– collinear 190–191, 196, 352
– coupled mode analysis 179–187
– critical thickness 176
– figure of merit 182, 184
– number of resolvable directions 186
– phase asynchronism 185, 192, 230
– tangential 191
– with SAW 193–196
Actuator, piezoelectric 358–362
Admittance
– BAW resonator 26–28
– harmonic 127–130, 133, 388
– matrix (SAW) 155, 404, 407
– SAW transducer 90, 117–119, 138
– SAW transducer 76
Aluminar *see* Sapphire
Aluminium 35, 36, 44, 61, 145, 213, 263, 297
Aluminium nitride
– BAW constants 40
– SAW application 147
Aluminium phosphate 277
– BAW resonator 277
Antiresonance frequency 26, 49, 54
Array factor 87
Attenuation (SAW) 125, 340, 366
Autocorrelation
– of rectangular pulse 309
– of SAW transducer impulse response 60
– signals 301, 314, 317

Bandshape factor 33–40, 51
Barker
– code 300
– sequence 300
BAW filter 269–280
BAW resonator 44
– admittance 26
– antiresonance frequency 26, 49, 54
– BVA 274
– electrical impedance 27, 54
– equivalent circuit 28
– high-frequency 275
– miniature 276
– piezocomposite 210
– QAS 274
– quartz 28, 269–278
– resonance frequency 26, 49
– structure 273–278
BAW transducer 5–55
– bandshape factor 33–40, 51
– crossed-field model 14
– efficiency 30
– electrical impedance 26–30, 46
– equivalent circuit 14, 29–30
– frequency response 20–21
– high-frequency 30–40
– impedance matrix 10–12
– impulse response 23–26
– in-line field model 14
– KLM model 16–30
– low-frequency 19–30
– Mason and Redwood circuit 12
– matching 30, 39, 41
– materials 40–45, 277
– Nyquist plot 49
– one-dimensional model 6
– piezocomposite 41
– radiation conductance 33, 37

– structure 7, 20, 32, 36–38
– technology 40–45
– time response 23
– transmitted power 21–23
Berlinite *see* Aluminium phosphate
Bessel function 137, 173
Biaxial crystal 161
Bilinearity factor 307
Binary coding
– by amplitude modulation 299
– by phase modulation 300
Bleustein–Gulyaev wave 262
Bragg
– angle, diffracted intensity 229
– cell 195, 356
– condition 109, 160, 177, 193, 393
Brillouin effect 158
$B\Theta$ product 287, 295, 355
BVA resonator 274

Cadmium sulfide 34
Capacitance, of finger pair 69, 111, 385
Cell
– Bragg 195, 356
– of ladder filter 266
– of monolithic filter 279
– of transducer 122, 123, 125, 131, 133, 383
Charge distribution 67, 83, 387
– measurement 376–378
Chebyshev polynomials 86, 136
Clausius-Mossotti relation 165
Coded signal 283
Coherent electronic detection 203
Components
– acousto-optic 348–357
– elastic wave 237
Composite
– resonator 209
– transducer 41
Compressed pulse 285–287, 301, 316
Continuous level gauge 347
Contrast 364
Convolver (SAW) 306–310
Coordinate sensor 338–339
Coupled mode method 103–123, 179–187
– theory 391–397
Coupling
– acoustic, of SAW resonators 265
– by evanescent waves (BAW) 278
– electrical, of SAW resonators 265
– fluid 364
– of vibration modes 52, 362, 393
– torsion 360
Critical thickness (acousto-optic) 176
Crossed-field model
– BAW 14
– SAW 139–143, 155
Crystal
– acoustic signature 365
– biaxial 161
– uniaxial 162, 187, 190, 228, 231
Cylindrical guide 344

DART transducer 121
Deflector, acousto-optic 355
Delay line
– BAW 245–249
– – elasto-optic interaction 248
– – frequency response 247
– – structure 245
– echo 247
– SAW 249–251, 311
Delta function method 70–72, 149
Detector
– dew 334
– gas 334
– liquid level 339–343
Detuning parameter 105, 112
Dew detector 334
Diffraction
– Bragg incidence 177–193
– normal incidence 172–177
Dioptric
– plane 225
– system 413–416
Directionality factor 116, 410
Directivity pattern 219
Discrete source method 70–72, 149
Dispersive
– effect 291, 295
– structure 78
Doppler
– effect 206, 379
– velocimetry 205–206, 369
Dummy finger 63, 293

Echo
– continuous level gauge 347
– delay line 247
– point level sensor 345
Echography 368
Effective distance 110
Elasto-optic tensor 164
– of materials 168

Electro-optic tensor 171
Electrode lithography 143
Electromechanical coupling
– BAW 40, 42, 50
– SAW 84
Electrostatic potential 82
Equivalent circuit
– BAW transducer 12–19
– for transmission line section 54
– SAW transducer 139–143, 155
Evanescent wave coupling 278

Fan-shaped transducer 312
Figure of merit (acousto-optic) 182, 184
Filter
– acousto-optic (BAW) 232, 351–354
– acousto-optic (SAW) 195
– BAW 269–280
– dispersive reflection grating 295–298
– dispersive transducer 290–295
– elasto-optic interaction 301
– frequency response 255
– impulse response 255
– matched to discrete code 299–300
– matched to medium 373
– miniature 280
– monolithic 278–280
– SAW 251–269
– signal-matched 280–303
– stationary wave 261–269
– transversal 72, 252
– with bidirectional transducers 253–258
– with multistrip coupler 258
– with unidirectional transducers 259–261
Finger
– dummy 63, 293
– slanted 258, 311
Fluorescence 353
Focussing, elastic wave 223
Force dipole 216
Fourier transform 23, 80, 82, 88, 148, 150, 235, 252, 304, 312, 366, 387
Fourier transform processor 304–306
Frequency modulation 283
Frequency response
– of BAW transducer 20–21
– of SAW transducer 113–115

Gallium arsenide 195
Gallium phosphate 277
Gas detector 334
Generation
– centre 115, 122, 134
– coefficient 131
– factor 112
– photothermal 210–225
Golay code 317
Gold electrode 34, 36
Green function 84, 134–139, 216, 389, 416
Groove reflection grating 298
Group
– delay 109, 289
– velocity 336, 344
GSM system 268

Head wave 211
Hexapole 10, 94, 125, 407
Hilbert transform 78, 150

Impedance matrix (BAW) 10–12
Impulse response model 72–80
In-line field model
– BAW 14
– SAW 139
Index
– ellipsoid 161–163
– surface 163
Insertion losses 120
Interaction length 182
Interactive window display 343
Interdigital-electrode transducer 58
Interferometer
– Fabry–Pérot cavity 206
– Michelson 200
Interferometric probe 200–206, 326

Klein–Cook parameter 193
KLM model 16–19, 54

Ladder structure transducer 257
Lamb wave 290
– antisymmetric mode 336, 343
– coordinate sensor 339
– dispersion curves 336
– generation 45, 224, 342
– in tube 224
– mechanical displacement 336
– sensor 336–344
– symmetric mode 336
Langasite 277
Laplace's equation 383
Laser printer 349

Layer, piezoelectric 43, 146
Lead molybdate 355
– elasto-optic constants 168
– figure of merit 184, 230
Legendre
– function 65, 125, 399–401
– polynomial 68, 87, 384, 399–401
Level sensor oscillator 342
Lift-off technique 144
Liquid level detector 339–343
Lithium niobate
– BAW constants 40
– BAW transducer 23, 355
– elasto-optic constants 168
– figure of merit 184
– filter 260
– multistrip coupler 242
– SAW constants 144
– SAW permittivity 81
– SAW transducer 63, 76, 80, 88, 134
Lithium tantalate
– BAW resonator 277
– elasto-optic constants 168
– figure of merit 184
– filter 258, 279
– SAW constants 144
– SAW transducer 138
Love wave 334

Materials
– for BAW transducers 40, 277
– for SAW transducers 143–148
Maxwell's equations 225
Michelson interferometer 200
Microbalance 320
Miniature resonator (BAW) 276
Mirage effect 228
Mirror, time reversing 372
Mixed matrix 94, 99–103, 119, 123, 407
Modulator, acousto-optic 348–351
Monolithic filter 278–280
Multistrip coupler 240–243, 258, 310

Nelson–Lax tensor 169
Non-destructive testing 43, 209, 223, 368, 375
Numerical model, SAW transducer 134–139
Nyquist plot 50

Optical probe
– deflection 196
– diffraction 198
– heterodyne 202–205
– interferometric 200–206
Orthogonality relation 181, 228, 392
Oscillator
– BAW (quartz) 270
– SAW 251
– ultrastable 273
Osteodensitometer 368

P-matrix 94, 99–103, 119, 123
Parametric effect 379
Parametric interaction probe 379–382
Paratellurite 191, 231, 349, 354
– acousto-optic figure of merit 183
Parseval's theorem 48, 78
Passive array
– dispersive 295–298
– elementary cell 124
– modes 153
– reflection coefficient 107, 124
– reflective 130
– transmission coefficient 107
Phase asynchronism 185, 192, 230
Phased source array 223
Photothermal generation 210–225
Piezoceramic 41
– resonator 53
– transducer 22, 24, 41, 207, 344, 350, 370
Piezocomposite
– resonator 209
– transducer 41
Piezoelectric permittivity
– method 80–94, 102
– surface 80
Piezopolymer 42
– BAW constants 40
Platinum 44, 245
Pockels tensor 164, 166
Point level sensor 344
Point source model 216–220
Poisson ratio 220, 344
Poisson's equation 8, 136
Potential distribution 67, 83, 383–384, 389
Poynting vector 111, 225, 340
Pressure pulse 378
Pressure sensor 321–324, 334
Pressure wave method 376–378
Printing, by laser 349
Profilometer 351
Pseudo-surface wave 138, 144, 264
Pulse compression ratio 287, 290, 298

PZT4 ceramic (properties) 40

QAS resonator 274
Quality factor 29, 50, 269
Quartz
– BAW constants 40
– BAW resonator 27, 28, 269–278
– elasto-optic constants 168
– figure of merit 184
– SAW constants 144
– SAW delay line 60
– SAW filter 260–261
– SAW transducer 134
– thermometer 330
Quartz crystal cuts 270–272
– doubly-rotated 272
– singly-rotated 272
Quasi-static approximation 85

Radar 280, 314, 347, 357
Radiation conductance
– BAW transducer 33, 37
– SAW transducer 76–80, 90, 118
Radio astronomy 354–357
Raman–Nath
– parameter 173
– regime 160, 174, 193, 228
Ray model 365
Rayleigh wave
– generation 59, 210, 220, 364
– potential 83, 85
– transported power 100
– velocities 144
Reciprocity
– principle 100, 404, 406, 408
– relations 126, 154, 407
Reemission, SAW transducer 62, 91, 97
Reflection
– centre 115, 122, 134
– coefficient
–– reflective array 105, 109, 149
–– single-finger 112
Resonance frequency 26, 49, 53
Resonant cavity (SAW) 122, 264
RSPUDT transducer 122, 260

SAM microscope 368
Sapphire 34, 43, 245, 328, 364
SAW filter 251–269
SAW resonator 122
– associations of 264
– longitudinal coupling 265
– one or two-port 262–264
– transverse coupling 265
SAW transducer 57–156
– admittance matrix 155
– antisymmetric 96, 410
– charge distribution 67
– converted power 93
– coupled mode method 110–120
– crossed-field model 139–143
– discrete source method 70–72, 149
– dispersive 295
– electrical admittance 76
– electrode lithography 143
– electrostatic potential 82
– equivalent circuit 139–143, 155
– frequency response 59, 74, 89, 113–115
– generation factor 112
– harmonic admittance 127, 133
– impulse response 59, 74, 89, 90
– impulse response model 72
– interdigital electrode 58
– interdigitated interdigital 257
– ladder structure 257
– matching 79
– materials 143–148
– mixed matrix 99
– numerical model 134–139
– P-matrix model 123
– permittivity method 80–94, 102
– potential distribution 67
– radiation conductance 76–80, 90, 118
– receiver 91, 96, 151
– reemission 62, 91, 97
– reflected power 94, 151
– reflection 65
– RSPUDT 122, 260
– sampled 313
– scattering matrix 95
– slanted finger 258, 311
– SPUDT 101, 121, 260
– symmetric 96, 151, 410
– technology 143–148
– transmitted power 75
– unidirectional 98, 120–123, 146, 259, 410
Scanning antenna 357
Scattering matrix 95, 101, 404
Schaefer and Bergmann method 229
Seismic prospection 225
Sensor
– cylindrical guide 344–347
– guided wave 331–347

– Lamb wave 336–344
– pressure 321–324, 334
– SAW 331–336
– temperature 321, 330
– thickness shear mode 320–324
– vibrating beam 324–330
Sezawa mode 309
Sidelobes 286, 287, 294
Signal-to-noise ratio, matched filter 281–283, 287, 314
Signature, of crystal 365
Silica
– acousto-optic interaction 176, 177, 350
– elasto-optic constants 168
– figure of merit 184
Silicon 290, 309
SLAM microscope 368
Slanted finger 258, 311
Solid–liquid interaction 346
Son et lumière 354
Sorting plastic objects 353
Spectrometer, for radioastronomy 354–357
Spectrum
– fluorescence 353
– of material 353
– of variable frequency signal 287, 315
Spectrum analyser 304
SPUDT transducer 101, 121, 260
Standing wave ratio 247
Stationary phase method 287, 315, 316
Stationary wave filter 261–269
Stop band 114, 130, 407
Synchronism frequency 59, 125

Technology
– BAW transducer 40–45
– reflection grating filter 297
– SAW transducer 143–148
Tellurium oxide *see* Paratellurite
Temperature
– distribution 213
– sensor 321, 330
Tensor
– elasto-optic 164
–– of materials 168
– electro-optic 171
– Nelson–Lax 169
– Pockels 164, 166
Thermoelastic regime 212–220
Thermometer, quartz 330
Time reversal 371–376
Time reversing mirror 372
Titanium 375
Tobolka matrix 94
Torsion coupler 360
Touch-sensitive screen 332–334
Transformer
– electromechanical 13
– Fourier 304–306
Transition matrix 418
Transmission
– coefficient (reflective array) 105, 149
– matrix 107, 126, 405–407
Transversal filter 72, 252
Travelling wave motor 359
Triple-transit echo 97, 119, 151, 258
Tuning fork thermometer 330

Ultrasonic
– motor 358–362
– probe, parametric 379
Uniaxial crystal 162, 187, 190, 228, 231

Velocimetry, Doppler 205–206, 369
Velocity in materials
– bulk wave 40
– surface wave 144

Wave vector diagram 179, 187, 352
Window function 254

YAG laser 212, 223, 328, 377
Yttrium aluminium garnet *see* YAG laser

Zinc oxide
– BAW constants 40
– deposition 43, 146
– transducer 30–34, 37, 44, 146, 245, 309

Printing: Mercedes-Druck, Berlin
Binding: Buchbinderei Lüderitz & Bauer, Berlin